Environmental Pollution, Climate Change and Altered Lifestyle during COVID

THE EDITORS

Prof. Arun Arya, Ex- Head, Department of Botany and Environmental Studies in The Maharaja Sayajirao University of Baroda. He is a Botanist, Phytopathologist, Philatelist and a popular Science writer. Serving as hon. President of IAAPC, he is Fellow of I B S, I P S and IAAPC. He has actively served as Head of the Department of Botany during 2008 – 2013.

Prof. Vinod S. Patel, B.E.(Civil) M.S. (USA), MIE, ASCE, FIWWA, FIAAPC, PE (MASS.) is Consulting Engineer and Environmental expert has been working in this field for the past 40 years. He has made significant contribution in the field of radial well and ground water supply management. Visiting Prof. in the Faculty of Tech. and Engg., The Maharaja sayajirao University of Baroda. He has contributed in the field of air and soil pollution and global climatic change and has published 50 technical papers and 4 books on pollution and their management issues.

Professor Madhoolika Agrawal, FNA, FNASc, FNAAS is Professor of Botany and presently the Coordinator, Interdisciplinary School of Life Sciences at Banaras Hindu University, Varanasi, India. She has worked on various projects with MoEn., UGC, DST, CSIR, DIFD, U.K., Research council of Norway, SEI, U.K and Asia Pacific Network, Singapore. Dr. Agrawal has published about 258 research papers.

Professor C. N. Murthy, M.Tech. Ph.D. Prof. C. N. Murthy is an alumnus of Indian Institute of Technology, Mumbai where he obtained his M.Sc. (Chemistry), M. Tech. (Materials Science) and Ph.D. degrees. He has about three years' industrial experience with a multinational company, M/s Johnson & Johnson, and 23 years of teaching & research experience. He was the Head of the Applied Chemistry Department (2004-2010) at Faculty of Technology & Engineering, The M.S. University of Baroda. He has about 50 publications in peer reviewed journals of international repute.

Prof. Bindu Bhatt, M.Sc.(Geography) Ph.D. is currently Head of the Dept. of Geography. She has 21 years of teaching & research experience. Appointed as Hon. Secretary (2000-2006) and joint coordinator of Env. Science in the Maharaja Sayajirao University of Baroda till 2012. She is visiting faculty of Env. Science and P.G. Diploma in Disaster Manangement, and master of Urban and regional Planning (Fac. of Engg. And technology). She has published 35 research paper in journals of repute.

Dr. Geeta Padate is Professor in Department of Zoology, Faculty of Science, The M.S. University of Baroda, Vadodara. She is an expert in Bird studies. She Headed the Department for more than 4 years. She has been involved in organizing various mass awareness programme related to environment, birds and wildlife. She has published 61 research papers in journals of National and International reputes. 13 students have completed their Ph.D. under her guidance while 4 are working for the same.

Environmental Pollution, Climate Change and Altered Lifestyle during COVID

Arun Arya

Vinod S. Patel

Madhoolika Agrawal

C. N. Murthy

Bindu Bhatt

Geeta Padate

2022

Daya Publishing House®

A Division of

Astral International Pvt. Ltd.

New Delhi – 110 002

ISBN: 9789354614590

Publisher's Note:

Every possible effort has been made to ensure that the information contained in this book is accurate at the time of going to press, and the publisher and author cannot accept responsibility for any errors or omissions, however caused. No responsibility for loss or damage occasioned to any person acting, or refraining from action, as a result of the material in this publication can be accepted by the editor, the publisher or the author. The Publisher is not associated with any product or vendor mentioned in the book. The contents of this work are intended to further general scientific research, understanding and discussion only. Readers should consult with a specialist wherever appropriate.

Every effort has been made to trace the owners of copyright material used in this book, if any. The author and the publisher will be grateful for any omission brought to their notice for acknowledgment in the future editions of the book.

Published by : **Daya Publishing House®**
A Division of
Astral International Pvt. Ltd.
– ISO 9001: 2015 Certified Company –
4760-61/23, Ansari Road, Darya Ganj,
New Delhi-110 002
Ph. 011-43549197, 23278134
e-mail: info@astralint.com
Website: www.astralint.com

Dedicated to

The Young Scientists and Nature Lovers

With the Blessings of

Swami Gyan Swarup Sanand Ji

Engineer, Prof. (Dr.) G. D. Agrawal, Ph.D. Berkley, USA

Professor of Env. Engineering, IIT, Kanpur

Leading Environmental Expert, and founder member of IAAPC

and

Dr. D.N. Shukla, D.Sc.

Ex Head, Dept. of Botany, University of Allahabad
Founder of Ganga Pollution Exhibition, Prayagraj

Two crusaders of our country, who sacrificed their lives for the purity and sanctity of Ma Ganga so the pristine water can flow perpetually in the holy river year after year.

"I believe in an ideal life"

I believe that, in a little flower,
there is a divine power hidden in beauty
which is more potent than a gun.
I believe that in the birds' note
Nature expresses herself with a force
which is greater than that revealed
in the deafening roar of a cannon.

– by Rabindranath Tagore

For Rabindranath Tagore, Nature was his favorite school, as he recorded in "My Life": "I had a deep sense, almost from infancy, of the beauty of nature, an intimate feeling of companionship with the trees and the clouds, and felt in tune with the musical touch of the seasons in the air. All these craved expression, and naturally I wanted to give them my own expression."

Dr. Kamlesh Lulla, ***Ph.D.; Ph.D.; NASA Medalist***

- ☆ *Recipient of Government of India Pravasi Bhartiya Samman Medal and Award from the President of India 2015*
- ☆ *Recipient of NASA Medal for Exceptional Achievement for Science 2005*
- ☆ *Recipient of NASA Medal for Exceptional Service 2011*
- ☆ *Recipient of Astronaut Onizuka Medal for Science and Technology 2012*
- ☆ *Chief Scientist for Earth Observations-Space Shuttle*
- ☆ *Current Director, University Research and Technology Development*

NASA Johnson Space Center, Houston, Texas, USA.

Foreword

The current pandemic COVID-19 and its impact on global communities has brought into focus one self-evident fact: we, all of us humans, are connected, all ecosystems are interconnected, all organisms are interconnected and we are part of interacting complex planetary processes and cycles. Recent scientific reports indicate that as the COVID-19 pandemic brought global communities and their daily life as well as commercial activities to a grinding halt and this has resulted in global greenhouse gas emissions to reduce by 10 to 30 percent in the early months of 2020. This reduction in emissions is expected to be temporary and the resumption of our activities could spur the rise in emissions again. Experts have suggested that this temporary occurrence will not slow down the climate change. However, this unexpected planetary level experiment has revealed the resilience of our ecosystems to recover from environmental issues and pollution. This information could guide decision makers and planners to device better and environmentally responsible responses as the nations reopen their economies, reopen global travel, trade and industrial activities.

Environmental pollution, climate change, biodiversity, and human well-being and health are all interconnected. Experts have surmised that COVID-19 virus jumped from some wild animal/s to humans: a clear warning of the need to provide adequate habitats for animal populations and to maintain biodiversity. Some years ago, when I was a senior university professor in the U.S.; I traveled to Indiana University to hear the eminent Harvard biologist Professor E O Wilson who suggested *"Biological diversity is the key to the maintenance of the world as we know it. Eliminate one species, and another increase to take its place. Eliminate a great many species and the local ecosystem starts to decay"*. We, as human societies, are a part of "nature" and must understand our role in altering ecosystems around the globe.

Our planetary systems are exhibiting symptoms: global temperature rise, warming oceans, ocean acidification, shrinking icesheets, glacial retreats, decreased snow covers, sea level rise, declining arctic sea ice, extreme weather events-floods, heat waves, storms and hurricanes. These vital signs of our "planetary health" are our call to action to find science and technology-based solutions so we can serve as better stewards of our habitat "Our Earth"! At NASA, we are devoting our remote sensing satellites to monitor both rapid and long-term changes that are revealing the state of our planet Earth's health.

As scientists, we are not alarmists or fear-mongers, but we are equipped to gain insights into the complex functioning of our ecosystems and innovate long term solutions. I am reminded of one of my favorite scientists- Nobelist Marie Curie who stated, *"Nothing in life is to be feared. It is only to be understood. Now is the time to understand more, so that we may fear less."* We must not fear COVID-19 or Climate crisis. This COVID-19 crisis has provided us with an opportunity to apply our collective knowledge to overcome these challenges and use our wisdom to make life better for all inhabitants of this planet we call home!

I sincerely congratulate Professors Arun Arya, V.S.Patel, C.N.Murthy, M.Agrawal, Bindu Bhatt, Geeta Padate for serving as editors of this very timely and very interdisciplinary book. I am impressed with their collaboration and their vision to bring diverse but inter-related topics under one publication. I am positive that this book will serve as an important source of information and reference for both the researchers and the practitioners. I wish this book a great success!

Finally, I fondly recall my tenure as an Assistant Professor of Ecology in the Faculty of Science at the Maharaja Sayajirao University of Baroda in the seventies (1971-1978). Professor S J Bedi, Professor and Pro Vice Chancellor S D Sabnis, Professor S N Padate and Professor and Head Chandravadan H Pathak were my mentors. They introduced me to the science of ecology and environmental pollution and empowered me to apply newly emerging technologies of remote sensing for ecological research during my Ph D work at the M S University. I was awarded a prestigious fellowship to the United States where I received my second Ph D in Remote Sensing and later joined Indiana State University as a Professor of Remote Sensing and Geosciences. After ten years in the US academia, I joined NASA as Mission Scientist and Chief for Earth Observations for the Space Shuttle flights. This scientific journey has enabled me to view our planet Earth both from Space and from the ground/field research: it has confirmed our ancient Bhartiya wisdom "Vasudev Kutumbkam" "We are one global family"!

Dr. Kamlesh Lulla

Preface

ओउम् यस्य सूर्य श्चक्षु श्चन्द्र माश्च पुनर्णवः
अग्निं यश्चक्र आस्यं तसमै ज्येष्ठाय नमः

— Atharva Veda 10-7-33

Whose, the sun and the ever new moon are the eyes,
fire is the mouth; to such a great one, we pay our homage.

We require food and water to survive but role of clean air is still more important than water and soil. Each healthy person requires 11500 liters of air per day. The industrialization has affected these pristine air, water and natural resources forcing us to suffer with phenomenon like pollution and climate change. Recent lockdown in countries world over resulting into no or very restricted activities recorded drastic change in quality of air and water. Air is most essential for life. Man cannot survive without fresh air. The beauty of Planet and celebration of life can't be imagined in polluted atmosphere. Air the most important life supporting element, sometimes becomes our bitter enemy as and when it gets polluted, since it causes a number of diseases in our body. The harmful effects caused to human body due to the polluted air present in the form of dust, fumes, gas, mists, pollen, microbes, odour, smoke or vapours for a long duration is injurious to human, plant, or animal life as well as to buildings and properties. Nature is an essential component of ecosystem, describing it Rashtra kavi Shri Maithili Sharan Gupt wrote in Panchwati....

चारूचंद्र की चंचल किरणें, खेल रहीं हैं जल थल में,
स्वच्छ चाँदनी बिछी हुई है अवनि और अम्बरतल में।
पुलक प्रकट करती है धरती, हरति तृणों की नोकों से
मानों झूम रहे हैं तरू भी, मन्द पवन के झोंकों से।।

British poet W.H. Auden once noted," Thousands have lived without love, not one without water." Yet while we all know water is crucial for life, we trash it anyway. Some 80 per cent of the world's waste water is dumped-largely untreated-back into the environment, polluting rivers, lakes and oceans. Flowing for about 2,525km, the Ganges is the longest river in country. The government of India has set up National Mission for Clean Ganga (NMCG) under the Union Ministry of Jal Shakti for maintaining its cleanliness. Urbanization and industrialization have resulted in a profound deterioration of India's air quality. Anthropogenic sources (human activity) are mostly related to burning different kinds of fuel, vehicular pollution and "stationary sources" including smoke stacks of power plants, manufacturing facilities and waste incinerators, as well as furnaces and other types of fuel-burning heating devices. The traditional biomass burning includes wood, crop waste and dung. Fumes from paint, hair spray, varnish, aerosol sprays and other solvents and waste deposition in landfills, which generate pollutant methane an asphyxiant.The CO_2 released from soil carbon in the top matter of soil can increase the temperature by 0.03°C per year. Secondary pollutants result due to chemical or photochemical reactions of the primary pollutants with elements or compounds present in the atmosphere. These reactions depend on factors like concentration of reactants, moisture, photo activation, meteorological parameters and topography. The secondary pollutants are Sulfur trioxide, nitrogen dioxide, aldehydes, ketones, various sulfates and nitrate and smog. Organic Pollutants mainly consist of carbon and hydrogen. However, some may contain oxygen, nitrogen, phosphorous and sulfur. Hydrocarbons, aldehydes, ketones, organic compounds of sulfur are some of the other examples.

Air pollution is increasing at an alarming rate in big cities where there is rapid industrial growth and number of automobiles is also on an increase. Therefore it is necessary to gather information about the baseline data on existing air quality. The physicochemical monitoring may be carried out by suitably selecting sampling sites in the specified area and collecting periodical samples and analyzing them for different air pollutants. The biological monitoring can be carried out by systematic study of plants, humans and animals as regards damage or adverse impacts on them. Government of India enacted the Air (Prevention and Control of Pollution) Act 1981 to arrest the deterioration in the air quality. The act prescribes various functions for the Central Pollution Control Board at the apex level and State Pollution Control Board at the state level. It is interesting to note that during the rule of Maharaja Sayajirao Gaekwad III sprinkling of water was practiced on roads of Vadodara to reduce the particulate pollution.

Fortunately not all things have changed,

Not the leaves on the trees, Not the birds in the breeze.

Not the water in the sea, Not the rocks of you and me.

Not the sun, actually a star, Not those fearful pits of tar.

Not the bones of the dinosaurs, Not the land you thought was yours.

Hopefully we can change our Earth, Make it once again green,

Make it what's its worth, Make it once again clean.

Air pollution and climate change are two major global problems. To generate awareness among masses and to provide current know how about various aspects of air pollution to the industries, scientists and planners an association was formed in Benaras Hindu University, Varanasi on 18th September 1976. Through different workshops, conferences and webinars organized by IAAPC the knowledge is shared between various stakeholders. The climate change is now a reality and is already affecting the environment. The climate change is a change in the long-term weather patterns that characterize the regions of the world. It has put adverse impact on biogeochemical pathways. According to general consensus, the global climate is changing, which may also affect agricultural and livestock production. This time potential impact of climate change on food production is widely debated. Climate change induced by increasing greenhouse gases is likely to impact directly on food production across the world, and that can vary from region to region. In the long run, the climatic change could affect the agriculture affecting the quantity and quality of crops. This will cause threat for food security in the country. Every year there is loss of life and wealth due to floods and cyclones in different parts of the country. There is no simple solution for the abatement and adaptation to climate change. An all out effort is required to promote renewable energy and develop innovative technologies. The loss of biodiversity needs to be checked and ecological services used effectively. Prof. C.K. Varshney from JNU, New Delhi, in his inaugural address delivered at National Symposium in Vadodara in 2020 quoted the words of UN General Secretary Antonio Guterres, "Humanity is facing nothing less than a direct existential threat."

Protozoa, bacteria, viruses, fungi, pollens and spores also cause biological polluiton. The Corona pandemic has made our life miserable. We have to cage ourselves in our houses and a lot of life style changes were automatically adopted in industries, offices and academic institutions. The war against invisible virus is on and we are facing challenges on daily basis. Millions of People are affected and many lost their lives because of poor immunity or lack of proper medical attention. But how this virus entered in our body raises many unanswered questions. Did the virus jumped from animal to the human being in search of a stable host for its survival ? Or this was a monster created by Chinese scientists in Wuhan's viral research lab. It was an accidental release of deadly Corona. Any how! We have the responsibility of protecting our mother nature and devise ways to live in harmony with other creatures.

काश हम जान पाते
सूक्ष्म जीवों की यह विनाश लीला
रोक पाते बढ़ते संक्रमण को
न होती यह विनाश लीला
कराहते लोग और
दम तोड़ता पर्यावरण हमारे आस-पास॥

The present volume contains 32 chapters on pollution, climate change and topics related to altered life style due to COVID-19. We hope that contributors have brought new knowledge to the Science. Few of these were presented in a National

Symposium on 'Climate change, pollution and Harmony with nature' organized by IAAPC in The M.S. University of Baroda, Vadodara, on 25th January, 2020. The chapters will provide new knowledge to enthusiastic readers and will pave way to policy makers in keeping the Blue Planet "Earth" Clean and Green. It is said:

The past no longer belongs to us,

And the future does not exist;

The present is what we have, and

It is where we must live.

The editors acknowledge the help and support received from experts and reviewers for making good suggestions to be incorporated in the text. Although due care has been taken it is important to mention that information presented in the volume is opinion of the authors only, not of IAAPC or the publishers. We are sincerely thankful to Dr. Kamlesh Lulla from NASA, Houston, Taxas, for writing the Foreword and to all the contributors for sparing their valuable time and submitting their papers for this volume. We highly appreciate Prof. Anil Kane, Prof. S.D. Sabnis, Shri Prasoon Gargava, Shri B.R. Naidu, Prof. H.V. Bhavnani and Prof. S.J. Bedi for their encouragement and good wishes. Our sincere thanks are due to Astral International, New Delhi for nicely publishing this book.

Arun Arya

V.S. Patel

Madhoolika Agrawal

C.N. Murthy

Bindu Bhatt

Geeta Padate

Contents

— PART 2 —
Pollution: Management Strategies

— PART 3 —
Environmental Issues: Effects on Climate Change

— PART 4 —
General Topics on Environment: Including Changes in Lifestyle during COVID

Part 1

Pollution: Causes and Effects

साहस भरी उड़ान

निकलती है सुनहरी धूप हर दिन
सूरज जब चमकता है,
अपनी स्वर्ण रश्मियों से हमें प्रकाशित करता है।
हर सुबह रफ्तार देती है,
जिन्दगी जीने की चाह देती है।
तहे दिल से स्वागत करो सुबह का
यही तो उड़ान भरने का जज्बा देती है
पक्षी उड़ान भरते हैं घोसलों से
लाते हैं चोंच में दाना अपने बच्चों के लिए।
होती है उड़ान दूर-दूर पर
ज्ञात है उन्हें उनकी सीमाएं।
हम भी ऊंची उड़ान का हौंसला रखें,
पर लक्ष्य पर नजर साफ होनी चाहिए।
आओ हम सब मिलकर सुबह सुन्दर बनाएं,
समाज को दें नई दिशा।
और मानव जीवन को इन्द्रधनुषी बनाएं।।

Chapter 1

Interactive Effects of Air Pollutants on Plants: A Review of Indian Studies

Durgesh Singh Yadav*, S.B. Agrawal** and Madhoolika Agrawal

Laboratory of Air Pollution and Global Climate Change, Department of Botany, Institute of Science, Banaras Hindu University, Varanasi – 221005

e-mail: *durgeshsinghy@gmail.com, **sbagrawal56@gmail.com

ABSTRACT

Gaseous air pollutants do not exist singly in the atmosphere. Therefore, researches on interactive effects of air pollutants must be encouraged to understand the role of air pollutants in combination. The aim of this review is to evaluate the responses of different Indian agricultural crops to a combination of ozone (O_3) and other major air pollutants such as sulphur dioxide (SO_2), carbon dioxide (CO_2), nitrogen dioxide (NO_2) and aerosols. Amongst the air pollutants, O_3 is the most reactive and phytotoxic, therefore it was selected as a common pollutant in all the combinations. Air pollutant levels and trends in India as well as plant responses under combined exposure of air pollutants based on Indian studies are reviewed. Carbon dioxide in a limited dose of exposure improves the plant performances, however, other combinations of air pollutants caused damaging effects on productivity of the plants. We describe here physiological and biochemical changes along with growth and yield responses under combined exposure of air pollutants which varied with pollutant combination, their levels and plants under consideration. In combinations of air pollutants, responses of agricultural crops are mostly synergistic with respect to physiological, biochemical, yield and quality traits. The review clearly indicates that combinations of air pollutants have potential to cause significant unfavorable impact on agricultural crops at comparatively lower concentrations.

Keywords: *Air pollutants, Trends, Combined effects, Agriculture, Crops, Phytotoxicity.*

1.1 Introduction

Air pollution is established as a significant threat to crop production throughout the world. However, it is a serious challenge to agricultural productivity in rapidly developing countries (Sharma *et al.*, 2019; Mukherjee *et al.*, 2020).The negative impacts on plants due to air pollutants depend on their concentration, exposure duration and different combinations (Rai *et al.*, 2011; Devrajani *et al.*, 2020). India is experiencing an increase in the number of very high-pollution events from the past few decades (Pommier *et al.*, 2018) due to the high population and rapid urbanization and industrialization in the country (Gao *et al.*, 2019). Rising greenhouse gases (GHGs) such as carbon dioxide (CO_2) and ozone (O_3) due to anthropogenic activities, have become a cause of concern for air quality for this region (Pommier *et al.*, 2018; Gao *et al.*, 2019).

Over the last few years, there has been an intensive effort to address gaps in contemporary knowledge by exploring the plant responses (such as yield, growth, biochemical and physiological changes) to combined stresses (Mishra *et al.*, 2013a; Pandey *et al.*, 2018; Mina *et al.*, 2019; Yadav *et al.*, 2019a).Since many pollutants co-exist in the atmosphere, it is a challenge to identify different stress events and potential phytotoxicity elicited by individual pollutants (Papazian and Blande, 2020). The concomitant responses of different plant parts to multiple stresses may further increase the complexity of plant responses (Danner *et al.*, 2015).The co-exposure of air pollutants may have synergistic (effect is more worse than the simple sum of effects of individual pollutants), additive (effect is equal as the sum of effects of individual pollutants), potentiating (toxic substance made worse by presence of non-toxic substance) or antagonistic (substance reduced by presence of non-toxic substance toxic) effects on the extent of plant response as compared to the effects of individual pollutants (Heagle and Johnston, 1979; Lefohn *et al.*, 2018). In addition, different air pollutants can affect plants differently depending on their mode of action such as depositing on leaf surface and clogging of stomata by aerosols (Mina *et al.*, 2019), atmospheric sulphur and nitrogen deposition absorbed by the foliar tissues and the soil (MacKenzie and Dietrich, 2020) and CO_2 and O_3 absorbed through stomatal uptake of plants (Mishra *et al.*, 2013a; Yadav *et al.*, 2020). Plant exposure to various atmospheric pollutants leads to numerous symptoms depending upon the natural protection mechanisms operated in the plants. Furthermore, effects of air pollutants are mainly dose-dependent and result in detrimental effects on physiological and biochemical processes (Muneer *et al.*, 2014).

For the above reason, the main focus of this review is to explore the plant responses to combined air pollutants based on data available from Indian studies. Moreover, investigating their interactions is important in two key contexts: (1) current status and trends of atmospheric air pollutants in India and the risk of plant exposures; (2) interactive effects of air pollutants in different combinations on growth, physiology, metabolism and yield of crops.

1.2 Current Pollution Levels and Trends of Atmospheric Air Pollutants in India

For understanding the interactive effects of air pollutants on plants, it is important to have knowledge about currently encountered pollution levels and emission trends in different region and the country in particular. Presently, air pollution is prevalent in Indian urban areas where automobiles are the major contributors followed by commercial industries and thermal power plants (Gurjar *et al.*, 2016). Moreover, combustion of fossil fuel and seasonal burning of crop residues in India are also major sources of air pollution. In the past two centuries, anthropogenic emissions of air pollutants such as CO_2, NOx and SO_2 have constantly altered the atmospheric composition by increasing their concentrations in the atmosphere (Papazian and Blande, 2020).

The concentrations of air pollutants vary annually or seasonally and certain pollutants show high upsurge than the others (Koukouli *et al.*, 2018). Variations in weather system can also result in oscillation of pollutant levels. For instance, in Kanpur, diurnal CO, O_3 and NOx mixing ratios follow a similar trend for all seasons (Gaur *et al.*, 2014). However, daytime emissions of the primary pollutants (NOx, SO_2 and CO) were highest in winter and lowest in the monsoon season. Similarly, the O_3 mixing ratio is lowest in the monsoon season, but is the highest in the pre-monsoon (summer) season, when the weather conditions are presumably optimal for photochemical oxidative reactions (Gaur *et al.*, 2014). The monitoring of pollutant fluxes in different locations also highlights the substantial geographical and temporal variations and the wide ranging air pollution scenarios. Consequently, the responses of a particular plant to air pollution has to be examined and it is pertinent to consider the pollution scenarios to which plants will be exposed in their growth stages.

1.2.1. Carbon Dioxide (CO_2)

The CO_2 has the highest concentration in the atmosphere compared to other GHGs. The increasing CO_2 concentration in the atmosphere is mainly because of anthropic activities associated with land-use change and fossil fuel burning with the average increments of 7.8 PgC yr^{-1} and 1.1 PgC yr^{-1}, respectively (Nunes *et al.*, 2020). The release of sequestered CO_2 from forest is 11.6 PgC yr^{-1} primarily because of respiration and forest fire (Nunes *et al.*, 2020). The IPCC A1B emission scenario suggests that at the end of 21st century, the CO_2 concentration will reach up to 700-1000 ppm (IPCC, 2013). The emission of CO_2 from industry and fossil fuels burning continue in an upward direction, with a 2 per cent projected global emission rate and also a similar rate was found for India (Ainsworth *et al.*, 2019).

The atmospheric CO_2 levels are rising continuously over time and have increased from 280 to more than 400 ppm. The global monthly mean CO_2 concentration was recorded 413.61 ppm in May 2020 (https: //www.esrl.noaa.gov/gmd/ccgg/trends/global.html).

Figure 1.1 shows the annual average global CO_2 level from 1990 to 2019, which shows a constant increasing trend across the world.

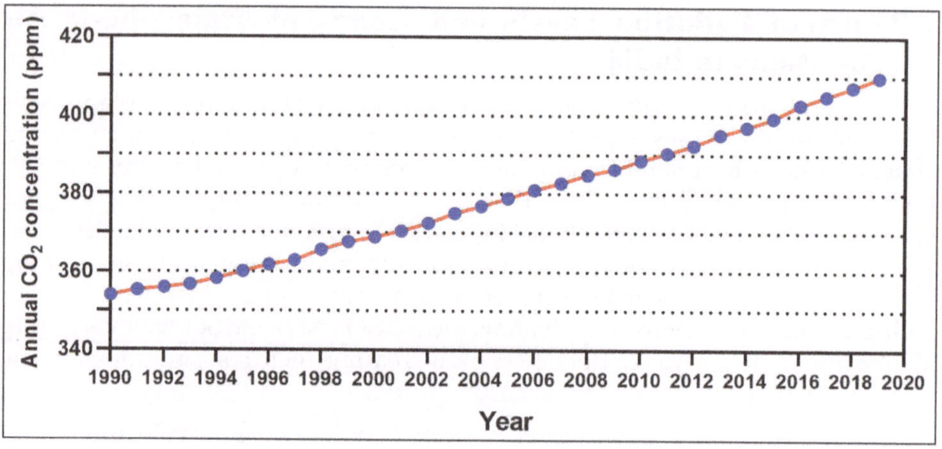

Figure 1.1: Global Annual Average CO_2 Concentration (ppm) Over the Period 1990 – 2019. Data extracted from Global Monitoring Laboratory (NOAA) (https: //www.esrl.noaa.gov/gmd/ccgg/trends/gl_data.html).

1.2.2. Ozone (O_3)

Tropospheric O_3 is a secondary air pollutant which is formed from the photochemical oxidation between O_3 precursors such as carbon monoxide (CO), nitrogen oxides (NOx) and volatile organic compounds (VOCs) (Rai and Agrawal, 2012; Zheng *et al.*, 2018). It is a strong oxidative air pollutant of phytotoxic nature and hence increase in ground level O_3 concentration is thought to be of a great concern in many regions of the world especially in developing countries like India (Ghude *et al.*, 2014; Sicard *et al.*, 2017). Since the start of industrialization (1750), concentration of tropospheric O_3 has increased and now contributes to+0.40 W m^{-2} in global average radiative forcing (IPCC, 2013), which makes O_3 the 3rd most potent anthropogenic GHG after CO_2 and CH_4 (Myhre *et al.*, 2013). The studies have revealed that most fertile and productive agricultural regions in India are facing high surface O_3 concentrations and showed the highest-burden of yield losses (Roy *et al.*, 2009; Ghude *et al.*, 2014). It was also reported that the increase of O_3 concentration in India is mainly because of biomass burning or fire emission which contributed to 4 to 10 per cent in Indo-Gangetic plain (IGP) region and 10 to 25 per cent in Central India during the months of March and April (Jena *et al.*, 2015). Surface O_3concentration in developing countries like India is projected to rise up to 13 per cent by 2050 mainly due to an increase in anthropogenic emissions of O_3 precursors (Pommier *et al.*, 2018), which could enhance the vulnerability of the agriculture sector. In a peri-urban area of Varanasi (in middle IGP region), the mean ambient O_3 concentration during wheat growing season was recorded as 52.7 ppb in year 2017 (Yadav *et al.*, 2019b). Satellite estimates showed increases in O_3 concentration (more than 2 per cent per annum) in Peninsular India, however, no clear trends were observed in peak concentration of mean monthly tropospheric $O_3$for year 2000-2009 in the IGP region of India (David and Nair, 2013). The Population-weighted average surface O_3 concentration from 1990 to 2015 is presented in Figure 1.2. The Increasing Trend of O_3 Levels in the India.

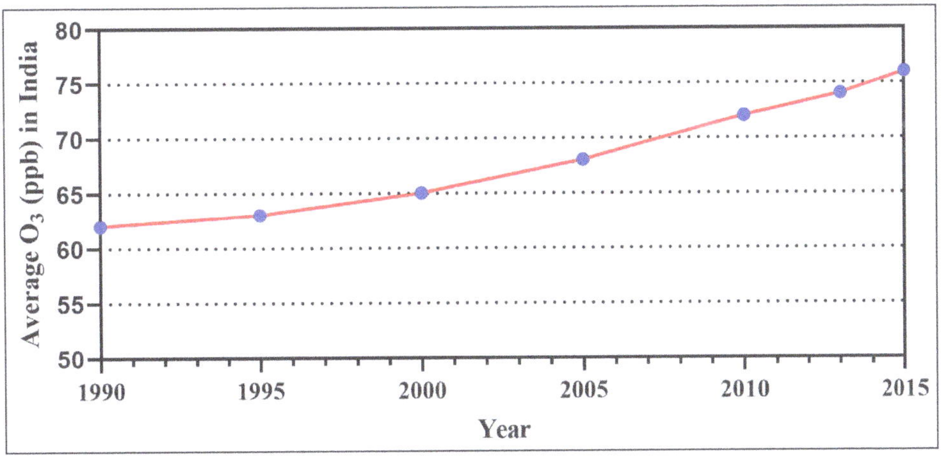

Figure 1.2: Annual Average Data of O_3 (ppb) in India Over the Period 1990 to 2015. Extracted from the State of Global Air (2017) available at https: //ourworldindata.org/grapher/ozone-o3-concentration-in-ppb?tab=chart and country=~IND.

1.2.3. Sulphur Dioxide (SO_2)

Atmospheric SO_2 is one of the major air pollutant, which is emitted in the atmosphere by both natural (*i.e.* volcanic activity)as well as anthropogenic (*i.e.* fossil fuel burning) activities (Sun *et al.*, 2018). SO_2 emissions during 1970s-80s were very high in many parts of Europe and USA. The situation gained attention very quickly and raised social and political awareness as it had affected people's health negatively. Thereafter, many preventive emission policies have come into regulation such as reduction in use of coal and improvement in refining processes. As a result, the global SO_2 emission has declined dramatically from the past thirty years. However, in some fast-developing Asian countries such as China and India, SO_2 is still present in a significant amount in the atmosphere (Smith *et al.*, 2011). Gurjar *et al.* (2016) observed a decreasing trend of SO_2 in all monitoring sites in mega cities of India due to the lowering of sulphur content in coal and diesel. SO_2 concentration has declined throughout the country in 2009 with concentrations varying between 1.5- 13.7 ppb (Rai *et al.*, 2011).

Figure 1.3 shows the annual average SO_2 concentration in India for 2007 to 2018, which is extracted from national ambient air quality monitoring programme. It also shows decline in concentration of SO_2 from past decades.

1.2.4. Nitrogen Oxides (NOx)

According to CPCB 2012, nitrogen dioxide (NO_2) has the second-highest exceeding rate in India after PM10. On a diurnal scale, emissions of NOx are highest in the morning and evening, coinciding with the rush hour of traffic (Tong *et al.*, 2017). The initial NOx peak is swiftly followed by an increase in O_3 concentration in afternoon as temperature and sunlight promote the photochemical reactions for O_3 formation. Therefore, diurnally, O_3 has a tendency to reach at peak after the early rise of NOx level and to decrease with declining sunlight (Tong *et al.*, 2017). In cities,

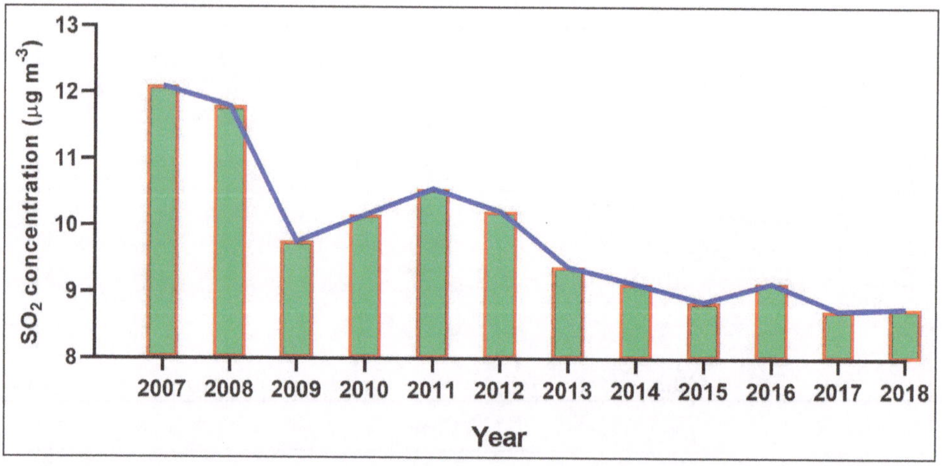

Figure 1.3: Annual Average SO_2 Concentration (µg m^{-3}) in India Over the Period, 2007–2018. Extracted from data of National Ambient Air Quality Monitoring Programme (NAMP) available at CPCB ENVIS (2020) (http: //www.cpcbenvis.nic.in/air_quality_data.html#).

vehicular emissions are the largest contributors of NOx emission in India (Pathak and Bhatia, 2017). The annual NOx emissions from fuel combustion are growing at 6.5 per cent a year in India. Additionally, Indian NOx emissions increased by 52 per cent from 1991 to 2001 and 69 per cent from 2001 to 2011(Sahu *et al.*, 2012). The NOx emission in India was estimated as 3.46 Tg during 1995 with an increase in the rate ~5.6 per cent yr^{-1} (Garg *et al.*, 2001). According to the monitoring data of NAMP, annual concentration of NO_2 went down in India from 2008 to 2015, but again increased in recent years from 2016-2018 **(Figure 1.4).**

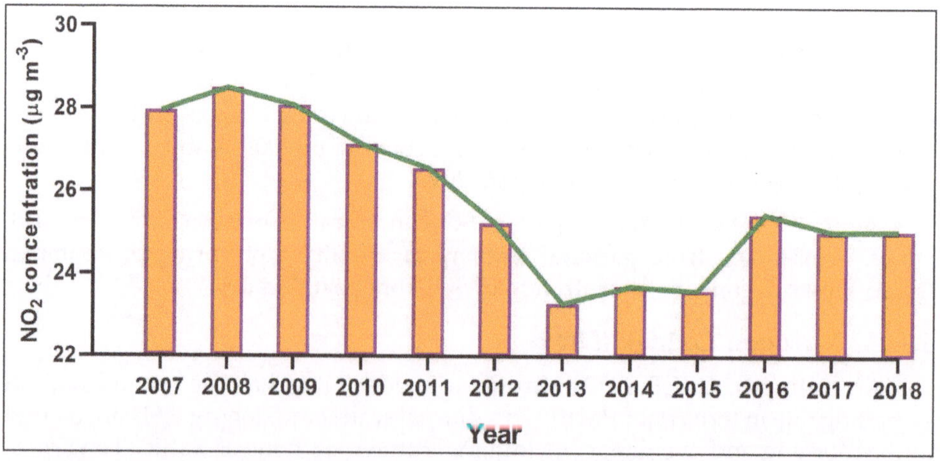

Figure 1.4: Annual Average NO_2 Concentration (µg m^{-3}) in India Over the Period, 2007–2018. Extracted from data of National Ambient Air Quality Monitoring Programme (NAMP) available at CPCB ENVIS (2020) (http: //www.cpcbenvis.nic.in/air_quality_data.html#).

1.2.5. Ultraviolet-B

Due to the stratospheric O_3 layer depletion, biologically active UV-B radiation level has been increased by an average of 7 per cent in northern mid-latitudes since last two decades (Yang *et al.*, 2007; Gilbert *et al.*, 2009). It was further estimated that there will be maximum 14 per cent annual increase in UV radiation during 2010–2020 in the northern hemisphere (Taalas *et al.*, 2000). Recently, the O_3 layer over the Arctic region has reached record-breaking low values in April 2020. Similarly, a strong chemical O_3 depletion was observed during spring 2011 over the same region and O_3 depletion in 2020 seems to be even stronger than the last time (Atmospheric Monitoring Service, 2020). Tropospheric O_3 level increases in parallel to UV-B flux densities on the Earth's surface (Ambasht and Agrawal, 2003). Therefore, tropospheric O_3 and UV-B radiation may affect the plants in combination (Tripathi and Agrawal, 2013a, 2013b; Tripathi *et al.*, 2019).

1.2.6. Aerosols

Atmospheric aerosols are solid or liquid particles present in the air. India is globally the largest source for ammonia emission and nearly double NOx emissions, which are important sources of atmospheric aerosols as the particle conversion is facilitated (Sharma *et al.*, 2007; Behera and Sharma, 2010). Aerosols can also be produced from dust storms, forest fires, volcanic eruptions, burning of fossil fuels and crop lands. Aerosols regulate weather conditions of a region by absorbing/emitting radiation, scattering incoming solar radiation and reflecting back from the Earth surface and by altering cloud patterns(Mina *et al.*, 2019). Aerosols, thus reduce the quantum of photosynthetic active radiation (PAR),which is crucial for plant growth and photosynthesis (Latha *et al.*, 2017).A thick layer of anthropogenically emitted absorbing aerosols is reported in the IGP region of India (Ramanathan and Ramana, 2005). Lal *et al.* (2012) reported that aerosol index (AI) was growing with 1.7 ± 1.2 per cent per year in North India over the IGP region during 1979-1992. However, aerosol levels are low in year 2020 from the past two decades due to the strict lock down during the COVID 19 Pandemic(NASA, 2020). Clogging of stomata and shading of stomatal pores are the primary effects of aerosol deposition, which cause abrasion on plants (Gupta *et al.*, 2016; Mina *et al.*, 2018). Small particles of aerosols (up to 2 µm) can enter into the leaf through stomatal aperture and cause internal leaf injury (Song *et al.*, 2015). Furthermore, aerosols also contain heavy metals like Mn, Zn, Pb and when deposited on leaf surfaces or settle down on soil surfaces, can cause metal toxicity to crops by affecting the mesophyll functions of the leaves (Burkhardt *et al.*, 2012).

1.3 Plant Responses to Combinations of Ozone and other Air Pollutants

In this review, we are summarizing current knowledge on agricultural crop responses to most studied pollutant combinations with O_3 such as CO_2, SO_2, NOx, UV-B and aerosols in Indian context. Most of the air pollutants (such as O_3, CO_2, SO_2, and NOx) interact with plants at the leaf level by entering through the stomatal pores and making a way into the intracellular spaces (Mukherjee *et al.*, 2020).

1.3.1. Crop Responses to Combination of O_3 and CO_2

Knowledge of crop responses to a combination of elevated CO_2 and O_3concentrations are an important issue for future global climate change as both the GHGs have strong effects on photosynthesis and crop production. Usually, higher CO_2 has a growth-stimulating effect, while O_3 tends to have the opposite effect on plants. In general, high CO_2 diminishes the O_3-induced damages and crop yield losses by reducing the stomatal O_3 uptake via a reduction in stomatal conductance and by increasing detoxification capacity of the plants (Mishra *et al.*, 2019). In a number of crop species, foliar O_3 injury was reduced significantly due to elevated CO_2, for instance, wheat (Mishra *et al.*, 2013a) and potato (Kumari and Agrawal, 2014).The studies conducted to evaluate the combined effects of CO_2 and O_3showed variable plant responses (**Table 1.1**). The protective effect of elevated CO_2 against O_3 injury involves increases in photosynthetic rate and antioxidant enzymes, which increases the ability of plants to tolerate O_3 toxicity, leading to increase in grain yield of crops, (Rao *et al.*, 1995; Mishra *et al.*, 2013a, 2013b; Yadav *et al.*, 2019a). High CO_2 and O_3treatments synergistically enhanced the oil content of Indian mustard compared to O_3 treatment (Singh *et al.*, 2013). Kumari *et al.* (2013) have conducted an experiment on beet plant by exposing individual and combined treatments of O_3 (ambient +20ppb) and CO_2 (550 ppm). The combined treatment increased beet biomass, lipid peroxidation, enzymatic and non-enzymatic antioxidants, while decreased root and shoot length, photosynthetic pigments and nutritional quality. Similar effects on growth, physiology and biochemical constituents at the same concentrations of $O_3$and CO_2were reported on potato plants (Kumari *et al.*, 2015; Kumari and Agrawal, 2014). In combination, high CO_2counteracted the damaging effects of O_3 by reducing oxidative stress and O_3flux inside the leaves and also by improving photosynthetic performance, thus increased carbon assimilation (Kumari *et al.*, 2015).

1.3.2. Crop Responses to Combination of O_3, NO_2 and SO_2

Many atmospheric gaseous substances are toxic to the plants. The three most important gases in the order of their phytotoxicity are O_3,SO_2, and NO_2.Only O_3in combination with SO_2 and NO_2, was found to be responsible for ~90 per cent of the crop losses caused from air pollutants in the United States (Reinert, 1984). The combined treatment of O_3+SO_2 and O_3+NO_2 treatments induced the defense mechanism by reducing stomatal conductance to inhibit the excess uptake in plants. Thus, it reduced the toxicity levels into the cellular compartments with an inadvertent lower CO_2 uptake, resulting in decline of net photosynthesis (Rao *et al.*, 1983). A biomonitoring study in peri-urban areas of Varanasi reported that the lowest photosynthetic rate, chlorophyll content, above ground biomass and yield in wheat and mustard during winter and in mung bean and spinach during summer period at a site showing highest concentrations of SO_2 and NO_2 and second highest concentration of O_3 (Agrawal *et al.*, 2003). Synergistic adverse effects on growth, physiology, oxidative and antioxidative properties of carrot plants under ambient air pollution having different concentrations of O_3, SO_2, and NO_2have been reported by Tiwari *et al.* (2006) (**Table 1.1**). Pandey *et al.* (2018) reported better antioxidant defense and improvement in the yield under combined treatments of O_3 and nitrogen compared to only O_3 treated wheat plants. The interactive effect of O_3 and SO_2 was synergistic causing greater adverse impact on leaf injury, pigments and yield than the addition of individual effects (Agrawal *et al.*, 1982; Khan and Khan, 1994).Low O_3 and SO_2 concentrations in combination exhibited foliar injury on leaves of rice plants (Agrawal *et al.*, 1982). Further, reductions in photosynthetic pigments and accessory pigments were higher under combined treatment of O_3+SO_2 compared to single treatments. Khan and Khan (1994) have also performed a study with O_3 and SO_2 on tomato plant and found a synergistic impact on reducing stomatal numbers, photosynthetic pigments and yield (**Table 1.1**). Studies also reported that at lower doses of SO_2 can be beneficial to the plants by contributing to the important mineral pool (Deepak and Agrawal, 1999), which is used as a source of sulphur in plant metabolism to synthesize amino acids, proteins, vitamins and cofactors (Muneer *et al.*, 2014; Chandra and Pandey, 2016).

1.3.3. Crop Responses to Combination of O_3 and UV-B

Increased tropospheric O_3 concentration and UV-B radiation may occur together and hence may affect plant physiology, biochemical characteristics and yield of the plants (Ambasht and Agrawal, 2003; Tripathi and Agrawal, 2013a; Tripathi *et al.*, 2019). In India, tropospheric O_3 concentrations are above the critical level. The interactive effects of O_3 and UV-B have been focused on plant growth and of wheat, linseed and sunflower (**Table 1.1**). Ambasht and Agrawal (2003) have found less than additive interactive effects of O_3 and UV-B on photosynthetic pigments, antioxidants, biomass and yield. Similarly, the effect ofO_3 and UV-B together was less than individual effects on linseed plant. The combined treatment also altered metabolite properties, antioxidants and leaf proteome and genome of linseed (Tripathi *et al.*, 2011). Additionally, combined treatment promoted the ROS production, lipid peroxidation, secondary metabolites, but caused reductions in plant biomass and seed yield of linseed and sunflower along with deterioration of oil quality (Tripathi and Agrawal, 2013a, 2013b;Tripathi *et al.*, 2019) (**Table 1.1**).

1.3.4. Crop Responses to Combination of O_3 and Aerosol

Tropospheric O_3 and aerosols both have detrimental effects on agricultural productivity. Model studies have confirmed that O_3 and aerosols contributed to large yield gaps in various crops in IGP region (Burney and Ramanathan, 2014). Mina *et al.* (2019) have observed that the exposure of both elevated O_3 and aerosol simultaneously on a wheat cultivar (HD 2967), caused adverse effects on growth, biochemical and physiological attributes. Further, combined treatment of O_3 and aerosols enhanced stress synergistically on photosynthesis rate, stomatal conductance, transpiration and eventually on grain yield(up to 17.3 per cent loss) in wheat cultivar HD 2967 (**Table 1.1**). Burney and Ramanathan (2014) calculated 18.9 per cent yield loss in wheat during 2010 due to combined effect of O_3 and aerosols in India through a simulation-based study.

1.4. Conclusion

Plant responses to a combination of air pollutants are least studied and under-represented in the literature. We have outlined here what is known regarding

Table 1.1. Summary of Indian Studies on Interactive Effects of Air Pollutants on Plants

Air Pollutant Interaction	*Experimental Condition*	*Plant Species*	*Responses*	*References*
O_3 (120 ppb) + CO_2 (800 ppm)	CSTR	Wheat (*Triticum aestivum* L.)	Ameliorating effects on shoot biomass compared to O_3 treatment. Preferential changes in the composition of antioxidant enzymes; synthesis of a new isoform of glutathione reductase.	Rao *et al.* (1995)
O_3 (Ambient +10ppb) + CO_2 (700 ppb)	OTCs		Increases in plant height, number of leaves, total biomass, growth indices, grain yield and quality compared to ambient treatment	Mishra *et al.* (2013a)
O_3 (Ambient +10ppb) + CO_2 (700 ppm)	OTCs		Increases in chlorophyll content, photosynthesis rate, ROS levels and antioxidant enzymes, while decrease in stomatal conductance and carotenoid content.	Mishra *et al.* (2013b)
O_3 (70 ppb) + CO_2 (550 ppm)	FAOCE		Increases in leaf area index, photosynthesis rate and grain yield parameters compared to ambient treatment	Yadav *et al.* (2019a)
O_3 (ambient +20ppb) + CO_2 (550 ppm)	OTCs	Beet (*Beta vulgaris* L.)	Increases in plant biomass, enzymatic and non-enzymatic antioxidants and lipid peroxidation. Decreases in root and shoot length, photosynthetic pigments and nutritional quality.	Kumari *et al.* (2013)
O_3 (ambient +30ppb) + CO_2 (550 ppm)	OTCs	Mustard (*Brassica juncea* L.)	Synergistic action on yield and nutritional quality. Increase in oil content of seeds compared to only O_3 treated plants.	Singh *et al.* (2013)
O_3 (ambient +20ppb) + CO_2 (550 ppm)	OTCs	Potato (*Solanum tuberosum* L.)	Increases in plant growth, biomass, starch, sugar and organic carbon content. Decreases in nitrogen, amino acid and protein contents.	Kumari and Agrawal (2014)
O_3 (ambient +20ppb) + CO_2 (550 ppm)	OTCs	Potato (*Solanum tuberosum* L.) Chickpea (*Cicer arietinum* L.)	Increases in photosynthetic pigments and photosynthesis rate, antioxidative enzymes, total phenolics and ascorbic acid content Reductions in stomatal conductance, antioxidant enzymes and protein contents.	Kumari *et al.* (2015)
O_3 (70 ppb) + CO_2 (550 ppm)	FAOCE		Combined effects eliminate the negative impact of O_3 on growth, biomass and WUE.	Singh *et al.* (2017)
O_3 (40 ppb) + SO_2 (0.250 ppm)	Exposure chambers	Rice (*Oryza sativa* L.)	Synergistic impact on leaves in terms of leaf injury, reductions in chlorophyll and carotenoid contents	Agrawal *et al.* (1982)

Air Pollutant Interaction	*Experimental Condition*	*Plant Species*	*Responses*	*References*
O_3 (50 ppb) + SO_2 (50 ppb)	Exposure chambers	Tomato (*Lycopersicon esculentum* L.)	Synergistic impact on reduction in number of stomata and photosynthetic pigments and yield of plants decline and increment in size of stomatal pores, reduction in	Khan and Khan (1994)
O_3 + SO_2 + NO_2 (ambient)	In situ exposure	Wheat, mustard, mung bean (*Vigna radiata* L.) and palak (*Spinacia oleracea* L.)	Reduction in photosynthetic rate, stomatal conductance, chlorophyll content, plant height, above ground biomass and yield	Agrawal *et al.* (2003)
O_3 + SO_2 + NO_2 (ambient)	OTCs	Carrot (*Dacus carota* L.)	Synergistic negative effects of combined pollutants on dry weight, photosynthesis rate, stomatal conductance; shoot length, leaf number and leaf area reduced. Increase in antioxidants.	Tiwari *et al.* (2006)
O_3 (ambient +20 ppb) + N (1.5 × 120 kg ha^{-1})	OTCs	Wheat (*Triticum aestivum* L.)	Enhanced antioxidant defense, reduced the yield losses due to O_3.	Pandey *et al.* (2018)
O_3 (70 ppb) + UV-B (ambient +7.1kJ m^{-2})	OTCs	Wheat (*Triticum aestivum* L.)	Less than additive effects on biomass, yield, photosynthetic rate, chlorophyll, carotenoid and ascorbic acid contents, catalase activity and total phenol content.	Ambasht and Agrawal (2003)
O_3 (ambient +10 ppb) + UV-B (ambient +7.2kJ m^{-2})	OTCs	Linseed (*Linum usitatissimum* L.)	Less than additive effects compared to individual UV-B or O_3. Alterations in metabolites, antioxidants, leaf proteome and genome	Tripathi *et al.*, 2011)
O_3 (ambient +10 ppb) + UV-B (ambient +7.2kJ m^{-2})	OTCs	Linseed (*Linum usitatissimum* L.) Sunflower (*Helianthus annuus* L.)	Less than additive effects on yield, oil quality and seeds.	Tripathi and Agrawal (2013a)
O_3 (ambient +10 ppb) + UV-B (ambient +7.2kJ m^{-2})	OTCs		Less than additive effects on induction of ROS, membrane damage, secondary metabolites, biomass and seed yield.	Tripathi and Agrawal (2013b)
O_3 (ambient +10 ppb) + UV-B (ambient +7.2kJ m^{-2})	OTCs		Reductions in leaf area, number of heads, weight of heads, oil content and quality.	Tripathi *et al.* (2019)
O_3 (70 ppb) + Aerosol	FAOCE	Wheat (*Triticum aestivum* L.)	Reductions in air pollution tolerance, gaseous exchange, clogging of stomata and crop yield	Mina *et al.* (2019)

CSTR: Continuous Stirred Tank Reactor; OTC: Open Top Chamber; FAOCE: Free Air Ozone and Carbon dioxide Enrichment facility.

the plant responses to combination of air pollutants through Indian studies. Carbon dioxide is not considered as a harmful air pollutants for plant health as CO_2 is a substrate for photosynthesis. However, O_3 exposure always affects plant performance negatively. It is evident that the magnitude of negative effects of individual pollutants are largely different from combined effects. The damaging effects of O_3 on plants are often mitigated by simultaneous exposure to high CO_2, while combination of O_3+SO_2 showed synergistic negative effects on plants. The combination of O_3+UV-B treatment caused less than additive negative effects.

In order to develop a clear understanding of how plants respond to combined air pollutants, it is important to take site specific measurements of pollutant concentrations. The predictions of future pollution levels based on emissions will be helpful to estimate their future impacts on plants. Recent trends to explore the effects of multiple stresses on plants and vegetation should be extended to take a broad range of biotic and abiotic factors. Responses of plants under combinations of stresses will enable the understanding of the harmful effects of air pollution in a holistic manner and will allow the development of better guidance for essential environmental policies.

Acknowledgements

Authors are thankful to the Head, Department of Botany and the Coordinators, CAS in Botany and Interdisciplinary School of Life Sciences of Banaras Hindu University, Varanasi for providing necessary facilities for conducting systematic literature review. Durgesh Singh Yadav is grateful to the University Grant Commission, New Delhi for providing research fellowship.

REFERENCES

Agrawal, M. Singh, B. Rajput, M. Marshall, F. and Bell, J. N. B. (2003). Effect of air pollution on peri-urban agriculture: A case study. *Environmental Pollution*, 126(3): 323–329. https: //doi.org/10.1016/S0269-7491(03)00245-8

Agrawal, M. Nandi, P. K. and Rao, D. N. (1982). Effect of ozone and sulphur dioxide pollutants separately and in mixture on chlorophyll and carotenoid pigments of *Oryzasativa*. *Water, Air, and Soil Pollution*, 18(4): 449–454. https: //doi.org/10.1007/BF02426052

Ainsworth, E. A. Lemonnier, P. and Wedow, J. M. (2019). The influence of rising tropospheric carbon dioxide and ozone on plant productivity. *Plant Biology*, 22(S1): 5–11. https: //doi.org/10.1111/plb.12973

Ambasht, N. K. and Agrawal, M. (2003). Effects of enhanced UV-B radiation and tropospheric ozone on physiological and biochemical characteristics of field grown wheat. *BiologiaPlantarum*, 47(4): 625–628. https: //doi.org/10.1023/B: BIOP.0000041076.95209.c3

Atmospheric Monitoring Service (2020). CAMS tracks a record-breaking Arctic ozone hole | Copernicus. https: //atmosphere.copernicus.eu/cams-tracks-record-breaking-arctic-ozone-hole

Behera, S. N. and Sharma, M. (2010). Investigating the potential role of ammonia in ion chemistry of fine particulate matter formation for an urban environment. *Science of the Total Environment*, 408(17): 3569–3575. https: //doi.org/10.1016/j.scitotenv.2010.04.017

Burkhardt, J. Basi, S. Pariyar, S. and Hunsche, M. (2012). Stomatal penetration by aqueous solutions-an update involving leaf surface particles. *New Phytologist*, 196(3), 774–787. https: //doi.org/10.1111/j.1469-8137.2012.04307.x

Burney, J. and Ramanathan, V. (2014). Recent climate and air pollution impacts on Indian agriculture. *Proceedings of the National Academy of Sciences of the United States of America*, 111(46), 16319–16324. https: //doi.org/10.1073/pnas.1317275111

Chandra, N. and Pandey, N. (2016). Role of sulfur nutrition in plant and seed metabolism of *Glycine max* L. *Journal of Plant Nutrition*, 39(8): 1103–1111. https: //doi.org/10.1080/01904167.2016.1143495

CPCB ENVIS (2020). National Air Quality Monitoring Programme (NAMP). http: //www.cpcbenvis.nic.in/air_quality_data.html#

Danner, H. Brown, P. Cator, E. A. Harren, F. J. M. van Dam, N. M. and Cristescu, S. M. (2015). Aboveground and belowground herbivores synergistically induce volatile organic sulfur compound emissions from shoots but not from roots. *Journal of Chemical Ecology*, 41(7): 631–640. https: //doi.org/10.1007/s10886-015-0601-y

David, L. M. and Nair, P. R. (2013). Tropospheric column O_3 and NO_2 over the Indian region observed by Ozone Monitoring Instrument (OMI): Seasonal changes and long-term trends. *Atmospheric Environment*, 65: 25–39. https: //doi.org/10.1016/j.atmosenv.2012.09.033

Deepak, S. S. and Agrawal, M. (1999). Growth and yield responses of wheat plants to elevated levels of CO_2 and SO_2, singly and in combination. *Environmental Pollution*, 104(3), 411–419. https: //doi.org/10.1016/S0269-7491(98)00171-7

Devrajani, S. K. Qureshi, M. Imran, U. and Nisa, T. U. (2020). Impact of gaseous air pollutants on agricultural crops in developing countries: A review. *International Journal of Research in Environmental Science*, 6(2): 71–82. https: //doi.org/10.20431/2454-9444.0602001

Gao, M. Gao, J. Zhu, B. Kumar, R. Lu, X. Song, S. Zhang, Y. Wang, P. ... McElroy, M. (2019). Ozone pollution over China and India: Seasonality and sources. *Atmospheric Chemistry and Physics Discussions*, 2000: 1–29. https: //doi.org/10.5194/acp-2019-875

Garg, A. Shukla, P. R. Bhattacharya, S. and Dadhwal, V. K. (2001). Sub-region (district) and sector level SO_2 and NO(x) emissions for India: Assessment of inventories and mitigation flexibility. *Atmospheric Environment*, 35(4): 703–713. https: //doi.org/10.1016/S1352-2310(00)00316-2

Gaur, A. Tripathi, S. N. Kanawade, V. P. Tare, V. and Shukla, S. P. (2014). Four year measurements of trace gases (SO_2, NOx, CO, and O_3) at an urban location,

Kanpur, in Northern India. *Journal of Atmospheric Chemistry*, 71(4): 283–301. https: //doi.org/10.1007/s10874-014-9295-8

Ghude, S. D. Jena, C. Chate, D. M. Beig, G. Pfister, G. G. Kumar, R. and Ramanathan, V. (2014). Reductions in India's crop yield due to ozone. *Geophysical Research Letters*, 41(15): 5685–5691. https: //doi.org/10.1002/2014GL060930

Gilbert, M. Pörs, Y. Grover, K. Weingart, I. Skotnica, J. Grimm, B. Seidlitz, H. K. Langebartels, C. and Wilhelm, C. (2009). Intra- and interspecific differences of 10 barley and 10 tomato cultivars in response to short-time UV-B radiation: A study analysing thermoluminescence, fluorescence, gas-exchange and biochemical parameters. *Environmental Pollution*, 157(5): 1603–1612. https: //doi.org/10.1016/j.envpol.2008.12.023

Gupta, G. P. Kumar, B. Singh, S. and Kulshrestha, U. C. (2016). Deposition and impact of urban atmospheric dust on two medicinal plants during different seasons in NCR Delhi. *Aerosol and Air Quality Research*, 16(11), 2920–2932. https: //doi.org/10.4209/aaqr.2015.04.0272

Gurjar, B. R. Ravindra, K. and Nagpure, A. S. (2016): Air pollution trends over Indian megacities and their local-to-global implications. *Atmospheric Environment*, 142, 475–495. https: //doi.org/10.1016/j.atmosenv.2016.06.030

Heagle, A. S. and Johnston, J. W. (1979). Variable responses of soybeans to mixtures of ozone and sulfur dioxide. *Journal of the Air Pollution Control Association*, 29(7): 729–732. https: //doi.org/10.1080/00022470.1979.10470855

IPCC. (2013). In: Stocker TF, Qin D, Plattner G-K, Tignor M, Allen SK, Boschung J, Nauels A, Xia Y, Bex V, Midgley PM (eds) Climate Change 2013: The Physical Science Basis. Contribution of Working Group I to the Fifth Assessment Report of the Intergovernmental Panel on Climate Change. *Cambridge University Press*, Cambridge, UK\New York, NY.

Jena, C. Ghude, S. D. Pfister, G. G. Chate, D. M. Kumar, R. Beig, G. Surendran, D. E. Fadnavis, S. and Lal, D. M. (2015). Influence of springtime biomass burning in South Asia on regional ozone (O_3): A model based case study. *Atmospheric Environment*, 100: 37–47. https: //doi.org/10.1016/j.atmosenv.2014.10.027

Khan, M. R. and Khan, M. W. (1994). Single and interactive effects of O_3 and SO_2 on tomato. *Environmental and Experimental Botany*, 34(4), 461–469. https: //doi.org/10.1016/0098-8472(94)90029-9

Koukouli, M. Theys, N. Ding, J. Zyrichidou, I. Mijling, B. Balis, D. and Johannes Van Der A, R. (2018). Updated SO_2 emission estimates over China using OMI/Aura observations. *Atmospheric Measurement Techniques*, 11(3), 1817–1832. https: //doi.org/10.5194/amt-11-1817-2018

Kumari, S. and Agrawal, M. (2014). Growth, yield and quality attributes of a tropical potato variety (*Solanumtuberosum* L. cv*Kufri chandramukhi*) undor ambient and elevated carbon dioxide and ozone and their interactions. *Ecotoxicology and Environmental Safety*, 101(1): 146–156. https: //doi.org/10.1016/j.ecoenv.2013.12.021

Kumari, S. Agrawal, M. and Singh, A. (2015). Effects of ambient and elevated CO_2 and ozone on physiological characteristics, antioxidativedefense system and metabolites of potato in relation to ozone flux. *Environmental and Experimental Botany*, 109, 276–287. https: //doi.org/10.1016/j.envexpbot.2014.06.015

Kumari, S. Agrawal, M. and Tiwari, S. (2013). Impact of elevated CO_2 and elevated O_3 on *Beta vulgaris* L.: Pigments, metabolites, antioxidants, growth and yield. *Environmental Pollution*, 174, 279–288. https: //doi.org/10.1016/j.envpol.2012.11.021

Lal, D. M. Ghude, S. D. Patil, S. D. Kulkarni, S. H. Jena, C. Tiwari, S. and Srivastava, M. K. (2012). Tropospheric ozone and aerosol long-term trends over the Indo-Gangetic Plain (IGP), India. *Atmospheric Research*, 116: 82–92. https: //doi.org/10.1016/j.atmosres.2012.02.014

Latha, R. Murthy, B. S. Lipi, K. Srivastava, M. K. and Kumar, M. (2017). Absorbing aerosols, possible implication to crop yield - a comparison between IGB stations. *Aerosol and Air Quality Research*, 17(3), 693–705. https: //doi.org/10.4209/aaqr.2016.02.0054

Lefohn, A. S. Malley, C. S. Smith, L. Wells, B. Hazucha, M. Simon, H. Naik, V. Mills, G. ... Gerosa, G. (2018). Tropospheric ozone assessment report: Global ozone metrics for climate change, human health, and crop/ecosystem research. *Elementa*, 6, 1. https: //doi.org/10.1525/elementa.279

MacKenzie, M. D. and Dietrich, S. T. (2020). Atmospheric sulfur and nitrogen deposition in the Athabasca oil sands region is correlated with foliar nutrient levels and soil chemical properties. *Science of the Total Environment*, 711: 134737. https: //doi.org/10.1016/j.scitotenv.2019.134737.

Mina, U. Chandrashekara, T. K. Kumar, S. N. Meena, M. C. Yadav, S. Tiwari, S. Singh, D. Kumar, P. and Kumar, R. (2018). Impact of particulate matter on basmati rice varieties grown in Indo-Gangetic Plains of India: Growth, biochemical, physiological and yield attributes. *Atmospheric Environment*, 188: 174–184. https: //doi.org/10.1016/j.atmosenv.2018.06.015.

Mina, U. Kandpal, A. Bhatia, A. Ghude, S. Singh, D. and Kumar, P. (2019). Quantitative assessment of individual and combined impact of ozone and aerosols on wheat cultivar: Growth, development, biochemical, physiological and yield attributes. *Agricultural and Forest Meteorology*. https: //doi.org/10.1016/j.agrformet.2019.107817.

Mishra, A. K. Agrawal, S. B. and Agrawal, M. (2019). Rising atmospheric carbon dioxide and plant responses: Current and future consequences. *In Climate Change and Agricultural Ecosystems: Current Challenges and Adaptation*. https: //doi.org/10.1016/B978-0-12-816483-9.00011-6.

Mishra, A. K. Rai, R. and Agrawal, S. B. (2013a). Differential response of dwarf and tall tropical wheat cultivars to elevated ozone with and without carbon dioxide enrichment: Growth, yield and grain quality. *Field Crops Research*, 145: 21–32. https: //doi.org/10.1016/j.fcr.2013.02.007.

Mishra, A. K. Rai, R. and Agrawal, S. B. (2013b). Individual and interactive effects of elevated carbon dioxide and ozone on tropical wheat (*Triticumaestivum* L.) cultivars with special emphasis on ROS generation and activation of antioxidant defence system. *Indian Journal of Biochemistry and Biophysics*, 50(2): 139–149.http: //nopr.niscair.res.in/handle/123456789/17098.

Mukherjee, A. Agrawal, S. B. and Agrawal, M. (2020). Responses of tropical tree species to urban air pollutants: ROS/RNS formation and scavenging. *Science of the Total Environment*, 710: 136363. https: //doi.org/10.1016/j.scitotenv.2019.136363.

Muneer, S. Kim, T. H. Choi, B. C. Lee, B. S. and Lee, J. H. (2014): Effect of CO, NOx and SO_2 on ROS production, photosynthesis and ascorbate-glutathione pathway to induce *Fragariaannasa* as a hyperaccumulator. *Redox Biology*, 2(1), 91–98. https: //doi.org/10.1016/j.redox.2013.12.006.

Myhre, G. Shindell, D. Bréon, F. M. Collins, W. Fuglestvedt, J. Huang, J. Koch, D. Lamarque, J. F. … Zhang, H. (2013). Anthropogenic and natural radiative forcing, In Climate Change 2013: The Physical Science Basis. Contribution of Working Group I to the Fifth Assessment Report of the Intergovernmental Panel on Climate Change. (pp. 659–740). *Cambridge University Press*,Cambridge, UK and New York, USA. https: //doi.org/https: //doi.org/10.1017/CBO9781107415324.018.

NASA. (2020). Airborne particle levels plummet in Northern India. *NASA Earth Observatory*. https: //earthobservatory.nasa.gov/images/146596/airborne-particle-levels-plummet-in-northern-india.

Nunes, L. J. R. Meireles, C. I. R. Gomes, C. J. P. and Ribeiro, N. M. C. A. (2020). Forest contribution to climate change mitigation: Management oriented to carbon capture and storage. *In Climate*, 8(2): 21. https: //doi.org/10.3390/cli8020021.

Pandey, A. K. Ghosh, A. Agrawal, M. and Agrawal, S. B. (2018). Effect of elevated ozone and varying levels of soil nitrogen in two wheat (*Triticumaestivum* L.) cultivars: Growth, gas-exchange, antioxidant status, grain yield and quality. *Ecotoxicology and Environmental Safety*, 158: 59–68. https: //doi.org/10.1016/j.ecoenv.2018.04.014.

Papazian, S. and Blande, J. D. (2020). Dynamics of plant responses to combinations of air pollutants. *Plant Biology*, 22(S1), 68–83. https: //doi.org/10.1111/plb.12953.

Pathak, H. and Bhatia, A. (2017). Reactive Nitrogen and Its Impacts on Climate Change: An Indian Synthesis. *In The Indian Nitrogen Assessment: Sources of Reactive Nitrogen, Environmental and Climate Effects, Management Options, and Policies* (Issue 3). https: //doi.org/10.1016/B978-0-12-811836-8.00024-0

Pommier, M. Fagerli, H. Gauss, M. Simpson, D. Sharma, S. Sinha, V. Ghude, S. D. Landgren, O. … Wind, P. (2018). Impact of regional climate change and future emission scenarios on surface O3 and PM2.5 over India. *Atmospheric Chemistry and Physics*, 18(1): 103–127. https: //doi.org/10.5194/acp-18-103-2018.

Rai, R. and Agrawal, M. (2012). Impact of tropospheric ozone on crop plants. *Proceedings of theNational Academy of Sciences India Section B - Biological Sciences*, 82(2), 241–257. https: //doi.org/10.1007/s40011-012-0032-2.

Rai, R. Rajput, M. Agrawal, M. and Agrawal, S. B. (2011). Gaseous air pollutants/ : A review on current and future trends of emissions and impact on agriculture. *Journal of Scientific Research*, 55(771): 1. http: //citeseerx.ist.psu.edu/viewdoc/summary?doi=10.1.1.487.785.

Ramanathan, V. and Ramana, M. V. (2005). Persistent, widespread, and strongly absorbing haze over the Himalayan foothills and the Indo-Gangetic Plains. *Pure and Applied Geophysics*, 162(8–9), 1609–1626. https: //doi.org/10.1007/s00024-005-2685-8.

Rao, I. M. Amundson, R. G. Alscher Herman, R. and Anderson, L. E. (1983). Effects of SO_2 on stomatal metabolism in *Pisumsativum* L. *Plant Physiology*, 72(2): 573–577. https: //doi.org/10.1104/pp.72.2.573.

Rao, M. Hale, B. A. and Ormrod, D. P. (1995). Amelioration of ozone-induced oxidative damage in wheat plants grown under high carbon dioxide: Role of antioxidant enzymes. *Plant Physiology*, 109(2), 421–432. https: //doi.org/10.1104/pp.109.2.421.

Reinert, R. A. (1984). Plant response to air pollutant mixtures. *Annual Review of Phytopathology*, 22(1): 421–442. https: //doi.org/10.1146/annurev.py.22.090184.002225.

Roy, S. D. Beig, G. and Ghude, S. D. (2009). Exposure-plant response of ambient ozone over the tropical Indian region. *In Atmospheric Chemistry and Physics*9(14), 5253–5260). https: //doi.org/10.5194/acp-9-5253-2009.

Sahu, S. K. Beig, G. and Parkhi, N. S. (2012). Emerging pattern of anthropogenic NO_X emission over Indian subcontinent during 1990s and 2000s. *Atmospheric Pollution Research*, 3(3): 262–269. https: //doi.org/10.5094/APR.2012.021.

Sharma, A. Ojha, N. Pozzer, A. Beig, G. and Gunthe, S. S. (2019). Revisiting the crop yield loss in India attributable to ozone. *Atmospheric Environment*: 1, 100008. https: //doi.org/10.1016/j.aeaoa.2019.100008.

Sharma, M. Kishore, S. Tripathi, S. N. and Behera, S. N. (2007). Role of atmospheric ammonia in the formation of inorganic secondary particulate matter: A study at Kanpur, India. *Journal of Atmospheric Chemistry*, 58(1): 1–17. https: //doi.org/10.1007/s10874-007-9074-x.

Sicard, P. Anav, A. De Marco, A. and Paoletti, E. (2017). Projected global ground-level ozone impacts on vegetation under different emission and climate scenarios. *Atmospheric Chemistry and Physics*, 17(19): 12177–12196. https: //doi.org/10.5194/acp-17-12177-2017.

Singh, R. N. Mukherjee, J. Sehgal, V. K. Bhatia, A. Krishnan, P. Das, D. K. Kumar, V. and Harit, R. (2017). Effect of elevated ozone, carbon dioxide and their interaction on growth, biomass and water use efficiency of chickpea (*Cicerarietinum* L.). *Journal of Agrometeorology*, 19(4): 301–305.

Singh, S. Bhatia, A. Tomer, R. Kumar, V. Singh, B. and Singh, S. D. (2013). Synergistic action of tropospheric ozone and carbon dioxide on yield and nutritional quality of Indian mustard (*Brassica juncea* (L.) *Czern.*). *Environmental Monitoring and Assessment*, 185(8): 6517–6529. https: //doi.org/10.1007/s10661-012-3043-9

Smith, S. J. Van Aardenne, J. Klimont, Z. Andres, R. J. Volke, A. and Delgado Arias, S. (2011). Anthropogenic sulfur dioxide emissions: 1850-2005. *Atmospheric Chemistry and Physics*, 11(3): 1101–1116. https: //doi.org/10.5194/acp-11-1101-2011

Song, Y. Maher, B. A. Li, F. Wang, X. Sun, X. and Zhang, H. (2015). Particulate matter deposited on leaf of five evergreen species in Beijing, China: Source identification and size distribution. *Atmospheric Environment*, 105, 53–60. https: //doi.org/10.1016/j.atmosenv.2015.01.032.

State of Global Air. (2017). A special report on global exposure to air pollution and its disease burden. https: //www.stateofglobalair.org/sites/default/files/soga_2017_report.pdf

Sun, W. Shao, M. Granier, C. Liu, Y. Ye, C. S. and Zheng, J. Y. (2018). Long-term trends of anthropogenic SO_2, NOx, CO, and NMVOCs emissions in China. *Earth's Future*, 6(8): 1112–1133. https: //doi.org/10.1029/2018EF000822.

Taalas, P. Kaurola, J. Kylling, A. Shindell, D. Sausen, R. Dameris, M. Grewe, V. Herman, J. … Steil, B. (2000). The impact of greenhouse gases and halogenated species on future solar UV radiation doses. *Geophysical Research Letters*, 27(8): 1127–1130.

Tiwari, S. Agrawal, M. and Marshall, F. M. (2006). Evaluation of ambient air pollution impact on carrot plants at a sub urban site using open top chambers. *Environmental Monitoring and Assessment*, 119(1–3): 15–30. https: //doi.org/10.1007/s10661-005-9001-z.

Tong, L. Zhang, H. Yu, J. He, M. Xu, N. Zhang, J. Qian, F. Feng, J. and Xiao, H. (2017). Characteristics of surface ozone and nitrogen oxides at urban, suburban and rural sites in Ningbo, China. *Atmospheric Research*, 187: 57–68. https: //doi.org/10.1016/j.atmosres.2016.12.006.

Tripathi, R. Sarkar, A. Pandey Rai, S. and Agrawal, S. B. (2011). Supplemental ultraviolet-B and ozone: Impact on antioxidants, proteome and genome of linseed (*Linum usitatissimum* L. cv. *Padmini*). *Plant Biology*, 13(1): 93–104. https: //doi.org/10.1111/j.1438-8677.2010.00323.

Tripathi, R. and Agrawal, S. B. (2013a). Evaluation of changes in lipid peroxidation, ROS production, surface structures, secondary metabolites and yield of linseed (*Linum usitatissimum* L.) under individual and combined stress of ultraviolet-B and ozone using open top chambers. *Indian Journal of Biochemistry and Biophysics*, 50(4): 318–325.http: //nopr.niscair.res.in/handle/123456789/20881

Tripathi, R. and Agrawal, S. B. (2013b). Interactive effect of supplemental ultraviolet B and elevated ozone on seed yield and oil quality of two cultivars of linseed (*Linum usitatissimum* L.) carried out in open top chambers. *Journal of the Science of Food and Agriculture*, 93(5): 1016–1025. https: //doi.org/10.1002/jsfa.5838

Tripathi, R.Rai, K. Singh, S. Agrawal, M. and Agrawal, S. B. (2019). Role of supplemental UV-B in changing the level of ozone toxicity in two cultivars of sunflower: growth, seed yield and oil quality. *Ecotoxicology*, 28(3): 277–293.

Yadav, A. Bhatia, A. Yadav, S. Kumar, V. and Singh, B. (2019a). The effects of elevated CO_2 and elevated O_3 exposure on plant growth, yield and quality of grains of two wheat cultivars grown in north India. *Heliyon*, 5(8): e02317. https: //doi. org/10.1016/j.heliyon.2019.e02317.

Yadav, D. S. Mishra, A. K. Rai, R. Chaudhary, N. Mukherjee, A. Agrawal, S. B. and Agrawal, M. (2020). Responses of an old and a modern Indian wheat cultivar to future O_3 level: Physiological, yield and grain quality parameters. Environmental Pollution, 259, 113939. https: //doi.org/10.1016/j.envpol.2020.113939.

Yadav, D. S. Rai, R. Mishra, A. K. Chaudhary, N. Mukherjee, A. Agrawal, S. B. and Agrawal, M. (2019b). ROS production and its detoxification in early and late sown cultivars of wheat under future O_3 concentration. *Science of the Total Environment*, 659: 200–210. https: //doi.org/10.1016/j.scitotenv.2018.12.352.

Yang, L. Wang, Y. Dong, G. Gu, H. Huang, J. Zhu, J. Yang, H. Liu, G. and Han, Y. (2007) The impact of free-air CO_2 enrichment (FACE) and nitrogen supply on grain quality of rice. *Field Crops Research*, 102(2): 128–140. https: //doi. org/10.1016/j.fcr.2007.03.006.

Zheng, B. Tong, D. Li, M. Liu, F. Hong, C. Geng, G. Li, H. Li, X. Zhang, Q. (2018) Trends in China's anthropogenic emissions since 2010 as the consequence of clean air actions. *Atmospheric Chemistry and Physics*, 18(19): 14095–14111.

Chapter 2

Evaluation and Assessment of Urban Exposure on Water Quality of Perennial River: A Case Study of Vishwamitri River, Vadodara, India

*Pruthviraj Parmar, Sazia Barodawala and Bablu Prasad**

Department of Environmental Studies, Faculty of Science, The Maharaja Sayajirao University of Baroda, Vadodara – 390 002
**email: akashganga812@gmail.com*

ABSTRACT

Monitored ions were observed to be significantly increased with increase in urban exposure. In addition, a reduced amount of DO and significant increase in free CO_2 were observed in downstream as compared to upstream of river. Oil and grease also follow the similar trend as it passes through the urban area and recorded to be higher than the survival limit of aquatic organisms. Increase in oil and grease is often linked with discharge of domestic grey water. Based on water quality index analysis, high potential of Vishwamitri River in pollution load handling were estimated.

Keywords: *Anthropogenic activities, Environmental impact, Urbanization, Vadodara, Water quality, Assessment, Water quality index (WQI).*

2.1 Introduction

Quality and quantity of water in a river ecosystem is directly linked with human health and biodiversity of flora and fauna. Deterioration of water quality of river is dependent on various activities such as urbanization, industrialization, hydropower projects, untreated wastewater discharge, surface runoff, lack of awareness and seasonal variations (Anderson *et al.*, 2015; Chadetrik *et al.*, 2015; Trivedi, 2010). In addition, quality of water also depends on their origin and change in climate (Arnell *et al.*, 2015; Gregory, 2006). River ecosystems are one of the most vulnerable and most exploited environmental compartment of our planet since last 100 years (Aneta, *et al.*, 2017; Bunn and Davies, 2000; Gregory, 2006). The dramatic decline in biological diversity is a serious threat to the hydrological and biological sustainability of freshwater ecosystems. With an increased understanding of the importance of drinking water quality to public health and raw water quality to aquatic life, there is a great need to assess surface water quality. This is true for Vishwamitri River which flows through Vadodara city. Similar to the other rivers, it plays a major role in assimilating or carrying off industrial and municipal effluent, manure discharges and runoff from agricultural fields, roadways and streets. Flow of rivers is a function of many factors including precipitation, surface runoff interflow, groundwater flow and pumped inflow and outflow (Vega, *et al.*, 1998). Seasonal variations of these factors have a strong effect on flow rates and hence on the concentration of pollutants in the river water.

Local, regional, and global assessments of water quality monitoring are the key component of environmental management of aquatic resources. Furthermore, it is crucial to understand influence of various activities on water resources both in terms of negative and positive impact (Aartsen *et al.*, 2018). In current investigation, physicochemical and biological parameters namely temperature, pH, conductivity, total dissolved solids (TDS), total suspended solids (TSS), total solids (TS), dissolved oxygen (DO), free CO_2, oil and grease, turbidity, alkalinity, acidity, total hardness, calcium hardness, magnesium hardness, calcium, magnesium, sodium, potassium, sulphate, phosphate, nitrate, chloride, chemical oxygen demand (COD), biological oxygen demand (BOD) of Vishwamitri River flowing through Vadodara city is assessed for a period of four months (December 2016 to March 2017). Water quality index (WQI) has been often used to monitor water quality of various water resources and also used for water resources management (Banerjee and Srivastava, 2009; Ghosh *et al.*, 2013; House, 1989; Kaur *et al.*, 2014; Khosravi *et al.*, 2017; Kumar *et al.*, 2015; Reza and Singh, 2010; Saeedi *et al.*, 2010; Singh and Hussian, 2016; Stigter *et al.*, 2006; Vasanthavigar *et al.*, 2010; Yisa *et al.*, 2012). In this study, WQI were calculated using multivariate analysis methods developed by Saeedi *et al.* (2010).

2.2 Materials and Methods

2.2.1. Sampling Locations

The Vishwamitri basin is a part of Dhadhar basin which originate from Pavagadh hills and extends over an area of 3423 km^2 in the state of Gujarat, India. The total length of the river from the head to its outfall into the Dhadhar is about 80 km. The climate of Vishwamitri river watershed is semi-arid and sub-tropical having a temperature range of 21°C to 41°C in summer and 7°C to 29°C in winter while, most of rainfall is received during the mid-June to early October (Pancholi *et al.*, 2015). Water samples were collected from 10 sampling stations distributed alongside Vishwamitri River at Vadodara city during December 2016 to April 2017. The 1st and 10th sampling stations were located just before river enters urban area of Vadodara and point where it leaves urban area, respectively. The sampling stations fall between latitude N 22°28′ to N 22°14′ and longitude E 073°24′ to E 073°10′. The sampling stations were selected on the basis of their contamination vulnerability, *i.e.*, the stations upstream to the city were supposed to be less impaired by urban contamination and the stations downstream to the city were supposed to be more impaired by contamination. Grab sampling was carried out at each sampling point to collect water samples.

2.2.2. Analysis of Water Parameters

The monitored water quality parameters, their units and the methods of analysis are summarized in Table 2.1. The temperature, pH, electrical conductivity and dissolved oxygen of each water sample were measured *in situ*. The other parameters, namely total suspended solids (TSS), total solids (TS), total dissolve solids (TDS), acidity, alkalinity, sodium (Na^+), potassium (K^+), calcium (Ca^{2+}), magnesium (Mg^{2+}), nitrate (NO_3^-), sulphate (SO_4^{2-}) total phosphate (PO_4^{3-}), chloride (Cl^-), Hardness, oil and grease, biological oxygen demand (BOD_3), chemical oxygen demand (COD) and total coli form (MPN) were analysed in laboratory using the analytical techniques following the standard methods (APHA 2012). All the samples were analysed in triplicate and mean were compared to the Bureau of Indian Standards of drinking water specification (IS 10500: 2012; BIS 2012).

Table 2.1: List of Water Parameters, their Methods of Analysis and Indian Standards along with Assigned Weight and Relative Weight

Parameters	*Analytical Methods*	*Indian Standards (BIS 10500 2012)*	*Weight (Wi)*	*Relative Weight (Wr)*
Temperature	Temperature probe	-	-	-
pH	Potentiometry/pH probe	6.5-8.5	-	-
DO	Portable Oximetry	-	-	-
Free CO_2	Base Titration	-	-	-
Conductivity	Conductometry	-	-	-
TDS	Gravimetric	500	5	0.140
TSS	Gravimetric	-	-	-
TS	Gravimetric	-	-	-
Turbidity	Nephelometry	1	2	0.056
Acidity	Base Titration	-	-	-
Alkalinity	Acid Titration	200	3	0.084
Total Hardness	EDTA Titrimetric Method	200	3	0.084
Ca Hardness	EDTA Titrimetric Method	-	-	-
Mg Hardness	EDTA Titrimetric Method	-	-	-
Ca	EDTA Titrimetric Method	75	3	0.084
Mg	EDTA Titrimetric Method	30	3	0.084

Parameters	*Analytical Methods*	*Indian Standards (BIS 10500 2012)*	*Weight (Wi)*	*Relative Weight (Wr)*
Na and K	Flame Emission/Flame Photometry	-	-	-
Sulphate	Colorimetry	200	5	0.140
Phosphate	Stannous Chloride/ Spectrophotometry	-	-	-
Nitrate	Spectrophotometry	45	5	0.140
Chloride	Argentometric titration	250	5	0.140
COD	Close reflux	-	-	-
BOD_3	Dilution method	-	-	-
Oil and Grease	Gravimetric	-	-	-
Total Coliform	Decimal dilution techniques	-	-	-

2.2.3. Water Quality Index

WQI were calculated by assigning a weight (W_i) of each selected parameters according to its relative importance in the overall quality of water for drinking purpose. The maximum weight of 5 were assigned to the parameters like total dissolved solids, nitrate, chloride and sulphate due to their major importance in water quality assessment (Srinivasamoorthy *et al.*, 2008; Vasanthavigar *et al.*, 2010). Turbidity was given the weight of 2 as it plays less significant role in the water quality assessment. Other parameters like alkalinity, total hardness, calcium and magnesium were assigned weight 3(Vasanthavigar *et al.*, 2010). QWI were computed using the following equation:

$$WQI = \sum SI_i$$

Equation 1

In Equation 1, SI_i is the sub index of *i*th parameters. Sub index of each parameters were computed using the following equation:

$$SI_i = W_r \times \frac{C_i}{S_i} \times 100$$

Equation 2

In equation 2, SI_i is the sub index of *i*th parameters, W_r is the relative weight of respective parameters, Ci is the concentration of *i*th parameters and S_i is the standards of *i*th parameters according to BIS 10500: 2012 standards of drinking water specification(BIS 2012). The relative weight (W_r) is computed from the following equation:

$$Wr = \frac{Wi}{\sum_{i-1}^{n} Wi}$$

Equation 3

In equation 3, W_r is the relative weight, W_i is the weight of each parameter and *n* is the number of parameters. Calculated relative weight (W_r) values of each parameter are presented in Table 2.1.

2.3 Results and Discussion

Result of water analysis showed a gradual increase in values of most of the analysed parameters as the river passes through Vadodara city except temperature and dissolved oxygen. The average temperature during this study was observed to be 28.80 ± 0.90°C. Temperature at sampling station S6 (Kalaghoda bridge) and S1 (Pratappura) were observed to be highest and lowest, respectively (Table 2.2). In current investigation, difference in water temperature were observed to be insignificant, possibly due to marginal variation in the atmospheric temperature at Vadodara city during mid-winter and pre summer season (Morrill *et al.*, 2005). Trivial increase in pH was observed as it passes through the centre of city (Table 2.2). This is possibly due to addition of dissolve solids in water and/or decomposition of organic matter discharge from different point (Deshkar *et al.*, 2014; Shiddamallayya and Pratima, 2008). In addition, a reduced amount of DO and significant increase in free CO_2 were observed in downstream as compared to upstream (Table 2.2). The reduced dissolved oxygen at downstream clearly indicated organic pollution as it passes through the urban area. Deshkar *et al.* (2014) has reported direct discharge of ~180 MLD treated sewage and ~25 MLD of untreated sewage into Vishwamitri River through the drains passing through slum areas of Vadodara city. At the same time, high concentration of DO at upstream signifies a condition favourable for the aquatic organisms to thrive in the water. The concentration of DO lower than 2 mg/L in a prolong period could be detrimental leading to asphyxiation of fishes.

Electrical conductivity, turbidity and alkalinity trail increasing trend due to increase in cation and anion such as Na^+, K^+, Ca^{2+}, Mg^{2+}, PO_4^{3-}, SO_4^{2-}, NO_3^- and Cl^- (Table 2.2). These parameters are regarded as pollution indicator in the water bodies (Agrahari and Kushwaha, 2012; Das *et al.*, 2006). In addition to total dissolved solids (TDS), total suspended solids (TSS) were reported to increase with the exposure of urban area of Vadodara. TSS is water insoluble materials, including organic and inorganic, that are suspended in water such as silt, plankton and industrial wastes (Kumar *et al.*, 2010). The Environmental Management Bureau stated that the higher the TSS value, the lower is the ability of the water to support aquatic life due to reduced light penetration affecting plant photosynthesis. The possible reason for these observations could be the release of sewage effluent, discharge of industrial effluent, solid waste disposal leachates from municipal solid waste transfer station and construction on the bank of river. In addition, there are other factors that contribute to the change in physical-chemical status of water bodies such as sedimentation, biological uptake, mineralization, transformation, decomposition, run-of-river hydropower and ionic exchange (Anderson *et al.*, 2015; Deshkar *et al.*, 2014; Duff *et al.*, 2009; Qualls and Richardson, 2003). Many mineral deposits have the potential to increase acidity, which dissolves metals that are toxic to many organisms such as algae, benthic invertebrates, fish, *etc.*, thus posing a threat to water resources (Medeiros *et al.*, 2017). Oil and grease also follow the similar trend as it passes through the urban area. Increase in oil and grease is often linked with discharge of domestic grey water (Friedler, 2004). Moreover, chemical oxygen demand (COD) was observed to be increased with increase in urban area exposure (Table 2.2). Concurrently, biological oxygen demand (BOD_3) was not detectable at

Table 2.2: Results of Water Parameters of Selected Sampling Station

Parameter	*Water Sampling Station*									
	S1	*S2*	*S3*	*S4*	*S5*	*S6*	*S7*	*S8*	*S9*	*S10*
Temperature (°C)	27.7	28.1	27.8	27.9	28.2	30.1	29.6	29.1	29.8	29.8
pH	7.3	7.5	7.7	7.6	7.8	7.8	7.5	7.7	7.7	8.1
DO (mg/L)	4.2	3.6	1.0	1.2	0.4	ND*	0.44	0.91	ND	ND
Free CO_2 (mg/L)	33.6	49.1	19.4	35.8	70.8	32.1	71.2	93.7	15.4	72.2
Conductivity (us/cm)	560	698	789	830	727	950	940	1317	1010	1572
TDS (mg/L)	300	397	463	518	720	712	499	796	903	1361
TSS (mg/L)	501	532	407	737	1181	1050	787	525	876	1665
TS (mg/L)	802	929	870	1255	1901	1762	1286	1321	1779	3027
Turbidity (NTU)	12.4	19.0	15.2	47.0	47.8	33.0	35.3	23.3	33.3	89.3
Acidity (mg/L as $CaCO_3$)	106.3	70.1	128.6	143.5	117.3	103.5	48.2	182.4	107.0	194.2
Alkalinity (mg/L as $CaCO_3$)	210.6	219.3	248.1	235.1	248.4	252.4	240.6	347.3	415.9	506.8
Total Hardness (mg/L as $CaCO_3$)	148.3	181.9	172.0	186.6	266.2	237.5	296.7	215.1	177.5	297.5
Ca Hardness (mg/L as $CaCO_3$)	100.3	148.1	116.5	125.6	172.6	169.7	122.6	126.6	125.7	211.6
Mg Hardness (mg/L as $CaCO_3$)	48.0	33.8	55.5	60.9	93.5	67.7	104.6	88.2	51.7	85.9
Calcium (mg/L)	40.8	58.3	47.5	50.3	69.1	68.0	54.9	50.7	50.3	84.7
Magnesium (mg/L)	11.3	8.0	13.2	14.5	22.2	16.1	16.7	20.6	12.1	20.1
Sodium (mg/L)	43.3	40.4	43.8	47.7	54.8	60.6	59.6	62.1	69.8	74.1
Potassium (mg/L)	0.14	0.28	5.62	5.69	7.42	8.07	8.21	8.14	8.64	9.22
Sulphate (mg/L)	4.10	22.72	18.72	22.72	30.32	26.72	29.52	30.72	34.32	34.72
Phosphate (mg/L)	0.97	3.59	3.74	3.89	3.94	4.09	4.29	4.19	4.44	4.44
Nitrate (mg/L)	1.94	2.92	2.43	3.37	3.30	4.20	4.58	5.03	7.02	9.18
Chloride (mg/L)	28.61	41.28	32.38	43.08	35.05	44.94	30.05	135.1	177.4	179.1
COD (mg/L)	76.67	246.7	200.0	296.2	236.7	296.4	166.7	323.8	393.3	403.6
BOD_3 (mg/L)	20.15	65.29	45.78	75.23	59.16	ND	42.05	81.39	ND	ND
Oil and Grease (mg/L)	4.33	4.2	11.5	5.33	17.33	12.7	16.89	15.06	19.67	22.05
MPN (CFU/100ml) x 10^3	17	35	160	17	>160	>160	160	>160	>160	>160

*ND = Not detectable.

numerous sampling stations. The above results clearly indicate disposal of non-biodegradable organic matter. The concentrations of total coli form in water of all the sampling stations including upstream and downstream were found to be high and failed to meet the BIS standards which suggested unsuitability of water for drinking even for recreational or bathing purposes. Untreated sewage discharge, combined with sewer overflows, solid waste disposal and storm water runoff are the largest causes of urban water pollution (Phillips and Chalmers, 2009). These results are consistent with results reported in the literature (Deshkar *et al.*, 2014; Pancholi *et al.*, 2015; Gosai *et al.*, 2016).

The results of water quality index analyses showed better water quality in the monitoring stations located in the upstream of the river than those stations in the midstream and downstream of the rivers. Gradual increase in WQI was observed up to S5 sampling station followed by decreasing trend up to S8 sampling station (Figure 2.1). Decrease in WQI is possibly due to physical and biological remediation of pollutants in river ecosystem. These results clearly indicated the potential of Vishwamitri River in pollution load management. Further increase in WQI is possibly due to further discharge of untreated sewage and solid waste disposal in unmanageable quantity. Based on above analyses, it can be predicted that the maximum discharge and disposal of solid waste is carried out between sampling station S2 to S5 and S8 to S10. Maximum remediation of pollutants by biological, physical and chemical processes in Vishwamitri River is carried out between sampling station S5 to S8. This might be possible due to high vegetation on the bank of river which provides suitable conditions for remediation of several pollutants. Vegetation on the bank of river also remediates various pollutants by phytodegradation, phytosequestration, rhizodegradation, phytohydraulics, phytoextraction and phytovolatilization (Lone *et al.*, 2008). These results are consistent with literature that suggest river as a plug flow bioreactor with high self-purification capacity (Vicente *et al.*, 2011).

Based on forgoing discussion, it is evident that the water in upstream is suitable for public supply after the conventional treatment procedure (coagulation and flocculation, sedimentation, filtration and disinfection) to meet the specification as per the Bureau of Indian Standards for drinking water (BIS, 2012). On the other hand, water quality in downstream of the river is threatened or impaired and conditions usually depart from desirable quality as required under public supply. However, all of the monitoring stations had excellent to good condition/status of water if utilized for agriculture, irrigation and other purposes.

The deterioration in the water quality of the Vishwamitri River is the cumulative result of human interferences with the natural vegetation through encroachment, localized development pressure, community expansion, extensive construction on the bank of river and large amount of organic load introduction by municipal wastes reducing the DO level and, in turn, affecting aquatic life. The water of the Vishwamitri is overburdened with inorganic and organic pollutants, in many sampling station, beyond permissible limits and guideline values. Although people seldom use the Vishwamitri water for drinking, a large number of domestic cattle and poultry are dependent on the same for their water needs which might enter human food chain. Hence, the comparison with the drinking water quality guidelines becomes relevant. The present work has highlighted the state of the quality of the water of the Vishwamitri River and hence the need for adopting appropriate management plans for restoring the river to its pristine state.

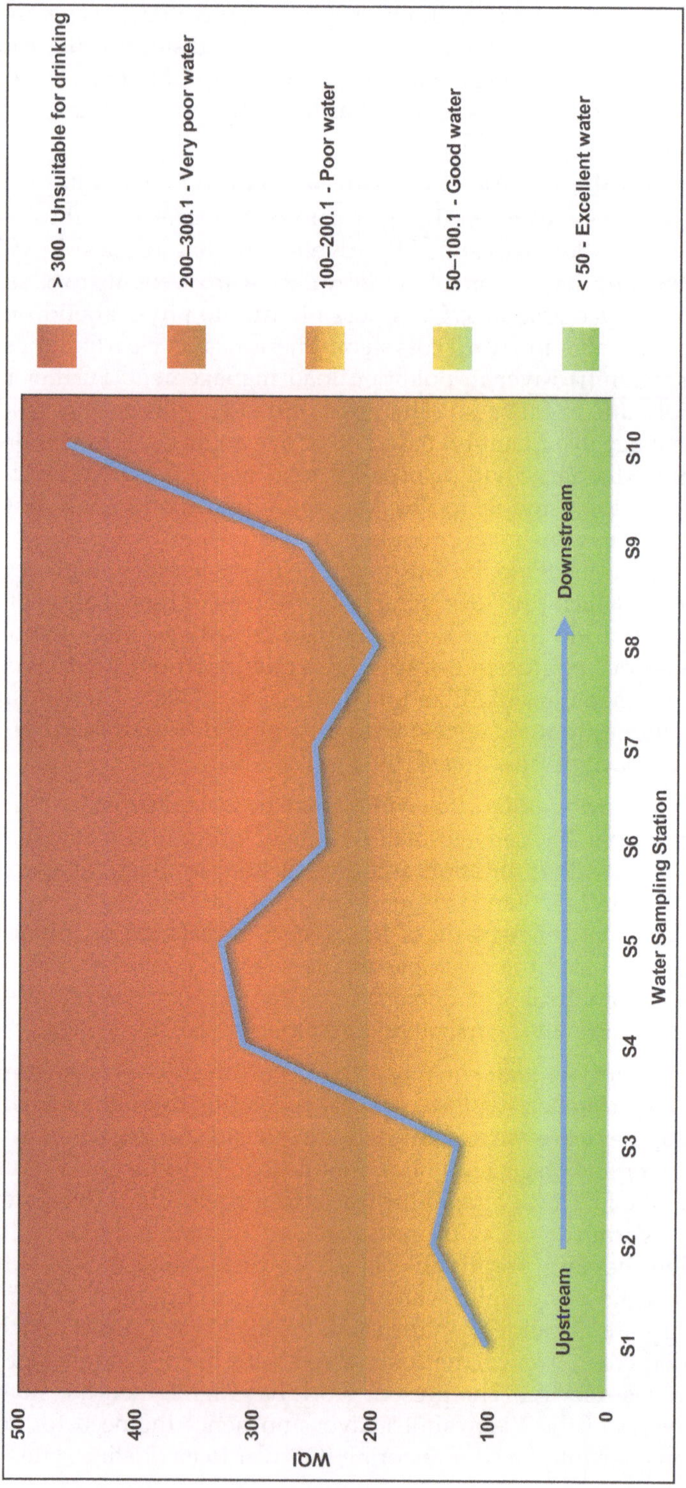

Figure 2.1: Impact of Urban Exposer on Water Quality Index of Vishwamitri River.

2.4 Conclusions

This assessment has found high and low quality conditions of the Vishwamitri River at upstream and downstream of Vadodara urban area, respectively, with a huge capacity to self-remediate pollutant. Electrical conductivity, TDS, and turbidity, an indirect measure of dissolve salts was found to be increasing as it passed through urban areas of Vadodara. The water in the downstream was harder as compared to water in the upstream and was dominated by calcium, magnesium, sulphate and chloride ions. In current investigation, the sources of dissolve inorganic salts and organic matter were not identified.

Based on current investigation, it can be concluded that the growth and development in urban area had negative effects on the riverine system. Further investigation is needed on the seasonal concentrations of chemical constituents in the river and its tributaries to understand the impact of development especially that associated with discharge of waste and construction on the bank of river. The information provided by the WQI is a useful tool for describing the state and trends of the river water quality. The WQI is useful as decision tool in decision-making related to water resource protection, improvement, utilization towards a sustainable water resource management and environmental planning. This study also recommends a continuous monitoring of water quality of freshwater bodies in the city of Vadodara.

Acknowledgement

The authors thank The Maharaja Sayajirao University of Baroda, Vadodara for providing the necessary infrastructure to conduct present research investigation.

REFERENCES

Aartsen, M., Koop, S., Hegger, D., Goswami, B., Oost, J., van Leeuwen, K. (2018) Connecting water science and policy in India: lessons from a systematic water governance assessment in the city of Ahmedabad. Reg. Environ. Chang. 1–13. https: //doi.org/10.1007/s10113-018-1363-1

Agrahari, M., and Kushwaha, V.B. (2012) Effect of Domestic Sewage on the Physico-Chemical Quality of River Rapti at Gorakhpur. The Bioscan 7: 135–138.

Anderson, D., Moggridge, H., Warren, P., Shucksmith, J. (2015) The impacts of "run-ofriver" hydropower on the physical and ecological condition of rivers. Water Environ. J. 29, 268–276. https: //doi.org/10.1111/wej.12101

Aneta S., Kubicka J., Strzelec M. (2017) The Use of Biological Indices for the Assessment of the River Quality (Ruda River, Poland). Ecol. Chem. Eng. 24: 285–298. https: //doi.org/10.1515/eces-2017-0020

APHA, (2012) Standard Methods for Examination of Water and Wastewater, 22nd ed. American Public Health Association, Washington, DC.

Arnell, N.W., Halliday, S.J., Battarbee, R.W., Skeffington, R.A., Wade, A.J. (2015) The implications of climate change for the water environment in England. Prog. Phys. Geogr. 39: 93–120. https: //doi.org/10.1177/0309133314560369

Banerjee, T., and Srivastava, R.K. (2009) Application of water quality index for assessment of surface water quality surrounding integrated industrial estate-Pantnagar. Water Sci. Technol. 60, 2041–2053. https: //doi.org/10.2166/wst.2009.537

BIS, (2012) Indian Standard Drinking Water Specification (Second Revision). Bur. Indian Stand. IS 10500, 1–11.

Bunn, S.E., and Davies, P.M. (2000) Biological processes in running waters and their implications for the assessment of ecological integrity. Hydrobiologia 422: 61–70. https: //doi.org/10.1023/A: 1017075528625

Chadetrik, R., Arun, L., Prakash, D.R. (2015) Assessment of Physico-chemical parameters of River Yamuna at Agra Region of Uttar Pradesh, India. Int. Res. J. Environ. Sci. 4, 25–32.

Das, R., Samal, N., Roy, P., Mitra, D. (2006) Role of electrical conductivity as an indicator of pollution in shallow lakes. Asian J. Water, Environ. Pollut. 3(1): 143-146.

Deshkar, S., Khyati, M., Gavali D. (2014) Spatial difference in pollution levels across Vishwamitri River. Int. J. Environ. Biol. 4, 51–56.

Duff, J.H., Carpenter, K.D., Snyder, D.T., Lee, K.K., Avanzino, R.J., Triska, F.J. (2009) Phosphorus and nitrogen legacy in a restoration wetland, Upper Klamath Lake, Oregon. Wetlands 29, 735. https: //doi.org/10.1672/08-129.1

Friedler, E. (2004) Quality of individual domestic grey water streams and its implication for on-site treatment and reuse possibilities. Environ. Technol. 25: 997–1008. https: //doi.org/10.1080/09593330.2004.9619393

Ghosh, M.K., Ghosh, S., Tiwari, R. (2013) A Study of Water Quality Index Assessment of Groundwater and Pond Water in Sirsakala Village of Bhilai-3, Chhattisgarh, India. Int. J. Civil, Struct. Environ. Infrastruct. Eng. Res. Dev. 3: 63–74.

Gosai D.P., Joshi G.S., Prakash I., Patel A. (2016) Assessment of Flood Potential in Vishwamitri River Basin Using Remote Sensing and GIS, Gujarat. *Proc. Natl. Conf. Recent Adv. Civ. Eng*. SVNIT Surat. 554–561.

Gregory, K. (2006) The Human Role in Changing River Channels. Geomorphology 79: 172–191. https: //doi.org/10.1016/j.geomorph.2006.06.018

House, M.A. (1989) A Water Quality Index for River Management. Water Environ. J. 3: 336–344. https: //doi.org/10.1111/j.1747-6593.1989.tb01538.x

Kaur, B.J., George, M.P., Mishra, S. (2014) Groundwater Quality and Water Quality Index of Delhi City, India 32: 865–871. https: //doi.org/10.5829/idosi.wasj.2014.32.05.1007

Khosravi, R., Eslami, H., Almodaresi, S.A., Heidari, M., Fallahzadeh, R.A., Taghavi, M., Khodadadi, M., Peirovi, R. (2017) Use of geographic information system and water quality index to assess groundwater quality for drinking purpose in Birjand city, Iran. Desalin. Water Treat. 67: 20458. https: //doi.org/10.5004/dwt.2017.20458

Kumar, K.S., Prasad, K.H., Rajesh, B., Prasad, R.S., Venkatesh, T. (2015) Assessment of Groundwater Quality Using Water Quality Index. Int. J. Innov. Res. Adv. Eng. 2: 103–108.

Lone, M.I., He, Z., Stoffella, P.J., Yang, X. (2008) Phytoremediation of heavy metal polluted soils and water: Progresses and perspectives. J. Zhejiang Univ. Sci. B 9: 210–220. https: //doi.org/10.1631/jzus.B0710633

Medeiros, A.S., Wood, P., Wesche, S.D., Bakaic, M., Peters, J.F. (2017) Water security for northern peoples: review of threats to Arctic freshwater systems in Nunavut, Canada. Reg. Environ. Chang. 17: 635–647. https: //doi.org/10.1007/s10113-016-1084-2

Morrill, C.J., Bales, C.R., Conklin, H.M. (2005) Estimating Stream Temperature from Air Temperature: Implications for Future Water Quality. J. Environ. Eng. 131, 139–146. https: //doi.org/10.1061/(ASCE)0733-9372(2005)131: 1(139)

Pancholi, V.H., Lodha, P.P., Prakash, I. (2015) Estimation of Runoff and Soil Erosion for Vishwamitri River Watershed, Western India Using RS and GIS 1: 7–14. https: //doi.org/10.11648/j.ajwse.20150102.11

Phillips, P., and Chalmers, A. (2009) Wastewater effluent, combined sewer overflows, and other sources of organic compounds to Lake Champlain. J. Am. Water Resour. Assoc. 45, 45–57. https: //doi.org/10.1111/j.1752-1688.2008.00288.x

Qualls, R.G., Richardson, C.J. (2003) Factors controlling concentration, export, and decomposition of dissolved organic nutrients in the Everglades of Florida. Biogeochemistry 62: 197–229. https: //doi.org/10.1023/A: 1021150503664

Reza, R., and Singh, G. (2010) Assessment of Groundwater Quality Status by Using Water Quality Index Method in Orissa, India 9: 1392–1397.

Saeedi, M., Abessi, O., Sharifi, F., Meraji, H. (2010) Development of groundwater quality index. Environ. Monit. Assess. 163, 327–335. https: //doi.org/10.1007/s10661-009-0837-5

Shiddamallayya, N. and Pratima, M. (2008) Impact of domestic sewage on fresh water body. J. Environ. Biol. 29: 303–308.

Singh, S., and Hussian, A. (2016) Water quality index development for groundwater quality assessment of greater Noida sub-basin, Uttar Pradesh, India. Cogent Eng. 3, 1–17. https: //doi.org/10.1080/23311916.2016.1177155

Srinivasamoorthy, K., Chidambaram, S., Prasanna, M. V, Vasanthavihar, M., Peter, J., Anandhan, P. (2008) Identification of major sources controlling groundwater chemistry from a hard rock terrain - A case study from Mettur taluk, Salem district, Tamil Nadu, India. J. Earth Syst. Sci. 117: 49. https: //doi.org/10.1007/s12040-008-0012-3

Stigter, T.Y., Ribeiro, L., Carvalho Dill, A.M.M. (2006) Application of a groundwater quality index as an assessment and communication tool in agro-environmental policies – Two Portuguese case studies. J. Hydrol. 327, 578–591. https: //doi.org/10.1016/j.jhydrol.2005.12.001

Trivedi, R.C. (2010) Water quality of the Ganga River – An overview. Aquat. Ecosyst. Health Manag. 13: 347–351. https: //doi.org/10.1080/14634988.2010.528740

Vasanthavigar, M., Srinivasamoorthy, K., Vijayaragavan, K., Rajiv Ganthi, R., Chidambaram, S., Anandhan, P., Manivannan, R., Vasudevan, S. (2010) Application of water quality index for groundwater quality assessment: Thirumanimuttar sub-basin, Tamilnadu, India. Environ. Monit. Assess. 171, 595–609. https: //doi.org/10.1007/s10661-009-1302-1

Vega M., Pardo R., Enrique B. L.D. (1998) Assessment of Seasonal and Polluting Effects on the Quality of River Water by Exploratory. Water Res. 32: 3581–3592.

Vicente, J., F., C.M., E., S., A., R., G., G.M., R., B., M., J.A. (2011) Evaluation of the water quality in the Guadarrama river at the section of Las Rozas Madrid, Spain. Water Environ. J. 25: 55–66. https: //doi.org/10.1111/j.1747-6593.2009.00188.x

Yisa, J., Jimoh, T.O., Oyibo, O.M. (2012) Under groundwater Assessment using Water Quality Index. Leonardo J. Sci. 21: 33–42.

Chapter 3

Fungal Flora Associated with Kitchen Waste and Management of Municipal and Agro Waste: Turning Garbage into Gold

*Suneet Dabke[1], Chandni A. Sejpal[2] and Arun Arya[3]**

[1]Concept Biotech, 3, Manu Smruti Complex, Near Navrachna School, Sama, Vadodara
[1]Biotechnology Engineering Department, VVP Engineering College, Rajkot
[2]Department of Environmental Studies, Faculty of Science, The Maharaja Sayajirao University of Baroda, Vadodara – 390 002
**e-mail: conceptbiotec@gmail.com, sarojarun10arya@rediffmail.com*

ABSTRACT

The municipal corporations being the responsible aulhority in India for Solid Waste Management are also having respoisibilities related to good health and sanitation in towns and cities. Collectlon and transportation of solid waste poses a big challenge and only a part of waste is collected and very small portion is treated or utilized back in day to day use. The paper gives a glimpse how this waste generated in our houses or waste generated in agricultural fields can be reutilized and various solutions are suggested in detail. The composition of MSW at generation sources and collection points was determined on a wet weight basis and it consists mainly of a large organic fraction (40–60 per cent), ash and fine earth (30–40 per cent), paper (3–6 per cent) and plastic, glass and metals (each less than 1 per cent). The C/N ratio ranges between 20 and 30, and the lower calorific value ranges between 800 and 1000 kcal/kg. In Vadodara, municipal waste site was started in Vadsar in 1974-1975. Waste was collected from this site and 20 different fungi were isolated. Many of these belonged to Ascomycetes and Fungi imperfecti group. Five species of Aspergillus were present, Rhizopus a member of Zygomycetes was also isolated. Mycoflora present in vegetable waste was isolated and identified. Results showed presence of one member of Zygomycetes, 6 of Ascomycetes and 7 different members of Fungi imperfecti on PDA slants. Five species of Aspergillus were present in kitchen waste.

Aerobic and anaerobic conversion of MSW and agro waste into compost or liquid fertilizer and other waste remediation processes including recycling of plastic waste and value addition are presented in the paper.

Keywords: *Fungal flora, Kitchen waste, Management, Agro waste, Plastic waste, Recycling, Mushroom cultivation.*

3.1 Introduction: Problems of Managing Municipal Solid Waste

Rapid industrialization and population explosion in India has led to the migration of people from villages to cities, which generate thousands of tons of MSW (Municipal Solid Waste) daily. The amount of MSW is increasing significantly day by day. Poor collection and inadequate transportation are responsible for the accumulation of MSW at every nook and corner. The management of MSW is going through a critical phase, due to the unavailability of suitable facilities to treat and dispose of the larger amount of MSW generated daily in metropolitan cities. Unscientific disposal causes an adverse impact on all components of the environment and human health (Rathi, 2006). Disposal of MSW has become increasingly difficult over the past several decades. Waste disposal rates have remained consistent while permitting and construction of new landfill sites have decreased due to scarcity of suitable sites, enhanced regulatory requirements, and citizen opposition for shifting the sites. The optimization of waste placement procedures and the physical and engineering properties of MSW are still not well understood. In 2011, Americans generated approximately 250 million tons of municipal solid waste (USEPA 2013a), which is equivalent to approximately 2kg of MSW per person per day. After recycling and re-use, 164 million tons of MSW is discarded into landfills around the United States. There is the need for optimizing waste placement procedures and maximizing landfill capacity. Generally, MSW is disposed of in low-lying areas without taking any precautions or operational controls.

Therefore, MSWM is one of the major environmental problems of Indian megacities. It involves activities associated with generation, storage, collection, transfer and transport, processing and disposal of solid wastes. But, in most cities, the MSWM system comprises only four activities, *i.e.*, waste generation, collection, transportation, and disposal. The management of MSW requires proper infrastructure, maintenance and upgrade for all activities. This becomes increasingly expensive and complex due to the continuous and unplanned growth of urban centers. The difficulties in providing the desired level of public service in the urban centers are often attributed to the poor financial status of the managing municipal corporations (Raje *et al.*, 2001). Covering an area of 328'7263 km, India is the seventh largest country in the world. Per day a lot of waste is generated in the country on account of its increasing population and increased GDP, although the civic services have not been expanding proportionately and hence are under tremendous pressure. The farmers have adopted more production of paddy in Punjab and Haryana and during Nov. to Jan. severe atmospheric pollution is recorded due to burning of stubble.

With increasing urbanization and changing life styles, Indian cities now generate eight times more MSW than they did in 1947. Presently, about 90 million t of solid waste are generated annually as byproducts of industrial, mining, municipal, agricultural and other processes. The amount of MSW generated per capita is estimated to increase at a rate of 1–1.33 per cent annually (Bhide, and Shekdar, 1998). It is reported that the MSW generation rates in small towns are lower than those of metro cities, and the per capita generation rate of MSW in India ranges from 0.2 to 0.5 kg/day. The urban population of India was approx. 341 million in 2010. The projected MSW quantities are expected to increase from 34 million tonnes in 2000 to 83.8 million tonnes in 2015 and 221 million tonnes in 2030. It is also reported that per capita per day production will increase to 1.032 kg, and urban population as 586 million in 2030 (Kaushal *et al.*, 2012).

The quantity of MSW generated (CPCB, 2000) and the per capita generation rate of MSW (CPCB,2004) are shown in Table 3.1 and Figure 3.1, respectively. It can be seen from Table 3.1 and Figure 3.1 that the per capita generation rate is high in some states (Gujrat, and Delhi) and cities (Delhi, Bhopal and Ahmedabad). This may be due to the high living standards, the rapid economic growth and the high level of urbanization in these states and cities. However, the per capita generation rate is observed to be low in other states (Meghalaya, Assam, Manipur and Tripura) and cities (Nagpur, Pune and Indore).

Table 3.1: Municipal Waste Generation in Certain States

Sl.No.	*State*	*Cities*	*Population*	*Solid Waste (t/d)*	*Per Capita Generation (Kg/day)*
1.	Andhra Pradesh	32	10,85,905	3943	0.364
2.	Bihar	17	5,278,361	1479	0.28
3.	Gujarat	21	8443,962	3805	0.451
4.	Haryana	12	2,254,353	623	0.276
5.	Delhi	1	8,419,084	4000	0.475
6.	M.P.	23	7,225,833	2286	0.316
7.	Maharashtra	27	22,727,186	8589	0.378
8.	Chandigarh	1	504,094	200	0.397
9.	Rajasthan	14	4,979,301	1768	0.355
10.	Punjab	10	3,209,903	1001	0.312

Source: Status of MSW generation CPCB 2000.

The composition and the quantity of MSW generated form the basis on which the management system needs to be planned, designed and operated. In India, MSW differs greatly with regard to the composition and hazardous nature, when compared to MSW in the western countries (Gupta *et al.*, 1998). The composition of

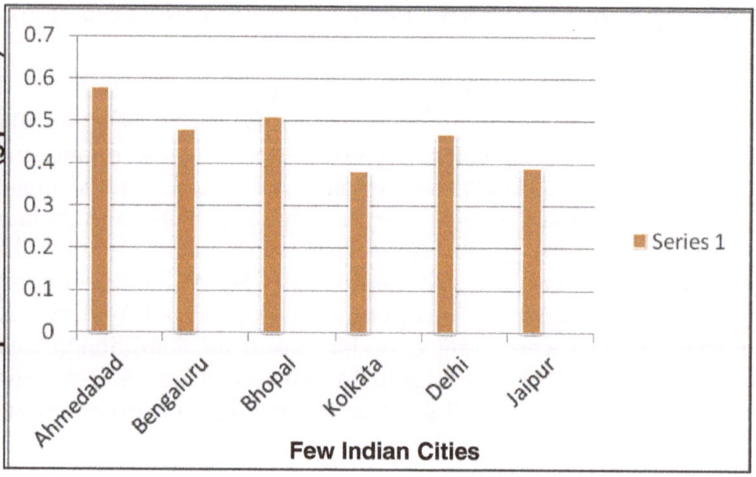

Figure 3.1: Per Capita Generation of MSW for Few Indian Cities (CPCB 2004).

MSW at generation sources and collection points was determined on a wet weight basis and it consists mainly of a large organic fraction (40–60 per cent), ash and fine earth (30–40 per cent), paper (3–6 per cent) and plastic, glass and metals (each less than 1 per cent). The C/N ratio ranges between 20 and 30, and the lower calorific value ranges between 800 and 1000 kcal/kg. The physical characteristics of MSW is presented in Figure 3.2. It has been noticed that the amount of Ash and compostable matter is more as compared to paper and plastic.

Improper MSW disposal causes pollution of air, soil, and water. Indiscriminate dumping of wastes contaminates surface and groundwater supplies. In urban areas, MSW clogs drains, creating stagnant water for insect breeding and floods during rainy seasons. Uncontrolled burning of MSW and improper incineration contributes significantly to urban air pollution. Greenhouse gases are generated from the decomposition of organic wastes in landfills, and untreated leachate pollutes surrounding soil and water bodies. The U.S. Public Health Service identified 22 human diseases that are linked to improper MSWM (Alam and Ahmade 2016). Workers and pickers in developing countries are exposed to direct contact and injury, and the co-disposal of hazardous and medical wastes with MSW causes serious health hazards. Exhaust fumes from waste collection vehicles, dust stemming from disposal practices and the open burning of waste also contribute to overall health problems. People know that poor sanitation affects their health, especially in developing and low-income countries, where the people are the most willing to pay for environmental improvements (Gupta *et al.*, 1998)

The decomposition of waste into constituent chemicals is a common source of local environmental pollution. Very few existing landfills in the world's poorest countries would meet environmental standards accepted in industrialized nations, and with limited budgets there are likely to be few sites rigorously evaluated prior to use in the future. The problem is again compounded by the issues associated with rapid urbanization. A major environmental concern is release of methane as by-product of the anaerobic respiration of bacteria, and these bacteria thrive in landfills with high amounts of moisture. Methane concentrations can reach up to 50 per cent of the composition of landfill gas at maximum anaerobic decomposition (Alam and Ahmade 2013). Liquid leachate management varies throughout the landfills of the developing world. Leachate poses a threat to local surface and groundwater systems. The use of dense clay deposits at the bottom of waste pits, coupled with plastic sheeting-type liners to prevent infiltration into the surrounding soil, is generally regarded as the optimum strategy to contain excess liquid. In this way, waste is encouraged to evaporate rather than infiltrate (Royal Commission on Environmental Pollution, 1984).

3.2 Characterization of Municipal Solid Waste

It is very essential to know the composition and the quantity of MSW generated in town and big cities in order to manage it. In India, MSW differs greatly with regard to the composition and hazardous nature, when compared to MSW in the western countries (Shannigrahi *et al.*, 1997). The composition of MSW at generation sources and collection points on a wet weight basis may consists mainly of a large organic fraction (40–60 per cent), ash and fine earth (30–40 per cent), paper (3–6 per cent) and plastic, glass and metals (each less than 1 per cent). The C/N ratio ranges between 20 and 30, and the lower calorific value ranges between 800 and 1000 kcal/kg. The physical characteristics of MSW in metro cities are presented in Figure 3.2.

Solid waste may be: Residential, institutional, industrial, municipal, construction *etc.* On the basis of reaction it can be of following types.

i) Corrosive: these are wastes that include acids or bases that are capable of corroding mental containers, *e.g.* tanks.

ii) Ignitabile: this is waste that can create fires under certain condition, *e.g.* waste oils and,

iii) Solvents Reactive: these are unstable in nature, they cause explosions, toxic fumes when heated.

iv) Toxic waste: which are harmful or fatal when ingested or absorb.

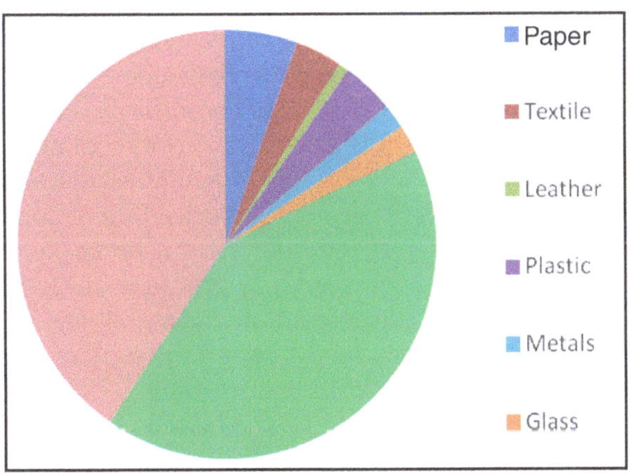

Figure 3.2: Pi Chart Showing Characteristics of MSW.

In the literature as well as in practice most often used definition of waste comes from Waste Framework Directive and is defined as any substance or object which the holder discards or intends or is required to discard (Directive 2008/98/EC). Another definition highlighting the characteristics of the waste determines it as a solid or liquid (but not being wastewater) by-product coming from production or consumption, useless in time and place where it was created and harmful or onerous for environment or people (Dobrzańska, 2010). There is a rather complicated definition of waste recovery existing in Polish law, which means any operation the principal result of which is waste serving a useful purpose by replacing other materials which would otherwise have been used to fulfil a particular function, or waste being prepared to fulfil that function, in the plant or in the wider economy (Ustawa, 2012). In practice, most often waste recovery means energy recovery. It is the conversion of non-recyclable waste materials into useable heat, electricity, or fuel through a variety of processes, including combustion, gasification, pyrolization, anaerobic digestion, and landfill gas recovery. This process is often called waste-to-energy (WTE), (Directive 2008/98/EC).

Waste management means collection, transport, treatment and disposal of waste as well as control, monitoring and regulation of the production and mentioned processes. It also covers prevention of waste production through in – process modifications, reuse and recycling. In the general waste management system three main elements can be distinguished: management institution, management instruments and managed object. In the process perspective objects of steering are following (Poskrobko and Poskrobko 2012):

- ☆ Preventing the creation of waste
- ☆ Generating waste
- ☆ Selective collection
- ☆ Recovery and recycling
- ☆ Waste neutralization
- ☆ Storing on landfills

In the case of tires mainly energy recovery is used. It involves burning tires in whole or after earlier shredding them. During combustion the heat is produced which is used in technological processes such as the clinker-burning in cement kilns or for the steam production in power plants. The calorific value of the rubber is comparable to or even higher than the calorific value of coal. In cement kilns tires may be for up to ¾ of the required fuel mass. In Poland, the biggest interest in energy recovery from tires is expressed by cement factories. The second most common method of waste tires processing is recycling. It means any recovery operation by which waste materials are reprocessed into products, materials or substances whether for the original or other purposes. It includes the reprocessing of organic material but does not include energy recovery and the reprocessing into materials that are to be used as fuels or for backfilling operations (Ustawa, 2012).

Table 3.2: Physical Characteristics of MSW in Certain Indian Metro Cities

City	Characteristics of MSW (Per cent by weight)							
	Paper	Textile	Leather	Plastics	Metal	Glass	Ash	Compostable Matter
Ahmedabad	6	1	-	3	-	-	50	40
Bengaluru	8	5	-	6	3	6	27	45
Bhopal	10	5	2	2	-	1	35	45
Delhi	6.6	4	0.6	1.5	2.5	1.2	51.5	31.78
Indore	5	2	-	1	-	-	49	43
Kanpur	5	1	5	1.5	-	-	52.5	40
Mumbai	10	3.6	0.2	2	-	0.2	44	40
Kolkata	10	3	1	8	-	3	35	40
Vadodara	4	-	-	7	-	-	49	40

Source: Status of solid waste, CPCB 2000 (Sharholy *et al.*, 2008).

3.3 Fungal Flora Associated with Kitchen Waste

Municipal solid waste contains household and other materials. A survey was conducted in Vadsar dumping site Kalai in Vadodara in Jan 2011 and isolation of fungi was undertaken. This site was started in 1974-1975. Table 3.3 gives a list of 20 different fungi isolated, many of these belonged to Ascomycetes and Fungi imperfecti. Five species of *Aspergillus* were present. *Rhizopus* a member of Zygomycetes was also isolated.

Table 3.3: Mycoflora Associated with different Articles Present in Municipal Waste

Sl.No.	Material Used for Isolation	Associated Mycoflora
1.	Broomsticks	*Aspergillus awamori* Nakaz. *Aspergillus niger* van Tiegh Aspergillus sp. *Drechslera halodes* (Drechsler) Subam. and B.L.Jain *Fusarium* sp. *Rhizopus stolonifer* (Ehrenb.) Vuill
2.	Coconut husk	*Aspergillus awamori* Nakaz. *Aspergillus flavus* var. flavus Link *Aspergillus fumigatus* Fresen *Aspergillus niger* van Tiegh Aspergillus sp. *Cladosporium herbarum* (Persoon)Link. *Curvularia prasadii* Mathur and Mathur *Fmericella nidulans* (Eldam)Vuill. *Penicillium citrinum* Thom.

Sl.No.	Material Used for Isolation	Associated Mycoflora
3.	Cloth material	*Aspergillus flavus* var. *flavus* Link *Aspergillus niger* van Tiegh Aspergillus sp. *Cladosporium herbarum*(Persoon)Link. *Curvularia prasadii* Mathur anad Mathur *Penicillium citrinum* Thom Rhizopus stolonifer (Ehrenb.) Vuill *Trichoderma harzianum* Rifai
4.	Feather waste	*Alternaria alternate* (Fr.) Keissl. *Aspergillus flavus* var. flavus Link Aspergillus sp. *Curvularia lunata* Boedijn *Curvularia* sp.
5.	Green vegetables	*Alternaria alternate* (Fr.) Keissl. *Aspergillus awamori* Nakaz. *Aspergillus fumigatus* Fresen *Aspergillus niger* van Tiegh *Curvularia lunata* Boedijn *Curvularia* sp. *Cladosporium herbarum* (Persoon)Link. *Curvularia prasadii* Mathur aad Mathur *Drechslera halodes* (Drechsler)Subam. and B.L.Jain *Fusarium* oxysporum Schlecht. *Rhizopus stolonifer* (Ehrenb.) Vuill *Trichoderma viride* Pers.
6.	Hairs	*Curvularia prasadii* Mathur anad Mathur *Curvularia lunata* Boedijn
7.	Leather waste	*Aspergillus niger* van Tiegh *Cladosporium herbarum*(Persoon)Link. *Trichoderma viride* Pers.
8.	News paper	*Aspergillus niger* van Tiegh *Curvularia lunata* Boedijn *Penicillium citrinum* Thom
9.	Plastic	*Alternaria alternata* (Fr.) Keissl. *Aspergillus flavus* var. *flavus* Link *Aspergillus terreus* Thom. *Trichoderma viride* Pers.

Sl.No.	*Material Used for Isolation*	*Associated Mycoflora*
10.	Rubber waste	*Alternaria alternate* (Fr.) Keissl. *Aspergillus flavus* var. *flavus* Link *Curvularia lunata* Boedijn *Cladosporium herbarum*(Persoon)Link. *Drechslera halodes* (Drechsler)Subam. and B.L.Jain *Fusarium oxysporum* Schlecht. *Fusarium* sp. *Rhizopus stolonifer* (Ehrenb.) Vuill
11.	Wood waste	*Aspergillus awamori* Nakaz. *Aspergillus fumigatus* Fresen *Aspergillus niger* van Tiegh *Fusarium oxysporum* Schlecht. *Penicillium citrinum* Thom *Trichoderma viride* Pers.
12.	Used cotton	*Aspergillus flavus* var. *flavus* Link *Aspergillus fumigatus* Fresen *Aspergillus niger* van Tiegh *Fusarium graminearum* Schwabe *Fusarium* sp. *Penicillium citrinum* Thom *Rhizopus stolonifer* (Ehrenb.) Vuill.

Source: Sejpal 2011.

Mycoflora present in peelings and waste vegetables in a kitchen were isolated and identified. Results are presented in Table 3.4. One member of zygomycetes, 6 of ascomycetes and 7 different members of Fungi imperfecti. Five species of *Aspergillus* were present in kitchen waste.

Table 3.4: Mycoflora Associated with different Vegetables and Fruits Present in Municipal Waste

Sl. No.	*Group*	*Fungi Isolated*	*Vegetables*					*Fruit*		
			1	*2*	*3*	*4*	*5*	*6*	*7*	*8*
1	Zygomycetes	*Rhizopus stolonifer* (Ehrenb.) Vuill.	+	-	+	-	-	-	-	+
2	Ascomycetes	*Aspergillus awamori* Nakaz	-	-	-	-	-	+	-	+
3	Ascomycetes	*Aspergillus flavus* var. *flavus* Link	-	-	-	+	+	-	-	-
4	Ascomycetes	*Aspergillus fumigatus* Fresen	-	-	-	-	-	-	-	-
5	Ascomycetes	*Aspergillus niger* van Tiegh	+	-	-	+	+	+	+	-
6	Ascomycetes	*Aspergillus terreus* Thom.	+	-	-	-	+	+	-	-
7	Ascomycetes	*Penicillium citrinum* Thom	-	-	-	+	-	+	+	+
8	Fungi imperfecti	*Alternaria alternata* (Fr.) Keissl.	-	-	+	-	-	+	-	-

Sl. No.	Group	Fungi Isolated	Vegetables					Fruit		
			1	2	3	4	5	6	7	8
9	Fungi imperfecti	*Fusarium oxysporum* Schlecht	-	-	-	-	-	-	-	-
10	Fungi imperfecti	*Cladosporium herbarum* (Persoon) Link.	-	-	-	-	-	-	-	-
11	Fungi imperfecti	*Curvularia lunata* Boedijn	-	-	-	+	-	-	+	-
12	Fungi imperfecti	*Drechslera halodes* (Drechsler) Subam. and B.L.Jain	-	+	+	-	-	-	-	-
13	Fungi imperfecti	*Trichoderma harzianum* Rifai	-	-	-	-	-	-	-	-
14	Fungi imperfecti	*Trichoderma viride* Pers.	-	-	-	-	-	-	+	-

1. Onion, 2. Carrot, 3. Potato, 4. Chilly, 5. Bottle gourd, 6. Banana, 7. Orange, 8. Pomegranate.

Source: Sejpal, 2011.

3.4 Collection and Disposal of Organic Waste

As per the Municipal Solid Waste (Management and Handling) Rules 2000, notified by the Ministry of Environment and Forests, Government of India, Vadodara Municipal Corporation has introduced Door-to-Door Garbage collection system from March-2006. The garbage from residential and commercial areas is directly collected from the door-step by specially designed closed body vehicles. All four zones of the city are covered by Door-to-Door Garbage collection system. More than 50 per cent of the total waste of the city is collected by this system. The Door to Door Garbage collection system vehicles are equipped with GPS monitoring system from November, 2010.

Storage of MSW at the source is substantially lacking in most of the urban areas. The bins are common for both decomposable and non-decomposable waste (no segregation of waste is performed), and the waste is disposed at a communal disposal center. Storage bins can be classified as movable bins and fixed bins. The movable bins are flexible in transportation but lacking in durability, while the fixed bins are more durable but their positions cannot be changed once they have been constructed (Nema, 2004; Malviya *et al.*, 2002). The collection of MSW is the responsibility of corporations/municipalities. The predominant system of collection in most of the cities is through communal bins placed at various points along the roads, and sometimes this leads to the creation of unauthorized open collection points. Efforts to organize house-to-house collection are just starting in many megacities such as Delhi, Mumbai, Bangalore, Madras and Hyderabad with the help of NGOs. It has been observed that many municipalities have employed private contractors for secondary transportation from the communal bins or collection points to the disposal sites. Others have employed NGOs and citizen's committees to supervise segregation and collection from the generation source to collection points located at intermediate points between sources and dumpsites. In addition, the welfare associations on specified monthly payment arrange collection in some urban areas. A sweeper who sweeps the roads manually is allotted a specific area (around 250 m^2). The sweepers put the road wastes into a wheelbarrow, and then transfer the waste to dustbins or collection points (Nema, 2004; Malviya *et al.*, 2002; Kansal *et al.*, 1998; Bhide and Shekdar, 1998). In most cities, a fraction of MSW generated remains uncollected on streets, and what is collected is transported to processing or disposal sites. The collection efficiency is the quantity of MSW collected and transported from streets to disposal sites divided by the total quantity of MSW generated during the same period. Many studies on urban environment have revealed that MSW collection efficiency is a function of two major factors: manpower availability and transport capacity. The average collection efficiency for MSW in Indian cities and states is about 70 per cent, (Rathi, 2006; Nema, 2004; Gupta *et al.*, 1998; Maudgal, 1995; Khan, 1994).

Generally, overcrowded low-income settlements do not have MSW collection and disposal services. The reason is that these settlements are often illegal and the inhabitants are unwilling or unable to pay for the services. They throw away the waste near or around their houses at different times, which makes the collection and transportation of waste very difficult in these areas. The Central Pollution Control Board (CPCB) has collected data for the 299 Class-I cities to determine the mode of collection of MSW. It is found that manual collection comprises 50 per cent, while collection using trucks comprises only 49 per cent (CPCB, 2000).

3.4.1 Disposal of MSW in Vadodara

The two leading innovative mechanisms of waste disposal being adopted in India include composting (aerobic composting and vermi-composting) and waste-to-energy(WTE) (incineration, pelletisation, biomethanation). WTE projects for disposal of MSW are a relatively new concept in India. Although these have been tried and tested in developed countries with positive results, these are yet to get off the ground in India largely because of the fact that financial viability and sustainability is still being tested. Different methods for the disposal and treatment of MSW have been discussed in the subsequent sections.

1. The first waste processing facility (compost plant) is located at Atladara STP premises running on BOOT basis. The capacity of this plant for treating un-segregated solid waste is 250 TPD. The biodegradable waste is segregated and composted in the plant and the reject is transported to the secured sanitary landfill site. Solid waste is processed by Mechanical Aerobic Windrow composting method. Compost/Manure is manufactured from this processed solid waste. At present with reference to the change in characteristics of waste and with intention to minimize processing reject waste to the secured sanitary landfill, revamping and upgrading the treatment process by implementing advanced technology and adopting profit sharing model is carried out which will result in the production of combination of easily marketable by products including green energy besides compost, resulting in enhancing the viability of the project and consequently its sustainability on long term basis. The tenure of the agreement is of 20 years which will expire in 2024.
2. Construction of Elevated Transfer Station for Centralized collection of Municipal Solid Waste under JnNURM Scheme is under progress.

3. Municipal Solid Waste treatment facility:
   a) The first waste processing facility of Atladara is under upgradation on Waste to Energy Concept.
   b) The second processing plant adjoining to the landfill cell phase I site of 300 MT/day (expandable up to 700 TPD) capacity based on Integrated Processing Technology with the intention to minimize waste load on the landfill site and to increase the life span of the landfill site.
4. Sanitary land fill cell: The Sanitary Land Fill cell (SLF) of capacity 4 Lakh Metric Tonns is in use with expected life of the same as 5 to 7 years adjacent to the processing facility near Jambuva.

3.4.2. Recycling of Municipal Solid Waste

If the organic waste is left unattended, it will tend to decompose by natural process giving rise to odors, hosting and feeding a variety of insects and pests, which in turn, form the carriers of disease creating severe health problems. The segregation, decomposition and stabilization of the organic waste by biological action forms the basis of recycling through different natural cycles.

3.5 Composting of Kitchen Waste

Biodegradable waste is used for composting by individual house owners. The MSW can be separated and biodegradable waste can be converted into useful compost. Two methods are available.

3.5.1 Aerobic Composting

The bacterial conversion of the organics present in MSW in the presence of air under hot and moist conditions is called composting, and the final product obtained after bacterial activity is called compost (humus), which has very high agricultural value. It is used as fertilizer, and it is non odorous and free of pathogens. As a result of the composting process, the waste volume can be reduced to 50–85 per cent. The composting methods may use either manual or mechanical means and are accordingly termed as a manual or mechanical process. Manual composting is carried out in smaller urban centers and mechanical composting plants have been set up in big Indian cities (Bhide and Shekdar, 1998).

Concept Biotech has developed aerobic composting tumbler of 50 L with a daily capacity of 500g. The container is made up of recycled HDPE material (Figure 3.3), waste handling is easy. With the incorporation of digesting inoculums the compost is available for mulching after 30 days. It is good for kitchen and garden waste. For societies a bigger double window composter with a daily capacity of 5-10 kg is developed. Total capacity is 250 liters. A person has to add the starter inoculums in the beginning phase only then churning of waste is done by rotating the drum manually twice daily. No foul odour is felt as bacterial growth is restricted by adding dry leaves or sawdust.

Figure 3.3: Aerobic Composter Designed by Concept Biotech, Vadodara.

Composting was encouraged in the early initiatives of the Government of India (GOI) regarding MSWM focused primarily on promoting composting of urban MSW. In the 1960s, the Ministry of Food and Agriculture offered soft loans to urban local bodies for this purpose. The 4th five year plan (1969–1974), block grants and loans were provided to state governments for setting up MSW composting plants. Finally, in 1974, GOI introduced modified scheme to revive MSW composting, particularly in cities with a population over 0.3 million. As far as large-scale composting is concerned, many mechanical compost plants with capacities ranging from 150 to 300 t/day were set up in the cities of Bangalore, Baroda, Mumbai, Calcutta, Delhi, Jaipur and Kanpur during 1975–1980 under the central Govt. scheme of MSW disposal. Indore city was a famous center for MSW composting, and the name was used to describe the composting process. The composting was done successfully for many years up to 1980, but after that the compost from MSW was not used for soil enrichment due to many problems.

The first large-scale aerobic composting plant in the country was set up in Mumbai in 1992 to handle 500 t/day of MSW by Excel Industries Ltd. However, only 300 t/day capacity is being utilized currently due to certain problems, but the plant is working very successfully and the compost produced is being sold at the rate of 2 Rs./kg. Another plant with 150 t/day capacity has been operated in the city of Vijaywada, and over the years a number of other plants have been implemented in the principal cities of the country. Now, about 9 per cent of MSW is treated by composting.

3.5.2 Anaerobic Digestion (Biomethanation)

If the organic waste is buried in pits under partially anaerobic conditions, it will be acted upon by anaerobic microorganisms with the release of methane and carbon dioxide; the organic residue left is good manure. This process is slower than aerobic composting and occurs in fact naturally in landfills. However, thermophilic digestion for biomethanation is much faster and has been commercialized. Anaerobic digestion leads to energy recovery through biogas generation. The biogas, which has 55–60 per cent methane, can be used directly as a fuel or for power generation. It is estimated that by controlled anaerobic digestion, 1 t of MSW produces 2–4 times as much methane in 3 weeks in comparison to what 1 t of waste in landfill will produce in 6–7 years (Khan, 1994).

3.5.3 Anaerobic Digestion (Production of liquid fertilizer)

Two alternatives are available to treat the waste anaerobically with15 and 250 liter capacity composter. The bigger one has the capacity of 25 kg per day. No mixing or churning is required, this works with the with solar power and bacterial culture powder. It provides high quality liquid fertilizer. Ideal for houses and flats, can be placed in a small space. The consortium of anaerobic bacteria is used in the process.

Figure 3.4: Rocket Composter Used to Produce Liquid Fertilizer through Anaerobic Digestion.

3.6 Production of Vermicompost

Vermicomposting involves stabilization of organic waste through the joint action of earthworms and aerobic microorganisms. Initially, microbial decomposition of biodegradable organic matter occurs through extra cellular enzymatic activity (primary decomposition). Earthworms feed on partially decomposed matter, consuming five times their body weight of organic matter per day. The ingested food is further decomposed in the gut of the worms, resulting in particle size reduction. The worm cast is a fine, odorless and granular product. This product can serve as a biofertilizer in agriculture. Vermicomposting has been used in Hyderabad, Bangalore, Mumbai and Faridabad. Experiments on developing household vermicomposting kits have also been conducted. However, the area required is larger, when compared to dry composting (Khan, 1994).

3.7 Management of Agrowaste

The common farming system in the Indo-Gangetic Plain (IGP) is the rice-wheat rotation system. The IGP is an important region located in South Asia blessed with fertile agricultural farmlands and a diverse ecosystem. Geographically, it occupies about 20 per cent of the total land area of India, Pakistan, Bangladesh, and Nepal. Out of approximately 66 million hectares (total area of the IGP), about 12 million hectares are utilized for the wheat-rice (Crop rotation) production system (Sharma *et al.*, 2010). In India an estimated 9.6 million hectares of land are utilized annually for the rice-wheat cropping system (Sharma *et al.*, 2010). Farmers use combine harvesters for harvesting grains which generate a huge amount of stubble consisting of tall stalks, about 15cm high, which are difficult to be incorporated into the soil (Pratika and Sandhu, 2020). A significant amount of the stubble generated is set to fire on the field. According to the Indian Agricultural Research Institute (IARI), approximately 14 million tons (Mt) out of the 22 Mt of the rice stubble (about 63.6 per cent) generated each year in India is set to fire (IARI, 2012). Haryana and Punjab, two of the key agricultural states of India, alone contribute 48 per cent to this amount (Gadde *et al.*, 2009). In the Punjab region, rice and wheat account for about 85.91 per cent of the total cultivation with other crops produced in relatively minor quantities. From the farmers' perspective, it is easier to burn the crop stubble after harvest to quickly prepare the farmland for the next plantation.

Burning of firecrackers during Diwali, (Arora *et al.*, 2018) is the cause of poor winter air quality in India. However, the air quality became poor even before the event (Barman *et al.*, 2008; Ghei and Sane, 2018). The Diwali activities coincide with the crop stubble burning periods and thus results in further deterioration of the air quality. Ghei and Sane (2018) reported that the burning of firecrackers during the Diwali ceremony results in a quantitatively small and statistically significant rise in air pollution. They compared the stubble burning periods with and without Diwali and found an increase of about 40 µg/m^3 $PM_{2.5}$ (Particulate matter with aerodynamic diameter less than 2.5 microns) concentration during Diwali days in 2018. It is worth noting that stubble burning is not the only cause of post-monsoon deteriorated air across the IGP; meteorological conditions such as ambient temperature, relative humidity, wind speed, wind direction and ambient pressure also play important roles (Ravindra *et al.*, 2019).

A number of studies have established a link between the post-monsoon poor quality of air in India and the variation in meteorological conditions (Banerjee *et al.*, 2011). Variations in the $PM_{2.5}$ concentrations is dependent upon specific meteorological parameters (Sembhi *et al.*, 2020). In other words, the post-monsoon meteorological conditions favor more stable atmosphere rendering the pollutants to accumulate and reside longer in the atmosphere leading to severe levels of pollution. The impact of stubble burning is more severe during the rice stubble burning season as the lower winter temperature leads to a more stable atmosphere (Inversion conditions) (Ghei and Sane, 2018). The smoke generated, cause damage to the environment (Pratika and Sandhu, 2020). Air pollution poses a grave threat to human health and wellbeing leading to a rise in mortality and morbidity rates in many parts of the world. About 7 million people in the world annually are affected (WHO 2008). The Energy and Resources Institute (TERI) reported that in 2012, around 22 per cent of the total deaths in the region occurred due to air pollution that is about 5 million deaths in South Asia. It claims approximately 5 trillion dollars from the global economy annually (Ghosh *et al.*, 2019). The poor visibility in the NCR as a result of air pollution has caused many people to migrate out of the region especially during severe pollution episodes which conversely affects tourism, especially in Delhi.

Stubble burning creates a negative externality in the form of emissions, with implications for climate change and health costs to people in affected regions, as well as disruptions in economic activity (cancellation/delays in flights and trains, and slow road traffic and accidents). Stubble burning emits fine particulate matter ($PM_{2.5}$), an air pollutant that is a concern for people's health when levels in the air are high; the particles can get trapped inside the lungs and increase the risk of lung cancer by 36 per cent. The cost of air pollution due to stubble burning in India is estimated to be $30 billion annually. Burning 1 tonne of rice accounts for a loss of nitrogen (5.5kg), phosphorous (2.3 kg), potassium (25 kg) and sulphur (1.2 kg) in the soil. Moreover, the heat from burning crop residue kills critical bacterial and fungal populations in the soil, apart from reducing organic carbon.

3.7.1 Effective Policy Measures and Management

Various policy measures at the national and sub-national levels seek to resolve the problem of crop stubble burning in India. There are also schemes to promote in situ and ex situ crop residue management through such farm equipment as the "happy seeder", rotavator and baler. However, there are many gaps in terms of policy design, implementation and awareness. In terms of policy design, the national programme on crop diversification does not have clear provisions on outreach activities to inform farmers about alternate crop options. Similarly, there is insufficient convergence with other programmes, such as the National Rural Employment Guarantee Scheme, National Rural Livelihood Mission and agro-enterprise related schemes, which could help with the management of paddy stubble or crop diversification. In terms of implementation, much-needed equipment is still unaffordable to many farmers despite subsidy provisions.

A holistic approach is required to address crop residue burning.

In the short term, misconceptions among farmers regarding paddy straw management needs to be resolved. These misconceptions include improvement in soil fertility due to stubble burning and reduction in yield due to use of in situ machines. In the medium term (the next seven years), there is a need to encourage crop diversification and rotation. While technological interventions for the management of crop residue may be useful in the short term, crop diversification as a policy intervention needs to be emphasized by the government given the multiple environmental externalities of present cropping patterns, such as depleting groundwater, poor soil quality and air pollution. Crop diversification can improve resilience from the effects of greater climate variability and extreme events. It can be implemented in various ways such as crop rotation, poly-cultures, increased structural diversity, or agro-forestry. A detailed study involving all stakeholders is required along with the awareness campaigns, economic incentives such as minimum support prices for alternative crops, along with infrastructure support such as agricultural inputs for identified alternative crops, cold storage facilities and market promotion mechanisms.

3.7.2 Use of Agrowaste as Feed and Fermentation

Waste of agricultural crops like wheat and many grasses are used as fodder for cattle. However there are a number of crops and paddy is not equally utilized as fodder.

The rice straw can be converted after pretreatment with alkali into fermentable sugars as glucose and xylose with the help of fungus *Mucor circinelloides*. Bioethanol can be produced from lignocelluloses. Collection from fields and cutting and mixing into small pieces adds to the cost hence not a viable commercial practice.

3.7.3 Use of Agrowaste as Fuel

It just might be time to redefine the word "trash." That's because if a Canadian startup has its way, trash will no longer be worthless. Enerkem, a company backed by giants such as Waste Management and Valero Energy, has a patented process that converts wood, plants, plastics, textiles, and other materials–just about anything except metal, glass and stone–into fuel that we can use to power our cars. The process can also convert garbage into feedstock for industrial chemicals.

Enerkem takes unrecyclable waste from dumps and converts it into a syngas and then converts the gas through multiple chemical steps into methanol or ethanol. It chemically convert the waste to syngas. Says Marie-Hélène Labrie Vice President at Enerkem: "We think our economics are quite attractive, and at full capacity our fuel will be competitive with corn ethanol and gasoline." Enerkem estimates that North America produces enough garbage each year to generate the equivalent of about 14 billion gallons of fuel a year–or about 10 per cent of U.S. gasoline demand.

Another advantage is that Enerkem is locating its plants next to urban landfills, which gives it a huge economic advantage over other biofuel makers. The Montreal firm, is scheduled to open its first commercial scale plant in Edmonton, Alberta. After that it plans to build at least two others: one in Montreal and the other in Pontotoc, Mississippi, which is backed by $130 million in taxpayer funds from the U.S. Department of Agriculture and Department of Energy. The plants will each produce only about 10 million gallons a year–a pittance compared to gasoline demand. But if the plants work, Enerkem will have proven that the technology is viable on a commercial scale–a major tripping point for startups in the fuel business.

It produces an improved solid fuel or pellets from MSW. The refuse derived fuel (RDF) plant reduces the pressure on landfills. Combustion of the RDF from MSW is technically sound and is capable of generating power. RDF may be fired along with the conventional fuels like coal without any ill effects for generating heat. Operation of the thermal treatment systems involves not only higher cost, but also a relatively higher degree of expertise. Many RDF plants are in operation in India, in Bangalore RDF plant was established and has had regular production of fuel pellets since October, 1989, compacting 50 t/day of garbage, converting into 5 tonne of fuel pellets. The plant is owned by Excel India at present. The Hyderabad RDF plant was commissioned in 1999 near the Golconda dumping ground with a 1000 t/day capacity (but receiving only 700 t/day at present).The RDF production is about 210 t/day as fluff and pellets, and it is going to be used for producing power (about 6.6 MW)(Yelda, and Kansal, 2003).

3.7.4 Use of Agrowaste as Building Material

Agrowaste like rice husk, paddy straw and saw dust can be used to make hard board which can be used for construction material. The straw can be placed in a linear fashion and members of basidiomycetes like *Pleurotus* is grown, this can be used to make small blocks and then the blocks can be pressed and coated with polyvinyle like substance to make them water proof. The boards are light in weight and can be ideally used as partition walls or for roofing purpose. The boards are extremely light in weight.

3.7.5 Composting of Agrowaste

Composting is the most responsible technical solution for many developing countries especially, where the climate is arid and the soil is in serious need of organic supplements. The composting process usually follows 2 basic steps as shown in Figure 3.5. Preparation of fungal inoculum as starter. This step involves mixing of saw dust with jaggery and wheat flour (10: 1: 1). Fungus *Trichoderma harzianum* Rifai is sprayed and water is sprinkled. The starter is ready after 15 days. Second step is mixing the starter with other agro waste. Good results have been obtained with leaves of mango, straw of castor, sugarcane bagasse, straw of paddy *etc.* This is an aerobic process and humidity is maintained. Crushing is required if the material is very large in size. Paddy straw can be collected in heaps or in pits and starter can be mixed to give desired results in 30-50 days. It can be mixed to soil as compost and is good for plant growth.

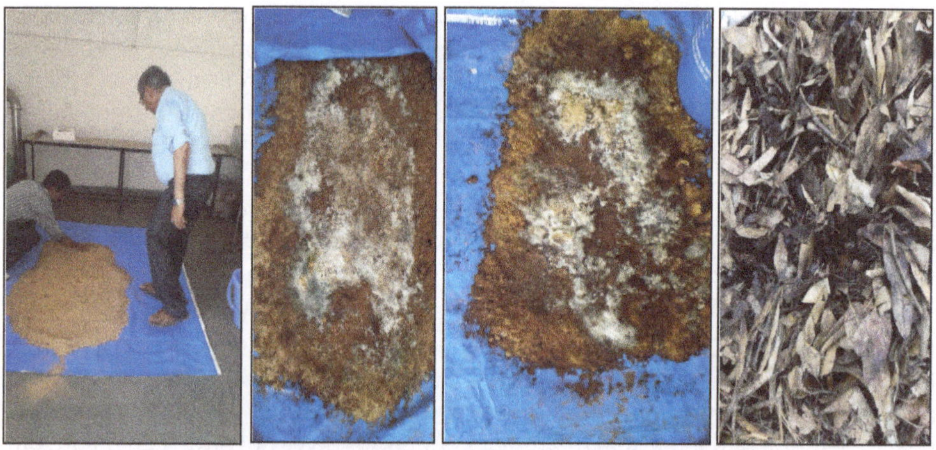

Figure 3.5 (a,b,c,d): Production of Inoculum using *Trichoderma* for Seeding the Organic Waste.

3.8 Production of Mushrooms from Biodegradable Waste

Production of Oyster mushroom (*Pleurotus sajor caju* Fr. Singer) was tried on cotton and waste paper. The mixing of 50 per cent cotton or paper with substrate of paddy straw resulted into better biological efficiency. Complete cotton or paper when used alone were not economical. The use of biodegradable waste into production of edible protein *i.e.* species of *Pleurotus* may be tried on commercial scale. After harvesting of mushroom the spent mass can be used as compost.

Many agricultural wastes can be utilized as substrates for production of *Pleurotus* species (Zadrazil and Brunnert, 1981), saw dust and rice straw in

Bangladesh (Moonmoon *et al.*, 2010), rice straw, corncob and rice bran, mango saw dust (Khan 2012).

Figure 3.6A: ***Pleurotus*** **Grain Spawn.**

Figure 3.6b: ***Pleurutus florida*** **Grown on Paper Waste.**

Figure 3.6C: ***Pleurotus*** **Grown on Paddy Straw.**

It is evident from Table 3.4 that biological efficiency of *P. florida* (Mont.) Singer was higher in straw and paper (Mix 75: 25) and straw and cotton (75: 25). By mixing the paper or cotton, number of flush *i.e.* fruiting was noticed more times.

Table 3.4: Biological Efficiency of *Pleurotus florida* (Mont.) Singer on different Substrates

Sl.No.	*Treatment*	*Number of Flush*	*Total Weight*	*Per cent Biological Efficiency*
1.	Straw	3	530	53
2.	Straw + Paper 50: 50	4	625	62.5
3.	Straw + Paper 75: 25	4	944	94.4
4.	Straw + cotton 50: 50	5	750	75
5.	Straw + cotton 75: 25	3	674	67.4
6.	Straw + cotton 25: 75	2	192	19.2

Source: Nisha B.Sc. Env. Science dissertation 2011.

3.9 Management of Plastic Waste

America's first plastic recycling mill in Conshohocken, Pennsylvania, opened in 1972. It took several years and a concerted effort for average citizens to embrace the recycling habit, but embrace it they did, and they have continued to do so in increasing numbers–but is this enough?

Plastic recycling may have come to the fore during the late-mid-20th century, Mother Earth-loving, hippie counter-culture revolution–but the idea was nothing new even then. The concept of repurposing and reusing products is as old as hand-me-downs. For thousands of years, household products were made with the idea that if they broke, they could be repaired–not simply replaced. Paper was being recycled in Japan as far back as the year 1031. A little closer to current history, plants for recycling aluminum cans were opened in Chicago and Cleveland in 1904. During World War II, the U.S. government asked the public to recycle and reuse products, a list that included tires, steel, and even nylon. Prior to today's disposable containers, fleets of milkmen home-delivered milk and cream in glass bottles that were collected when empty. They were then cleaned, sterilized, and filled back up to start the cycle all over.

Recycling plastic is unlike glass or metal processes due to the greater number of steps involved and the use of dyes, fillers, and other additives used in virgin plastics (resin produced directly from a petrochemical or biochemical feed-stock). The process begins with sorting the various items by their resin content. There are seven different plastic recycling symbols marked on the bottoms of plastic containers. At recycling mills, plastics are sorted by these symbols (and are sometimes sorted an additional time based on the color of the plastic). Once sorted, the plastics are chopped up into small pieces and chunks, and then cleaned to further remove debris such as paper labels, contents residue, dirt, dust, and other contaminants. After the plastic is cleaned, it's melted down and compressed into tiny pellets called nurdles that are ready to be reused and fashioned into new and completely different products. (Recycled plastic is hardly ever used to create the same or identical plastic item as its original form.)

- Polyethylene Terephthalate (PET, PETE): Known for superior clarity, strength, toughness, and as an efficient barrier to gas and moisture. Commonly used in the bottling of soft drinks, water, and salad dressing, and for peanut butter jars.
- High-Density Polyethylene (HDPE): Known for its stiffness, strength, toughness, resistance to moisture, and permeability to gas. HDPE is commonly used in the bottling of milk, juice, and water, as well as for trash and retail bags.
- Polyvinyl Chloride (PVC): Known for its versatility, clarity, bendability, strength, and toughness. PVC is commonly used in juice bottles, cling films, and PVC piping.
- Low-Density Polyethylene (LDPE): Known for its ease of processing, strength, toughness, flexibility, ease of sealing, and as an efficient moisture barrier. It's commonly used for frozen food bags, freezable bottles, and flexible container lids.

3.9.1 Recycling of Used Plastic

The plastic recycling process is fraught with flaws. Some dyes used in creating plastic products can be contaminated, causing entire batches of potential recycling material to be scrapped. Another issue is that producing recycled plastic does not reduce the need for virgin plastic. However, due to its use in the manufacture of composite lumber and many other products, plastic recycling can and does reduce the consumption of other natural resources, such as timber. While it's true that there's still a large percentage of people who refuse to recycle (the actual numbers of plastic being returned for reuse is roughly only 10 per cent of what is purchased as new by consumers), there are a lot of plastic items–like drinking straws and children's toys, and multi layered packaging (MLT) –that aren't considered recyclable at all. In addition, over the last few years, overwhelmed by the sheer volume and rising costs, many communities no longer offer recycling options or have added restrictions (washing and drying containers, and disallowing certain grades of plastic) for items that could have been recycled in the past.

3.9.2 Use of Plastics in Road and Construction Work

The plastic is used with tar and fly ash in building roads. LDPE is melted with tar (bitumen) to construct roads. The construction is also done by making plastic bricks and other way of using plastic is by mixing with foundry dust or waste sand to make bricks and paver block. The plastic bricks are used for construction purpose (Figure 3.7A). The alternative bricks are made using PET bottles filled with waste plastic. The waste plastic can be used with sand and other waste to produce bricks. Each brick has 20 per cent waste plastic. A wall of Unity with black coloured bricks is shown in Figure 3.7B.

Figure 3.7A: A Bench Made of Plastic Bottle Bricks.

Figure 3.7B: Wall of Bricks made from Plastic Waste.

With the support of L and T Technological Services a project was undertaken in Dhumad village of Vadodara to collect the waste plastic and reuse it in different ways. It can be used in construction of building as well as in making school benches. These are termite and water resistant. Waste plastic collected from 11 different villages was processed and a 5 feet wall of Unity was constructed at Dhumad, waste collection site in Vadodara.

3.10. Methane from Land Fill Sites

There are also facilities that convert landfill gas into lower BTU gas, which the EPA says are used to produce enough electricity to power the equivalent of 1.1 million homes per year. According to the EPA's Landfill Methane Outreach Program, there are 594 landfills around the country that have some capacity for turning methane into energy. (As of 2011, according to the EPA, there were 1,908 municipal solid waste landfills in the United States.) The conversion process at Fresh Kills works like this: The landfill gas is extracted from the ground through high-vacuum pressure and then cooled; it is then piped to a purification facility, where the methane is isolated, scrubbed and cleaned. Because methane is odorless and colorless, it is then mixed with a chemical deodorant called mercaptan so it can be detected. From there, the resulting natural gas is piped directly to about 22,000 households on Staten Island.

"It's amazing how you can take captured gas from rotting garbage and actually produce a valuable commodity out of it," said Vito Turso, Deputy Commissioner for Public Information and Community Affairs at the New York City Department of Sanitation. New York City is paid $12 million each year for the gas. That's not the only benefit to the process: It also means that the methane generated when trash decomposes - a greenhouse gas - is no longer released into the atmosphere."Not only are there environmental benefits, but also economic benefits and also the reuse of the solid waste within the local community," said Ted Nabavi, director of waste management engineering for the Department of Sanitation.

Fresh Kills, which was opened in 1948, stopped accepting waste in 2001: The city is in the process of turning what was once the largest landfill in the country, at 2,200 acres, into a park. (The city now compacts its waste and ships it to landfills in other states.) But the decomposing waste that has been dumped here for decades is still generating methane, and it is expected to do so at a significant rate for at least another decade.

At one place when we can generate energy from waste the conversion of waste into compost can be utilized for organic farming. It is termed as black gold. Trials conducted in Nagpur indicated that after third year, organic plot of cotton produced as much cotton as cultivated with pesticides and chemicals. The waste can be used for agriculture. Organic farming has a bright future and so agri-waste and compost from household waste.

3.11. Production of Electricity from Solid Waste

Gasification is a process in which partial combustion of MSW is carried out in the presence of oxygen, but in lesser amount than that is required for complete combustion, to generate a combustible gas (fuel gas) rich in carbon monoxide and hydrogen *e.g.* the conversion of coal into town gas. When a gasifier is operated at atmospheric pressure with air as the oxidant, the end products of the gasification process are a low-energy gas typically containing (by volume) 20 per cent CO, 15 per cent H_2, 10 per cent CO_2 and 2 per cent CH_4.

3.12 Landfill

In many metropolitan cities, open, uncontrolled and poorly managed dumping is commonly practiced, giving rise to serious environmental degradation. More than 90 per cent of MSW in cities and towns are directly disposed of on land in an unsatisfactory manner. In the majority of urban centers, MSW is disposed of by depositing it in low-lying areas outside the city without following the principles of sanitary land filling. Compaction and leveling of waste and final covering by earth are rarely observed practices at most disposal sites, and these low-lying disposal sites are devoid of a leachate collection system or landfill gas monitoring and collection equipment (Bhide and Shekdar, 1998; Gupta *et al.*, 1998). In Vadodara, tree garden is built now on Vadsar dumping site and walking track is made. There is no dumping of waste now.

3.13 Conclusion

Keeping good health of urban environment both long and short term planning and design of municipal solid waste management systems are required. Models are needed for accurate prediction of solid waste generation. Collection and safe disposal of municipal waste needs a fresh outlook. A successful long term planning will require the correct characteristics of the solid waste and estimation of future quantities. In foreign countries waste is segregated at source and collected in different parts. Recyclable material is processed and used. Decisions related to treatment, choices and disposal options for solid waste management will get affected by the composition of solid waste material introduced in the future. Plastic once very essential is problematic now. The paper suggests various alternatives available and used in different parts of the country or abroad. These practical approaches not only reduce the bulk of waste generated everyday but methods are also suggested to treat the agro-waste which is a cause of concern in Delhi NCR region. An integrated approach will be a right solution to generate wealth out of waste and limit the curse of pollution and ill health.

Acknowledgements

We thank L and T Technological Services for adequate financial support to conduct trials with waste plastic in form of bricks and use of PET bottles bricks to make the sitting bench, and a wall of Unity, which was inaugurated on31st Oct. 2020, at Dhumad village, Vadodara.

REFERENCES

Alam P. and Ahmade K. (2013) Impact of solid waste on health and environment. Special Issue of International Journal of Sustainable Development and Green Economics (IJSDGE), 165-168pp. ISSN No.: 2315-4721, V-2, I-1.

Arora, A, Kumari, A, Kulshrestha, U. (2018) Respirable mercury particulates and other chemical constituents in festival aerosols in Delhi. Curr. World Environ. 13 (1). https: //doi.org/10.12944/CWE.13.1.02.

Banerjee, T., Singh, S.B., Srivastava, R.K. (2011) Development and performance evaluation of statistical models correlating air pollutants and meteorological variables at Pantnagar, India. Atmos. Res. 99, 505–517. https: //doi.org/10.1016/j.atmosres.2010.12.003.

Bhide, A.D., Shekdar, A.V. (1998) Solid waste management in Indian urban centers. International Solid Waste Association Times (ISWA) (1), 26–28.

Barman, S.C., Singh, R, Negi, M.P.S., Bhargava, SK. (2008) Ambient air quality of Lucknow city (India) during use of fireworks on diwali festival. Environ. Monit. Assess. 137, 495–504. https: //doi.org/10.1007/s10661-007-9784-1. PMID: 1756220.

CPCB, (2000) Status of Municipal Solid waste Generation, Collection,Treatment and Disposal in Class I Cities, Series: ADSORBS/31/1999–2000.

CPCB, (2004) Central Pollution Control Board. Management of Municipal Solid Wastes, New Delhi, India.

Directive 2008/98/EC of the European Parliament and of the Council of 19 November 2008 on waste and repealing certain directives, http: //eur-lex.europa.eu (retrieved 05.09.2016).

Dobrzańska B, Dobrzański G, Kielczewski D. (2010) Ochrona œrodowiska przyrodniczego [Natural environment protection]. Warszawa: PWN; 2010.

Elliot, V. "Milkmen are making a comeback as families shun plastic milk containers in favor of recyclable glass bottles and organic farm food delivered to the doorstep.

Gadde, B., Bonnet, S., Menke, C., Garivait, S. (2009) Air pollutant emissions from rice straw open field burning in India, Thailand and the Philippines. Environ. Pollut. 157, 1554–1558. https: //doi.org/10.1016/j.envpol.2009.01.004.

Ghei, D., Sane, R. (2018) Estimates of air pollution in delhi from the burning of firecrackers during the festival of diwali. Edited by Krishna Prasad Vadrevu. Plos One 13 (8), e0200371. https: //doi.org/10.1371/journal.pone.0200371.

Ghosh, P., Sharma, S., Khanna, I., Datta, A., Suresh, R., Kundu, S., Goel, A., Datt, D. (2019) Scoping study for South Asia air pollution. Energy Resour. Inst. 153. https: //www.gov.uk/dfid-research-outputs/scoping-study-for-south-asia-air-pollution. Ghude, S.D., Chate, D.M., Jena, C., Beig, G., Kumar, R., Barth, M.C., P.

Gupta S, Mohan K., Prasad R., Gupta S. (1998) Solid waste management in India: options and opportunities. Resoures. conservation and Recycling 24: 117-154.

Huang, Y., Shen, H., Chen, Y., Zhong, Q., Chen, H., Wang, R., Shen, G., Liu, J., Li, B., Tao, S. (2015) Global organic carbon emissions from primary sources from 1960 to 2009. Atmos. Environ. 122, 505–512. https: //doi.org/10.1016/j.atmosenv.2015.10.017.,.

IARI (2012). Crop residues Management with Conservation Agriculture: Potential, Constraints and Policy Needs. Indian Agricultural Research Institute, New Delhi, pp. 7–32.

IPCC, (2006) Guidelines for National Greenhouse Gas Inventories. IPCC. https: //www.ipcc-nggip.iges.or.jp/public/2006gl/.

Jalan R.K., Srivastava, V.K. (1995) Incineration, land pollution control alternative – design considerations and its relevance for India. Indian Journal of Environmental Protection 15 (12), 909–913.

Kansal, A., Prasad, R.K., Gupta, S. (1998) Delhi municipal solid waste and environment – an appraisal. Indian Journal of Environmental Protection 18 (2): 123–128.

Kaushal R.K., Varghese GK, and Chabukdhara M. (2012) Municipal Solid Waste Management in India-Current State and Future Challenges: A Review. International Journal of Engineering Science and Technology (IJEST) 4 (4): 1473-1489.

Khan, R.R. (1994). Environmental management of municipal solid wastes. Indian Journal of Environmental Protection 14 (1), 26–30.

Lazarus, S. (2019) "Could Asia's Passion for Tofu Solve the Plastics Crisis?" CNN. December 9, 2019.

Malviya, R., Chaudhary, R., Buddhi, D. (2002) Study on solid waste assessment and management – Indore city. Indian Journal of Environmental Protection 22 (8), 841–846.

Maudgal, S. (1995) Waste management in India. Journal of Indian Association for Environmental Management 22 (3): 203–208.

Moonmoon M. M., Uddin N., Ahmed S. Shaelly NJ and Khan MA (2010) Cultivation of different strains of king oyster mushroom (Pleurotus eryngii) on saw dust and rice straw in Bangladesh 17: 341-345.

Nema, A.K., 2004. Collection and transport of municipal solid waste. In: Training Program on Solid Waste Management. springer, Delhi, India.

Moeller, D. W. (2005). Environmental Health (3rd ed.). Cambridge, MA: Harvard University Press.

Malviya, R., Chaudhary, R., Buddhi, D., 2002. Study on solid waste assessment and management – Indore city. Indian Journal of Environmental Protection 22 (8), 841–846.

Nema, A.K. (2004) Collection and transport of municipal solid waste. In: Training Program on Solid Waste Management. springer, Delhi, India.

Nisha (2011) Conversion of paper and cotton waste into protein B.Sc. dissertation. M.S. University of Baroda pp.101.

Pratika, C., Sandhu, H.A.S. (2020) Stubble burn area estimation and its impact on ambient air quality of Patiala and Ludhiana District, Punjab, India. Heliyon 6 (1), e03095. https: //doi.org/10.1016/j.heliyon.2019.

Poskrobko B, Poskrobko T. (2012) Environment management in Poland. Bialystok: Bialystok University Press; 2012.

Ravindra, K., Singh, T., Mor, Sahil, Singh, V., Mandal, T.K., Bhatti, M.S., Gahlawat, S.K., Dhankhar, R., Mor, Suman, Beig, G., 2019. Real-time monitoring of air pollutants in seven cities of North India during crop residue burning and their relationship with meteorology and transboundary movement of air. Sci. Total Environ. 690, 717– 729. https: //doi.org/10.1016/j.scitotenv.2019.06.216.

Raje, D.V., Wakhare, P.D., Despande, A.W., Bhide, A.D., 2001. An approach to assess level of satisfaction of the residents in relation to SWM system. Journal of Waste Management and Research 19, 12–19.

Rathi, S. (2006) Alternative approaches for better municipal solid waste management in Mumbai, India. Journal of Waste Management 26 (10), 1192–1200.

Royal Commission on Environmental Pollution. (1984) 10th Report tackling pollution– experience and prospects London: HMSO; 1984. Feb.

Sharholy M. Ahmad K. Mahmood G and Trivedi R.C. (2007) Municipal solid waste management in Indian cities- a critical review. Waste Management 28(2008)459-467.

Sharma, A.R., Kharol, S.K., Badarinath, K.V.S., Singh, D. (2010) Impact of agriculture crop residue burning on atmospheric aerosol loading–a study over Punjab State, 661 India. Ann. Geophys. 28 (2). (09927689) https: //doi.org/10.5194/angeo-28-367.

Sembhi, H., Wooster, M.J., Zhang, T., Sharma, S., Singh, N., Agarwal, S., Boesch, H., Gupta, S., Misra, A., Tripathi, S.N., Mor, S., Khaiwal, R. (2020) Post-monsoon air quality degradation across Northern India: Assessing the impact of policyrelated shifts in timing and amount of crop residue burnt. Environ. Res. Lett. https: //doi.org/10.1088/1748-9326/aba714.

Sedaghat, L. (2018)"7 Things You Didn't Know About Plastic (and Recycling)." National Geographic Society Newsroom. April 4, 2018.

Sejpal CA. (2011) Studies on fungal flora of kitchen and municipal solid waste. B.E.Tech dissertation VVP Engineering College, Rajkot pp 86pp.

Shannigrahi, A.S., Chatterjee, N., Olaniya, M.S. (1997) Physico-chemical characteristics of municipal solid wastes in mega city. Indian Journal of Environmental Protection 17 (7), 527–529.

TERI (1997). Looking Back to Think Ahead: Green India 2041 lndia: Tata Energv R'sedrc! Insntute' (1997)346.

Ustawa z dnia 14 grudnia (2012) r. o odpadach [The act of 14 December 2012 on waste], (Journal of Laws 2013, item 21).

WHO (2008). Health risks of ozone from long-range trans-boundary air pollution. https: //www.euro.who.int/data/assets/pdf_file/0005/78647/E91843 (Accessed on June 27th 2020). World Bank, 2016.

USEPA. (2013). U.S. Greenhouse Gas Inventory Report: 1990-2013. USEPA.

Yelda, S., Kansal, S. (2003) Economic insight into MSWM in Mumbai: a critical analysis. International Journal of Environmental Pollution 19 (5): 516–527.

Zadrazil, F. and Brunnert F. (1981) Investigation of physical parameters important for the solid state fermentation of straw by white rot fungi. 11: 183-188.

Chapter 4

Increasing Concentration of Sulphur Dioxide in Ambient Air and its Health Impacts

Arun Arya

Department of Environmental Studies, The Maharaja Sayajirao University of Baroda, Vadodara
e-mail: sarojarun10arya@rediffmail.com

ABSTRACT

A quantitative description of the spatial distribution of biofuel and open field burning is required to assess the impact of this burning on the budgets of trace gases. From an end-use perspective, emissions from energy conversion, largely from electricity generation, have steadily increased over the 20th century to nearly 50 per cent of SO_2 emissions. Global emissions from domestic and industrial combustion peaked around 1970 as these sectors switched to electricity and natural gas. While energy conversion now comprises the largest emission sector, the remaining emissions are distributed over a range of end-uses. The burning of woodfuel was tied directly to forestry statistics published by the Food and Agriculture Organization (FAO) of the United Nations, while burning of agricultural residues in the developing and developed world was estimated as a fraction of the available residues. Gaseous air pollutants do not exist singly in the atmosphere. The monitoring of pollutant fluxes in different locations also highlights the substantial geographical and temporal variations and the wide ranging air pollution scenarios. Biomass burning has a significant impact on global atmospheric chemistry since it provides large sources of sulphur dioxide, carbon monoxide, nitrogen oxides, and hydrocarbons, primarily in the tropics. These gases are precursors of tropospheric ozone and influence the chemistry of the OH radical. Two notable components of biomass burning are the incineration of wood, charcoal and agricultural waste as household fuel, and the combustion of crop residue in open fields. As the developing world population continues to raise, the contributions from these types of biomass burning increases. The micromorphological and biochemical changes brought along with growth and yield responses under combined exposure of air pollutants varied with pollutant combination, their levels *and plants under consideration. Crops like mango and drumstick are more susceptible. In combinations of air pollutants, responses of agricultural crops are mostly synergistic with respect to physiological, biochemical, yield and quality traits. Higher conc. of SO_2 may result into respiratory troubles and affect human health at the same time it brings acid rains.*

***Keywords**: Sulphur dioxide, Atmosphere, Human health, Plant health, Monuments, Acid rains.*

4.1 Introduction

The analysis drives home the fact that air pollution is a huge public health concern. Total 91 per cent of the world's population lives in areas where outdoor air pollution exceeds guideline limits by the World Health Organization (WHO) and as a result, 4.2 million people die prematurely every year only due to ambient air pollution. The concentrations of air pollutants vary with time and place. Central Pollution Control Board has set up Ambient Air quality standards. These are measured regularly and places are designated as normal or Residential and Eco-sensitive zones. Earlier there were three categories: Industrial, residential and sensitive. The schools colleges and hospitals were considered as sensitive. The change was needed because the conditions needed for better health can not vary in house and industries for human beings. People are spending 8-12 hours in offices and industries and these are not supposed to have harmful pollutant gases.

Source SO_2 Emission 2018	*(kt/yr)*
Volcano	19868
Coal	15217
Oil and Gas	9385
Smelter	5216

Table 4.1: Recommended Max. Limits of SO_2 (Present in air) as Primary and Secondary Exposure

Sulphur dioxide (SO_2)	Primary	1 hour	75 ppb	99th percentile of 1-hour daily maximum concentrations, averaged over 3 years
	Secondary	3 hours	0.5 ppm	Not to be exceeded more than once per year

Based on Clean air Act USA 1990.

The fraction of emissions from coal and petroleum have remained remarkably constant since 1980, with around 50 per cent of emissions from coal and 30 per cent from petroleum products. While terrestrial petroleum emissions have declined, emissions from international shipping have increased. This is a sector where high sulphur bunker fuel could, until recently, be used without significant restriction in most of the world. Trends in emissions from coal combustion include a steady decline since the 1970s in Europe and North America combined with large changes in coal combustion in the states of the Former Soviet Union after 1990, and a large increase in coal combustion in China in recent years. Emissions from metal smelting have decreased since their peak around 1970 due to increased sulphur removal, in part due to the widespread introduction of flash smelter technologies. From an end-use perspective, emissions from energy conversion, largely from electricity generation, have steadily increased over the 20th century to now and comprise nearly 50 per cent of SO_2 emissions. Surface transportation contributes a relatively small portion of total sulfur dioxide emissions. The largest portion of emissions from petroleum combustion arise from use of heavy fuel oils, including residual fuel, which are not often used in surface transportation other than for domestic shipping. The only end-use sectors that have increased emissions between 1970 and 2005 are the energy sector and international shipping. Total process emissions (smelting, fuel processing and extraction, pulp and paper, *etc.*) appear to have exceeded emissions from industrial combustion since 1990. These trends point to the central role of energy-sector emissions policies to reduce sulphur emissions overall, but also the increasing importance that other sectors, such as international shipping can play, once energy sector emissions come under emission control regimes and begin to decrease (Miola *et al.*, 2010; Eyring *et al.*, 2010).

4.2 Increasing Concentration of SO_2

The Norilsk smelter site in Russia continues to be the largest anthropogenic SO_2 emission hotspot in the world, followed by the Kriel area in Mpumalanga province of South Africa and Zagroz in Iran. Other places with high coal consumption or oil and gas refining and combustion such as Rabigh in Saudi Arabia and Singrauli in India have been catching up with the top three hotspots in the last decade and have increased their pollution dramatically. This is mainly caused by expanding capacity of coal combustion and oil refining/consumption and in part due to slow implementation and uptake of stringent emission standards. Some countries, such as China, have enforced more stringent emission regulation for coal combustion and other industrial processes, leading to a decrease of SO_2 emissions. Singrauli, Neyveli, Talcher, Jharsuguda, Korba, Kutch, Chennai, Ramagundam, Chandrapur and Koradi are the major SO_2 emission hotspots in India. In some fast-developing Asian countries such as China and India, SO_2 is still present in a significant amount in the atmosphere(Smith *et al.*, 2011). Gurjar *et al.* (2016) observed a decreasing trend of SO_2 in all monitoring sites in megacities of India due to the lowering of sulphur content in coal and diesel. The SO_2 concentration has declined throughout the country in 2009 with concentrations varying between 1.5- 13.7 ppb (Rai *et al.*, 2011). Thermal power plants are major users of coal in India, accounting for more than 25 per cent of total emissions from 1973 to 1997. Similarly petroleum sector accounts for more than 40 per cent of total emission during 1973-1997 (Mukhopadhyay and Forssell, 2005). Among all sources, coal combustion is the major source of SPM, CO_2 and NOx emissions whereas oil combustions are responsible for SO_2 emission.

The total emissions of SO_2, in the YRD (Yangtze River Delta region, China) region for the year of 2017 were 1,552, Gg (An *et al.*, 2020). If ship emissions were not included, the air pollutant emissions above would be 1,437 Gg. Emission reductions on coal-fired facilities including power plants and boilers were the main reason for the significant decline in SO_2 emissions (Zheng *et al.*, 2018). The emissions of primary pollutants such as SO_2 and NOx from power plants and boilers have dropped significantly, but the contribution of mobile sources has become increasingly prominent, and the emissions of reactive organic compounds from industrial sources are still at a high level, resulting in outstanding secondary pollution issues (Sun *et al.*, 2018).

Emissions from agricultural waste burning on fields were from EDGAR v4.0 (JRC/PBL, 2009). Production statistics for 24 crop types from FAO (FAOSTAT Crop Production) were combined with information on the fraction burned on the end emission factors (Andreae and Merlet, 2001). Emissions from waste burning were calculated in a similar manner, although these are small and the data for this sector are incomplete.

Table 4.2: Geographic Distribution of Available Residue for Developing Continents (Percent of Total)

	Maize and M&S[a]	*Wheat and Barley*	*Cereals Rice*	*Total Cereals*	*Cotton*	*Sugar Cane*	*Minor Agro-Industry*	*Total*
Asia	13	13	39	**66**	5	6	1	**78**
China	8	6	15	**29**	3	1	0	**33**
India	3	3	8	**14**	1	3	0	**18**
S. America	5	1	1	**7**	1	5	0	**13**
Brazil	2	0	1	**3**	1	4	0	**8**
Africa	5	1	1	**7**	1	1	0	**9**
Total	**24**	**16**	**40**	**80**	**7**	**12**	**1**	**100**

[a]M&S = Millet and Sorghum.

Source: Yevich, and Logan, 2003.

Billions of tons of agricultural waste are generated each year in the developing and developed countries. Agricultural residue includes all leaves, straw and husks left in the field after harvest, hulls and shells removed during processing of crop at the mills, as well as animal dung. The types of crop residue which play a significant role as biomass fuels are relatively few. The single largest category of crops is cereals, with global production of 1800 Tg in 1985 (FAO, 1986a). Wheat, rice, maize, barley, and millet and sorghum account for 28 per cent, 25 per cent, 27 per cent, 10 per cent, and 6 per cent, respectively, of these crops. The waste products which are the main contributors to biomass burning are wheat residue, rice straw and hulls, barley residue, maize stalks and leaves, and millet and sorghum stalks. Sugarcane (0.95 gigatons) provides the next sizeable residue with two major crop wastes: barbojo, or the leaves and stalk, and bagasse, the crop processing residue. The cotton crop also gives nonnegligible residue in the form of stalks and husks, both of which are used as biofuels. Four minor crops provide residue from processing that is frequently used as fuel: palm empty fruit bunch and palm fiber, palm shells, coconut residue, groundnut shells, and coffee residue. Geographical distribution of crop residue (Table 4.2) is skewed by large crop productions in India and China (FAO, 1986a). The other countries of southeast Asia have rice and sugar cane as dominant crops.

In the Middle East, the crop mixture is more diverse with more cereals and less rice and sugar cane. In the dry lands of the Near East and Mediterranean northern Africa, wheat and barley predominate. In the sub Saharan Sahel in Africa, millet and sorghum are the main crops (Table 4.2).

Crop residues are especially important as cattle feed in the semi arid regions where much of the land is cultivated and little grassland and pasture land remains (Rao, 1985). In the northern wheat and barley growing zones and in mountain villages, the straw is used for fodder (Negi, 1994). The farmers of the northeastern provinces prefer to grow traditional rice with long straw as opposed to the short straw modern varieties because the straw is needed for buffalo fodder. Similarly, farmers throughout India grow wheat varieties which give good returns in straw to provide fodder (Pal, 1966). In the northwest, the rice straw is mixed with cow dung for use as fuel. Elsewhere, sorghum straw provides a major share of cattle feed (*von Oppen and Rao*, 1982). In some parts of India, millet is grown exclusively as a forage crop. However, residue is burned in the fields in India; for example, in Punjab (Meelu *et al.*, 1991; Jenkins *et al.*, 1992). Rice straw in the central region around Hyderabad is also burned in the fields.

According to an estimate, 20 million tonnes of rice stubble is produced every year in Punjab alone, 80 per cent of which is burnt. A study estimates that crop residue burning released 149.24 million tonnes of CO_2, over 9 million tonnes of CO, 0.25 million tonnes of oxides of sulphur. One metric tonne straw burnt can produce 2 kg of SO_2, 60 kg, CO, 1460 kg of CO_2 199 kg Ash and 3 kg PM.

4.3 Vehicular Sources as Pollution of SO_2

Sulphur is ubiquitous in the biosphere and often occurs in relatively high concentrations in fossil fuels, with coal and crude oil deposits commonly containing 1–2 per cent sulphur by weight. The widespread combustion of fossil fuels has, therefore, greatly increased sulphur emissions into the atmosphere, with the anthropogenic component now substantially greater than natural emissions on a global basis (Smith *et al.*, 2011). The recent data shows that in India the number of motor vehicles has increased from 0.3 million in 1951 to 159.5 million in 2012 (Government of India, 2015). Out of these, 32 per cent are concentrated in 23 metropolitan cities. For instance, Delhi itself accounts for about 8 per cent of the total registered vehicles and the number is more in comparison to the other metro and megacities (*e.g.* Mumbai, Kolkata, and Chennai) taken together. It was estimated that the total pollution load from the 260 transportation sector in India increased 68 times from 1947 (150 Tg) to 1997 (10300 Tg). India is the largest emitter of SO_2 in the world with more than 15 per cent of all the anthropogenic SO_2. Other sources of SO_2 emissions include processes such as extracting metal from ore; running of locomotives, ships and other vehicles that burn fuel with a high sulphur content.

The resulting estimate of global petroleum sulphur is consistent with the global inventory value for petroleum emissions for most years. The mass-balance estimate indicates that there may be an underestimate on the order of 5000 Gg SO_2 in the inventory data during the late 1970s. The 10 per cent retention mass-balance case results in a lower global estimate than the inventory data past 1990. The 5 per cent retention case matches well with the inventory during this time, perhaps indicating the impact of increased refinery efficiency. While this comparison validates the overall approach, significant errors could still be present at the country level. The global mass balance estimate is also subject to significant uncertainties, particularly in regard to the estimates of the amount of sulphur removed at refineries, which is not always reported separately from sulphur removed from natural gas. As noted by Smith *et al.* (2004), the aggregate emissions factor from petroleum products in the US as implied by current estimates (Gschwandtner *et al.*, 1986; USEPA, 1996) before about 1980 appeared to be lower than those for other world regions. In order to better estimate petroleum-related emissions in the United States, a mass balance estimate for the United States was constructed, accounting for imports and exports of crude oil by region and imports and exports of petroleum products (MEPC. EIA, 2009). This mass balance indicated that sulphur emissions from petroleum were substantially underestimated prior to 1980 by 30–50 per cent.

4.4 Reduction in Pollution of SO_2 during Covid-19

Concentrations of SO_2 in polluted areas in India have decreased by around 40 per cent between April 2019 and April 2020. Using data from the Copernicus Sentinel-5P satellite, from the European Union Copernicus programme, scientists have produced new maps which show the drop in concentrations across the country in times of COVID-19. In a report by Greenpeace last year, India was named the world's largest emitter of anthropogenic SO_2 – a significant contributor to air

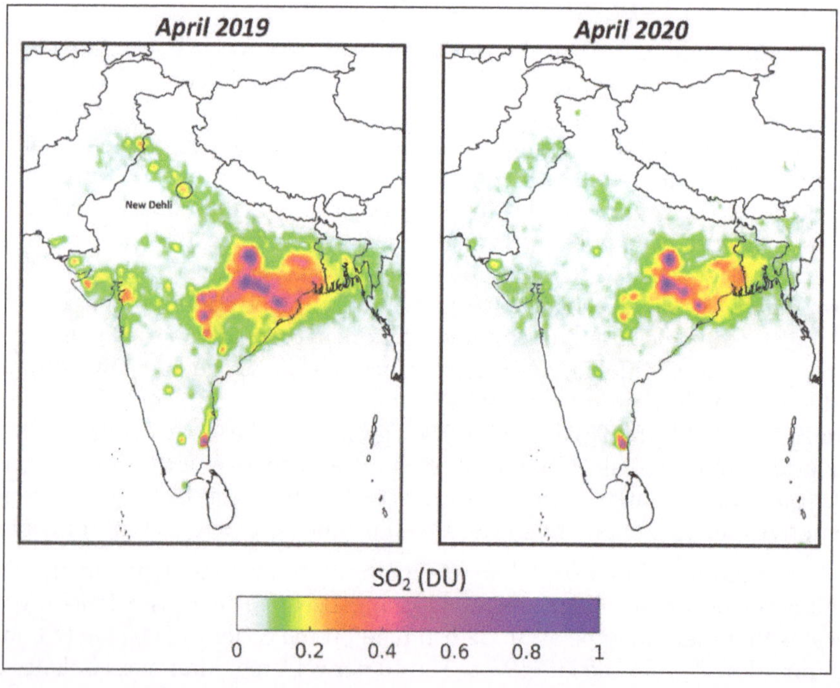

Figure 4.1: Map Showing Comparison of SO_2 Pollution in Two different Years.

pollution. While some atmospheric sulphur dioxide is produced from natural processes, such as volcanoes, a substantial amount is produced by human activities – predominantly from power plants burning fossil fuels. In India, emissions of sulphur dioxide have strongly increased over the last ten years, exacerbating haze problems over large parts of the country. However, owing to the COVID-19 pandemic, human and industrial activity dropped considerably since the beginning of its lockdown on 25 March 2020.

The Figure 4.1 shows the averaged SO_2 concentrations in April 2019, compared to April 2020. The darker shades of red and purple depict greater concentrations of SO_2 in the atmosphere, while the black dots indicate the locations of the large, coal-fired power plants. Sulphur dioxide concentrations have dropped significantly compared to the previous year, notably over New Delhi, over many large coal-fired power plants as well as other industrial areas. Some large plants in the northeast states of Odisha, Jharkhand, and Chhattisgarh have maintained a substantial level of activity, while others appear to have ceased activities entirely. This analysis was produced by using data from the Tropomi instrument on the Copernicus Sentinel-5P satellite. A recent algorithm improvement, completed by the Royal Belgian Institute for Space Aeronomy (BIRA-IASB), allows the team to better picture the evolution of anthropogenic SO_2 emissions over the country.

4.5 Effect of Sulphur Dioxide Pollution on Plants

Gaseous air pollutants do not exist singly in the atmosphere. Therefore, researches on interactive effects of air pollutants must be encouraged to understand the role of air pollutants in combination. The monitoring of pollutant fluxes in different locations also highlights the substantial geographical and temporal variations and the wide ranging air pollution scenarios. The micromorphological and biochemical changes along with growth and yield responses under combined exposure of air pollutants which vary with pollutant combination, their levels and plants under consideration. Crops like mango and drumstick are more susceptible. Figure 4.3 shows less fruiting in drumstick (*Moringa oleifera* Lam.) and black tip of mango (Arya, 2004; Figure 4.4). An accidental release of SO_2 caused burning in boot leaf of wheat crop (Figure 4.2). In combinations of air pollutants, responses of agricultural crops are mostly synergistic with respect to physiological, biochemical, yield and quality traits. Higher concentration of SO_2 may bring acid rains.

The Government of National Capital Territory of Delhi has taken several steps to reduce the level of air pollution in the city during the last 10 years. The use of Metro rail can be encouraged by provision of an adequate number of feeder buses at Metro stations that ply with the desired frequency. More frequent checking of Pollution Under Control Certificates needs to be undertaken by the civic authorities to ensure that vehicles are emitting gases within permissible norms. In order to develop a clear understanding of how plants respond to combined air pollutants, it is important to take site specific measurements of pollutant concentrations. The predictions of future pollution levels based on emissions will be helpful to estimate their future impacts on plants. Recent trends to explore the effects of multiple stresses on plants and vegetation should be extended to take a broad range of biotic and abiotic factors (Bell *et al.*, 2006).

Figure 4.2: Effect of SO_2 Pollution on Wheat Leaves.

Figure 4.3: Reduced Fruit Setting in Drumstick due to SO_2 Pollution.

Figure 4.4: Black Tip Occurs in Orchards Located Close to Brick-kilns.

Gases like carbon monoxide, SO_2 and ethylene constituting the fumes from brick kiln are known to damage growing tip of mango fruits (Arya, 2004).

4.6 Impact of Sulphur Dioxide Pollution on Buildings and Monuments

Deposition of particles can also stain and damage stone and other materials, including culturally important objects such as statues and monuments. The effects of acid deposition depend on the fragility of the materials, plants, soils and waters involved. In those instances where there is not the natural alkalinity to neutralise this acidity or alternatively the capability to withstand such attack, acid deposition has been linked with the acidification of ground and surface water, deforestation, reduction – even elimination – of aquatic life and building decay. The exposed surface of limestone ($CaCO_3$) used for the fabric of many historical buildings turns to gypsum ($CaSO_4$), which has a lower density and is more water soluble, hence suffers from sapling. The impacts on the natural environment can cause further problems for example, acidic water leeches out heavy metals. Soil no longer bound by tree roots, may be washed away, leaving a denuded landscape and potentially adding to the flood risk downstream where this soil is subsequently deposited as silt.

An analysis of water-soluble samples collected from marble and sandstone of monuments for different ions have been done. The combustion, manufacturing and other polluting operations existing within Agra area have been investigated. The measurements of flue gases amounting to 3.63×10^9 S.C.F. indicate atmospheric contamination and deterioration of archaeological monuments of Agra. It has been found that the principal sources of air contamination are the 325 iron foundries and 3 railway shunting yards located within 0.3 to 3.0 km. of the main monuments. The topographical and micrometeorological conditions of the city have tended to favour and aggravate the concentration of effluents in the surrounding air of the monuments. The annual average existing level of SO_2 ranges from 16 to 20 micrograms/m^3. The seasonal distribution of SO_2 and suspended particulate matter in the air at Taj Mahal, Red Fort and Sikandra have been discussed and illustrated. It has been observed that there is substantial sulphur dioxide contamination existing at Agra. The maximum concentration of SO^{2-}_4 and NO^-_3 amounting 0.46 and 0.38 respectively by weight percentage, found existing at Red Fort cause efflorescences of sandstone (Sharma and Sharma 1982).

4.7 Effect of Sulphur Dioxide Pollution on Human Health

Short-term exposures to SO_2 can harm the human respiratory system and make breathing difficult. People with asthma, particularly children, are sensitive to these effects of SO_2. The emissions that lead to high concentrations of SO_2 in the air generally also lead to the formation of other sulfur oxides (SO_x). SO_x can react with other compounds in the atmosphere to form small particles. These particles contribute to particulate matter (PM) pollution. Small particles may penetrate deeply into the lungs and in sufficient quantity can contribute to health problems.

Total 95 per cent of the SOx emitted from the combustion of fossil fuel is sulphur dioxide. SO_2 is a toxic gas, which is directly harmful to human health. It is heavier than air and has a suffocating odour at an atmospheric concentration of around 500 parts per billion (ppb), at which level it can be fatal. At lower levels, chest pains, breathing problems, eye irritation and a lowered resistance to heart and lung diseases can be experienced. At 20 ppb or lower concentration there should be no ill effects to a healthy person. The normal atmospheric background concentration of SO_2 is generally less than 10 ppb. A secondary effect is the formation of sulphates (and nitrates), in the form of aerosols or very fine airborne particles, which can comprise a significant proportion of the particulate matter and have been linked to increased asthma attacks, heart and lung disease and respiratory problems in susceptible population groups (Guan *et al.*, 2016).

A third effect can occur further away from the emission source where the sulphur oxides will have converted to acids by aqueous phase reactions in the atmosphere. These acidic aerosols are eventually precipitated as acid rain, snow, sleet or fog but only when they encounter the right meteorological conditions. In the absence of man made pollution rain water would be slightly acidic, around pH 5, due to the presence of carbonic acid from the interaction of water vapour and naturally occurring levels of CO_2. Acid rain on the other hand has been measured with pH levels below corresponding to vinegar.

Designation of this ECA (Emission Control Area) will significantly reduce emissions from ships, and deliver substantial benefits to large segments of the population, as well as to marine and terrestrial ecosystems. Air pollution from ships occurs not just in the United States and Canadian ports and coastlines, but is also carried hundreds of kilometres inland. When people breathe this polluted air, their health is adversely affected, leading to lost productivity due to increased illnesses, hospitalizations and even premature deaths. As of early this decade, more than 100 million people in the United States and Canada lived in areas with air pollution at levels exceeding our respective national ambient air quality standards.

Ships subject to the proposed ECA generate emissions that elevate on-land concentrations of harmful air pollutants such as PM2.5 and ozone, as well as SOx and NOx. Human exposure to these pollutants results in serious health impacts such as premature mortality and aggravation of heart and lung disease. However, we do not separately quantify the health impacts from exposure to sulphur oxides and nitrogen oxides alone, due to the difficulty of discerning those effects that are due solely to SOx versus its contribution to PM (*i.e.*, sulphate particles) or effects due solely to NOx versus its contribution to ozone and PM (*i.e.*, nitrate particles for PM).

In their 2009 joint proposal to the IMO, the USA and Canada stated that by designating the eastern and western seaboards of North America an Emissions Control Area, *"as many as 8,300 lives will be saved and over three million people will experience relief from acute respiratory symptoms each year"*(USEPA,2009).

REFERENCES

An J. Huang,Y. Huang,C. Wang, X. Yan R., Wang,Q. Wang, H. Jing S., Zhang Y., Liu Y., ChenY A.J., Xu C., Qiao L., Zhou M., Zhu,S. Hu,Q. Lu,J. Chen C. (2020) Emission inventory of air pollutants and chemical speciation for 2 specific anthropogenic sources based on local measurements in 3 the Yangtze River Delta region, China

Andreae, M. O., and P. Merlet, (2001) Emission of trace gases and aerosols from biomass burning, *Global Biogeochem. Cycles*, 15, 955– 966.

Arya, A. (2004) Tropical fruits: Diseases and Pests, Kalyani Publishers, Ludhiana,217pp.

Bell J.N.B. Davis L. and Jarraud N.S. (2006) In: *Plant response to Env. Stress* (Tripathi R.D. *et al.*, eds) Int. Book Dist Co. Lucknow India 103-112.

Eyring, V., Isaksen, I. S. A., Berntsen, T., Collins, W. J., Corbett, J. J., Endresen, O., Grainger, R. G., Moldanova, J., Schlager, H., and Stevenson, D. S. (2010) Transport impacts on atmosphere and climate: Shipping, Atmos. Environ., 44: 4735–4771.

Food and Agriculture Organization, Production Yearbook 1985, Rome, (1986a).

Gschwandtner, G., Gschwandtner, K., Eldridge, K., Mann, C., and Mobley, D. (1986) Historic emissions of sulfur and nitrogen oxides in the United States from 1900 to 1980, *Journal of Air Pollution Control Association*, 36: 139–149.

Guan, W. J., Zheng, X. Y., Chung, K. F., and Zhong, N. S. (2016) Impact of air pollution on the burden of chronic respiratory diseases in China: time for urgent action, Lancet, 388: 1939–1951.

Gurjar, B. R. Ravindra, K. and Nagpure, A. S. (2016). *Atmospheric Environment*, 142: 475–495. https: //doi.org/10.1016/j.atmosenv.2016.06.030

Jenkins, B., S. Q. Turn, and R. B. Williams, Atmospheric emissions from agricultural burning in California: Determination of burn fractions, distribution factors, and crop specific contributions, *Agric. Ecosyst. Environ.*, **38**, 313– 330, 1992.

MEPC (2009) 59/6/5 Proposal to Designate an Emission Control Area for Nitrogen Oxides, Sulphur Oxides and Particulate Matter. United States and Canada. International Maritime Organization, 2009

Meelu, O. P., Y. Singh, and B. Singh, (1991) Green manure and crop residue management in rice wheat rotation, in *Extended Summary, Proceedings International Symposium Rice Research: New Frontiers*, edited by K. Muralidharan *et al.*, pp. 355– 356, Dir. of Rice Res., Hyderabad, India, 1991.

Miola, A. Ciuffo, B., Giovine, E., and Marra, M. (2010) The State of the Art on Methodologies, Technologies and Policy Options, Publications Office of the European Union, Luxembourg, 2010.

Mukhopadhyay, K., and Forssell, O. (2005) An empirical investigation of air pollution from fossil fuel combustion and its impact on health in India during 1973–1974 to 1996–1997 *Ecological Economics* 55: 235– 250.

Negi, G. C. S. (1994) High yielding vs. traditional crop varieties: A socio agronomic study in a Himalayan village in India, *Mt. Res. Dev.*, 14: 251– 254.

Pal, B. P. (1966) *Wheat*, Indian Counc. of Agric. Res., New Delhi, 370 pp.,

Rai, R. Rajput, M. Agrawal, M. and Agrawal, S. B. (2011). *Journal of Scientific Research*, 55(771): 1.http: //citeseerx.ist.psu.edu/viewdoc/summary?doi=10.1.1.487.785.

Rao, P. P., Marketing of fodder in rural and urban areas of India, in *Agricultural Markets in the Semi Arid Tropics: Proceedings of the International Workshop at ICRISAT Center, India, 24–28 October 1983*, Int. Crops Res. Inst. for the Semi Arid Trop., Patancheru, Andhra Pradesh, India, 1985.

Sharma JS.Sharma D.N. (1982) Atmospheric contamination of archaeological monuments in The Agra region (INDIA) *Studies in Environmental Science*. 20: pp. 31-40

Smith, S. J., Andres, R., Conception, E., and Lurz, J. (2004) Sulfur Dioxide Emissions: 1850–2000 (JGCRI Report. PNNL-14537).

Smith, S. J. Van Aardenne, J. Klimont, Z. Andres, R. J. Volke, A. and Delgado Arias, S. (2011). Anthropogenic sulfur dioxide emissions: 1850-2005, *Atmospheric Chemistry and Physics*, 11(3): 1101–1116.

Sun, W. Shao, M. Granier, C. Liu, Y. Ye, C. S. and Zheng, J. Y. (2018). Long-term trends of anthropogenic SO_2, NOx, CO, and NMVOCs emissions in China. *Earth's Future*, 6(8): 1112–1133. https: //doi.org/10.1029/2018EF000822

USEPA (1990) Proposal to Designate an Emission Control Area for Nitrogen Oxides, Sulphur Oxides and Particulate Matter. Technical Support Document. USEPA

USEPA (2009) Proposal to Designate an Emission Control Area for Nitrogen Oxides, Sulphur Oxides and Particulate Matter. Technical Support Document. USEPA

von Oppen, M., and P. P. Rao, (1982) Sorghum marketing in India, in *Sorghum in the Eighties: Proceedings of the International Symposium on Sorghum*, vol. **2**, pp. 659– 674, Int. Crops Res. Inst. for the Semi Arid Trop., Patancheru, Andhra Pradesh, India.

Yevich, R. and Logan, J. (2003) An assessment of biofuel use and burning of agricultural waste in the developing world, *Global Biogeochem. Cy.*, 17: 1095–1034.

Zheng, B., Tong, D., Li, M., Liu, F., Hong, C. P., Geng, G. N., Li, H. Y., Li, X., Peng, L. Q., Qi, J., Yan, L., Zhang, Y. X., Zhao, H. Y., Zheng, Y. X., He, K. B., and Zhang, Q. (2018) Trends in China's anthropogenic emissions since 2010 as the consequence of clean air actions, Atmos. Chem. Phys., 18: 14095–14111.

Chapter 5

Role of Heavy Metal Zinc in Plants and its Effect on Germination of Wheat Grains

*Minu Kumaraswami, K. Shruti, Nirmal Chaudhary, Vraj Shah and Arun Arya**

Department of Environmental Studies, Faculty of Science, The Maharuja Sayajirao University of Baroda, Vadodara – 390 002

****e-mail: sarojarun10arya@rediffmail.com***

ABSTRACT

According to extensive surveys the zinc (Zn) content of soils, is generally in the range of 10-300 ppm. Certainly Zn, because of its concentration, can be considered as a trace element in soil. It occurs most frequently in the lithosphere as the mineral ZnS (sphalerite). Zn appears to be scattered throughout the mineral fraction of soils. The critical Zn level found in youngest expanded leaves for 90 per cent maximum yield was 16 mg kg^{-1} dry biomass. Zn-efficient Excalibur had greater net Zn uptake rates as compared to Zn-inefficient variety but they were not different at the 6-week harvest. Zinc-deficient plants had greater net uptake rates of Cu, Mn, B, P, and K but a reduced uptake rate of Fe. It is concluded that higher seed Zn content acted similar to a starter-fertilizer effect by improving vegetative growth and dissipating differences in Zn efficiency of wheat genotypes. Despite the importance of zinc as a micronutrient for plant growth, there have been relatively few studies of the mechanism of zinc uptake. It is clear from a recent review of the literature by Kochian (1993) that there is currently little agreement on how zinc crosses the plasmalemma of plant cells. The questions which have not been satisfactorily answered are whether zinc enters via ion channels or via a divalent-cation carrier. Zinc functions as an enzyme activator in carbohydrate metabolism and protein formation. Deficiency symptoms usually appear first on relatively young leaves early in growing season. On corn, a broad band of bleached tissue appears on either side of midrib. The deficiency begins at the base of the leaf and usually stays in the lower half of the leaf. In broadleaf plants, zinc deficiency results in a *shortening of internodes and a decrease in leaf size. Snap bean develops a yellowing between the leaf veins. However, it is very difficult to distinguish between zinc and manganese deficiencies in this crop.*

Zn in the leaf is associated with low molecular weight complexes, storage metallo-proteins, free ions, and insoluble forms associated with the cell wall. Zn can become inactivated within the cell either by legend formation or by forming different compounds with phosphorous. High amount of zinc concentration in soil have an adverse effect on seed germination. When the concentration of Zn is too high, the permeability of the seed membrane of the seed is destroyed, the conductivity of the seed increases with the increase of zinc concentration, and the electrolyte leakage rate increases with the increase of zinc concentration. The reduced germination was recorded in Cucumis, rhubarb and cluster beans at higher concentration. To understand the extent of zinc toxicity on seed germination, wheat grains have been exposed to different concentration of zinc sulphate and percentage germination, and young seedling growth was observed. Inhibitory effect of Zn was observed at higher concentration.

Keywords: *Zn uptake, Heavy metal, Plants, Germination, Wheat, Triticum aestivum.*

5.1 Introduction

Foods derived from plants are poorer sources of Zn for humans and monogastric animals when compared to those from animals because they can contain substances which interfere with the absorption and/or utilization (*i.e.*, bioavailability) of Zn. Examples of these substances include phytic acid and certain types of fibre, especially fibre from whole cereal grains. However, these substances are also known to play important roles in either the life cycle of higher plants or possibly, in the prevention of several human diseases including heart disease and certain forms of cancer. Therefore, it is not wise to reduce the content of these substances in food crops or in diets without a more thorough understanding of their role(s) in plant growth or human health (Welch, 1993). According to WHO, about 2 billion (33 per cent) of the world population is affected by Zn deficiency (Stein, 2010), which causes 450,000 death of children under 5 years of age every year (Black *et al.*, 2008). In developing countries Zn related deficiency is fifth major cause of human diseases and mortality (Hussain *et al.*, 2012). In recent times intake of Zn with vitamin C has been suggested to prevent from COVID disease.

Most of the zinc (Zn) in soils exists in unavailable forms. This essential micronutrient occurs in plants either as a free ion, or as a complex with a variety of low molecular weight compounds. Zinc may also be incorporated as a component of proteins and other macromolecules. As a component of proteins, zinc acts as a functional, structural, or regulatory cofactor of various enzymes. It plays an important role in auxin formation and in other enzyme systems. Presently, Zn is recognized as an essential component in several dehydrogenases, proteinases, and peptidases. High concentrations of exchangeable zinc may be toxic to many agronomic crops. It is probably held in crystal lattices, by isomorphous substitution and as occluded ions. Since it is a trace element, it is usually surrounded, by many other solid phases. Zn can also be held, by exchange sites, and adsorbed to solid surfaces. Crops differ in their sensitivity to zinc deficiency. Zn deficiencies are frequently found on soils, with restricted root zones. The movement of Zn to plant roots is dependent on the concentration and on the capacity factors (ability to replenish). Increasing the pH decreases the solubility of zinc in soils, and thereby reduces the concentration, the concentration gradient, and, hence, the uptake of Zn to plants.

Nene (1966) was the first to show that a foliar application of zinc sulfate corrects a disorder of the rice plant long known as Khaira disease in North-Central India. In August of 1966, a visit was made to the Nene's experimental fields at the Uttar Pradesh Agricultural University in Uttar Pradesh, India just as his experiments were being completed. Since, at that time, adequate analytical data on zinc status in the rice plant, as well as in the soil, were not available, the authors collected samples of soil and rice plants from the affected field for chemical analysis. The results of the analysis fully support the conclusion that Khaira disease was indeed an example of zinc deficiency (Yoshida and Tanaka, 1969). Usually crops take up less than 0.5 lb/a of zinc, yet when zinc is deficient, crops need external supply of zinc. Crops with high zinc requirements include corn, onion and spinach. Those with medium requirements are barely, beans, beet, canola, cucumber, lettuce, lupine, potato, radish, sorghum-sudan, soybean, tobacco, and tomato. Other crops have low zinc requirements and seldom exhibit zinc deficiency. In Wisconsin, zinc deficiencies have been observed on corn, snap bean and a few other vegetable crops (Schulte, 2004).

Many of the physiological activities associated with the zinc deficiency are disruption of normal enzyme activity, thus inhibition of photosynthesis was coincident with a decrease in activity of key photosynthetic enzymes. Zinc deficiency also increases membrane leakiness by inhibiting the activity of enzymes involved in the detoxification of membrane damaging oxygen radicals. Recent evidence suggests that zinc plays a key role in stabilizing RNA and DNA structure, in maintaining the activity of DNA synthesizing enzymes and controlling the activity of RNA degrading enzymes. Thus, zinc may play a role in controlling gene expression. It is interesting to note that though our understanding of the function of zinc has increased greatly in the last thirty years, there are still many facets of zinc metabolism that remain controversial.

Pahlavan Rad and Pessarakli (2009) working with wheat in Iran found that when four levels of Zn (soil applications of 0, 40, and 80 kg ha^{-1} and foliar application of 0.5 per cent $ZnSO_4$ solution), two levels of iron sulfate ($FeSO_4$; 0 and 1 per cent) as foliar application, and two levels of Mn (0 and 0.5 per cent) also as foliar applications were used in this study. Results showed that the interactive effects of Zn and Mn were significant on the number of grains in each spike. The highest number of grains resulted from the application of 80 kg $ZnSO_4$ ha^{-1} and foliar Mn. The interactive effects of Zn and Fe were significant on weight of 1000 grains. The highest weight of 1000 grains resulted from application of 80 kg Zn and foliar Fe. Application of 80 kg $ZnSO_4$ ha^{-1} alone and 80 kg $ZnSO_4$ ha^{-1} with foliar application of Mn significantly increased grain yield in 2003. Discussing the role of Zn in plants Stobart and others believed that the reason for the decrease in chlorophyll content was due to inhibition in the activity of chlorophyllate reductase enzyme, which in turn affected the synthesis of amino-r-pentanic acid, both of which are necessary for the synthesis of chlorophyll, resulting in a decrease in chlorophyll content (Stobart *et al.*, 1985). The exact concentration of zinc required for the better root length of the seedlings, stem length, chlorophyll content and seed germination is not fully understood. The toxic effects of Zn were reduced in *Brassica napus* by application of potassium, calcium and magnesium (Xu *et al.*, 2007).

5.2 Presence of Zinc in Soil

Zinc ions are held on the surface of clay and organic matter particles. Soil organic matter holds Zn in a chelated form. Chelation is a process by which certain metals are held within the structure of large organic molecules. Because zinc is held on soil particles by chelating, it does not move through the soil and is not leached under most conditions. The primary factors affecting zinc availability are soil texture, soil pH, soil phosphorus, and weather conditions. Mining activities result in extensive soil damage, causing drastic disturbances in landscape, altering the ecological environment of soil microorganisms, thereby disrupting the functional stability of the microbial community. The ultimate goal of mine land reclamation means the reestablishment of productive healthy and sustainable ecosystem for post mining activity. Currently criteria for successful restoration have largely been restricted to soil erosion, physicochemical status and vegetation characteristics. Microbial community can proceed detectable changes in soil physicochemical properties, thereby providing early signs of environmental stress or ecological environment evolution in the mining area.

Microorganisms differ greatly in their tolerance for pH, oxidized versus reduced environment. The physical and chemical conditions become more restrictive, for the diversity of microbial types that can maintain themselves. The ecosystem so developed is dominated by acidophilus sulphur and iron oxidizers. Some prokaryotic acidophilus heterotrophic organisms are present. Due to open cast mining a large area is disturbed and requires reclamation. To suggest suitable plantation a preliminary study was undertaken. Heavy metal stress, contributed by elements like cadmium, manganese or zinc is one of the crucial factors that limit the distribution and productivity of major healthy crop plants posing severe health hazards to human beings.

Restoration has been attempted on an experimental scale in various parts of the country and its implementation for the whole area has also started at some locations. In some cases the plantation of exotic and horticulture species has also been considered as restoration, while in other cases, *e.g.* Neyveli, the degraded land has been made productive and changed to agricultural land. One possible approach is to make use of the hyper-accumulator plants and the phytoremediation techniques. Furthermore, a general solution to this problem is chelation, which is generally understood as carbon binding to a compound resulting in a neutrally charged complex that can move more freely through a variety of substrates. Several chelators are known to perform this function in soil plants (Rauser 1999). The farming land is declining gradually and the main reasons include, intensive use of agricultural practices, urbanization, biotic and abiotic stresses *etc.* Among the abiotic stresses the salinity problem is increasing at an alarming rate throughout the world. Use of Cyanobacterial and other biofertilizer may help to reclaim the soil and reduce the effect of NaCl stress.

5.3 Deficiency of Zn in Plants

The biological role of Zn was first identified by Raulin in 1869, who observed that common bread mold (*Aspergillus niger*) did not grow in the absence of Zn. Soon thereafter, Zn was identified as an ubiquitous component of both animal and plant tissue. This observation stimulated Zn research in crops, and in 1914 the first demonstration of Zn deficiency in maize plants was made (Maze, 1914). The evidence that the Zn was essential for plants was further known for barley and sunflower in 1926 (Sommer and Lipman, 1926), while the first identification of Zn deficiency in field conditions was reported in 1937 in the deciduous orchards of California by Chandler (1937).

Zinc deficiency is now recognized as one of the most common micronutrient deficiencies and is becoming increasingly significant in crop production. A specific role for Zn in plants was not identified only in the late 1960's. Since then a series of Zn-containing enzymes have been identified and considerable progress has been made in detection of various chemical forms and physiological effects of Zn deficiency in plants. In the last 20 years several reviews have dealt with aspects of Zn in soils and plants. The two most comprehensive studies of these provide a good account of the reactions of Zn in soils (Lindsay, 1972) and forest ecosystems (Boardman and McGuire, 1990). The chapter discusses the chemical forms of Zn present in the plant and the role of Zn in plant metabolism and physiology. Particular attention is paid to the physiological and morphological effects of Zn deficiency in higher plants. In India alone, 50 per cent of the soils that groundnut is grown show Zn deficiency, which is causing considerable yield loss. Zinc is required for chlorophyll production, pollen function, fertilization and germination (Cakmak 2008). Zinc plays an important role in biomass production (Kaya *et al.*, 2000). Zinc in normal soil ranges from 10-300 mgkg^{-1}. The concentration of Zn present in Indian soil varies from 30 to 72 mgkg^{-1} depending on the type of soil (Katyal and Sharma 1991; Sharma *et al.*, 2013).

In plants, Zn does not undergo valance changes, and its chemistry differs from Mg^{2+}, Ca^{2+}, or Mn^{2+} in that it forms more stable complexes with a particular affinity for tetrahedral complexes. The majority of Zn in the leaf is associated with low molecular weight complexes, storage metalloproteins, free ions, and insoluble forms associated with the cell wall. Zn can become inactivated within the cell either by legend formation (Leece, 1978) or by complexation with phosphorous (Olsen, 1972). Depending on plant species, anywhere from 58 per cent to 91 per cent of plant Zn may be soluble (Welch *et al.*, 1976). This water-soluble Zn fraction is usually considered to be the physiologically active fraction and is regarded as a better indicator of Zn status than is total Zn content (Cakmak and Marschner, 1987). Among these soluble Zn forms, low molecular weight complexes are frequently in the greatest abundance and are probably the most significant form of active Zn. Though more than 70 Zn metalloenzymes have been identified, functional metalloproteins represent only a small proportion of the total Zn present in the plant.

Identification of the dominant low molecular weight Zn complexes has been performed with a range of electrophoretic and chromatographic techniques and more recently has been estimated by using chemical equilibrium computer programs (CHELATE, GEOCHEM). Most of the soluble Zn in the cell is associated with low molecular weight anionic complexes, with only a small percent present as free ions. In plant leaves, soluble Zn exists largely as an anionic compound possibly associated with amino acids. In lettuce (*Lactuca sativa* L.) the major Zn fraction had a molecular weight of 1250 and contained S, reducing sugars, and amino acids (Walker and Welch, 1987). The isolated low molecular weight fractions represented 73 per cent of total soluble Zn, which is 58 per cent of the total Zn in lettuce leaf. Similar results have been observed in seed, where 62 per cent to 70 per cent of seed Zn was soluble (Khan and Weaver, 1989). Free Zn ion present in leaf tissues is generally low and ranges from 5.8 per cent of total Zn in tomato (Bowen *et al.*, 1962) to 6.5 per cent in alfalfa (Johnson and Schrenk, 1964). The cell wall may also play a role in controlling free Zn activity. A relationship between the affinity of cell wall extracts for free Zn, and the relative tolerance of diverse species to excess Zn, suggests a role for the cell wall in controlling free Zn in the cell (Turner, 1970; Turner and Marshall, 1971). As reviewed by Torre *et al.* (1991) various cell wall compounds *i.e.* lignin, cellulose, hemicellulose, *etc.* possess a high binding affinity to Zn. Accordingly, it is assumed that 90 per cent or more of the total Zn in roots is adsorbed in the apoplast of rhizodermal and cortical cells (Schmid *et al.*, 1965). The significance of cell wall binding of Zn, however, remains controversial, particularly as the total binding capacity of the cell wall may be too small to be physiologically significant (Wainright and Woolhouse, 1978).

5.4 Zinc Uptake by Plant Cells

Zinc functions as an enzyme activator in carbohydrate metabolism and protein formation. Deficiency symptoms usually appear first on relatively young leaves early in growing season. On corn, a broad band of bleached tissue appears on either side of midrib. The deficiency begins at the base of the leaf and usually stays in the lower half of the leaf. In broadleaf plants, zinc deficiency results in a shortening of internodes and a decrease in leaf size. Snap bean develops a yellowing between the leaf veins. However, it is very difficult to distinguish between zinc and manganese deficiencies in this crop. In most crops, the typical leaf Zn concentration required for adequate growth approximates 15-20 mg Zn/kg DW (Marschner, 1995; Singh and Prasad, 2014).

Despite the importance of zinc as a micronutrient for plant growth, there have been relatively few studies of the mechanism of zinc uptake. It is clear from a recent review of the literature by Kochian (1993) that there is currently little agreement on how zinc crosses the plasmalemma of plant cells. The questions which have not been satisfactorily answered are whether zinc enters via ion channels or via a divalent-cation carrier, the link between uptake and metabolic energy transduction, the existence of an active efflux mechanism, whether fluxes can be described by Michaelis-Menten kinetics with Vm and K, and the possible involvement of phytosiderophores. Most studies of zinc uptake have used solution cultured roots and radioactive $^{65}Z_n$ either to measure short-term influxes or to estimate fluxes from compartmental analysis of efflux kinetics of tissues equilibrated in $^{65}Z_n$ (Schmid *et al.*, 1965; Chaudhry and Loneragan 1972). However, it is difficult to assess the reliability of the tracer flux measurements because in none of these studies a clear distinction has been made between extracellular binding and actual membrane influx. In complex tissues such as roots it is difficult to do so, although implicit faith in the effectiveness of desorption treatments to remove extracellularly bound radiotracer seems to be widespread in studies of divalent- and trivalent-cation uptake. The experiments with charophyte cells, in which it is indeed possible to assess the effectiveness of desorption (Reid and Smith 1992a,b; Reid *et al.*, 1993), have led us to question the validity of the application of such practices to short-term uptake experiments with more complex tissues. We demonstrate here that the characteristics of cation binding in cell walls of *Chara* are not fundamentally different from apoplasmic binding in cereal roots, and that by adopting the same techniques used to measure Ca^{2+} (Reid *et al.*1993, Pifieros and Tester,1995).

In *Chara,* the general properties of membrane fluxes of zinc can be described. The giant-celled alga *Chara cora* Uina Klein ex Will. R.D. Wood was grown in the laboratory in large plastic tanks in tap water on a substrate of garden soil. The cultures were illuminated on a 16 h/8 h light/dark cycle at a photon flux of approximately 35 ~tmol. m^{-2}. s^{-1} at the surface of the solution. Before experiments, individual internodal cells (40-90 mm long and approximately 1 mm in diameter) were isolated from the plant and stored overnight in an artificial pond water (APW) containing 1 mol^{-1} m^{-3} NaC1, 0.1 mol. m^{-3} K_2SO_4 and 0.5 mol^{-1} m^{-3} $CaCI_2$. Unless otherwise stated, experimental solutions were buffered at pH 6.2 with 1 mol. m^{-3} 2-0N-morpholino) ethanesulfonic acid. Wheat (*Triticum aestivum* L. cv. Halberd) was grown on a 10 per cent Hoagland's solution for 10 d. Influx of zinc was measured with $^{65}Z_n$ using the methods described in Reid and Smith (1992a,b) for 4SCa. The advantage of using charophyte internodal cells lies in their large size and cylindrical shape which enables easy separation of cell wall from intracellular contents after the radioactive influx period. Two different techniques were used, one to obtain short-term influxes and a second to measure longer-term accumulation of zinc. In the latter method, individual internodal cells were incubated in 6~Zn-APW in a Petri dish (10^{12} cells/200ml), then desorbed for 30min in 1 tool-m^{-3} LaC13 in APW, the solution being changed six times during this period. The cells were then blotted, allowed to wilt slightly, the ends were excised and a hypodermic needle inserted into one end. The vacuole was expelled by gently injecting an air bubble through the cell, and the cytoplasm was then flushed out by rapidly injecting 1 ml of deionised water, leaving a clear sleeve of cell wall, whose $^{65}Z_n$ activity could be measured separately from that in the cytoplasmic and vacuolar fractions. For the measurement of short-term (*i.e.* unidirectional) zinc influx, short rinse times were used to reduce the possibility of efflux of $^{65}Z_n$ during the desorption of the cell wall. The increased risk of contamination of the cell contents during the separation of the cell wall was overcome by mounting the cell in a three compartment chamber in which only the centre compartment contained $^{65}Z_n$.

The ends of the cells remained unlabelled and there was therefore much less likelihood of contamination during the surgical procedures required for separating the cell walls from the cell contents. Influx of ^{45}Ca was measured using the same techniques as for zinc. Influx of ^{42}K was measured using normal tracer techniques with several 1-min rinses to remove extracellular activity. Large volumes of influx solution were needed to prevent depletion of $^{65}Z_n$ by adsorption to cell walls. Influx solutions were gently agitated and cells were illuminated with dim light (approx. 35 ktmol^{-1} m^{-2}-s^{-1}). The electrical potential difference (PD) between the vacuole and the outside solution was varied by an external voltage-clamp technique. Individual internodal cells were mounted in a chamber in which the centre portion of the cell was isolated externally from the two end portions with silicone grease. A KCl- filled glass micropipette was inserted into the vacuole to measure the PD, which was then controlled by passing hyperpolarizing or depolarizing current between external Ag/AgCl electrodes located in the centre compartment and one end compartment. When a stable PD was obtained, $^{65}Z_n$ was added to the solution flowing through the centre compartment. After 20 rain the $^{65}Z_n$ was rinsed away and the influx was determined using the cell-fractionation techniques described above. Action potentials were generated by applying a voltage pulse between the Ag/AgC1 electrodes of approximately 3 V for 0.2s. Isolated cell wall sleeves were obtained by removing the ends of internodal cells and injecting demonized water through the lumen. This process appeared to be quite efficient in removing the cell contents since the chloroplasts, which are known to be embedded in a cytoplasmic gel adjacent to the plasma membrane, were completely displaced.

Cell wall material was obtained from wheat roots by boiling 50-mm apical root segments for 20 min in deionised water. Washed and dried cell wall material was incubated in 10 mmol'm^{-3} $^{65}Z_n$ in APW for 2 h then desorbed in a solution containing 1 mol.m^{-3} LaC13 in APW. The rinse solutions were periodically changed and sampled to determine the $^{65}Z_n$ desorbed in each rinse period. Initial binding and the time-course of desorption were calculated by summing the $^{65}Z_n$ remaining in the tissue after 60 min and the preceding rinses. The zinc concentration in culture solution, cells and cell walls was measured by graphite furnace atomic absorption spectrometry (GFAAS) after extraction in 1000 tool" m^{-3} nitric acid. Zinc speciation in solution was calculated using GEOCHEM-PC (Parker *et al.*, 1994).

Rauser (1999) reported that plants produce a range of legends for cadmium (Cd), copper (Cu), nickel (Ni), and Zn. Cd- and Zn-citrate complexes are prevalent in leaves, even though malate is more abundant. In the xylem sap moving from roots to leaves, citrate and histidine are the principal legends for Cu, Ni, and Zn. Phosphorus-rich globular bodies in young roots are probably Zn-phytate. The Ec protein from wheat embryos has Cys in three domains, binds Zn, and disappears with seedling development. One complex is cytosolic, the dominant one is vacuolar. Together, they can bind a large proportion of cellular Cd; other legends may also function as complexes with tetrahedral coordination and a Cd-sulfur inter atomic distance of 2.52 A. Little is known about the counterpart situation for Cu and Zn.

5.5 Role of AM Fungi in Metal Hyperaccumulator Plants

In nature, some plants hyperaccumulate heavy metals. For example, *Viola calaminaria* and *Thlapsi calaminare* grow over calamine deposits in Aachen, Germany and contain over 1 per cent (dry weight) zinc in their tissues. Also, some *Alyssum* species like *A. bertolinii* grow on serpentine soils in Tuscany, Italy and contain over 1 per cent (dry weight) nickel. These species are respectively called calamine and serpentine flora. *Thlaspi caerulescens* from the Brassicaceae family can also hyperaccumulate both Zn and Cd (Brooks, 1998). Heavy metal complexes in hyperaccumulator plants are usually associated with carboxylic acids like citric, malic and malonic acids. These organic acids help in the storage of heavy metals in leaf vacuoles. Amino acids like cysteine, histidine, glutamic acids, and glycine also forms heavy metal complexes in hyperaccumulators (Homer *et al.*, 1997). These complexes are more stable than those with carboxylic acids. They are mostly involved in heavy metal transport through xylem. Moreover, hyper-accumulator plants can increase availability of metals like Fe and also Zn, Cu and Mn by releasing chelating phytosiderophores. This mechanism may thus be related to rhizosphere processes such as to the release of chelating agents (phytosiderosphores and organic acids) and/or to differences in the number or affinity of metal root transporters (Lombi *et al.*, 2001).

Although hyperaccumulator plants are widely used in phytoextraction, they are generally of low biomass, inconvenient for phytoremediation. However, arbuscular mycorrhizae fungi (AMF), especially *Glomus intraradices*, colonized *Festuca* and *Agropyron* species have shown higher heavy metal (Zn, Cd, As and Se) content than non-colonized controls (Giasson *et al.*, 2006). As for hyperaccumulators, fungi can synthesize cysteine-rich metal binding proteins called metallothioneins (Gadd and White, 1989), AMF might therefore, be directly implicated in heavy metal hyperaccumulation in plants.

Many researchers have shown that mycorrhizal colonization can have an impact on heavy metal assimilation by plants (Bradley *et al.*,1981; Gildon and Tinker, 1983a, b). Dehn and Schuepp (1990) have found that mycorrhizal infection enhances heavy metal accumulation in lettuce roots but not in shoots. However, Jamal *et al.* (2002) have shown that mycorrhizae enhance heavy metal accumulation in legume shoots like soybean, alfalfa and lentils, Killham and Firestone (1983) have found similar results with grasses. In the case of cesium (Cs) and strontium (Sr), Entry *et al.* (1999) have indicated that mycorrhizal plants produce higher biomass and accumulate higher Cs and Sr content in plant tissues than non-mycorrhizal plants. Moreover, Turnau and Mesjasz-Przybylowicz (2003) have found that *Berkheya coddii*, a hyperaccumulator from the Asteraceae family, cultivated with well-developed mycorrhization, which includes arbuscule formation, increased not only the shoot biomass of the plant but also strongly increased the Ni content of shoots. Ni shoot content of *B. coddii* colonized with *Glomus intraradices* was 1.3 per cent of dry weight, while in nonmycorrhizal plants it was below 0.5 per cent.

In a glasshouse experiment, Giasson *et al.* (2006) studied four commonly found AMF species well adapted to North American soils: Glomus *intraradices, G. mossae, G. etunicatum*, and *Gigaspora gigantean*. *Glomus* spp. and *Gigaspora* spp. are AMF species identified in metal rich soils (Chaudhry *et al.*, 1999). A grass mixture of *Festuca rubra* and *F. eliator* (70 per cent), *Agropyron repens* (25 per cent), and *Trifolium repens* (5 per cent) was used. This vegetation mix is used in land reclamation in Eastern Canada to revegitate mine tailings. *Festuca* species like *F. rubra* are considered characteristic species on metalliferous soils and can accumulate excessive amounts of metals (Smith and Bradshaw, 1979; Pichtel and Salt, 1998).

5.6 Effect of $ZnSO_4$ on Germination of Wheat

Germination rate and germination potential are often used to evaluate the commonly used indicators of seed germination, reflecting the seed germination rate and germination uniformity. Zinc sulphate solution of 100ppm 500ppm and 1000ppm concentrations were prepared. Ten wheat seeds were taken in 10 Petri dishes in each treatment. These seeds were exposed for 6 hour to the different zinc sulphate solutions prepared. Then they were allowed to germinate and readings were taken after every 2 days and observations were recorded in Table 5.1. Seeds in control solution germinated in expected time period. After 4 days 60 per cent seeds germinated in 100 ppm zinc sulphate, 20 per cent seeds germinated in 500 ppm, while no seed germination was observed in 1000ppm $ZnSO_4$ solution. It is evident from the table that it took longest time to germinate a seed in 1000ppm only 10 per cent seed germination was observed after 10 days. Figure 5.1A shows no germination of seeds in 500 and 1000ppm of $ZnSO_4$ after 2 days. After 4 days 80 per cent seed germination was recorded in control, while it was 60 at 100 ppm (Figure 5.1C) and 20 at 500 ppm. After 8 days at 500 ppm concentration of $ZnSO_4$, 30 per cent germination of wheat was recorded. Higher concentration of Zn salt showed longer germination time and the germination percentage was very poor. Figure 5.1D shows reduced growth of radical in higher conc. of zinc sulphate.

Table 5.1: Percentage Germination of Wheat Grains at different Concentrations of $ZnSO_4$

Sl.No.	*Days of Treatment*	*Percentage Seed Germination in different Con. of $ZnSO_4$*			
		Control	*100 ppm*	*500ppm*	*1000 ppm*
1.	02	20	00	00	00
2.	04	80	60	00	00
3.	06	100	60	20	00
4.	08	100	80	30	00
5.	10	100	80	40	10

Note: Data based on 100 wheat grains.

Possible roles of zinc in protecting plant cells from damage by reactive oxygen species and its effect on plant metabolism has also been well reviewed (Cakmak 2000; and 2008, Broadley, *et al.*, 2006). However, little information about the effect of zinc nano particle induced biochemical responses in rice seeds is available. The role of Zn in protecting plant cells from damage by reactive oxygen species may be an important response in seed germination and not simply affecting seed water content, dry matter accumulation and changes in antioxidant balance during germination. Zinc is an indispensible micronutrient which regulates abiotic stress responses in plant. Zinc mitigates arsenic toxicity in plant by modulating ROS and antioxidant function in plant (Das *et al.*, 2016). Zinc nutrition also helps in mitigation

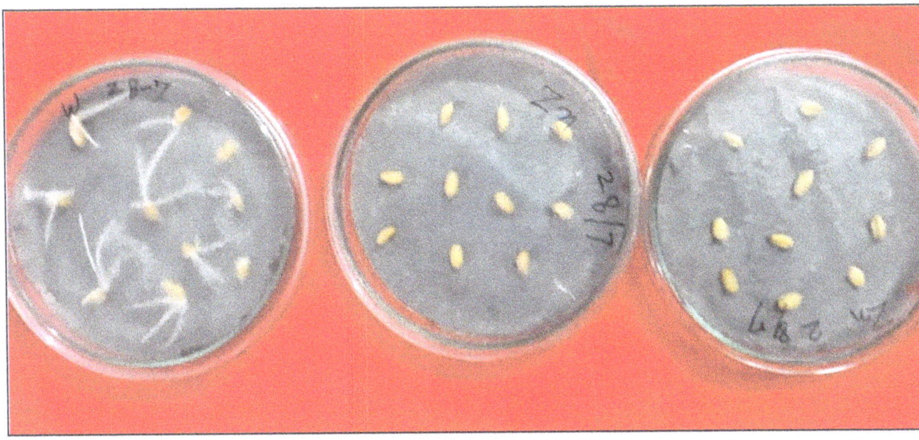

Figure 5.1A: No Germination of Wheat Grains after 4 Days at 500 and 1000 ppm of $ZnSO_4$.

Figure 5.1B: Hundred Percentage Germination in Control after 10 Days.

of iron toxicity in rice (Shahid, 2014). Zinc nutrition also modulates drought induced growth and antioxidant dysfunction in plant and mitigates dehydration induced damages and improves post stress rehydration responses in plant (Upadhyaya *et al.*, 2013 and 2016).

Contrary to present findings, it was reported by Liu *et al.* (2018) that the germination rate of Rhubarb seeds was different with the concentration of Zn^{2+} solution. And germination potential was higher than the control. Among them, when the Zn^{2+} solution was at 100mg/l, the germination rate of the seeds was at the highest level, 86 per cent and the germination of the seeds of the rhubarb was promoted effect. When the concentration of Zn^{2+} was 200mg/l, the germination index of the seeds was lower than that of the control, and the germination index was gradually decreased with the increase of the concentration of Zn^{2+}. The results

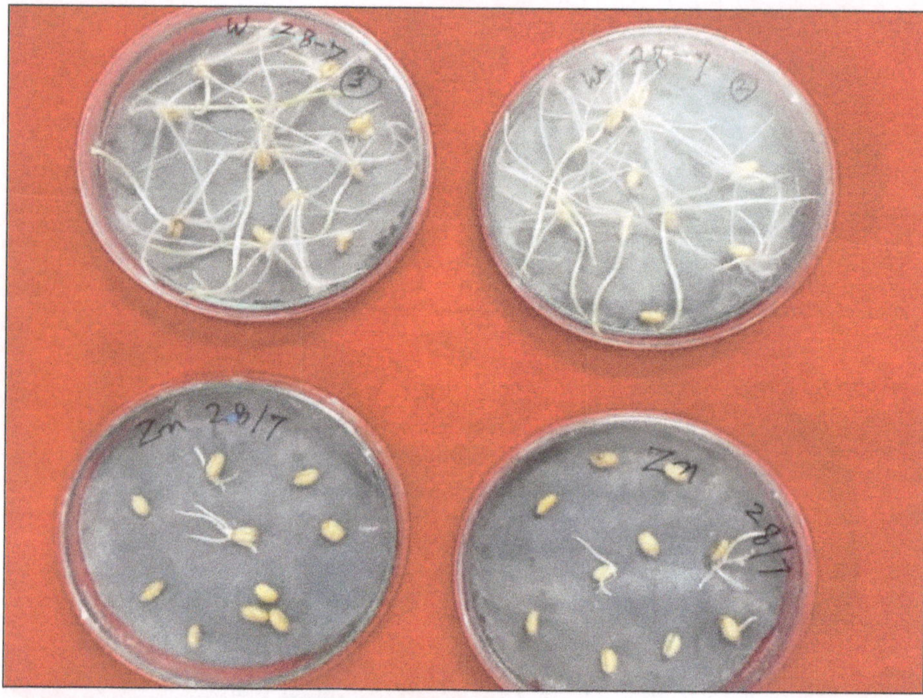

Figure 5.1C: Germination After 6 Days in Control and 500 ppm $ZnSO_4$.

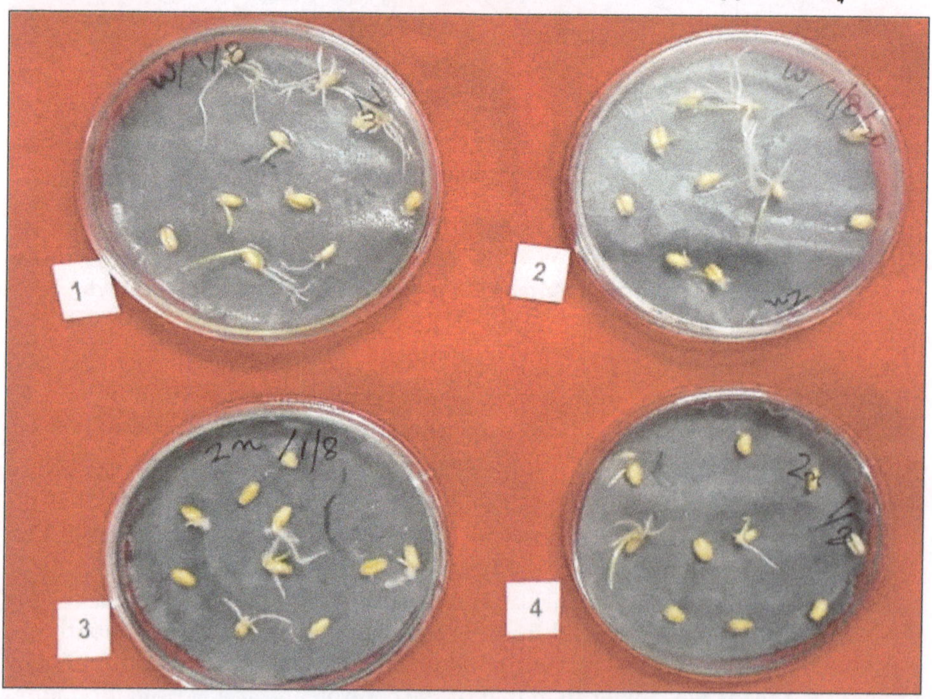

Figure 5.1D: Germination in Wheat Showing Reduced Growth of Radical after 6 Days in Petri Dish 1 and 2 at 100 ppm and 2 and 3 at 500 ppm Conc. of $ZnSO_4$.

showed that low concentration of Zn^{2+} was beneficial to seed germination. High concentration of Zn^{2+} could inhibit the seed germination.

Rengel and Graham (1995) found that seed nutrient reserves may be important for an early establishment of crops on low-fertility soils. The glasshouse pot study evaluated effects of seed Zn content on vegetative growth of two wheat (*Triticum aestivum* L.) genotypes differing in Zn efficiency. Low-Zn (around 250 ng Zn per seed) and high-Zn seed (around 700 ng Zn per seed on average) of Excalibur (Zn efficient) and Gatcher (Zn inefficient) wheats were sown in a Zn-deficient siliceous sand fertilized with 0, 0.05, 0.2, 0.8 or 3.2 mg Zn kg^{-1} soil. After 3 weeks, plants derived from the high-Zn seed had better root and shoot growth; the cv. Excalibur accumulated more shoot dry matter than the cv. Gatcher. Zinc-deficient plants had greater net uptake rates of Cu, Mn, B, P, and K but a reduced uptake rate of Fe. It is concluded that higher seed Zn content acted similar to a starter-fertilizer effect by improving vegetative growth and dissipating differences in Zn efficiency of wheat genotypes.

Zn can effectively reduce the oozing of the seeds in the germination, and the permeability of the seeds during germination. The damage of the seeds after low concentration of Zn^{2+} treatment was lower than that of the control, and the integrity of the membrane was also repaired. When the concentration of Zn^{2+} is too high, the permeability of the seed membrane of the seed is destroyed, the conductivity of the seed increases with the increase of zinc concentration, and the electrolyte leakage rate increases with the increase of zinc concentration. The results obtained by Liu *et al.* (2018) showed that the proper amount of Zn^{2+} could promote the germination of the seeds by the determination of the conductivity of the seeds. The results also showed that zinc could inhibit the root length, bud length and chlorophyll content of rhubarb seedlings. The seeds of *Cucumis* germinated in the presence of some moderate and high concentrations of zinc in substrate. The concentrations used for the treatment do not influence negatively the seed germination, but affect the seedlings growth (especially the root elongation), the formation and growth process of the lateral roots and the seedlings vigour. The effect of delay of the growth process is very pronounced in the case of high concentration (Stratu and Costică, 2015).

Seeds of cluster bean treated with low level of zinc (10 and 25 mgl^{-1}) showed a significant increase in germination, seedling growth, fresh and dry weight over the control (Manivasagaperumal *et al.*, 2011). This value indicates that zinc at lower level had a significant stimulatory, beneficiary and nutritional effect (Reichman 2002). The growth process beyond these levels (50–200 mgl^{-1}) has an adverse effect as is observed in case of wheat in the present experiment. The decrease in seed germination of cluster bean can be attributed to the accelerated breakdown of stored food materials in seed by the application of zinc. Reduction in seed germination can also be attributed to alterations of selection permeability properties of cell membrane. The decrease in seed germination of cluster bean due to zinc treatment is in conformity with the previous findings (Mahalakshmi and Vijayarengan, 2003; Mahmood, *et al.*, 2005). The reduced seedling length at higher concentrations of zinc could be the reduction in meristematic cells present in this region and some enzymes contained in the cotyledon and endosperms. Cells become active and begin to digest and store food which is converted into soluble form and transported to the radical and plumule tips for enzyme amylase which converts starch into sugar and protease act on protein. So when the activities of hydrolytic enzyme are affected, the food does not reach to the radical and plumule affecting the seedling length (Shafiq *et al.*, 2008). The reason for different response of seedling and root growth to heavy metals is not known but might be due to rapid accumulation of heavy metals in root than shoot or to faster rate of detoxification in shoot than root. The reduced growth of radical was obtained at higher conc. of zinc sulphate (Figure 5.1C and 5.1D).

5.7 Conclusion

Heavy metals in growth media can function as stresses, causing physiological constraints that decrease plant vigor and inhibit plant growth. Numerous studies conducted on the detrimental effects of metallic pollutants have shown that beyond a particular concentration, the metals can prove to be toxic to plants and animals. Heavy metals like zinc are required as structural and catalytic components of protein and enzymes for normal growth and development. But when present in elevated levels in soils, they are toxic and can ultimately cause the death of plants. Hence efforts have been made to establish the toxic level of zinc on seed germination, growth and various biochemical content of cluster bean seedling in the present study. In present experiment the germination of wheat grains at 500ppm and 1000ppm was poor. At 1000 ppm not only germination was delayed but it was 10 per cent and at different conc. of $ZnSO_4$ the length of radical was much less than control seedlings of wheat.

REFERENCES

Asher C (1987) Effects of nutrient concentration in the rhizosphere on plant growth. In: XIII th Congress Int Soc Soil Sci Symposium, 5. ISSS, Hamburg, pp 209-216

Black RE Lindsay HA Bhutta ZA Caulfield LE De Onnis M, Ezzati M, Mathers C,Rivera J. (2008) Maternal and child under nutrition: global and regional exposures and health consequences. Lancet 371: 243-260.

Bostrom TE, Walker NA (1975) Intercellular transport in plants. I. The rate of transport of chloride and the electric resistance. *J Exp Bot* 26: 767-782

Boardman R. and McGuire D.O. (1990) The role of zinc in forestry. II. Zinc deficiency and forest management: Effect on yield and silviculture of *Pinus radiata* plantations in South Australia. *Forest Ecology and Management* 37 (1-3): 207-218

Bowen H J M, Cause P A and Thick J (1962) The distribution of some inorganic elements in plant tissue extracts. *J. Exp. Bot.* 13, 257–267.

Broadley MR, White PJ, Hammond JP, Zelko I, Lux A. (2006) Zinc in plants. *New Phytol.* 2006; 173: 677-702. Ref.: https: //goo.gl/5VYYXR

Brooks RR. (1998) *Plants that hyperaccumulate heavy metals.* Wallingford, UK: CAB International.

Cakmak I. (2008) Enrichment of cereal grains with zinc: agronomic or genetic biofortifi cation? *Plant Soil.* 2008; 302: 1-17. Ref.: https: //goo.gl/q9KNHp

Cakmak I and Marschner H (1987) Mechanism of phosphorous-induced zinc deficiency in cotton III. Changes in physiological availability of zinc in plants. *Physiol. Plant.* 70, 13–20.

Carroll MD, Loneragan JF (1972) Responses of plant species to concentrations of zinc in solution. I. Growth and zinc content of plants. *Aust J Agric Res* 19: 859-868

Chaudhry FM, and Loneragan JF (1972) Zinc absorption by wheat seedlings and the nature of its inhibition by alkaline earth cations. *J Exp Bot* 23: 552-560

Chaudhry, T.M., Hill, L., Khan, A.G., Kuek, C. (1999) Colonization of iron and zinc-contaminated dumped ®ltercake waste by microbes, plants and associated mycorrhizae. In: Wong, M.H., Wong, J.W.C., Baker, A.J.M. (Eds.), *Remediation and Management of Degraded Land.* CRC Press LLC, Boca Raton, Chap. 27, pp. 275±283.

Chandler W.H. (1937) Zinc as a nutrient for plants. *Bot. Gaz.*, 98: 625-46.

Das I, Sanyal SK, Ghosh K, Das DK. (2016) Arsenic mitigation in soil-plant system through zinc application in West Bengal soils. *Bioremediation Journal.* 20: 24-37. Ref.: https: //goo.gl/AH1BkH

Dehn B. Schüepp H. (1990) Influence of VA mycorrhizae on the uptake and distribution of heavy metals in plants. *Agriculture, Ecosystems and Environment* 29, (1–4): 79-83

Entry, J.A. Watrud L.S. M. Reeves M. (1999) Accumulation of 137Cs and 90Sr from contaminated soil by three grass species inoculated with mycorrhizal fungi' *Environmental Pollution* 104 (1999) 449-457

Gadd, G.M. and C. White. (1989) The removal of thorium from simulated acid process-streams by fungal biomass. *Biotechnol. Bioeng.* 33: 592–597.

Giasson E. Clark R.T. Inda A.V. and Merte G.H. (2006) Digital soil mapping using multiple logistic regression on terrain parameters in southern Brazil. *Scientia Agricola* 63(3): 262-268 DOI: 10.1590/S0103-90162006000300008

Gildon A. and Tinker P. B. (1983a) Interactions of vesicular-arbuscular infection and heavy metals in plants I. The effect of heavy metals on the development of vesicular-arbuscular mycorrhixas. *New Phytologkt* 95, 247-261.

Gildon A. and Tinker P. B. (1983b) Interactions of vesicular-arbuscular mycorrhixal infections and heavy metals in plants II. The effects of infection on uptake of copper. *New Phytologist* 95, 263-268

Homer FA, Morrison RS, Brooks RR, Clemens J, Reeves RD (1991) Comparative studies of nickel, cobalt and copper uptake by some nickel hyperaccumulators of the genus Alyssum. *Plant Soil* 138: 195-205

Hope AB, Walker NA (1975) The physiology of giant algal cells. Cambridge University Press, London

Hussain S. Maqsood MA Rengel Z and Aziz T. (2012) Biofortification and estimated human bioavailability of zinc in wheat grains as influenced by methods of zinc application. *Plant Soil.* 361: 279-290.

Jamal, A., Ayub, N., Usman, M and Khan, A. G. (2002). Arbuscular mycorrhizal fungi enhance Znand Ni uptake from contaminated soil by soybean and lentil. Int. J. Phytoremed. 4: 205-221.

Kaya C, Higgs D, Burton A, (2000) Phosphorus acid phosphates enzyme activity in leaves in leaves of tomato cultivars in relation to Zn supply. *Commun Soil Science Plant Ana.* 31: 3239-3248. Ref.: https: //goo.gl/wDBAEq

Katyal JC, and Sharma BD. (1991) DTPA extractable and total Zn, Cu, Mn and Fe in Indian soils. *Geoderma.* 49: 165-179. Ref.: https: //goo.gl/QmVjTA

Katsuhara M, Tazawa M (1992) Calcium-regulated channels and their bearing on physiological activities in Characean cells. Phil Trans R Soc London Ser B 338: 19-29

Khan A and Weaver C M (1989) Pattern of Zn65 incorporation into soybean seeds by root absorption, stem injection and foliar application. *J. Agric. Food Chem.* 37, 855–860.

Killham K, Firestone MK (1983) Vesicular-arbuscular mycorrhizal mediation of grass response to acidic and heavy metal deposition. *Plant Soil* 72: 39–48

Kochian LV (1993) Zinc absorption from hydroponic solutions by plant roots. In: Robson AD (ed) Zinc in soils and plants. Kluwer, Dordrecht Boston London, pp 45 57

Kochian LV, Norvell WA, Shaft JE, Chaney RL (1991) Characterizing root iron uptake using a ferrous chelate to buffer free Fe 2+ ion activity in solution. (Abstr) *Plant Physiol* 96: Suppl 142

Leece, D. R. 1978 Effects of boron on the physiological activity of zinc in maize. *Aust. J. Agric.* Res.**29**, 739–747.

Lindsay, W. L. 1972 Inorganic phase equilibria of micronutrients in soils.*In* Micronutrients in Agriculture. Ed. J. J. Mortvedt, *et al.*, Soil Sci. Soc. Am. Madison, Wisconsin, pp 41–57.

Lombi E Zhao,F.J. Dunham, S J McGrath SP (2001) Phytoremediation of heavy metal-contaminated soils: natural hyperaccumulation versus chemically enhanced phytoextraction. *J Environ Qual*;30(6): 1919-26. doi: 10.2134/jeq2001.1919.

Mahalakshmi, G. and P. Vijayarengan. (2003) Effects of zinc on germinating seeds of three plants species. Nature Environ. *Pollut. Tech.* 2: 117-119.

Mahmood, S., A. Hussain, Z. Saeed and M. Athar. (2005) Germination and seedling growth of corn (*Zea mays* L.) under varying levels of copper and zinc. *Int. J. Environ. Sci. Tech.* 2: 269-274.

Manhart JR, Palmer JD (1990) The gain of two chloroplast tRNA introns marks the green algal ancestors of land plants. *Nature* 345: 268-270

Pifieros M, and Tester M (1995) Characterisation of a voltage-dependent Ca^{2+} selective channel from wheat roots. *Planta*, 195: 478-488

Maze, P. (1914). Influences respectives des 6l ements de la solution mineral a sur le developpement du maTs. *Annales de l'Institut Pasteur*, 28: 21-69.

Nene, Y. L. (1966) Zinc deficiency of the rice plant in calcareous soils, *Bull. Ind. Phytopath. Soc.*, 3: 97

Norvell WA, Welch RM (1993) Growth and nutrient uptake by barley (Hordeum vulgare L. cv. Herta): Studies using an N-(2- hydroxyethyl) ethylenedinitrilotriacetic acid-buffered nutrient solution technique. *Plant Physiol* 101: 619-625

Olsen S R (1972) Micronutrient interaction. In *Micronutrients in Agriculture*. Eds. J J Mortvedt, P M Giordano, W L Lindsay, pp 199–227. American Society of Agronomy, Madison.

Pahlavan Rad M.R. and Pessarakli M (2009) Response of Wheat Plants to Zinc, Iron, and Manganese Applications and Uptake and Concentration of Zinc, Iron, and Manganese in Wheat Grains. *Communications in Soil Science and Plant Analysis* 40 (7-8): 1322-1332 https: //doi.org/10.1080/00103620902761262

Parker DR, Norvell WA, Chaney RL (1994) GEOCHEM-PC: A chemical speciation program for IBM and compatible computers. In: Loeppert RH, Goldberg S, Schwab AP (eds) *Chemical equilibrium reaction models*. American Society of Agronomy,

Pichtel J, and Salt CA (1998) Vegetative growth and trace metal accumulation on metalliferous wastes. *J Environ Qual* 27: 618–624

Patrick H. Brown P.H. Cakmak I and Zhang Q (1993) Form and Function of Zinc Plants. In: *Zinc in Soils and Plants* pp 93-106.

Rauser W.E. (1999) Structure and Function of Metal Chelators Produced by Plants: The Case for Organic Acids, Amino Acids, Phytin, and Metallothioneins. *Cell Biochem Biophys*,31(1): 19-48. doi: 10.1007/BF02738153.

Reichman, S. M. (2002) The response of plant to metal toxicity: A review of focusing on copper, manganese and zinc. *Austr. Minerals Energy Environ*. Found. Melbourne, Australia, pp. 7.

Reid R J, and Smith FA (1992a) Measurement of calcium fluxes in plants using 45Ca. *Planta* 186: 558-566

Reid RJ, and Smith FA (1992b) Regulation of calcium influx in Chara. Effects of K +, pH, metabolic inhibition and calcium channel blockers. *Plant Physiol* 100: 637-643

Reid RJ, Tester M, and Smith FA (1993) Effects of salinity and turgor on calcium influx in *Chara*. *Plant Cell Environ* 16: 547-554

Rengel Z. and Graham R.D. (1995) Importance of seed Zn content for wheat growth on Zn-deficient soil *Plant and Soil* 173: 259–266.

Robson A.D. (1993), *Zinc in Soils and Plants* © Kluwer Academic Publishers 1993-94.

Santa Maria FE, and Cogliatti DH (1988) Bidirectional Zn-fluxes and compartmentation in wheat seedling roots. *J Plant Physiol* 132: 312-315

Schmid WE, Haag HP, and Epstein E (1965) Absorption of zinc by excised barley roots. *Physiol Plant* 18: 860-869

Schulte, E.E. (2004). Understanding plant nutrients, soil and applied Zn. Available at http: //learningstore.uwex.edu/Assets/pdfs/A2528.pdf.

Sela M, Garty J, and Tel-Or E (1989) The accumulation and the effect of heavy metals on the water fern Azolla filiculoides. *New Phytol* 112: 7-12

Shafiq, M., I. M. Zafar and M. Athar. (2008) Effect of lead and cadmium on germination and seedling growth of *Leucaena leucocephala*. *J. Appl. Sci. Environ.* Manage. 12: 61- 66.

Shahid M, Nayak AK, Shukla AK, Tripathi R, and Kumar A, (2014) Mitigation of Iron Toxicity and Iron, Zinc, and Manganese nutrition of wetland rice cultivars (*Oryza sativa* L.) grown in Iron Toxic Soil. *Clean Soil. Air, Water*. 2014; 42: 1604-1609.

Sharma A, Patni B, Shankhdhar D, and Shankhdhar SC. (2013) Zinc-an indispensable micronutrient. *Physiology and Molecular Biology of Plants*. 19: 11-20. Ref.: https: //goo.gl/fi uP85

Singh M.K. and Prasad S.K. (2014) Agronomic aspects of Zinc biofortification in rice (*Oryza sativa* L.) *Proc. Natl. Acad. Sci. India Sect. B. Biol. Sci.* 84(3): 613-623.

Smith RAH, and Bradshaw AD (1979) The use of metal tolerant plant populations for the reclamation of metalliferous wastes. *J Appl Ecol* 16: 595–612

Stein A.J. (2010) Global impacts of human mineral malnutrition. *Plant Soil*. 335: 133-154.

Sommer A L and Lipman C B (1926) Evidence on the indispensable nature of zinc and boron for higher plants. *Plant Physiol*. 1, 231–249.

Stobart A K, Griftths W T, and Bukhari A. (1985) The effects of Cd2+ on the biosynthesis of chlorophyll in leaves of barley [J].*Physiologia Plantarum*, 63: 293-298.

Stratu, A. and Costică N. (2015) The influence of zinc on seed germination and growth in the first ontogenetic stage in the species *Cucumis melo* L. PESD. 9(2): 215-227. DOI 10.1515/pesd-2015-0038 PESD,

Torre M, Rodriguez A R and Saura-Calixto F (1991) Effect of diet on fiber and phytic acid on mineral availability. *Crit. Rev. Food Sci. Nutr*. 1, 1–22.

Turner R G (1970) The subcellular distribution of zinc and copper within the roots of metal-tolerant clones of *Agrostis ienuis sibth*. *New Phytol* 69,725–731

Turner R G and Marshall C (1971) The accumulation of 65Zn by root homogenates of zinc-tolerant clones of *Agrostis tenuis sibth*. *New Phytol*. 70, 539–545

Turnau K. and Mesjasz-Przyby³owicz J. (2003) Arbuscular mycorrhiza of Berkheya coddii and other Ni-hyperaccumulating members of Asteraceae from ultramafic soils in South Africa. Mycorrhiza 13(4): 185-90 DOI: 10.1007/s00572-002-0213-6

Upadhyaya H, Dutta BK, and Panda SK. (2013) Zinc modulates drought induced biochemical damages in tea [*Camellia sinensis*(L) O Kuntze]. *J Agric Food Chem.* 2013; 61: 6660-6670. Ref.: https: //goo.gl/LwkPxK

Upadhyaya H, Dutta BK, and Panda SK. (2017) Impact of zinc on dehydration and rehydration responses in tea. *Biologia Plantarum.* 2017; 61: 1-4. Ref.: https: // goo.gl/hfRnG2

Wainright S J and Woolhouse H W (1978) Inhibition by zinc of cell wall acid phosphatase from roots in zinctolerant and non-tolerant clones of Agrostis tenuis sibth. *J. Exp. Bot.* 29, 525–531.

Walker NA, and Pitman MG (1976) Measurement of fluxes across membranes. In: Luttge U, Pitman MG (eds) *Encylopedia of plant physiology*, NS, vol. 2. Springer-Verlag, Berlin Heidelberg New York, pp 93-126

Walker C D and Welch R M (1987) Low molecular weight complexes of zinc and other trace metals in lettuce leaf. *J. Agr. Food Chem.* 35, 721–727.

Welch RM, Norvell WA, Schaefer SC, Shaft JE, and Kochian LV (1993) Induction of iron(Ill) and copper(II) reduction in pea (*Pisum sativum* L.) by Fe and Cu status 1 Does the root-cell plasmalemma Fe(III)-chelate reductase perform a general role in regulating cation uptake. *Planta* 190: 555-561

Welch R.M. (1993) Zinc concentrations and forms in plants for humans and animals In: *Zinc in Soils and Plants* (ed. E.D. Robson) Proceedings of the International Symposium on 'Zinc in Soils and Plants' held at The University of Western Australia, pp 183-195

Wu W-H, Assmann SM (1995) Is ATP required for K § channel activation in *Vicia* guard cells? *Plant Physiol* 107: 101-109

Xu J., Wei W. Han L. Guan Z-Q., Zheng H. C. and Chai T.Y. (2007) Effects of heavy metal ions on seed germination and radicle growth of *Brassica napus*. Zib. Zhi. Xue 27: 2263-2268.

Yoshida S. and Tanaka A. (1969) Zinc deficiency of the rice plant in calcareous soils, *Soil Science and Plant Nutrition*, 15: 2, 75-80, DOI: 10.1080/00380768.1969.10432783

Chapter 6

Biopollutants Posing Threat to Paper and Palm Leaf Manuscripts

Arun Arya[1]* *and N.R. Shah*[2]

[1]*Department of Environmental Studies, Faculty of Science,*
[2]*Department of Museology,*
The Maharaja Sayajirao University of Baroda, Vadodara – 390002
**e-mail: sarojarun10arya@rediffmail.com*

ABSTRACT

Scientists have studied aeromycoflora of cities like Baroda, Jabalpur, Lucknow, Bhopal etc. Arya and Arya (2007) reported numerous fungi from different vegetable and fruit markets of Baroda. Study of aeromycoflora of Agra city resulted into occurrence of 102 fungal types. In the indoor survey of hospitals in Agra, maximum concentration of fungal spores was obtained during rainy season. A large number of fungi are reported to damage the valuable cellulosic material. Pigments produced by fungi cause brown or black spots on paper and cloth objects.

Museums are centres of non-formal education, a good museum attracts, entertain, arouses curiosity, leads to questioning and this promotes learning. Study on damage to anthropogenic objects like Miniature paintings, palm leaf manuscript and damage to French paintings was studied. Hydrolyses of cellulose occurs which leads to degradation of paper. Enzymes cellodextrins and cellobiose break down cellulose. Damaged Miniature paintings in Baroda Museum showed the presence of C. gangligerum, A. versicolor, A. terreus, Penicillium sp. and Streptomyces. Palm leaf manuscript showed the occurrence of Nigrospora. Percentage loss of different papers was studied. Use of certain botanicals in form of Vacha rhizome powder and Cymbopogon oil was found useful.

The paper discusses aeromycoflora damaging the paper objects placed in Baroda Museum and picture gallery, museum of Botany Department of the M.S. University of Baroda and Maharaja Fatehsingh Museum, Vadodara. Deterioration of valuable objects due to fungal organisms and certain control methods are described.

Keywords: *Biopollutants, Fungi, Aspergillus, Threat paper, Deterioration, Palm leaf manuscripts.*

6.1 Introduction

Sculptures, paintings, manuscripts and historical monuments present all over the country constitute the rich cultural wealth of our country. Everyone knows about Tajmahal one of the seven wonders of the world but our country has still older monuments than this. Angkor watt 's Hindu temples are world famous. Many churches of Goa are quite old as well as many palaces in Rajasthan are very beautiful and quite old. Earlier rocks and metallic plates were used for writing. The use of leather parchment and Bhoj patra was done in India. *Cyperus papyrus* was the plant growing along river Nile which yielded the present day source of paper from its peduncle (Figure 6.8). Paper is a cellulosic material obtained from a large number of plants and even from cloth (Figure 6.2). Discoloration and damage to paper material may be caused by bacteria fungi and insects like silverfish (*Lepisma*).

The ASI has divided the entire country into 24 Circles. The ASI has a large work force of trained archaeologists, conservators, epigraphist, architects and scientists for conducting archaeological research projects. Earlier a lot of laws and acts had been passed by the government to protect these monuments, but majority of them were done on structures that were beneficial to the contemporary society. Also, the work that was carried out had a dearth of funds, enthusiasm and awareness. Later the '*Ancient Monuments and Preservation Act, 1904*', was passed with the prime objective to ensure the proper upkeep and repair of ancient buildings in private ownership excepting such as those used for religious purposes. Under this program, the conservation work is carried out in three main broad categories:

1. **Chemical Preservation** – The ASI's Science Branch is responsible mainly for the chemical conservation treatment and preservation of some three thousand five hundred ninety-three protected monuments besides chemical preservation of museum and excavated objects countrywide. Cellulose degrading fungi like *Chaetomium* may degrade paper and cloth. Fungal agents like *Alternaria* and *Curvularia* produce black pigments and discolors the objects. The fungal organisms may show the fast growth at increased temperature. They show increased production of enzymes at higher temperature. The fungi which grow at higher temperature are termed thermophilic fungi.
2. **Structural Conservation** – The workers in the field are acquiring cumulative knowledge of several generations and gaining expertise on the ways to improve and stabilize the structures by maintaining their pristine looks. The structures are given additional strength and reinforced to undo the harms done by pollution, acid rains, and other chemicals over the years. The Cyanobacteria like *Oscillatoria, Lyngbya, and Gloeocapsa* cause damage to old heritage buildings. The black crust coating the surfaces of building materials located in urban (polluted) environment may contain all kinds of organic compounds present in aerosols and particulate matter.
3. **Contemporary Awareness Program** – The citizens of India in general and students in specific are being roped in by the government to spread awareness and advertise about the preservation of the heritage. Many

seminars are being organized every year where the students are given lectures about the basic steps each can take individually on these issues.

Some of the steps that we can follow to do our share in their protection are:

i) Prevent ourselves and others from scribbling on the walls.

ii) Participate in the regular Cleanliness Drives. Being a part of the 'Adopt a Heritage' initiative.

iii) Spreading awareness around about these monuments and their importance.

In today's competitive world, we have to preserve the monuments as well as museum materials and showcase the valuable heritage to the next generation as the contributions or achievements of our ancestors. Damp surfaces are readily colonized by microbial settling from the air on other museum materials.

6.2 The World's Art under Attack by Microbes

☆ India is known for its rich, diversified, cultural properties kept in World heritage sites, National monuments, State monuments UNESCO, ICOMOS, ICCROM, ICOM *etc.*

Figure 6.1: Baroda Museum and Picture Gallery.

Cellulose

☆ Is a polysaccharide of glucose units in unbranched chains.

☆ Has β-1,4-glycosidic bonds.

☆ Cannot be digested by humans because humans cannot break down β-1,4-glycosidic bonds.

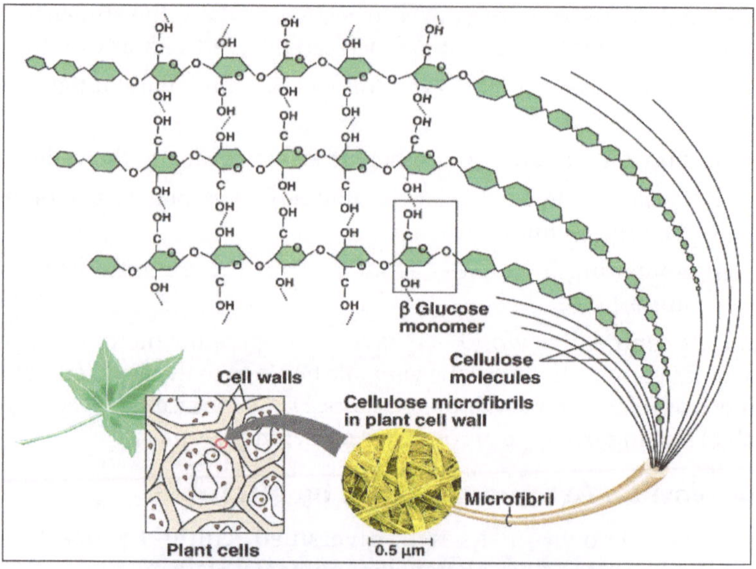

Cotton produced in plants

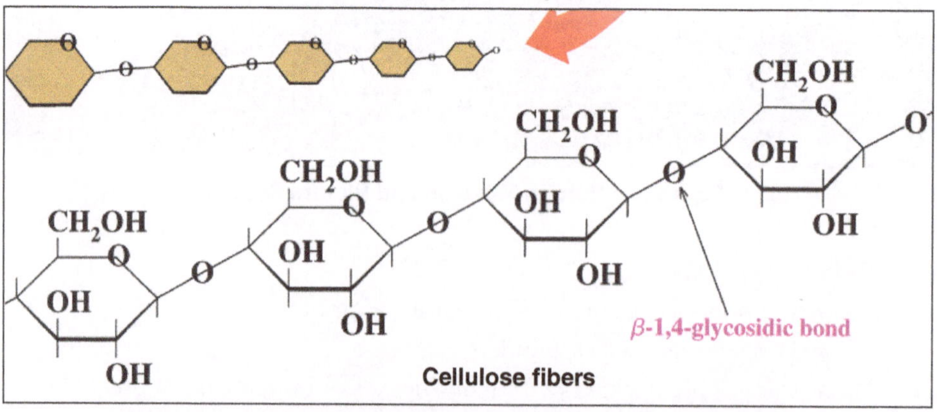

Figure 6.2: Cellulose Fibres and Cotton Produced in Plants.

- Gujarat is famous for Somnath and Akshardham temples, historical monuments (Figure 6.1), artistic sculptures, Adalaj vav and antique properties. Need for conservation and benefit of these requires special strategies to find out the deterioration and reduce it.

The bacterial and fungal colonies are a nuisance because their dark pigments leave stains like ink drippings along the walls on monuments and cloth and paper objects. The moist environment attracts other microbes that damage and weaken it. "That humidity really sets up the conditions where everything can get a toe hold," says Paula De Priest, deputy director of the Smithsonian's Museum.

Figure 6.3: Fungal Damage to Pastel Colour Painting B: Back, C: Front Side of Painting.

Biodeterioration phenomena represent a complex of physical and chemical alteration processes in various materials, such as those constituting the objects that represent our cultural heritage. The biodegradation of paper is conditioned by several variables such as the materials from which cellulose is obtained, the manufacturing processes employed, the occurrence of other affecting substances such as lignin or metallic compounds, and by the environmental conditions in which papers are conserved. In this study, biodeterioration of paper was artificially induced in order to evaluate the role of a range of chemical and physical variables on damage caused by cellulolytic fungi. A variable pressure SEM instrument was used to characterize paper samples with different fibre origins, and alterations obtained *in vitro*. Two fungal strains, *Aspergillus terreus* Thom and *Chaetomium globosum* Kunze, which are cellulolytic species frequently associated with paper spoilage, were used to produce stains with characteristics close to those observable on art objects made from paper. The stains obtained on the different samples of paper were compared at both low and high magnification, in order to visualize the macro and microscopic characteristics of paper fibers, inorganic constituents, impurities, and the deteriorating agents related to the spoiled areas. During this survey it was observed that single paper characteristics can strongly influence the intensity and the results of the fungal action. For example, the activity of a fungal strain on paper grades containing fibers of the same origin, but with different sizing, led to the formation of profoundly different stains and alterations. Moreover fungal structures, analyzed by low vacuum SEM, in areas on paper corresponding to the stains appeared in different physiological states suggesting an important effect of paper constituents on fungal growth and their sporulating ability. According to Nair (1972) fungi can grow on any substrata and play an important role in the deterioration of museum materials.

6.3 Fungi Recovered

The fungi most commonly recovered from indoor air were *Aspergillus niger*, and *Penicillium citrinum*, the percentage occurrence of these two fungi was more in 2018.

Figure 6.4: Damaged Miniature painting in Baroda Museum showed the Presence of *Chaetomium gangligerum, Aspergillus versicolor, A. terreus Penicillium* sp. and *Streptomyces*

Literature has suggested association between damp environments, microbial exposure, and higher prevalence of respiratory symptoms and diseases. The study began by evaluating the airborne fungal concentrations at urban and suburban areas of a typical metropolitan city in southern Taiwan for the estimation of related health risks. A group of representative homes, based on the housing characteristics questionnaires completed earlier, were selected from two parts of the city; urban and suburban. Burkard sampler (BURKARD, Rickmansworth, England) was used

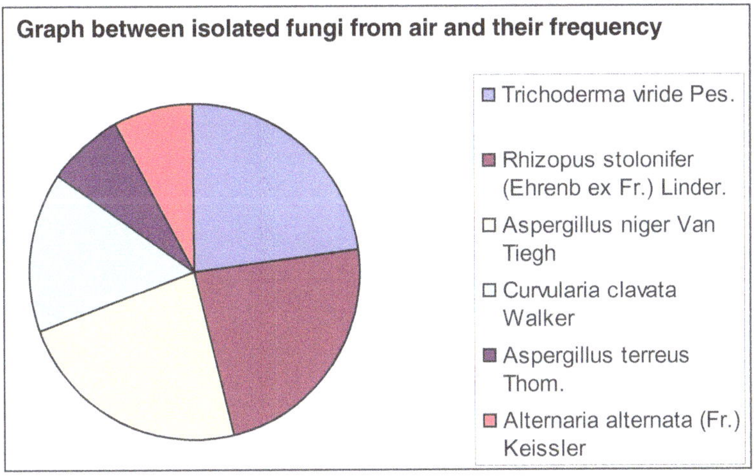

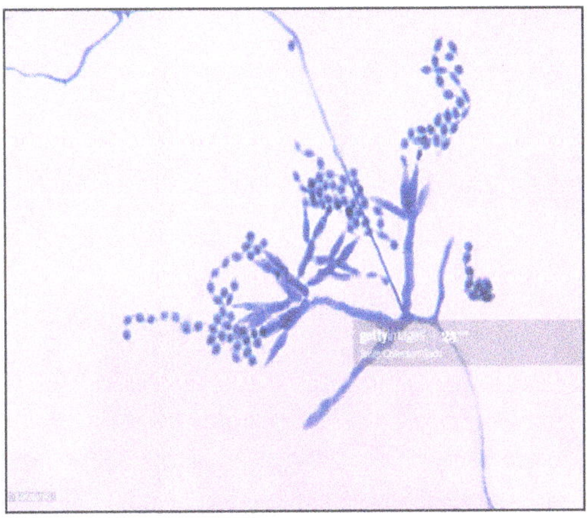

Penicillium citrinum

to collect airborne fungi onto agar plates with malt-extract. The species belonging to the first group are common both in the soil and in the air, they generally develop in moderate and humid climate where organic substances are available. Hyphomycetes include species with colourless or brightly coloured colonies (hyaline Hyphomycetes) and species producing dark brown, green-black, or black colonies (dematiaceous Hyphomycetes). Hyphomycetes excrete organic acids and can actively dissolve carbonates, they produce different kinds of pigments, but only a few species are melanine producers (Sterflinger, 2000) (Figure 6.9).

Damage to paper (sulfate pulp, cotton half-stuff, and flax half-stuff) caused by the *Aspergillus niger*, *A. sclerotiorum*, and *Penicillium chrysogenum* fungi is investigated by Raman spectroscopy, Fourier-transform infrared spectroscopy, and scanning electron microscopy. It is shown that the use of application of infrared Fourier-transform absorption spectroscopy allows one to identify the initial stages of damage from a decrease in the degree of crystallinity of the cellulose contained in paper.

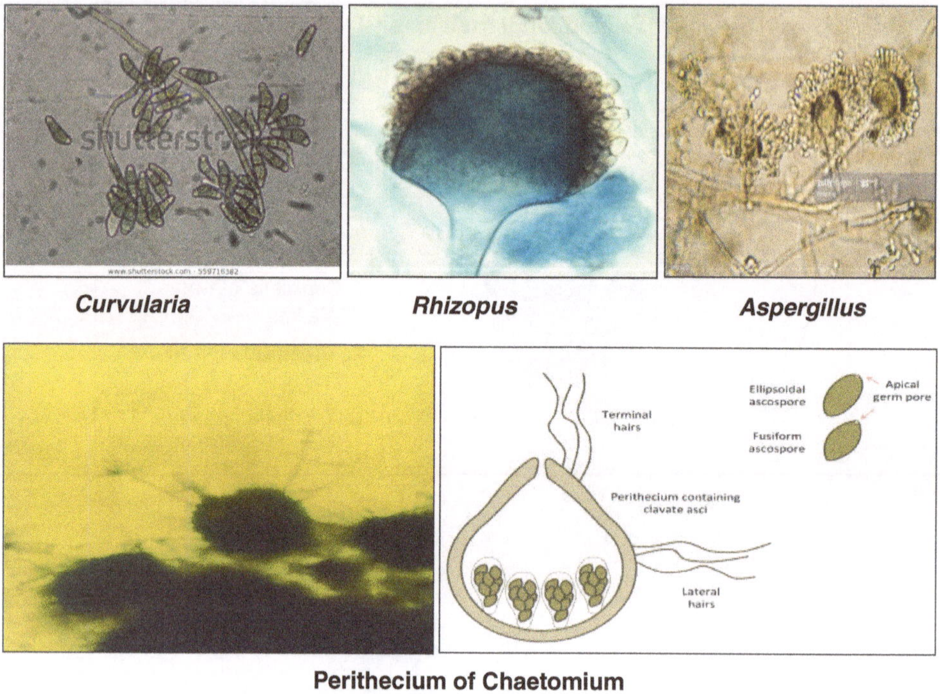

Figure 6.5: Photographs of Fungal Organisms as Seen under Microscope.

The absorption band near 900 cm^{-1} is used as an indicator of early stages of damage. An increase in the amide II peak at 1550 cm^{-1} and spectral changes in the region of valence vibrations of the C–H bonds (2800–3000 cm^{-1}) are observed in the case of heavier damage. The obtained data indicate that the vibrational spectroscopy techniques are promising in the study of damage of archive documents

6.4 Biodeterioration of Historic Paper: Preliminary FTIR and Microbiological Analyses

According to Zotti *et al.* (2008) paper is subjected to numerous biodeterioration processes, which may cause the irreversible degradation of important documents and works of art. Many chemical and physical factors can affect these processes and their behaviour, and fungi seem to play a key role in biodeteriorating paper materials. This study is mainly aimed at verifying the presence of fungi in biodeteriorated 18th century etchings, and characterizing the paper surface by means of Fourier transform infrared (FTIR) spectroscopy and fluorescence under UV radiation. The laboratory tests highlight the presence of fungal entities from all the samples investigated. Specifically, 14 species were identified; three of them were never isolated from paper until now. Furthermore, the data gathered do not confirm the theory according to which there is a correspondence between fluorescence of the stains under UV radiation and the vitality of microfungi (Figure 6.5).

The decay of cotton, Whatman and chemical pulp caused by *Trichoderma harzianum* and *Paecilomyces variotii*. The structural changes of the paper were evaluated by Infrared Spectroscopy (FTIR) and Scanning Electron Microscope (SEM). The SEM results show differences in hyphae colonization and paper decay patterns between studied species under the current study; *P. variotii* caused an eroded structure in the cotton (cavity forming), whereas the initial *T. harzianum* colonization produced rupture and erosion (soft-rot decay type II) for the three types of paper, the gaps were elongated with sharp pointed ends, which consisted either of individual cavities or in chains. Moreover, FTIR results confirmed that there a relationship could be observed between fungal decay and crystalline cellulose content because the intensity of peaks at 1335 and 1111 cm^{-1} significantly decreased due to the fungal decay. Furthermore, the intensity of O–H stretching absorption slightly decreased, and this may be attributed to hydrolysis of cellulose molecules (Figure 6.6, Hassan and Mansour 2018).

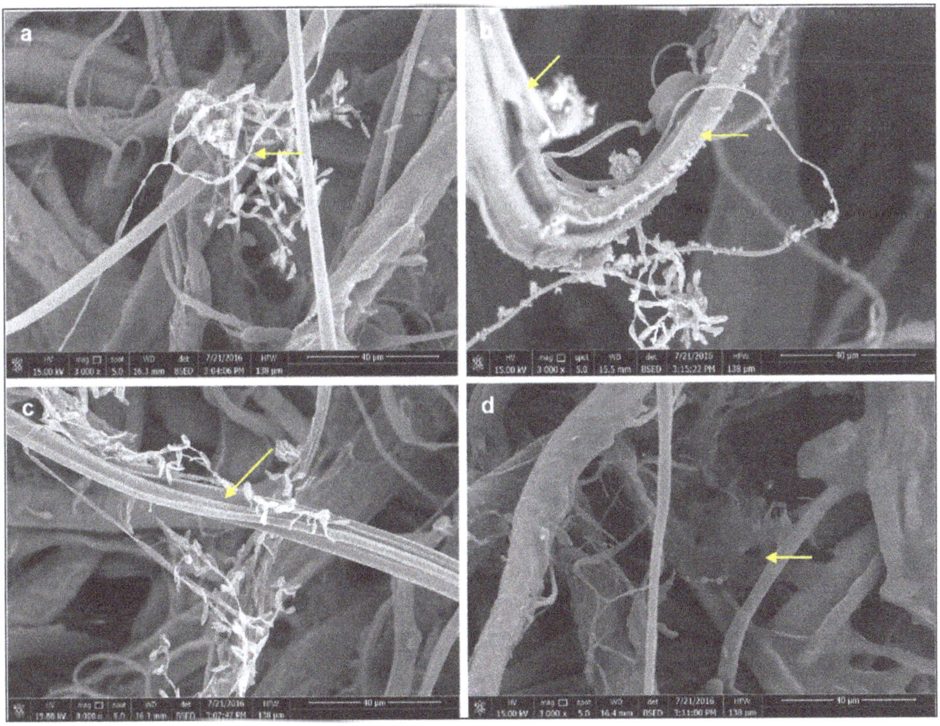

Figure 6.6: SEM Picture of Colonization of Cellulose Fibers by Fungal Hyphae and Conidia (Hassan and Mansour 2018).

6.5 Paper and Use of Adhesives

In paper conservation practice, adhesives are used for several purposes, such as mending tears and gaps, or paper consolidation. The criteria to choose one or another adhesive should be based on the knowledge of the properties and stability of those adhesives. However, the several adhesives available in the market still lack enough information to help the process of a rational decision-making. In the present work, five adhesives currently used in the paper conservation field (starch paste, unsupported *Archibond*, carboxymethylcellulose, hydroxypropylcellulose and methylcellulose) were analyzed for their chemical stability and fungal bioreceptivity (the ability of a material to be colonized by fungi). Bioreceptivity of products used in conservation and restoration is a still poorly explored subject, despite its great relevance for the preservation of objects.

The chemical and physical properties of the adhesives, before and after moist heat artificial ageing, were analyzed by thermogravimetry, capillary viscometry, measurement of water absorption capacity, colorimetery, and pH measurement.

Fungal bioreceptivity of the adhesives was tested on two different substrates (paper and glass) against three fungal species: *Aspergillus niger, Aureobasidium pullulans* and *Penicillium pinophilum*. Along 56 days of incubation, the colonization area on the adhesives was measured through digital photo analysis. Starch paste was the most bioreceptive adhesive, but on other hand was also the most stable adhesive to artificial ageing, regarding colour alteration, degree of polymerization and pH. Carboxymethyl cellulose and *Archibond* showed chemical deterioration with ageing. Nevertheless, these two adhesives presented only scarce bio receptivity to the tested fungi. Methylcellulose and hydroxypropyl cellulose showed the best relationship between higher chemical stability with artificial ageing and lower fungal bio receptivity.

6.6 Percentage Loss of Eight different Papers

Percentage loss of eight different papers are recorded in Table 6.1 given below.

Table 6.1: Percentage Loss of Eight different Papers by Four different Fungi

Sl.No.	*Type of Paper*	*Percentage Loss*			
		C. verruculosus	*P. citrinum*	*A. ustus*	*C. cymbiforme*
1.	Whatmans filter paper	15	11.4	9.5	12
2.	Common	14.2	48	2.5	14.3
3.	Brown	37	24	20.6	35.6
4.	News Paper	29	16	18.8	24.4
5.	Old Manuscript	36	14.4	15.8	28.4
6.	Card board	44.4	51.4	25	22
7.	Japenese Tissue paper	49	28	49	22
8.	Nepalese Tissue paper	48	12	20	10

Note: Source Shah and Arya (2000).

In 25 ml of Asthana and Hawker's modified medium 'A' 0.5 g different papers were added. After 25 days fungal mat and paper mass determined and percentage loss was calculated. It is evident that maximum percentage loss was observed in Japanese tissue paper by *Chchliobolus verruculosus* and *Aspergillus ustus.* In card board by *Penicillium citrinum* and brown paper by *Chaetomium cymbeforme* a cellulose degrading member of Chaetomiales group in Ascomycotina. Out of four fungi three caused negligible loss of Nepalese tissue paper and hence it can be safely used for restoration work. Ricelli (1999) studied the fungal growth on paper samples caused by sp. of *Aspergillus, Chaetomium, Penicillium* and *Stacybotrys* stored in high relative humidity.

Figure 6.7: An Old Egyptian Painting.

Figure 6.8: *Cynchrus papyrus* Grass.

6.7 Management Strategies

The good news is we can ward off microbial infestations by storing artefacts in conditions that don't encourage spores to settle and germinate. This means keeping things cool and dry. "In an indoor environment, like a museum or a library…you can control the temperature, humidity, and light," Palla says. Conservators also use UV-C fans to filter spores from the air and zap them with ultraviolet light.

Collections that are open to the public can't be chilled very much. "It's not very practical in museum settings, where you actually want people to see objects without being cold," DePriest says. One exception is Ötzi the Iceman, a natural mummy who froze in the Italian Alps 5,300 years ago. He's kept in a cold cell in the South Tyrol Museum of Archaeology in Italy that emulates the glacier he was discovered in Oriental Institute of the M.S. University of Baroda, Godavach (*Acorus calamus*) rhizome is used in cupboards in clean white cloth (Figure 6.10, Shah and Arya 2001). The oil was good for polishing the palm leaf manuscript (Figure 6.9). Eucalyptus leaves may be effective due to the presence of Cineole in them (Arya, 1988). It has been observed that the activity of extracts may be lost due to autoclaving (Lokesha *et al.*, 1986).

6.8 Conclusion

Likewise the population or community dynamics on cellulosic objects, before and after restoration, need more investigations and could give an important contribution to the conservation science in Cultural Heritage. SEM studies showed

Figure 6.9: Orrian Palm Leaf Manuscript damage by Fungus *Nigrospora spherical* Penz.

Figure 6.10: White Cloth Small Packets Containing Rhizome Powder, which Worked as Antifungal and Insect Repellent.

colonization of cellulose fibres by fungal hyphae. Table 6.1 shows that degradation of 8 different papers. *In vitro* it was observed by dry weight method in case of *Cochliobolus verruculosus, Aspergillus ustus, Penicillium citrinum* and by *Chaetomium cymbeforme.* Maximum reduction was observed in case of cardboard due to *P. citrinum.* Certain preventive and control methods are suggested.

REFERENCES

Allsopp D, Seal K, and Gaylarde C (2004) *Introduction to biodeterioration.* Cambridge University Press, Cambridge, pp 1–10

Arya A. (1988) Control of *Phomopsis* fruit rots by leaf extract of certain medicinal plants. In : Indigenous medicinal Plant Symp. (Kaushik M.P. ed) Pub. By today and Tomorrow's Printers and publishers New Delhi 41-46pp.

Arya A. (2006) Role of Fungi in Biodeteriorarion of museum objects In; *Recent Mycological Researches* by Sati S.C. (ed.) Pub by I.K. Int. New Delhi 260-271

Hassan R.A. and Mansour M.A. (2018) *Journal of Polymers and the Environment* 26, : 2698–2707

Lokesha S. Vasantha K. and shetty H.S. (1986) Effect of plant extracts on the growth and sporulation of *Aspergillus flavus.Pl. Dis. Reptr.* 1(1-2): 7981c f. Rev. Pl. Path. 67(1988): 4864

Nair, S.M. (1972) In : *Conservation in Tropics.* Proc. Of AST- pacific seminars on con. of cul. Properties (ed. Agrawal O.P.) Rome: 150-158.

Riecelli A. (1999) Fungal growth on samples of paper: Inhibition by new antifungals, *Restaurator* 20 (2): 97-107

Shah N.R. and Arya A. (2000) Studies on some fungal biodeteriogens by Bhartiya kala Prakashan, New Delhi pp202.

Sterflinger K. (2000) Fungi as geologic agents. *Geomicrobiology Journal* 17(2): 97–124. 10.1080/01490450050023791

Zotti M., Ferroni A., and Calvini P. (2008) *International Biodeterioration and Biodegradation* : 569-578

Part 2

Pollution: Management Strategies

खुशियों भरे दिन

इन हवाओं में पले प्यार
और मुहब्बत का सहारा हो
खुशियों भरे दिन हों, और
तारों की झिलमिलाहट हो।

सूरज की तपती गर्मी में
पसीने की बूंदे छलकें
मानों हर व्यक्ति को
उसकी मेहनत का पूरा भरोसा हो

घर-घर में बसी हो हलचल
हर मां को उसके बेटे का सहारा हो
हम मनायें खुशियां हर दिन
हमारे सपनों को बस सच्चाई का भरोसा हो।

क्यों बनायें साजो समान, हम लड़ाई का
भाई-भाई में जब बसा हो
पैगाम मोहब्बत का।
जीतें हम दिलों को प्यार से

कोई मंदिर गया कोई मस्जिद गया
फर्क कुछ भी नहीं ईश और खुदा में
ईश्वर समाया है कण-कण में
उसे बस अपनी इबादत पर भरोसा हो

हिन्दू हैं यहाँ मनाते होली
तो ईद मुसल्मां की भी हमें प्यारी है
हम दो भाई हैं एक वतन के
हमारी कहानी बस मोहब्बत की कहानी हो।

Chapter 7

Studies on the f-MWCNT/PES Mixed Matrix Membranes for Water and Wastewater Treatment

Priyanka Mistry and C.N. Murthy

Macromolecular Materials Laboratory, Applied Chemistry Department, Faculty of Technology and Engineering, The M.S. University of Baroda, Vadodara – 390001, Gujarat
e-mail: chivukula_mn@yahoo.com

ABSTRACT

Water shortages have become a major problem facing the world today. Membrane separation technology is commonly used in water treatment applications. The development of new materials for water treatment can reduce the energy required for water treatment, reduce cost, and improve efficiency. Ionic and molecular sieving membranes that enable fast solute separations from aqueous solutions are essential for processes such as water purification and desalination, sensing, and energy production. Nanomaterials exhibit excellent properties such as great chemical and physical stabilities, lower density, and large surface area. These excellent properties of nanomaterials make them popular candidates for the design and development of novel functional nanomaterial-based membranes. The combination of membranes with the unique properties of nanomaterials provides excellent physical and chemical stability, and also high rejection behavior of the nanomaterial-based membrane toward target compound/s. The design and development of novel membranes can be done by combining with nanomaterials such as graphenes, fullerenes, carbon nanotubes, and nanoparticles. Preparation methodologies of these nanomaterial-based membranes, as well as their potential in the effective separation of target compounds from environmental samples are discussed in detail. f-MWCNT/PES mixed matrix membrane were prepared via phase inversion method. It is also concluded that because of the tube-like structure CNTs are vertically aligned to the surface of the membrane, due to this pore volume reduces and number of pore increases. Azide MWCNT/PES membrane shows higher pure water flux and good rejection for the heavy metal like Pb, Cu, Mg, Cr. An improvement in hydrophilicity of the membrane was observed due to the presence of functional group on the nanotubes.

Keywords: *Mixed matrix membranes, Water treatment, Wastewater treatment, Multiwalled carbon nanotubes (MWCNT), Polyethersulfone (PES).*

7.1. Introduction

Wastewater treatment is the process from which contaminated water has been restored to good quality water (Englande *et al.* 2015). Most of wastewater discharged from the industries contains toxic heavy metals such as Cr, Cu, Pb, Cd, etc. and their compounds, which causes severe damage to humans as well as nature. Wastewater can be processed by membrane technology, which is supported by several recent reviews. Membrane technology has various applications because it has advantages like more energy efficient, its cost is low, easy to setup and easily operated. In membrane technology polymeric Membranes were widely used like polyethersulfone, polysulfone, polyvinylidene fluoride (PVDF), polyphenylene Oxide (PPO), polyamides, polyimides, polyurethanes etc. Using these polymers incorporating with solid particles, mixed matrix membrane have been prepared which have been further used for wastewater treatment.

Basically, mixed matrix membranes are prepared via the phase inversion technique in which solid particles, also called fillers, are added in to the polymeric dope solution and flat sheet membrane prepared. Various types of solid particles are used, like TiO_2, LiCl, CNTs, SiO_2, TiO_2, ZnO, GO, ZrO_2, and Al_2O_3. Other fillers such as zeolites, metal organic frameworks, chitosan, and silver nanoparticles have also been used. PES composites membranes prepared with five different additives like TiO_2, ZnO, GO, SiO_2, Al_2O_3, via non-solvent induced phase separation (NIPs) process, shows improved physiochemical properties, good hydrophilicity and 91.96-96.92 per cent dye rejection. The improvement in the membrane due to the addition of those five different solid particles, because they have porous structure, as well as, large surface area, high pore volume and also the presence of -OH group on the surface, which may give enhance hydrophilicity (Otitoju, 2020).

Nowadays, GO (Graphene Oxide) or f-GO have been used as an additive for the preparation of mixed matrix membranes. GO has a monolayer structure containing various oxygen containing functional groups like epoxy, hydroxy, carbonyls, phenols. PES/f-GO Mixed membrane shows higher removal efficiency of 97.1 per cent Fe^{2+}, 95.3 per cent Zn^{2+}, 92.7 per cent Cd^{2+}, 99.9 per cent Cr^{6+}, 98 per cent COD, and 99.9 per cent bacteria (Giwa, and Hasan 2020). In some recent articles GO has been used as a nanohybrid form, which is synthesized via blending of GO with other metal oxides. PES/GO-ZnO Membrane has higher hydrophilicity but lesser salt rejection, also because of the less membrane-solute hydrophobic attraction trac organic rejection increases. These membranes suitable for low-energy application due to the high fluxes, photocatalytic properties and rejection properties (Mahlangu, *et al.* 2017). For the Oily wastewater treatment PES/GO-SiO_2 Mixed matrix membrane prepared which gives increase water flux and 38 per cent increase in oil removal efficiency than the pristine PES membrane. This also shows excellent tensile strength, water permeability and hydrophilicity (Alkindy *et al.* 2020).

Membranes prepared via conventional methods has some disadvantages, these disadvantages can be resolved by the use of some additives as discussed above. Many additives also called as fillers in the form of solid particles have been used, which improves membrane properties by changing its morphology like pore structure, pore volume, and surface area. Among the various solid particles Carbon nano tubes (CNTs) has characteristics such as high aspect ratio, low density, and high mechanical strength and stiffness. Also, it has strong π-π stacking interaction with aromatic compounds which leads to irreversibly adsorbed contact between hydrophobic CNT and aromatic compounds. These properties prove CNT as an effective material in the fabrication of nanocomposite membranes (Sianipar *et al.* 2017).

Different types of functionalization of carbon nanotubes can be carried out, Among them Azide functionalized carbon nanotubes used for the preparation of Azide functionalized CNT/Polysulfone(PS) mixed matrix membrane, which shows higher flux and 90 per cent rejection Pb(II) and 93 per cent rejection for Cu(II) at lower pH and 4.06 bar pressure (Mangukiya *et al.* 2016). By varying the concentration of Azide functionalized MWCNTs from 0.1-1 per cent in to the polysulfone matrix, the higher rejection and flux obtained from the 1 per cent Azide functionalized MWCNTs /PS matrix at 0.49 MPa and acidic pH 2.6. Amide functionalized MWCNT/PS membrane also shows 94.8 per cent Cr (VI) rejection which much more better than the pristine Polysulfone membrane. These membranes are more hydrophilic, because of the reduced pore size, which suggest that a physical bond form between the functionalized MWCNT and polysulfone (Shah and Murthy, 2013).

With the 1 per cent oxidized MWCNT/Polysulfone membrane gave 96.8 per cent rejection of Cr (VI), 87.6 per cent rejection of As (III) and 94.2 per cent rejection of Pb (II). Similarly, 1 per cent oxidized MWCNT/PS membrane and 1 per cent amide MWCNT/PS membrane have 10 and 14 ml/min permeate flux at 2.06 bar pressure, while pristine polysulfone membrane has 16 ml/min permeate flux, this is due to the reduced pore size and porosity because of the greater hindrance on the surface of the membrane (Gupta, *et al.* 2015). Pore size, morphology of the membrane surface has been studied by the small angle neutron scattering (SANS). This study shows that by the increase the concentration of nanotubes in membranes, the pore dimensions decreasing up to 1wt per cent while the porosity of the membrane increases which is in agreement with the percentage of heavy metal removal (Nikita, *et al.* 2019). Figure 7.1 shows the SANS data of pristine as well as functionalized CNT impregnated membrane soaked in D_2O. The scattering intensities for the different membrane samples are presented in the arbitrary unit and shifted vertically for the clarity of the presentation. Two models are considered, first, where the pores were considered to be spherical and surrounded by a matrix material, which is called the polydisperse sphere model and second, where the first phase is the pores and the second phase the surrounding matrix material, called the random two-phase model.

This membrane SANS data follows the second model, which data are presented in Table 7.1 (Nikita *et al.* 2019b).

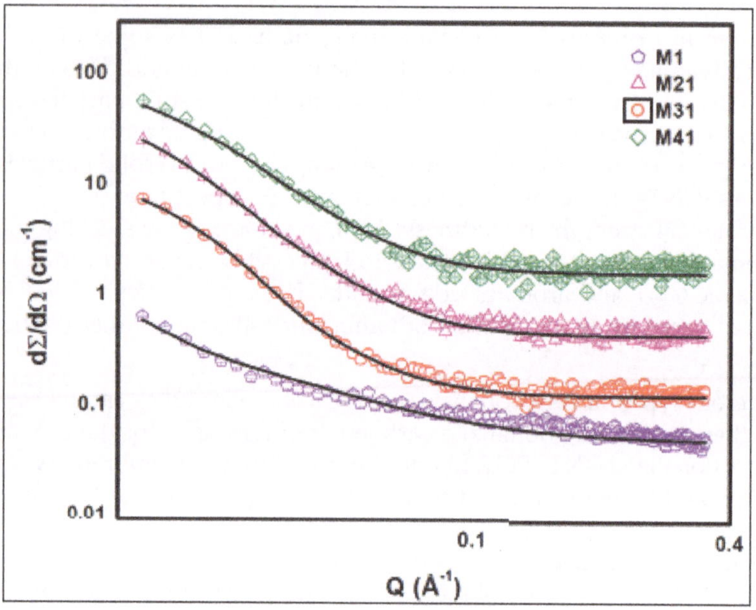

Figure 7.1: SANS Data for Pristine as well as Functionalized CNT Impregnated Membrane (Nikita *et al.*, 2019b).

Table 7.1: SANS Data of Membrane soaked in D_2O Containing 1 per cent of MWCNT, M1 (Pristine), M21 (Oxidized), M31 (Amide functionalized), M41 (Azide functionalized) (Nikita *et al.* 2019b)

Sample	*Pore Radius (nm)*	*Polydispersity (σ)*	*Correlation Length (nm)*
M1	14.8	0.27	3.9
M21	9.7	0.27	3.9
M31	8.3	0.29	3.9
M41	7.4	0.27	3.9

According to the above Figure 7.1 and Table 7.1, it supports that pore size decreases as the concentration of the azide functionalized MWCNT increases. The correlation length is same for all the membranes which suggests that pore size is reduced but the distance between the pores remains the same, which means that number of pores increases (Figure 7.2).

The Azide functionalized MWCNT containing MMM surface modified by click chemistry, in which Azide group on the surface of the membrane reacted with 1-pentyne in the presence of Cu(I) catalyst and converted into triazole group (Figure 7.3). After surface modification, the membrane becomes more hydrophilic, pure

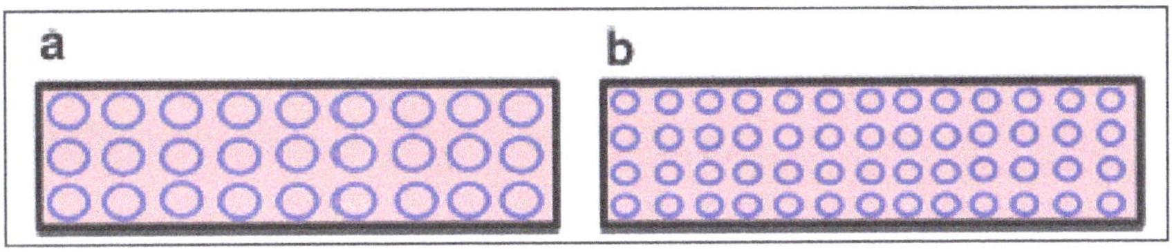

Figure 7.2: Schematic Representation of the Arrangement of Pores in (a) Pristine Membrane and (b) Functionalized CNT Impregnated Mixed-matrix Membrane (Nikita *et al.* 2019b).

water flux and heavy metal rejection increase slightly because of the more active sites on the surface of the membrane. Rejection of metal ions and flux recovery ratio improved due to the triazole group, which is a good ligand to complex metals. This modified membrane shows better rejection of Cu(II) due to the complexation of Cu(II) with the triazole group. This membrane also shows good antifouling property.

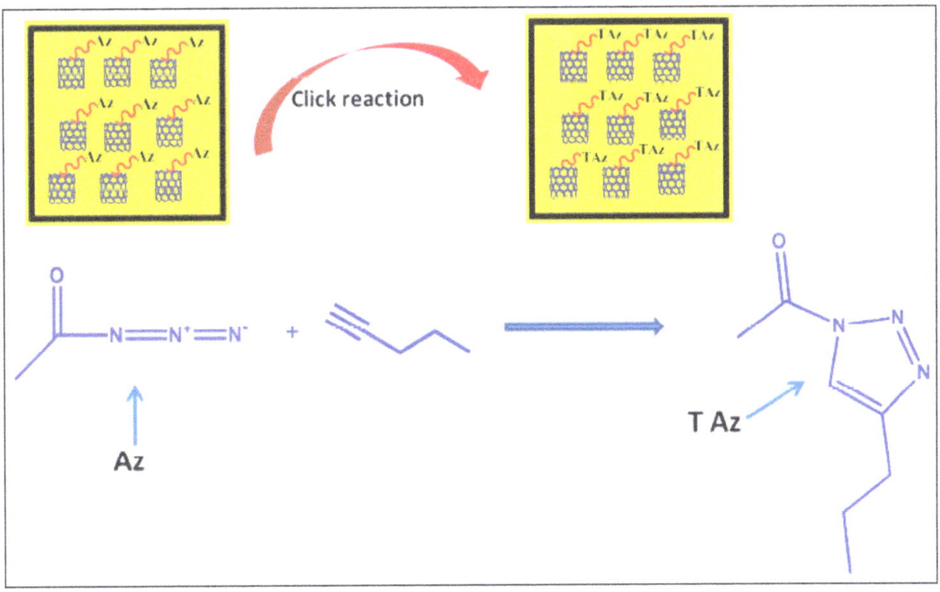

Figure 7.3: Surface Modification of Membrane via Click Reaction (Nikita *et al.* 2020).

7.2. Experiment

7.2.1 Materials

Polyethersulfone (PES), Multiwalled Carbon nanotubes (MWCNT), Concentrated sulphuric acid (Conc. H_2SO_4), Concentrated nitric acid (Conc. HNO_3), Concentrated hydrochloric acid (Conc. HCl), Thionyl Chloride ($SOCl_2$), Dimethyl formamide (DMF), Sodium Azide (NaN_3), Isopropyl alcohol.

7.2.2 Purification and Functionalization of MWCNTs

Commercially available Multiwalled carbon nanotubes (MWCNT) contains amorphous carbon and also metal impurities like Fe, Ni, Co, which were used as catalyst for the synthesis of CNTs via laser oven method or arc discharge evaporation method. These metal impurities removed by washing the commercial MWCNT with 50 per cent Conc. HCl. Metals react with hydrochloric acid and converted into metal chlorides, which then easily dissolved in water and removed. And then these purified MWCNTs were further undergoes Oxidation, in which Purified MWCNTs treated with the Mixture of (3:1) Conc. HNO_3 and Conc. H_2SO_4 for 24 hours at 33°C. After that Oxidized MWCNTs washed with water until the complete removal of acidity and dried. Oxidized MWCNTs further treated with thionyl chloride to convert oxidized MWCNTs into Acylated MWCNTs. For that Oxidized MWCNTs added into the thionyl chloride, DMF used as solvent, and keep it at 60°C for 24 hours. After completing the reaction, it washed with toluene for the complete removal of unreacted thionyl chloride and then dried.

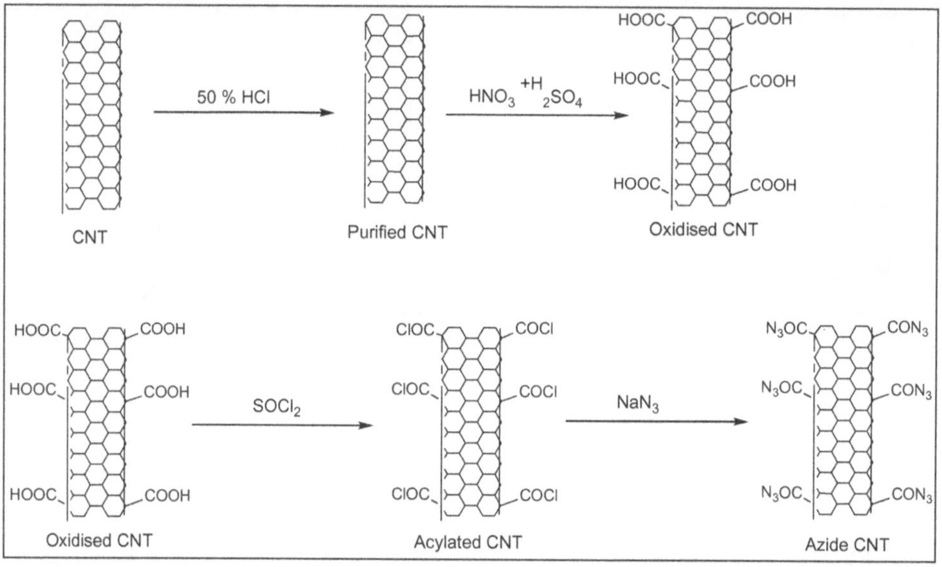

Acylated MWCNTs added in to the solution of Sodium Azide (NaN_3) in Dimethylformamide (DMF) and keep the reaction for 12 hours at room temperature. Azide MWCNTs were washed with distilled water and dried. This Azide functionalized MWCNT (f-MWCNT) used as additives for the preparation of f-MWCNT/PES Mixed matrix membrane.

7.2.3 Preparation of f-MWCNT/PES Mixed Matrix Membrane

f-MWCNT/PES mixed matrix membrane prepared via phase inversion method. In this method 18 per cent of PES dope solution prepared by dissolving PES into DMF and then various amount of f-MWCNTs were added. Sonicated the dope solution to assure the complete incorporation of f-MWCNTs in to the dope solution. After that using the dope solution a flat sheet membrane casted on the glass plate using the film applicator and put it into the coagulation bath containing coagulation solution, which is the mixture of Distilled water and isopropyl alcohol. Isopropyl alcohol was used as the pore former. The Membranes prepared with different percentage of f-MWCNT shown in the Table 7.2.

Table 7.2: Membranes Prepared with different Percentage of f-MWCNT

Membrane	*Per cent of f-MWCNT in PES Dope Solution*
M-1	0.0
M01	0.1
M03	0.3
M05	0.5
M07	0.7
M10	1.0

7.3. Characterization

The purified and functionalized CNT were characterized by FT-IR (8400 S Shimadzu Spectrophotometer). The prepared membrane surface morphology was studied by the SEM (Scanning electron microcopy). Analysis was done by taking the images of cross section of the membrane. And also, their permeability and flux results as well as rejection results taken by the membrane cell fabricated in the laboratory. Hydrophilicity of membranes was measured by the contact angle analysis.

7.3.1 Infrared Spectroscopy

Infrared spectroscopy is one of the most important analytical technique. This technique is based on the vibration of the atoms of molecules. An infrared spectrum is commonly obtained by passing infrared radiation through a sample and determining what fraction of the incident radiation is adsorbed at a particular energy. The energy at which any peak in an adsorption spectrum appears corresponds to the frequency of a vibration of part of a sample molecule. Fourier-transform infrared (FT-IR) spectroscopy is based on the idea of the interference of radiation between two beams to yield an interferogram. The latter is a signal produced as a function of the change of path length between the two beams. The two domains of distance and frequency are interconvertible by the mathematical method of Fourier-transformation.

7.3.2 Scanning Electron Microscopy (SEM)

A Scanning electron microscope scans a focused electron beam over a surface to create an image. The electrons in the beam interact with the sample producing various signals that can be used to obtain information about the surface topography and composition. Electrons are produced at the top of the column, accelerated down and passed through a combination of lenses and apertures to produce a focused beam of electrons which hits the surface of the sample. The sample is mounted on a stage in the chamber area and, unless the microscope is designed to operate at low vacuums, both the column and the chamber are evacuated by a combination of pumps. The level of the vacuum will depend on the design of the microscope.

7.3.3 Pure Water Permeability

The pure water permeability also known as the pure water flux is defined as the volume of water that passes through a membrane per unit time, per unit area and per unit of transmembrane pressure. This property indicates the effort required to generate permeate for membrane and is easy way to compare initial performance of a membrane. Flux is the flow rate of water applied per unit area of the membrane and has units of volume/ unit area/ time. J, which is the flow rate of permeate. Permeate is the water that has passed through the membrane. Pure water flux. J, is calculated by the following equation,

$$J = Q_p / A$$

where,

Q_p = Permeate flow rate (liter per hour), and A = area of the membrane (square meter).

7.3.4 Heavy Metal Rejection Studies

Rejection studies can be calculated by performing the experiment on membrane cell using different salt solution. The percentage of rejection was calculated by the following formula:

$$\text{Per cent Rejection} = = \frac{Cf - Cp}{Cf} * 100$$

where,

C_f is the conductivity of the feed solution and C_p is the conductivity of permeate.

The conductivity of the solution was measured by the use of Analab Conductivity meter.

7.3.5 Contact Angle

The angle between the substrate support surface and the tangent line at the point of contact of the liquid droplet with substrate is called contact angle. A liquid droplet *i.e.* sessile drop is placed on a solid surface and the contact angle has a single value for smooth homogeneous substrate surfaces. Contact angles measurements of liquid droplets on substrate surfaces are used to characterize surface wettability, surface cleanliness, and hydrophobic or hydrophilic nature of the surface. The value of contact angle of the liquid droplet will depend upon the surface energy of the substrate and surface tension of the liquid.

Half angle method is used for measuring contact angle which allows to take direct angle measurements and eliminates errors associated with arbitrary alignment of a hairline to the droplet image. Half angle method is based on the following formula:

$$\theta = 2^* \text{arc tan } (H/R)$$

where,

θ is the contact angle, H is the height of the droplet and R is the radius of the droplet base.

7.4. Results and Discussion

7.4.1 Functionalization CNTs

Functionalized MWCNTs were characterized by FT-IR analysis. The IR of Oxidized CNT and Azide CNT were shown in Figures 7.4 and 7.5.

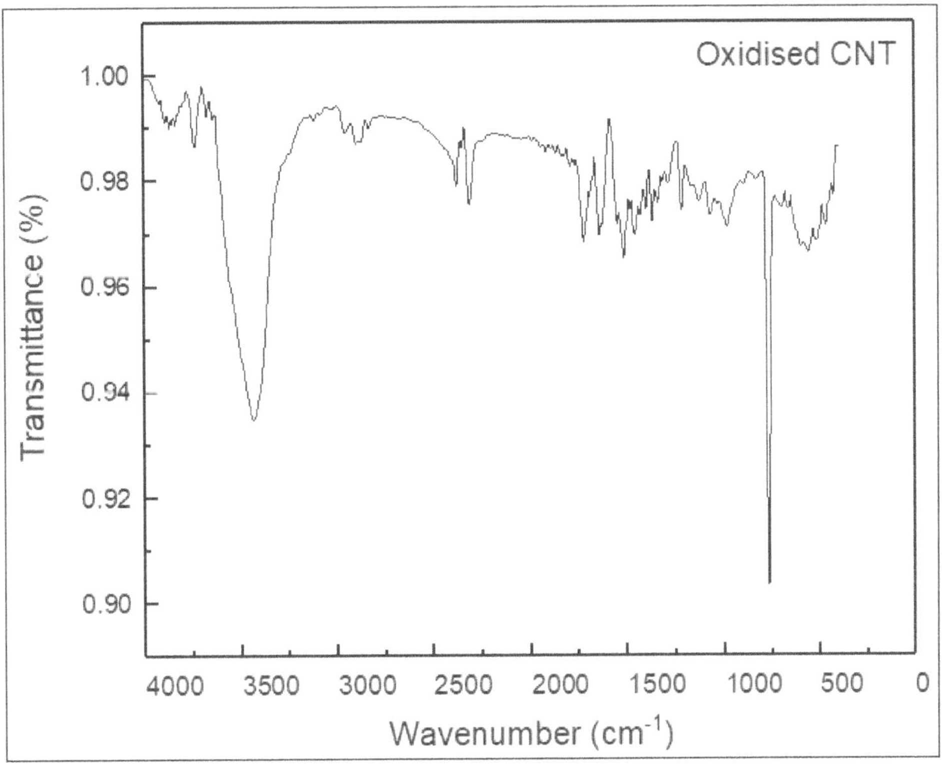

Figure 7.4: IR of Oxidized MWCNT.

In oxidized MWCNT sharp band at 3500 cm^{-1} shows the presence of -OH on the surface of the MWCNT. After azide functionalization this peak reduces and band between 2000-2500 cm^{-1} appears, which indicates the C-N stretching. These data confirm the functionalization of the MWCNTs and then these were further used for the preparation of membranes as additives.

7.4.2 Characterization of Membrane

From the SEM images membrane surface have porous structure and have voids, finger like structure. According to this result, f-MWCNT may aligned vertically into the membrane surface. The prepared membrane is asymmetric membrane which has

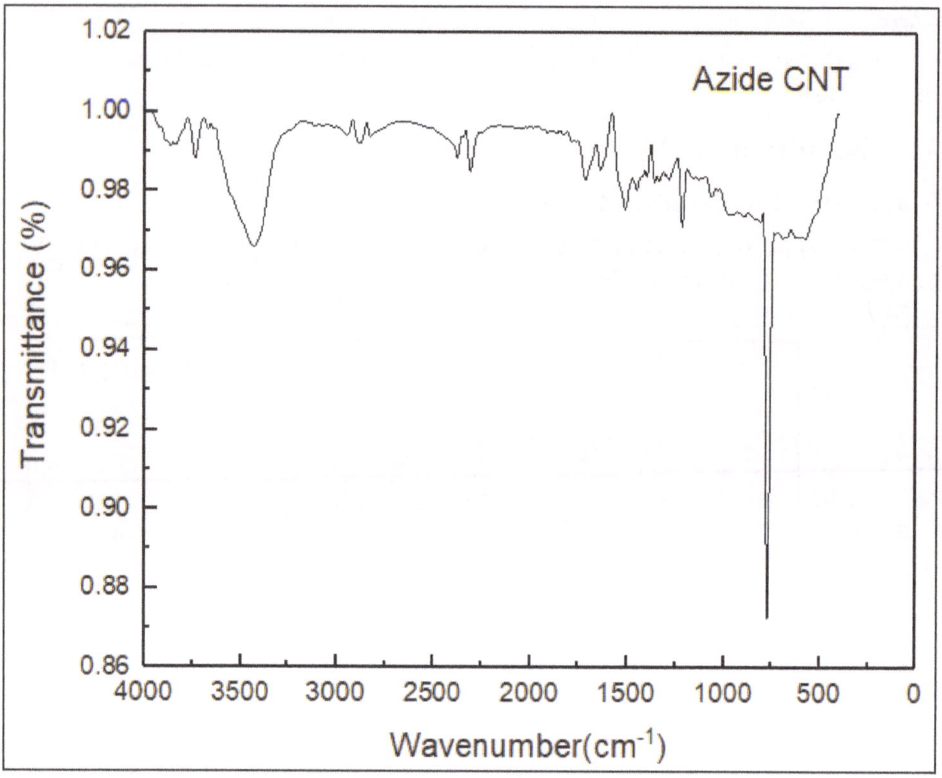

Figure 7.5: IR of Azide MWCNT.

small pores on upper surface and large pores on lower surface. After the addition of f-MWCNT membrane becomes smoother.

Pure water flux of the f-MWCNT/PES membrane have been higher than the pristine PES membrane. This improvement may be due to the incorporation of f-MWCNT into the PES polymer matrix. After addition of the f-MWCNT pore size reduces and number of pores increase, which supports the increase in the pure water flux. Also, as the concentration of the f-MWCNT increases the flux increases respectively. The Table 7.3 shows the data of the pure water flux.

Table 7.3: Pure Water Flux of Membranes

Membranes	*Pure Water Flux*
M-1	308
M01	377
M03	438
M05	450
M07	460
M10	595

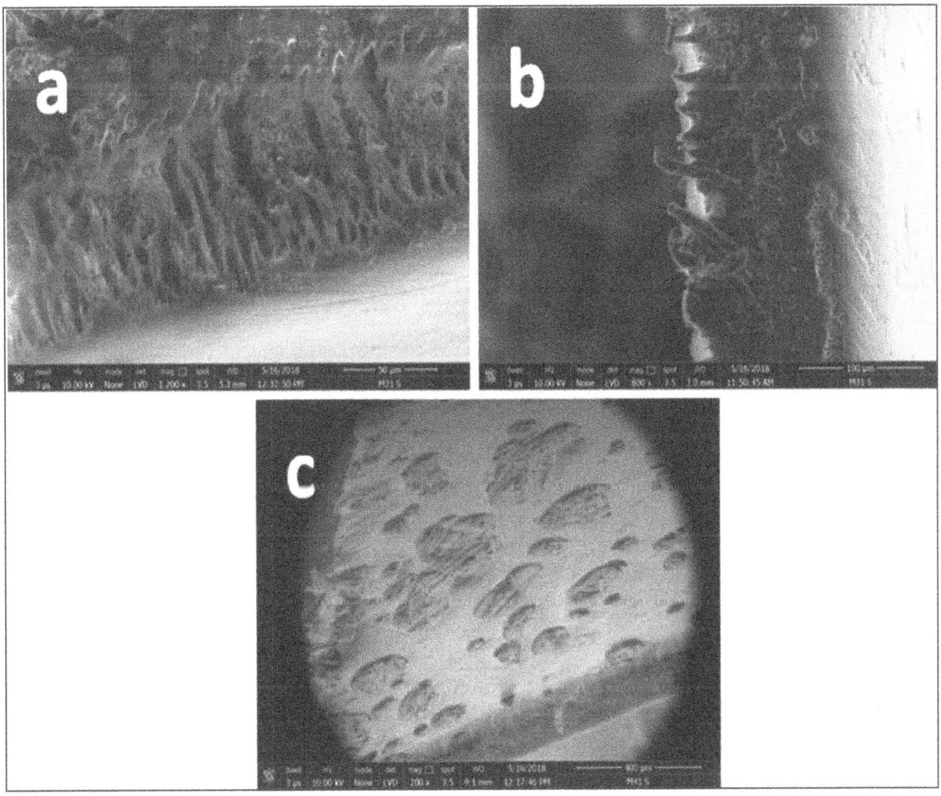

Figure 7.6: SEM Micrograph of Cross Section of Cold-fractured Membranes (a) Oxidized MWCNT/PES (b) Amide MWCNT/PES (c) Azide MWCNT/PES (Nikita *et al.*, 2019b).

Table 7.4: Metal Rejection Studies for different Membranes

Membranes	*Rejection*				
	$Pb(NO_3)_2$ (per cent)	*$Pb(NO_3)_2$ (At neutral pH) (per cent)*	*$K_2Cr_2O_7$ (per cent)*	*$CuSO_4$ (per cent)*	*$MgSO_4$ (per cent)*
M-1	60	60	57	65	60
M01	68	70	64	83	64
M03	71	71	65	85	68
M05	75	72	65	85	69
M07	73	75	66	86	69
M10	80	79	70	89	72

Heavy metal rejection also has good results with the incorporation of f-MWCNT in to the PES membrane. For this experiment stock solutions of the respective salts have been prepared, having the concentration of 2000 ppm. These prepared stock solutions were then passed through the membrane cell at 4 bar pressure. At first membrane cell should be stabilized for 1 hour and then collect the permeate and then measure its conductivity using conductivity meter, which is termed as the concentration of the permeate. The concentration of the feed measure before the solution had been passed through the membrane cell. According to the results membrane having the 1 per cent azide MWCNT content shows better rejection as well as pure water flux.

After the addition of the f-MWCNT membrane becomes more hydrophilic, this was supported by the contact angle measurement. This is due to the bonding between the function group on the surface of the MWCNT with the PES polymer, and also because of the increase in the porosity. As the percentage of f-MWCNT increases the hydrophilicity increases. f-MWCNT/PES membrane is more hydrophilic as compared to pristine PES membrane, because this membrane found to have aligned f-MWCNTs which are uniformly distributed.

Table 7.5: Contact Angle of Membranes

Membrane	*Contact Angle (in degree)*
M-1	49
M-2	46
M-3	44
M-4	43
M-5	47
M-6	57

7.5. Conclusions

From this study it has been concluded that the fillers play important role to enhance the properties of the membrane. Among all the fillers, Carbon nanotubes have special characteristics to improve the membrane surface morphology as it makes membrane smoother. It is also concluded that because of the tube-like structure CNTs are vertically aligned to the surface of the membrane, due to this pore volume reduces and number of pore increases. Azide MWCNT/PES membrane shows higher pure water flux and good rejection for the heavy metals like Pb, Cu, Mg, Cr. The Hydrophilicity of the membrane also improved because of the functional group on the MWCNTs.

REFERENCES

Alkindy, M.B. Naddeo, V. Banat, F. Hasan, S.W. (2020) Synthesis of polyethersulfone (PES)/GO-SiO_2 mixed matrix membranes for oily wastewater treatment, Water Sci. Technol. 81: 1354–1364. https://doi.org/10.2166/wst.2019.347.

Englande, A.J. Krenkel, P. Shamas, J. (2015) Wastewater Treatment &Water Reclamation, Elsevier Inc., https://doi.org/10.1016/b978-0-12-409548-9.09508-7.

Gupta, S. Bhatiya, D. Murthy, C.N. (2015) Metal Removal Studies by Composite Membrane of Polysulfone and Functionalized Single-Walled Carbon Nanotubes, Sep. Sci. Technol. 50: 421–429. https://doi.org/10.1080/01496395.2014.973516.

Giwa, A. Hasan, S.W. (2020) Novel polyethersulfone-functionalized graphene oxide (PES-fGO) mixed matrix membranes for wastewater treatment, Sep. Purif. Technol. 241 :116735. https://doi.org/10.1016/j.seppur.2020.116735.

Mahlangu, O.T. Nackaerts, R. Mamba, B.B. Verliefde, A.R.D. (2017) Development of hydrophilic GO-ZnO/PES membranes for treatment of pharmaceutical wastewater, Water Sci. Technol. 76: 501–514. https://doi.org/10.2166/wst.2017.194.

Mangukiya, S. Prajapati, S. Kumar, S. Aswal, V.K. Murthy, C.N. (2016) Polysulfone-based composite membranes with functionalized carbon nanotubes show controlled porosity and enhanced electrical conductivity, J. Appl. Polym. Sci. 133: 1–8. https://doi.org/10.1002/app.43778.

Nikita, K. Kumar, S.Aswal, V.K. Kanchan, D.K. Murthy, C.N. (2019a)Porous structure studies of the mixed-matrix polymeric membranes of polyether sulfone incorporated with functionalized multiwalled carbon nanotubes, Desalin. Water Treat. 146: 29–38. https://doi.org/10.5004/dwt.2019.23624.

Nikita, K. Karkare, P. Ray, D. Aswal, V.K. Singh, P.S. Murthy,C.N. (2019b) Understanding the morphology of MWCNT/PES mixed-matrix membranes using SANS: interpretation and rejection performance, Appl. Water Sci. 9: 1–9. https://doi.org/10.1007/s13201-019-1035-4.

Nikita, N. Ray, D.Aswal, V.K. Murthy, C.N. (2020) Surface modification of functionalized multiwalled carbon nanotubes containing mixed matrix membrane using click chemistry, J. Memb. Sci. 596: 117710. https://doi.org/10.1016/j.memsci.2019.117710.

Otitoju, T.A. Ahmadipour, M. Li, S. Shoparwe, N.F. Jie, L.X.Owolabi A.L. (2020), Influence of nanoparticle type on the performance of nanocomposite membranes for wastewater treatment, J. Water Process Eng. 36: 101356. https://doi.org/10.1016/j.jwpe.2020.101356.

Shah, P and Murthy, C.N. (2013) Studies on the porosity control of MWCNT/polysulfone composite membrane and its effect on metal removal, J. Memb. Sci. 437: 90–98. https://doi.org/10.1016/j.memsci.2013.02.042.

Sianipar, M. Kim, S.H. Khoiruddin,Iskandar, F. Wenten,I.G. (2017) Functionalized carbon nanotube (CNT) membrane: Progress and challenges, RSC Adv. 7: 51175–51198. https://doi.org/10.1039/c7ra08570b.

Chapter 8

Role of Plants in Attenuation of Dust Load with Special Reference to Open Cast Mining in Bundelkhand Region of Uttar Pradesh, India

Priyanka Singh and Amit Pal

Institute of Environment and Development Studies, Bundelkhand University, Jhansi – 284 128, Uttar Pradesh
e-mail: apu13@rediffmail.com

ABSTRACT

Mine and stone crushing industry in India has been growing rapidly due to increasing demand from the construction industry and the present emphasis on developing the country's infrastructure. The aim of the present study was to assess the effects of mining on air, especially dust particles and their effects on plant species in and around mining areas of Jhansi, Bundelkhand region, India. The mining and stone crushing activities have considerable effects on the environment and well-being of living organisms. The dust emissions and possibility of leaching of contaminants during the stone mining and its allied activities may contaminate the air and water therefore affecting the exposed living organisms. Stone crushing and associated activities mainly contribute SPM to surrounding environment. The variation in terms of dust deposition with species specific result are observed during the entire study. Decreasing of leaf pigment concentration indicate the positive impact of dust pollution. Species like Ficus hispida, Calotropis procera, Butea monosperma, Ficus benghalensis, showed the maximum deposition of dust on their leaf surface. Our observation may be helpful to find out some species which are resistant or to cope with open cast mining generated dust pollution in and around mining areas and adopt them also for the beautification of highways.

Keywords: *Bundelkhand region, Open cast mining, Dust pollution, Chlorophyll, SPM.*

8.1 Introduction

Mining is a major economic activity in many developing countries like our country as well. Operations whether small or large scale are inherently disruptive to the environment, producing enormous quantities of dust wastes that can have deleterious impacts for decades. Mine and stone crushing industry in India has been growing rapidly due to increasing demand from the construction industry and the present emphasis on developing the country's infrastructure (*e.g.* road, bridges buildings, *etc.*) though reliable statistics are lacking for this industrial sector due to its informal nature. Thousands of stone crushers in India provide direct employment to millions of rural migrant and unskilled workers. Stone crushing in India is basically a labor intensive small scale industry, where most of the operations are performed manually (Aslam *et al.*, 1992). Very little information is currently available regarding dust emissions from these units, associated with occupational exposures and the baseline respiratory health status of workers.

Mining and particularly, the extraction of rock and minerals in open mining have always been considered aggressive activities with a high and negative impact on the environment. The cutting operations of stone at quarries produce a large amount of fines that, in turn, cause a negative environmental impact over local surface waters. When it comes to defining a pollutant than a pollution can be any chemical *e.g.*, radio nuclide, organo-phosphorus compound or nay trace gas or geochemical (*e.g.*, dust, sediment), substance, biological organism or product, or physical property (*e.g.*, heat) that is released intentionally or inadvertently by man into the environment with actual or potential adverse, harmful or unpleasant and inconvenient effect.

Mining has an impact on all element of the environment even abandoned mining waste heaps and exaction spaces result in long-lasting, often historical impacts. Modern opencast mining involves a high degree of mechanization of the operations. Deep whole drilling, blasting of formations and sizing by crushing of the mineral are essential activities in most large open cast mines. Heavy earth moving machines (HEMM) is the essential feature of all large open mines. Operation of HEMM, crushers and blasting causes environmental degradation from dust, noise and ground vibration. For proposed mining projects, it is necessary to address the potential environmental impact issues that may arise due to proposed mining activities *i.e.* an assessment of the potential impacts of a project on the pre-mining environment.

It is clear that the impacts (both beneficial and adverse) of mining begin before a mine is even established and does not disappear with the closure of a mine. Due to lack of proper adequate precaution measures, granite mining has caused serious negative impacts on the environment and on human health (Singh *et al.*, 2009; Sheikh *et al.*, 2011). In Jhansi granite mining is done mainly through the open cast mining. Opencast mining is damaging the environment more than underground mining. Some potential impacts from blasting are ground vibrations, noise, dust, and fly rock (Langer, 2001). It starts within a natural ecosystem and it not only disturbs the existing ecosystems, but also generates an artificial one, which has its own factors including pollutants and contaminants. The important environmental problems that arise out of the opencast mining operation are water, soil, noise, land pollution *etc.* Mining operations not only affect the physical and biological structure of an area; it leads to an overall change of the surrounding *i.e.* the socio-economic status. Although opencast mining has various negative impacts on the environment but it has got positive impact even. It has also eliminated many traditional sources of income.

All major mining activity particularly opencast mining contributes to the problem of suspended particulate matter (SPM) directly or indirectly. SPM and RSPM (Respirable Suspended Particulate Matter) were the major sources of emission from various open pit mining activities, whereas emission of SO_x and NO_x were negligible. The suspended particulate matters are generated from the drilling, blasting, crushing and transportation operations. Gaseous pollutants are also released from the exhaust of HEMM and other vehicles used in mining areas. Control measures involve planning and implementation of a series of preventive and suppressive measures so that the levels of pollution are maintained within certain standards (Ghose and Sinha, 1990). However, there is no well defined method for predicting the air pollution impact due to mining. Increasing trend of opencast mining leads to the production of huge quantities of dust. The commonest dusts encountered in the industrial environments are those of mineral origin. Even after initial work has been carried out, and any dust has settled to the ground, it can be disturbed and put back into the air by traffic, wind or machinery. The digging, blasting and drilling required in the working of any mine, has the effect of throwing up concentrated levels of dust into the immediate atmosphere. Most of this dust is usually made up of silica (occurring as silicon dioxide SiO_2), and is present both in and around open-pit and underground mines. Among all the contaminants of the atmosphere, dust is probably the most abundant and ubiquitous. Silica dust gives rise to problems which tend to be even more acute when the dust which miners are exposed to is freshly cut. Opencast mining creates wider air-quality deterioration due to the dust and gaseous pollutants in and around the mining complexes. Therefore it is necessary to assess the impact on air quality due to mining and to suggest proper abatement measures for control of air pollution. Any mining activity would have a definite impact on the environment, especially on the quality of ground and surface waters (Singh and Pal, 2010). For economic growth of a country mineral resource plays an important role. Minerals and metals have played a crucial role in the development and continuation of human civilization. After agriculture it is the second largest industry at all scales and regions and has played a vital role in the development of civilization from ancient times. However Mining, like any other industrial activity tends to leave a strong negative impact on the environment unless it is meticulously planned and carefully executed. It is well established fact that mining is an environmentally destructive activity. The impact can be nullified or even better landscape and greenery can be developed by better efforts on re-vegetation of area after mining is over in that part of land.

Research on the effects of dust pollution on plant communities has never received the same level of attention as that given to phytotoxic pollutants such as SO_2, NO_2 and O_3. Results from research that has been undertaken, together with repeated observations of dust deposits on vegetation (Ahmad *et al.*, 1991), suggest that the effects of dust may be important and are worthy of greater research attention. Dusts consist of solid matter in a minute and fine state of subdivision so that the particles are small enough to be raised and carried by wind. They may originate from many sources. A large range of industrial processes can produce particulate emission. Measures to control these have, however, improved dramatically, so that there is now relatively little dust lost from closed processing plants in countries with strict pollution control. The main processes that regularly cause problems, however, are those concerned with mineral extraction. This ranges from the quarrying itself to the various processing operations (*e.g.* Cement manufacture or coal processing). Dust suppression in these operations is more difficult and dust levels can be very high. Mineral extraction is increasing in many countries in order to meet increased construction demand. Roads are a further common source of dust. Dusts may be produced from car exhausts, which are more controllable, and from the road surfaces. An increased reliance on road transport today is also evident. In this context, it is increasingly important that we understand the potential impacts that dust deposition may have on vegetation.

8.2 Characteristics of Dust

A number of characteristics of dust are important in considering its impacts. Dust can have both a physical and a chemical impact. Dust falling onto plants may physically smother the leaves. Thus the absolute level of deposition is important. This is affected by dust emission rates, meteorology and conditions on the leaf surface. Dusts of diverse origin have very different chemistries. The chemical effect of dust, either on soil or directly on the plant surface, may be more important than any physical effects. Thus before describing effects of dusts themselves it is necessary to consider how these characters vary. According to the International Standardization Organization (ISO 4225 - ISO, 1994), "Dust: small solid particles, conventionally taken as those particles below 75 μm in diameter, which settle out under their own weight but which may remain suspended for some time". According to the "Glossary of Atmospheric Chemistry Terms" (IUPAC, 1990), "Dust: Small, dry, solid particles projected into the air by natural forces, such as wind, volcanic eruption, and by mechanical or man-made processes such as crushing, grinding, milling, drilling, demolition, shoveling, conveying, screening, bagging, and sweeping. Dust particles are usually in the size range from about 1 to 100 μm in diameter, and they settle slowly under the influence of gravity." Levels of deposition of dust vary greatly. However, it is worth considering deposition from roads in more detail here. Unpaved roads produce higher dust levels than paved roads. Everett (1980) undertook a detailed study of an unpaved road in Alaska and found that in the summer about 10 gm per day was deposited at the roadside and that there was a logarithmic decline in deposition away from the road, with deposition still occurring 1km away.

A number of factors determine rates of dust deposition. Many of these are similar to those governing deposition of other pollutants. Belot *et al.* (1976) have modeled particulate deposition and showed that an increase in surface roughness causes a significant increase in deposition rates. The chemistry of dusts is varied.

Some dusts are relatively inert in their chemical effects, *e.g.* those from hard acidic rock quarries. However, limestone quarry dust, cement dust and that from many roads is highly alkaline (Everett, 1980). Darley (1966) described solutions of cement kiln dust having a pH of 12-0 and analysis showed that it contained a number of metals and bisulphate, all of which could have a direct toxic effect themselves. A number of mineral elements can be supplied in lime and cement dust. Cawse *et al.* (1989), for example, found that rainfall around a cement plant in south Wales was high in phosphorus and vanadium and had a high pH (>7.0). Many unpaved roads produce alkaline dusts, which have high calcium levels. The elements in road dust are concentrated in the smaller particles, so that Everett (1980) found that the decline in elemental deposition (excluding calcium) away from a road was not, therefore, as rapid as the decline in the mass of particulate matter. Coal also contains toxic substances, such as fluoride and sulphur compounds. These may be important in affecting vegetation if deposited as dust.

8.3 Stone Crushing Industry in India

A stone crushing industry is engaged in crushing mined stone into stone gravels of different sizes, which are used in activities such as construction of roads, building, dams and other miscellaneous purposes. The raw stones are mined and transported to crusher sites by trucks and unloaded in elevated storage hoppers constructed above the crushers. During unloading and crushing operations, substantial fugitive dust emissions occur from various locations, which pollute the surrounding environment (Figure 8.1). The fine silica dust is highly hazardous to human health and agriculture *etc.* and also adversely affects visibility in the nearby areas. Though environmental regulation exists and attempts have been made by crusher units to control the emissions, the dust control scenario in majority of the units is far from satisfactory. At national level the gravity of environmental problem with stone crusher is enormous. In India, there are over 15,000 crushers spread all over the country. As the stone crushing industry in India employs over 500,000 persons (mostly in rural areas where employment opportunities are limited) and has an estimated turnover of over Rs. 5000 crores (more than US $ 1 billion) and plays a key role in developmental activities such as road, dam and building construction and therefore, it is an important sector in the national economy.

Generally, the crushers are located in clusters. A cluster could have up to fifty crushers in one location. As the demand for the final products for road and building construction is primarily in large cities/towns, majority of the stone crushing clusters are located within two to twenty units.

The mining industry in India is a major economic activity which contributes significantly to the economy of India. The GDP contribution of the mining industry varies from 2.2 per cent to 2.5 per cent only but going by the GDP of the total industrial sector it contributes around 10 to 11 per cent. Even mining done on small scale contributes 6 per cent to the entire cost of mineral production.

From the national perspective, the important minerals of Bundelkhand are diaspore, used in paints and ceramics industries, glass sand, rock phosphate, a variety of stones and soils, and raw materials like lime for the cement industry.

(a) A typical stone crushing unit

(b) Stones are assembles for crushing

(c) Loading of stones for crushing unit

(d) Dust generation during crushing

(e) Dust generation during loading

(f) Dust generation during operation of crushing unit

Figure 8.1: Emission of Dust Particles from Opencast Mining Unit in Jhansi Area.

Lalitpur, Jhansi and Mahoba districts account for around 15 per cent of the country's reserves and nearly 40 per cent of the country's production of diaspore [Planning Commission, Vol, p 102]. The mineral is also found and extracted in large quantities in Chhatarpur district of Madhya Pradesh.

Cement plants, using locally available raw materials, namely cement grey limestone, clay and gypsum, are located in Damoh and Jhansi. The Diamond Cement factory of the Birla group, set up near Damoh town, provides direct employment to over a thousand people, and is Bundelkhand's only large industrial unit, apart from the public sector BHEL unit in Jhansi.

8.4 Quantification of Atmospheric Dust Removal by Plants

Investigations related to quantification of the dust capturing capacity of urban canopies have received attention in the recent past decades. The estimations of removal of atmospheric particulate matter by plants have been done by several researchers. For example, trees in Beijing, China captured 1,261.4 tons of pollutants that resulted in reduction of 772 tons of PM_{10} per annum from the air (Yang *et al.*, 2005). Nowak *et al.* (2006) demonstrated the variation in pollution removal by urban trees in several U.S. cities and estimated that the total annual air pollution removal by urban trees is approximately 71,000 metric tons ($3.8 billion value). Removal of approximately 4 per cent and 3 per cent of PM_{10} annually by urban plant canopies in West Midlands and Glasgow, United Kingdom, respectively, was reported by McDonald *et al.* (2007). Cavanagh (2008) reported removal of around 300 tons of air pollutants annually by trees in Christchurch, New Zealand.

Air quality improvement by urban greening is a sustainable and environmentally friendly and cost effective approach as has been discussed by Escobedo *et al.* (2008). Later on Cavanagh *et al.* (2009) again investigated the attenuation of particulate matter in an urbanized forest patch in New Zealand and concluded that at the local scale, interior parts of forest patches within urban areas can have substantially lower concentrations of particulate matter than forest edges. Reduction in NOx and PM_{10} concentrations by urban vegetations has been observed by several researchers (Currie and Bass, 2008; Escobedo *et al.*, 2009; Tallis *et al.*, 2011). An estimation of the attenuation of levels of air pollutants by urban vegetation especially in urban forest patches in Chicago suggested that these forests stores remove about 25,200 tons of carbon and 888 tons of air pollution per year (Nowak *et al.*, 2010). Effectiveness of green infrastructure in reducing air pollution and judicious use of vegetation has been discussed by Pugh *et al.* (2012). They have also suggested that urban vegetation can act as an efficient pollutant filter and may lead to rapid and sustained improvements in street-level air quality in dense urban areas. Nowak *et al.* (2013) through a modeled based study estimated the total amount of $PM_{2.5}$ removed annually by trees to be between 4.7 tons to 64.5 tons in different U.S. states.

8.5 Dust Deposition on to Vegetation

A wide variety of tree species have been studied in their response to dust (Ahmad *et al.*, 1991; Das *et al.*, 2006; Singh and Pal, 2017). Like much subsequent work, the dust studied was from cement industries. Dust may cause physical injury to tree leaves and bark, reduced fruit setting and a general reduction in growth, production of fruits, inhibited pollen germination. The dust forms a hard crystalline crust on the leaf surface, which dissolves releasing solutions of calcium hydroxide into the intercellular spaces that may causes cell plasmolysis and death. Finally, heavy cement/lime dust deposition can lead to growth reduction for many tree species (Brandt and Rhoades, 1973).

8.6 Foliar Features that Help in Dust Adsorption

Numerous studies have been carried out looking at the features that help the plants in capturing the airborne dust particulates and it has been suggested that the dust removal ability of different species can be influenced by the roughness of leaf surface, epidermal cell arrangement, and density of trichomes on leaf surfaces (Pal *et al.*, 1999, 2000, 2002). Foliar surface characters such as surface roughness and presence of trichomes has shown positive association with amount of dust on leaves and it has been suggested that such morphological features can enhance the dust filtering ability of the plant species (Khan *et al.*, 1989). Atmospheric pollutants get deposited on the canopy surface laterally (from the side) and vertically (from above the canopy). Lohr and Pearson- mims (1996) suggested that the particulate matters are removed from the ambient air due to the sedimentation, impaction and precipitation. Beckett *et al.* (1998) reviewed the role of urban woodlands in reducing the effect of particulate pollution and highlighted the role of leaf morphology in dust adsorption on urban vegetation. Foliar features that play important role in determining the ability of plants to capture airborne dust have been investigated during one growing season by Neinhuis and Barthlott (1998). The authors suggested that surfaces with wax crystals that erode very quickly are least effective in avoiding contamination and it can be expected that plants with such surfaces show greater damage in polluted areas.

Meusel *et al.* (1999) highlighted the role of roughness of leaf surfaces, frequency and length of trichomes, and size and frequency of stomata in determining the efficiency of plants to trap dust particulate. The effectiveness of tree species in capturing pollutant particles was quantified and it was observed that, finer and more complex structure of the plant foliage explained their much greater effectiveness at capturing particles (Beckett *et al.*, 2000).

In addition to leaf morphology of the plant species, deposition of dust on plant canopies depends on the physical characteristics of particles such as their size and shape (Harrison and Yin, 2000). The role of leaf surface character in dust capturing by the plant foliage and the significance of micro morphological leaf surface characters of plants in indication and mitigation of automobile exhaust pollution has been described by Pal *et al.* (2002) and Kulshreshtha *et al.* (2009) have also suggested that the morphological characteristics in combination with orientation of leaf on the main axis, shape, size, and wax deposition, play significant role in the interception of airborne dust by plants. Plant species differ significantly in their ability to mitigate traffic related pollution due to differences in their leaf surface characteristics such as epicuticular wax, cuticle, epidermis, stomata, and trichomes. Because of the roughness, large contact area, presence of certain foliar structures such as trichomes and stomata, the foliage of plants are able to filter numerous solid particles. Prusty *et al.* (2005) observed significant variations in dust interception ability among the plant species and implied that dust interception capacity of plants depends on their canopy shape and size, leaf phyllotaxy, and leaf surface characteristics such as hairs

(a) Dust deposition on leaves of *Ficus hispida*

(b) Leaves of *Ficus hispida* from controlled sites

(c) Dust deposition on leaves of *Calotropis procera*

(d) Leaves of *Calotropis procera* from controlled sites

(e) Dust deposition on leaves of *Datura metel*

(f) Leaves of *Datura metel* from controlled sites

Figure 8.2: Showing the Differences of some Leaves Surface Collected from Adjacent Opencast Mining Areas and Unpolluted Areas of Jhansi District.

and cuticles. They have also stated that except leaf morphology, dust interception ability of plants depends on leaf orientation and the sessile or semi sessile nature of leaves. Leaves with long petioles are easily put into motion by minor air movement and such leaves can hold lesser amounts of dust. In contrast leaves with short petioles can resist the air movement and can hold relatively more amounts of dust. Chakre (2006) assessed the contribution of canopy shapes toward dust filtration by plants and categorized thick and thin green belt based upon the plant canopy shape. Qiu *et al.* (2009) attributed the differences in dust retention potential of urban trees under same climatic conditions to tree canopy structure, branch density, leaf inclination angle, and variations in leaf area. The dust particulates adsorbed on the leaf surfaces may also affect the foliar morphology. Kayode and Otoide (2007) studied the epidermal variations in leaves of roadside plant and suggested that the alteration in foliar morphology induced by particulate pollution stress could serve as a gauge to assess the effects of air pollution on plants. Rai *et al.* (2010) reported the effect of particulate pollutants from automobile exhausts on the leaf morphology in terms of stomata closure, alteration in frequency of epidermal cells, stomata and trichome and changes in wax on epidermis. They have observed remarkable differences in micro morphological features in the dust-treated plants when compared to the respective controls. Particles of different sizes adsorbed on the leaf surfaces may be retained or gets responded by wind or may get washed off by rain water. Prajapati and Tripathi (2008) investigated the dust interception efficiency of roadside tree species and reported the seasonal variation in dust accumulation in plants and inferred that the deposition of dust on the leaves is a cumulative process followed by wash off with rain providing a cleaner surface for further deposition of dust. However, as the fine particulates are easily embedded in the leaf surface, the possibility of their resuspension was discarded by Liu *et al.* (2012). Except leaf morphology the structure and composition of wax particles may also contribute toward the dust capturing capacity of plants. Epicuticular wax layers of different thickness and composition influence the deposition of dust on plant foliage (Dzierzanowski *et al.*, 2011). Significant differences in accumulation of particulate matter were observed between tree species commonly cultivated in Poland and the variations were attributed to the dissimilarity in wax properties of different plant species (Popek *et al.*, 2013). Plants having suitable characters required for efficient dust adsorption must be selected for plantation during development of green belts around a proposed area. The areas for urban greening cannot be random instead these should be located at high priority zone. Morani *et al.* (2011) described the method to select strategic locations for greening, by using a planting priority index developed combining three main indicators such as pollution concentration, population density, and low canopy cover. Efforts to curb down the air pollutants have started several years ago and further advancement in this direction is being made. But pollution level is increasing and exceeding the ambient air quality standards in recent past. Therefore strategies to reduce the concentration of air pollutants must include reduction in emission of pollutant, increase in surfaces to accommodate such pollutants after their release into the atmosphere, or preventing setting up of high emitters near highly polluted areas. Having mentioned the capacity of urban vegetation to reduce air pollution, it has been suggested that the use of street trees must be considered cautiously. The specific combination of tree species, canopy volume, canyon geometry, and wind speed and direction must be scientifically approved via model-based studies (Pugh *et al.*, 2012) an extent that ultimately leads to yield losses (Calatayud and Barreno, 2004). Effects of air pollution on plants are reflected in terms of reduction in photosynthetic pigments, inhibition of certain physiological processes, alteration in metabolic functions, and enzymes activities (Heath, 2008). Wahid (2006) reported suppression of growth and reduction in yield of crops in response to elevated concentrations of air pollutants.

8.7 Impact of Dust Pollution on Plants

It has been established by several research groups that in response to abiotic stress, various biochemical changes occur in plants which include decreased chlorophyll and carotenoid content, and increased malondialdehyde content (Gavali *et al.*, 2002; Mandal and Mukherjee, 2000; Mashitha and Pise, 2001; Pandey *et al.*, 1999). It has been proved that reduction in chlorophyll pigments under stress may be due to several photochemical reactions such as oxidation, reduction and reversible bleaching. Rao and Le Blanc (1965) suggested that the reduction in chlorophyll content could be due to conversion of chlorophyll a to pheophytin and/or loss of phytol from chlorophyll b. Later, Pandey *et al.* (1991) observed relation of decrease in chlorophyll content with chloroplast damage while Ali (1993) attributed decrease in chlorophyll content to inhibition of chlorophyll biosynthesis. It is known that carotenoids have photo protective function and protect chlorophyll from photo oxidative damage (Siefermann- Harms, 1987). Under stress, the protective function of carotenoids may be hampered as it has been reported that carotenoids too are affected by air pollutants (Williams and Banerjee, 1995). Degradation of chlorophyll and discoloration of leaves are the visible injuries caused by air pollutants. Chlorophyll content indicates photosynthetic activity in plants (Larcher, 1995). Degradation in photosynthetic pigments caused by air pollution has been reported by Sandelius *et al.* (1995). Reduction in chlorophyll content has been observed in plants exposed to smoke pollution stress (Banerjee *et al.*, 2003).

Prusty *et al.* (2005) claimed that the photosynthetic pigments are most susceptible to damage by air pollution. In addition to reduced chlorophyll content, a group of researchers have reported increase in lipid peroxidation in plants exposed to pollution stress (Gavali *et al.*, 2002; Katiyar and Dubey, 2000; Mandal and Mukherjee, 2000; Masitha and Pise, 2001). The lipid peroxidation is one of the most damaging processes that occurs in both cell and organelle membranes when the reactive oxygen species (ROS) level is too high. Such high levels of ROS start affecting the normal cellular and/or physiological functioning and aggravate the oxidative stress condition through generation of lipid-derived radicals (Montillet *et al.*, 2005). Tripathi and Gautam (2007) reported pollution load dependent variations in biochemical parameters like chlorophyll, protein, ascorbic acid, nitrate reductase, superoxide dismutase and peroxides in the leaves of *Mangifera indica, Cassia fistula,* and *Eucalyptus hybrid.*

A study was undertaken to investigate the role of plants in filtering out dust particles from the atmosphere of Jhansi district especially opencast granite mined areas on the highway and the relative dust filtering potential of various plant species (Singh and Pal, 2017). Foliar samples of 15 selected roadside plant species in and around open cast granite mining areas were collected and dust load on the leaf surface was estimated (Table 8.1). Study showed that the maximum dust deposition occurred in leaves of *Ficus hispida* (18.44mg/cm^{2}) followed by *Calotropis procera*, *Butea monosperma*, *Ficus benghalensis*, *Ficus infectoria* respectively (Figure 8.2).

Table 8.1: Deposition of Dust on Leaves of Selected Plant Species

Sl.No.	*Vernacular Name*	*Scientific Name*	*Weight of Dust (mg/cm^{2})*
1	Akkaua	*Calotropis procera* Ait.	16.46
2	Pakhad	*Ficus infectoria* Roxb.	5.30
3	Banyan	*Ficus benghalensis* Linn.	9.11
4	Kathumbhar	*Ficus hispida* Linn.f	18.44
5	Dhakh	*Butea monosperma* Lam.	10.25
6	Chilbil	*Holoptelea integrifolia* Roxb.	1.99
7	Peepal	*Ficus religiosa* Linn.	0.85
8	Neem	*Azadirachta indica* A. Juss	1.06
9	Khajur	*Phoenix dactylifera* Linn.	3.13
10	Arru	*Ailanthus excelsa* Roxb.	4.04
11	Karanj	*Millettia pinnata* (L.) Panigrahi	3.35
12	Guava	*Psidium guajava* Linn.	3.40
13	Dhatura	*Datura metel* Linn.	8.78
14	Amaltas	*Cassia fistula* Linn.	1.49
15	Kanner	*Nerium oleander* Linn.	0.34

Source: Singh and Pal (2017).

Sharma *et al.* (2017) also studied the impact of dust accumulation on roadside plant species of NH-37, Assam. According to their study maximum amount of dust deposition was found to increase with time. Dust generated from day to day traffic activity will be accumulated in the leaf up to a limit. Therefore, irrespective of plant species dust deposition has been found to increase in different time intervals. The dust particles settled in the leaf surface regularly, but deposition was not uniform in all the 15 plant species (Singh and Pal, 2017). Settlement of dust particles on the leaf surface mainly depend on the phyllotaxy of leaf, smoothness of the leaf surface (*i.e.* presence of trichome), shape of leaf, petiole length as well as its position on the plant (Davison and Blackmore, 1976).

Roadside plant species showed the reduction of chlorophyll contents, protein and carotenoids in comparison to control site. Dust particles physically obstruct sunlight as well as block stomatal pore of leaf and thus dust deposition hampered photosynthetic activities of plant. Effect of dust as well as alkaline condition caused by dissolution of particles in cell sap degrade leaf pigment concentration has been reported earlier. Continuous exposure to dust on leaf surface leading to the formation of dense dust layers, which reduce light capturing capacity of plants and finally hamper plant photosynthetic activity. Decreased rate of photosynthesis and alteration of stomatal conductance are responsible for reduction of the stomatal index as well as dry matter content of leaf. Reduction of the stomatal index of the roadside plant may be due to shading effect of dust layers, which may block the stomata and reduce the photosynthesis rate of roadside plants (Sharma *et al.*, 2017). According to their study maximum amount of chlorophyll content was decreased considerably in the roadside plant species compare to control. It is evident that roadsides of open cast mining area plants produce less chlorophyll, protein and carotenoids than control sites (Table 8.2).

Table 8.2: Pigments Content in Leaves of some Selected Plant Species

Sl.No.	*Name of Plant Species*	*Chlorophyll Content (mg/g)*		*Protein Content (mg/g)*		*Carotenoid Content (mg/g)*	
		Un Polluted	*Polluted*	*Un Polluted*	*Polluted*	*Un Polluted*	*Polluted*
1	*Calotropis procera*	16.778	12.616	0.744	0.234	0.426	0.329
2	*Ficus infectoria*	8.653	4.702	0.138	0.076	0.126	0.284
3	*Ficus benghalensis*	14.641	7.511	0.292	0.151	0.284	0.324
4	*Ficus hispida*	19.227	13.632	1.261	0.504	0.477	0.366
5	*Butea monosperma*	15.122	8.508	0.344	0.184	0.358	0.325
6	*Holoptelea integrifolia*	5.475	4.748	0.107	0.045	0.121	0.146
7	*Ficus religiosa*	11.792	3.018	0.182	0.102	0.251	0.071
8	*Azadirachta indica*	12.203	2.464	0.055	0.030	0.276	0.075
9	*Phoenix dactylifera*	8.314	3.815	0.167	0.107	0.287	0.092
10	*Ailanthus excelsa*	11.685	7.184	0.187	0.116	0.296	0.140
11	*Millettia pinnata*	10.892	6.140	0.149	0.085	0.264	0.103
12	*Psidium guajava*	12.223	5.524	0.110	0.076	0.291	0.141
13	*Datura metel*	14.468	7.33	0.244	0.137	0.314	0.168
14	*Cassia fistula*	7.33	5.00	0.184	0.093	0.226	0.101
15	*Nerium oleander*	5.37	4.59	0.096	0.102	0.134	0.132

Source: Singh and Pal (2017).

The leaves are influenced by deposition of dust, which may lead to lower photosynthesis activity (Pourkhabbaz *et al.*, 2010). Degradation of chlorophyll contents in the leaf is a common phenomenon from dust deposition on leaf (Prajapati and Tripathi, 2008). A plenty of studies (Satao *et al.*, 1993; Prusty *et al.*, 2005; Gostin, 2009) reviewed declination of chlorophyll concentration due to deposition of dust in a number of annual non- leguminous crop. Continuous exposure to dust in leaf surface leading to the formation of dense dust layers, reduces light capturing capacity of plants (Prajapati and Tripathi, 2008; Pourkhabbaz *et al.*, 2010) and finally hamper plant photosynthetic activity. Alterations of stomatal conductance are responsible for reduction of the stomatal index as well as dry matter content of leaf (Khan *et al.*, 2015). Reduction of stomatal index of roadside plant may be due to shading effect of dust layer, which may block the stomata and reduce the photosynthesis rate of roadside plants.

Dust particles settled on the leaf surface regularly, but deposition was not uniform in all the 15 plant species. Settlement of dust particles on the leaf surface mainly depend on the phyllotaxy of leaf, smoothness of the leaf surface (*i.e.* presence of trichome) shape of leaf, petiole length as well as its position on the plant (Davison and Blackmore, 1976).

Based on extensive field observations and laboratory investigations, a few plant species have been identified which posses' high dust filtering capacity. These species may be raised in green belts around granite mining areas and both sides of national highway to lessen the dust load of the surrounding environment. The study revealed the definite correlation between the plant canopy and foliar morphological traits and amount of dust captured by them. Morphological characteristics which have been found to play a significant role in intercepting particulates are: orientation of the leaf on the main axis, size (leaf area), shape, nature of surface, venation pattern, presence and absence of trichomes, morphology and frequency of trichomes, wax deposition, *etc.*

The study found that the species with better dust filtering capacity include: *F. hispida, C. procera, B. monosperma, F. benghalensis, D. metel, Ailanthus excelsa* and could be considered as first choice species. Other species like : *P. dactylifera, N. oleander, F. religiosa, C. fistula, M. pinnata* are considered as second choice species and *P. guajava, F. infectoria, H. integrifolia, A. indica* are considered as third choice species for the designing of green belt in and around mining industries as well as beautification of road sides also.

8.8 Conclusion

In open cast mining especially in and around Jhansi district, a massive overburden activity which may require excavate, transports, loaders *etc.* leads to massive discharges of fine particulates from the overburden materials. Similarly normal operations also require excavation, transportation, loading, unloading, size reduction, stock piling *etc.* and all these activities generate particulate matter. The volume and variety of air borne dust particles in the ambient air is ever increasing and causing air pollution problems in the adjacent areas. Cowherd (1979) has reported that vehicular traffic on haul road of mechanized opencast mines can contribute as much as 80 per cent of the dust emitted. Chadwick *et al.* (1987) have estimated that about 50 per cent of total coal dust is released during journey time on the unpaved haul road, while 25 per cent of both during loading and unloading of dumper.

In present study emission from mining and vehicles are the prime source of dust generation for road side plant. Poor road infrastructure, the frequency of running vehicles in terms which determine the rate of dust generation. It has been noted that adaptive evolution of road side plant species is more visible in poor road infrastructure with more running vehicles. The response of plant to dust deposition may vary according to different species as deposition fluctuates with plant species due to leaf orientation, leaf surface character, phyllotaxy, epidermal and cuticular features height and canopy of road side plant (Davison and Blackemore 1976, Chaphekar *et al.*, 1980, Farmer 1993, Chaturvedi *et al.*, 2013). With the accumulation of dust roadside plant may exhibit adaptive response by changing morphological and physiological attributes (Pal *et al.*, 2002, Kulshreshtha *et al.*, 2009).

In an investigation carried out by Singh and Pal (2017), it was found that among 15 selected roadside plant species the deposition of dust particles was highest in *F. hispida* followed by *C. procera, B. monosperma, F. benghalensis, D. metel, A. excelsa* respectively, which support the above mentioned statement. Structural properties of leaf are influenced by – deposition of dust which may lead to lower photosynthesis activity. Study also showed reduction of chlorophyll, carotenoid and protein content as compared to dust free site (university campus). Studies showed that the dust particles physically obstruct sunlight as well as block stomatal pore and therefore, dust deposition hampered photosynthetic activities of plant. Significant variation in terms of dust deposition with species specific effect has been observed during the present investigation.

Plants on both sides of national highway or in the city not only have an ornamental function but they have also an important role in mitigation of air pollution in terms of gaseous pollutants and filtration of air-borne particles that allows characterizing an air pollutants and determine the source of their elemental and mineral profiles. Thus, plants with suitable traits for air pollution mitigation may be used for urban tree planting, roadside plantation programs, green belt in an around mining areas to improve air quality and at the same time such plants may be utilized for passive monitoring of air pollution.

Recommendations

On the basis of field data collected it has been observed that main pollution problem in the area is the dust. The ambient environmental conditions in and around the stone crusher in Jhansi are not congenial. Of course, mining and stone crushing activities have taken big strikes during the last century, contributing significant development and raising the living standards of mankind. But they have also brought pollution, degradation of natural resources, health risks and socio-economic instabilities. Mining and Stone crushing has caused potentially adverse impacts on natural environment, society and cultural heritage, the health and safety of workers and communities in close proximity to operations The workers engaged in stone crushers encounter with substantial exposure to dust and noise which may lead to manifestation of various occupational diseases in the long run. It has also caused serious social and environmental impacts, like displacement of local people, dust generation, air and water pollution, land degradation, harm to livestock and wild life, and reduction in the agricultural productivity. A control plan may be formulated involving the planning and implementation of a series of preventive and suppressive measures in addition to the dust extraction system. Although different abatement measures are enumerated but plantation of trees is one of the best measures for controlling air pollution. For inhibiting dust, trees with compact branches, foliose closely arranged broad leaves, shiny or waxy leaves and high proline content are preferred.

Air-pollution Control by Trees

A wide green belt of fast growing trees should be established around the mining industries. The following guidelines may be useful in increasing its efficiency.

- ☆ Trees should be planted in beds arranged perpendicular to the direction of wind.
- ☆ Trees should be planted in a concentric fashion around the source of pollution.
- ☆ Peripheral plantations may also be used as wind breaks on dry tailings.
- ☆ A judicious mixture of open and permeable planting with dense planting should be made to achieve maximum beneficial effect.
- ☆ Plants and trees of different heights should be planted on both sides of both traffic lanes.
- ☆ A wide green belt of 10-m width with six rows of tall f ast-growing trees should be raised around the mine complex.

Control measures involve planning and implementing a series of preventive and suppressive measures so that the levels of pollution are maintained within certain standards (Ghose and Sinha, 1990). For this, an air quality management strategy is formulated (Lohani, 1984). Four types of strategy have been developed: (1) air-quality management; (2) emission standards; (3) economic; (4) cost benefit. Maiti and Banerjee (1997) found that plants can act as filters of dust of pollutants in mining areas, reported that a 8-m wide road with green plants and buildings can reduce dust fall by two to three times; conifers reduce dust fall by up to 42 per cent in temperate urban areas. They also indicated that evergreen plants with shiny leaves like *Alstonia scholaris, Ficus lunea, F. benghalensis* and *Mangifera indica* are the best dust catchers. Plant species useful for combating air pollution are listed in Table 8.1.

Haul Road Dust Control

Haul roads are the most prolific sources of dust in opencast mines and watering of the roads is the most common methods of controlling dust emission. Various products are available in the market for consolidation of dust. Hygroscopic salts extract moisture from the atmosphere and keep the road surface damp. Surfactants/ wetting agents can reduce the surface tension of water helping wetting of dust particles and thus economize watering cost.

Chemicals for Dust Control

Water spraying requires a huge quantity of water. Suitable chemicals which act as wetting agents can therefore be added to water, helping to make wetting of finer dust particles more effective. The most efficacious chemicals include: (1) Aqueous solution $CaCl_2$, $MgCl_2$, hydrated lime sodium silicate *etc.* (2) Teepol, a laboratory grade product of Burmah Shell. (3) Kodak wetting agent. (4) Lessapol, producted by ICI. (5) Filset 50 and (6) Jalasakti, a hygroscopic salt.

Water Spraying by Ultrasonic Dry Fogging System

In this system water droplets with a diameter of 0.3-0.5 mm are created with help of compressed air. With increasing fineness of droplets, the surface area affinity for dust increases, as well as the number of droplets per volume unit. The droplets approximately the same size as the dust particles which are thereby agglomerated and suppressed at source.

There is no doubt however, that mining was and will continue to be crucial for the development of human societies. The main environmental problem of mining activity is the contamination of other vital resources, like water, soils and air. According to the United States Environmental Protection Agency (EPA), water contamination from mining is one of the top three ecological-security threats in the world.

Air pollution especially dust in atmosphere is the main problem in and around stone crushers in Bundelkhand region. Workers are prone to diseases such as skin problem, respiratory, eye problem, hearing disorder, fever, high B.P. *etc.* Immediate need is for a comprehensive occupational and environmental health management strategy.

- Remedial and control measures, can improve the environmental conditions.
- Proper maintenance of plant and good work practices, health and hygiene conditions, proper medical care and housekeeping, *etc.* are also to be improved in stone crusher units.
- Moreover, to prevent ill health and accidents the safety measures should be employed.
- Provide personal protective equipments.
- Construction of wind breaking walls
- Construction of metallic roads within the premises.
- Regular cleaning and wetting of the ground around mining sites.
- Establishment of a green belt.
- Proper environment impact assessment.
- Awareness about the health hazards caused by dust inhalation of people.
- Training to workers.
- Housing in a suitable structure with suitable roofing.
- Further research and action are needed to promote widespread occupational health improvements in the existing stone crusher units in this region.

REFERENCES

Ahmad KJ, Yunus M, Singh SN, Srivastava K, Singh N, Pandey V and Misra J (1991). Air pollution and plants. *CSIR News*, 15, 170 – 174.

Ali EA (1993). Damage to plants due to industrial pollution and their use as bioindicators in Egypt. *Environ. Pollut.* 81(3), 251-255.

Aslam M, Minocha AK, Kalra PD and Srivastava RS (1992). Fugitive dust emissios from stone crusher. *Indian J. Environ. Health*, 34(3), 186-191.

Banerjee S, Singh AK and Banerjee SK (2003). Impact of flyash on foliar chemical and biochemical composition of naturally occurring ground flora and its possible utilization for growing tree crop. *Indian Forester* 129, 964-977.

Beckett KP, Freer-Smith PH and Taylor G (1998). Urban woodlands: Their role in reducing the effects of particulate pollution. *Environ. Pollut.* 99, 347–360.

Beckett KP, Freer-Smith PH and Taylor G (2000). Particulates pollution capture by urban trees: effect of species and wind speed. *Global Change Biology*, 6, 995-1003.

Belot Y, Baille and Delmas JL (1976). *Atmos Environ* 10, 89-98.

Brandt CJ and Rhoades RW (1973). Effects of limestone dust accumulation on lateral growth of forest trees. *Environ. Pollut.* 3, 213-217.

Calatayud A and Barreno E (2004). Response to ozone in two lettuce varieties on chlorophyll a fluorescence, photosynthetic pigments and lipid peroxidation. *Plant Physiol. Biochem.* 42(6), 549-555.

Cavanagh JE (2008). Influence of urban trees on air quality in Christchurch; Preliminary estimates. LC0708/097, *Landcare Research*, Lincoln.

Cavanagh JAE, Zawar-Reza P and Wilson JG (2009). Spatial attenuation of ambient particulate matter air pollution within an urbanised native forest patch. *Urban Forestry and Urban Greening* 8(1), 21-30.

Cawse PA, Baker SJ and Page RA (1989). The influence of particulate matter on rainwater acidity and chemical composition. *UK Atomic Energy Authority Report* R 13478, HMSO, London.

Chadwick OA, Hendricks DM, Nettleton WD (1987). Silica in Duric soils – A depositional Model. *American Soc. Soil Sci. Journal* 51(4), 975 – 982.

Chakre OJ (2006). Choice of eco-friendly trees in urban environment to mitigate airborne particulate pollution. *J. Hum. Ecol.* 20(2), 135–138.

Chaphekar SB, Boralkar DB and Shetye RP (1980). Plants for air monitoring in industrial area. In: Furtado JI (ed) *Tropical Ecology* and *Development*. I.S.T.E. Kuala Lampur, pp. 669-675.

Chaturvedi RK, Prasad S, Rana S, Obaidullah SM, Pandey V and Singh H (2013). Effect of dust load on the leaf attributes of the tree species growing along the roadside. *Environ. Monitor. Assess.*, 185(1), 383-391.

Currie BA and Bass B (2008). Estimates of air pollution mitigation with green plants and green roofs using the UFORE model. *Urban Ecosyst*. 11, 409-422.

Darley EF (1966). Studies on the effect of cement-kiln dust on vegetation. *J. Air. Pollut. Contr. Assoc.* 16, 145-50.

Das TM, Pattanayak P, Bhaumik A and Ghosh A (2006). Dust filtering property of avenue trees and effect of air borne particulate matters on growth and yield of paddy plants. *Indian Biologist*, 38, 141-148.

Dzierzanowski K, Popek R, Gawronska H, Saebo A and Gawronski SW (2011). Deposition of particulate matter of different size fractions on leaf surfaces and waxes of urban forest species. *Int. J. Phytoremed.*, 13, 1037-1046.

Escobedo F, Seitz JA and Zipperer W (2009). Air pollution removal and temperature reduction by Gainesville's urban forest. Gainesville, Florida: School of Forest Resources and Conservation, Florida Cooperative Extension Service, Institute of Food and Agricultural Sciences, University of Florida.

Escobedo FJ, Wagner JE, Nowak DJ, De la Maza CL, Rodriguez M and Crane DE (2008). Analyzing the cost effectiveness of Santiago, Chile's policy of using urban forests to improve air quality. *J. Environ. Manage.* 86(1), 148-157.

Everett KR (1980). Distribution and properties of road dust along the northern portion of the haul road. In Environmental Engineering and Ecological Baseline Investigations along the Yukon River-Purdhoe Bay Haul Road, ed. J. Brown and R. Berg. US Army Cold Regions Research and Engineering Laboratory, *CRREL Report* 80-19, pp. 101-28.

Gavali JG, Saha D and Krishnayya NS (2002). Difference in sulfhur accumulation in eleven tropical tree species growing in polluted environs. *Indian J. Environ. Hlth*, 44, 88-91.

Ghose MK and Sinha DK (1990). Air pollution control plan in coal mining areas, *Industrial J. Environ. Protec.* 10, 752-756.

Gostin IN (2009). Air pollution effects on the leaf structure of some Fabaceae species. *Notulae Botanicae Horti Agrobotanici Cluj- Napoca* 37, 57-63.

Harrison RM. and Yin J (2000).Particulate matter in the atmosphere: which particle properties are important for its effects on health? *Sci Total Environ.* 249(1-3), 85-101.

Heath RL (2008). Modification of the biochemical pathways of plants induced by ozone: What are the varied routes to change? *Environ. Pollut.* 155(3), 453-463.

Katiyar V and Dubey PS (2000). Growth behavior of two cultivars of maize in response to SO_2 and NO_2. *J. Environ. Biol.* 21, 317-323.

Kayode J and Otoide JE (2007). Environmental pollution and leaf cuticular variations in Newbouldia laevis Seem. ex Bureau. *Asian J. Plant Sci.*, 6, 1024-1026.

Khan AM, Pandey V, Yunus M and Ahmad KJ (1989). Plants as dust scavengers. A case study, *Indian Forester*, 115, 670-672.

Khan ZS, Wolfram S, Yang P, Xiaoning Z, Hussein O, Xiongkui H and Muller J (2015). Effect of Dust Deposition on Stomatal Conductance and Leaf Temperature of Cotton in Northwest China. *Water*, 7, 116-131.

Kulshreshtha K, Rai A, Mohanty CS, Roy R.K and Sharma SC (2009). Particulate pollution mitigating ability of some plant species. *Int. J. Environ. Res.*, 3, 137-142.

Langer WN (2001). Environmental impacts of mining natural aggregate, in Bon, R. L., Riordan, R. F., Tripp, B. T., and Krukowski, S. T., eds., Proc. 35th Forum on the Geology of Industrial Minerals–*The Intermountain West Forum: Utah Geol. Survey Misc. Publ.*, pp. 127–138.

Larcher W (1995). Physiological Plant Ecology. Berlin: Springer.

Liu L, Guan D and Peart MR (2012). The morphological structure of leaves and the dust-retaining capability of afforested plants in urban Guangzhou, South China. *Environ. Sci. Pollut. Res.* 19(8), 3440-3449.

Lohr VI and Pearson-mims CH (1996). Particulate matter accumulation on horizontal surfaces in interiors: Influence of foliage plants. *Atm. Environ.* 30(14), 2565-2568.

Mandal M and Mukherji S (2000). Changes in chlorophyll context, chlorophyllase activity, Hill reaction, photosynthetic CO_2 uptake, sugar and starch contents in five dicotyledonous plants exposed to automobile exhaust pollution. *J. Environ. Biol.* 21, 37-41.

Mashitha PM and Pise VI (2001). Biomonitoring of air pollution by correcting the pollution tolerance index of some commonly ground trees of an urban area. *Pollut. Res.* 20, 195-197.

McDonald AG, Bealey WJ, Fowler D, Dragosits U, Skiba U and Smith RI (2007). Quantifying the effect of urban tree planting on concentrations and depositions of PM_{10} in two UK conurbations. *Atm. Environ.* 41(38), 8455-8467.

Meusel I, Neinhuis C, Markstadter C and Barthlott W (1999). Ultrastructure, chemical composition, and recrystallizationof epicuticular waxes: transverselyridged roadles. *Can. J. Bot.* 77, 706–720.

Montillet JL, Chamnongpol S, Rusterucci C, Dat J, van de Cotte B, Agnel JP, Battesti C, Inze D, Van Breusegem F and Triantaphylides C (2005). Fatty acid hydroperoxides and H_2O_2 in the execution of hypersensitive cell death in tobacco leaves. *Plant Physiol.*, 138(3), 1516-1526.

Morani A, Nowak DJ, Hirabayashi S and Calfapietra C (2011). How to select the best tree planting locations to enhance air pollution removal in the Million Trees NYC initiative. *Environ. Pollut.* 159(5), 1040-1047.

Neinhuis C and Barthlott W (1998). Seasonal changes of leaf surface contamination in beech, oak, and ginkgo in relation to leaf micromorphology and wettability. *New Phytol.* 138(1), 91-98.

Nowak DJ, Crane DE and Stevens JC (2006). Air pollution removal by urban trees and shrubs in the United States. *Urban forestry and Urban Greening.* 4, 115-123.

Nowak DJ, Hirabayashi S, Bodine A and Hoehn R (2013). Modeled $PM_{2.5}$ removal by trees in ten U.S. cities and associated health effects. *Environ. Pollut.* 178(0), 395-402.

Nowak DJ, Hoehn RE, Crane DE, Stevens JC and LeBlanc C (2010). Assessing urban forest effects and values: Chicago's urban forest. *Northern Resource Bulletin,* NRS-37. Newtown Square, PA: USDA Forest Service.

Pal A, Kulshreshtha K and Ahmad KJ (1999). Some Morphological Changes in two ornamental plant species due to automobile pollution. *Acta-Botanica Hungarica* 42(1-4), 39-47.

Pal A, Kulshreshtha K, Ahmad KJ and Behl HM (2002). Do leaf surface characters play a role in plant resistance to auto-exhaust pollution? *Flora*, 197(1), 47-55.

Pal A, Kulshreshtha K, Ahmad KJ and Yunus M (2000). Changes in leaf surface structures of two avenue tree species caused by auto-exhaust pollution. *J. Environ. Biol.* 21(1), 15-21.

Pandey DD, Nirala AK and Gaulam RR (1999). Impact of stone crusher dust pollution on maize crop. *Indian J. Environ. Ecoplann.*, 2, 43-46.

Pandey DD, Sinha CS and Tiwari MG (1991). Impact of coal dust pollution on biomass, chlorophyll and grain characteristics of rice. *Journal of Biology, 3*, 51-55.

Popek R, Gawronska H, Wrochna M, Gawronski SW and Saebo A (2013). Particulate matter on foliage of 13 woody species: deposition on surfaces and phytostabilisation in waxes-a 3-year study. *Int. J. Phytorem.*, 15, 245-256.

Pourkhabbaz A, Rastin N, Olbrich A, Langenfeld-Heyser H and Polle A (2010). Influence of Environmental Pollution on Leaf Properties of Urban Plane Trees, *Platanus orientalis* L. *Bull. Environ. Contam. Toxicol.*, 85, 251-255.

Prajapati, SK and Tripathi BD (2008). Seasonal variation of leaf dust accumulation and pigment content in plant species exposed to urban particulates pollution. *Journal of Environmental Quality*, 37, 865–870.

Prusty BAK, Mishra PC and Azeez PA (2005). Dust accumulation and leaf pigment content in vegetation near the national highway at Sambalpur, Orissa, India. *Ecotoxicol. Environ. Safety*, 60, 228-235.

Pugh TAM, MacKenzie AR, Whyatt JD and Hewitt CN (2012). Effectiveness of Green Infrastructure for Improvement of Air Quality in Urban Street Canyons. *Environ. Sci. Technol.* 46(14), 7692-7699.

Qiu Y, Guan D, Song W and Huang K (2009). Capture of heavy metals and sulfur by foliar dust in urban Huizhou, Guangdong Province, China. *Chemosphere*, 75, 447-452.

Rai A, Kulshreshtha K, Srivastava PK and Mohanty CS (2010). Leaf surface structure alterations due to particulate pollution in some common plants. *The Environmentalist*, 30, 18-23.

Rao DN and Leblanc F (1965). Effects of SO_2 on the lichens alga with special reference to chlorophyll. *Bryologist* 70, 69-75.

Sandelius AS, NaSlund K, Carlsson AS, Pleijel H and Sellden GUN (1995). Exposure of spring wheat (*Triticum aestivum*) to ozone in open-top chambers. Effects on acyl lipid composition and chlorophyll content of flag leaves. *New Phytol.*, 131, 231-239.

Satao RN, Kene HK, Nalamwar RV and Ulemale RB (1993). Effect of cement dust pollution on growth and yield of cotton. *Annual Review of Plant Physiology*, 7, 73-77.

Sharma B, Chanda SK and Bhuyan M (2017). Impact of dust accumulation on three road side plants and their adaptive responses at National Highway 37, Assam, India. *Tropical Plant Research*, 4(1), 161–167.

Sheikh A, Rana SVS and Pal A (2011). Environlmental health assessment of stone crushers in and around Jhansi, U. P., India. *J. Ecophysiol. Occup. Hlth,* 11, 107 – 115.

Siefermann-Harms D (1987). The light-harvesting and protective functions of carotenoids in photosynthetic membranes. *Physiologia Plantarum* 69(3), 561-568.

Singh G and Pal A (2010). Environmental impacts of mining on Bundelkhand region of Uttar Pradesh, India. *Rec. Res. Sc. Techno.,* 2(3), 50-57.

Singh G, Khoiyangbam RS and Pal A (2014). Assessment of Groundwater quality in Granite mining areas of Bijauli in Bundelkhand region of India. *Int. J. Sci. Res.,* 3(5), 1563–1567.

Singh G, Pal A and Khoiyangbam RS (2009). Impact of mining on human health in and around Jhansi, Bundelkhand region of Uttar Pradesh, India. *J. Ecophysiol. Occup. Hlth,* 9 : 47 – 54.

Singh G, Pal A and Singh A (2011). Status of air quality in open cast mining sites in Jhansi, Bundelkhand region of Uttar Pradesh, India. *J. Econ. Geol. Geores. Mgm.,* 8(1-2), 49 – 56.

Singh P and Pal A (2017). Response of dust accumulation on road side plant species due to open cast mining at Jhansi-Allahabad NH No. 76, U. P., India. *Trop. Pl. Res.,* 4(3), 461 – 467.

Tallis M, Taylor G, Sinnett D and Freer-Smith P (2011). Estimating the removal of atmospheric particulate pollution by the urban tree canopy of London, under current and future environments. *Landscape and Urban Planning, 103,* 129-138.

Tripathi AK and Gautam M (2007). Biochemical parameters of plants as indicators of air pollution. J. Environ. Biol. 28(1), 127-132.

Wahid A (2006). Productivity losses in barley attributable to ambient atmospheric pollutants in Pakistan. *Atm. Environ.* 40(28), 5342-5354.

Williams AJ and Banerjee SK (1995). Effect of thermal power plant emissions on themetabolic activities of *Mangifera indica* and *Shorea robusta*. Environ. Ecol. 13, 914–919.

Yang J, McBride J, Zhou J and Sun Z (2005). The urban forest in Beijing and its role in air pollution reduction. *Urban Forestry and Urban Greening,* 3, 65-78.

Chapter 9

Washcoated Monoliths for Automotive Emission Control

Vijay S. Narkhede, Rustom M. Cursetji, Dinesh Kumar and Touquire A. Siddiquie

Süd Chemie India (P) Ltd, A-1/2/3, Nandesari Industrial Area, P.O. Nandesari Vadodara – 391 340, Gujarat
e-mail: vijay.narkhede@sud-chemie-india.com

ABSTRACT

This chapter provides an overview of various emission control technologies employed to meet stringent vehicular emission standards worldwide. Emission exhaust gases from internal combustion engines consist of CO, hydrocarbons, NOx and particulate matters. The most common technologies used to convert the emission exhaust gases into environmentally benign gases are three-way catalyst, diesel oxidation catalyst, diesel particulate filter and selective catalytic reduction.

The industry for manufacture of emission control systems from automobile exhaust has evolved as a niche high technology area that is the largest market for precious group metals, platinum, palladium and rhodium. The key to efficient and optimal use of the precious metals is the formulation of the catalyst carriers and its application as a washcoat on metallic and ceramic substrates. This chapter attempts to describe the development as well as the challenges that lie ahead in meeting the ever increasing stringency of the norms for tail pipe emissions from automobiles.

Keywords: *Automotive emission control, Three-way catalyst, Diesel oxidation catalyst, Particulate filter, Selective catalytic reduction.*

9.1 Introduction

Vehicles equipped with internal combustion engines (ICE) that use fuels like gasoline, diesel, compressed natural gas (CNG) perform the majority of on-road transport. Road transport has overtaken rail transport in recent years and accounts today for ~85 per cent of all freight because of the advantage of door-to-door delivery of goods. The vehicles range from small motorbikes to heavy-duty trucks and the engines used are either spark ignited or compression-ignited. Although recent development in the area of batteries and fuel cell vehicles is promising, society will continue to rely on the ICE for mobility of people and goods in the foreseeable future (Lambert, 2019).

It may be borne in mind that unlike in conventional chemical processes, the catalytic converter has to perform under continuously varying conditions of space velocity > 100,000 h^{-1}, temperatures > 1000 °C and rapidly changing gas compositions. While carbon dioxide and water are the main products of fuel combustion that takes place in ICE, unburned hydrocarbons and intermediate oxidation products such as carbon monoxide (CO), aldehydes, alkenes are formed and emitted in exhaust gas. Some fuel also contains small amount of sulphur-containing and nitrogen-containing molecular species that result in emission of sulphur dioxides (SOx) and nitrogen oxides (NOx). The major source of NOx, however is the thermal reaction within air constituents nitrogen and oxygen, that occurs during flame combustion where the temperatures reach >1700 K. The detail composition of exhaust gas emission such as CO, hydrocarbons, NOx and other constituents as well as the fuel consumption and power output in a gasoline fuelled engine is a function of the air to fuel ratio (Lox, 2008).

While the increased mobility provided by automobiles has contributed vastly to economic development, it has also become a major contributor to air pollution. Exhaust gas emissions carry significant risks for human health and the environment. It is estimated that more than 50 per cent of the pollution in all major metropolis of the country is on account of vehicular air pollution that cause immediate and long term effects on the environment and human health. Automobile exhaust emit a wide range of gases and particulate matter causing global warming and acid rain. Since the industrial revolution of the nineteenth century the combustion of fossil fuels has caused an increase of around 0.74 C +/- 0.18° C in global temperature. A linear trend for past 50 years is 0.13° C per decade(Lox, 2008). If this continues it will seriously affect farming, wild life, sea levels and natural landscapes.

While poisoning effects of CO on humans are well known, reactive hydrocarbons and NOx participate in photochemical reactions when exposed to sunlight and generate smog and ozone. The secondary pollutant ozone is a powerful oxidising agent and injurious to the human respiratory systems. Sulfur dioxide and nitrogen dioxide mix with rain water to produce acid rain which damages crops, forests and buildings. Diesel engines emit a high level of soot which become air borne and causes skin and eye irritation and the very fine particles (*e.g.* PM 2.5µ) settle deep inside the lungs causing respiratory problems.

During the last five decades continuous worldwide efforts have been made to regulate obnoxious emission exhaust gases from internal combustion engines and several countries have introduced stringent emission standards which are periodically reviewed and mandated. This chapter provides an overview of an automotive emission control systems.

9.2 Regulatory Overview

In 1966, California took the initiative to introduce legislation to limit exhaust gas emission of CO and HCs from gasoline passenger cars.The automotive emission catalyst industry was born out of the US Clean Air Act of 1970 which mandated control of tail pipe emissions starting in 1975. Since then the driving force has been stringent legislations that have resulted in paradigm changes in both engine management and design of after-treatment catalysts for activity and durability. For instance closed loop engine system is where the engine operates with a feedback loop as shown in Figure 9.1. A sensor such as an oxygen sensor (also known as lambda sensor or air/fuel ratio mixture sensor) closes a loop. In closed loop mode, the engine management system gets feedback from the sensors to adjust fuel/air mixture close to stoichiometry where the catalyst efficiency is optimum for both oxidation of CO and hydrocarbons as well as for the reduction of NOx.

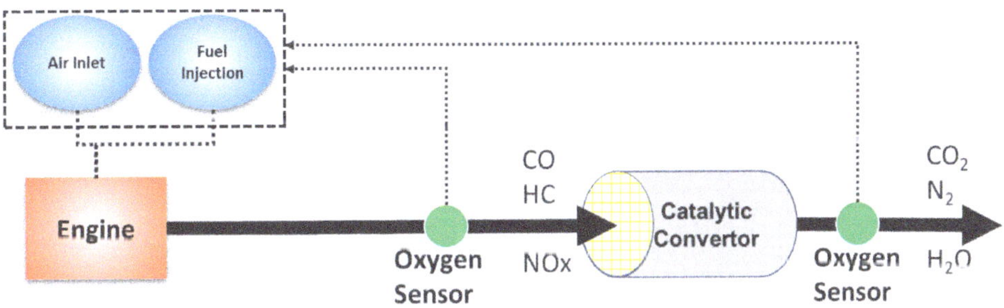

Figure 9.1: Schematic of Closed Loop Engine.

India introduced automotive emission standards for passenger cars starting from 1991 and progressively made it more stringent in subsequent years (Table 9.1). Similar emission standards have been introduced for two-wheelers equipped with small two-stroke or four-stroke single-cylinder gasoline engines, heavy duty vehicles (*e.g.* buses and trucks), larger construction equipment (*e.g.* cranes and bulldozers, small utility equipment (*e.g.* chain saws, lawn mowers). Emission legislation is a continuous process and variety of novel measures are constantly being introduced worldwide to further improve air quality. As per current state of the art in the design of after treatment devices, it has been demonstrated that the SULEV (Super Ultra Low Emission Vehicle) emits 90 per cent lower emission than the average gasoline powered vehicles (https: //en.wikipedia.org/wiki/ Super_ultra-low_emission_vehicle).

It is noteworthy that though emission legislation in India was introduced several years later in 1991 it is at par with worldwide standards as of today. One of the features specific to India is that the production figures per annum for 2-3 wheelers is higher than for 4 wheelers by a factor of more than 5x (Figure for FY 2018-19). Two and Three wheelers in India are the biggest consumers of petrol and account for more than 60 per cent of the total petrol sales.

Table 9.1: Emission Standards for Passenger Cars in India

Legislation	Year	Fuel Type	CO	HC	HC+NOx	NOx	PM	PN
Unit			g/km	g/km	g/km	g/km	g/km	Numbers/km
	1991	Gasoline	14.3	2	-	-	-	
	1992	Diesel						
	1996	Gasoline	8.68	-	3.00	-	-	
		Diesel	5.00	-	2.00			
	1998#	Gasoline	4.34	-	1.50	-	-	
India 2000	2000	Gasoline	2.72	-	0.97	-	-	
		Diesel	2.72	-	0.97	-	0.14	
BS II	April 2005*	Gasoline	2.20	-	0.50	-	-	
		Diesel	1.00	-	0.70	-	0.08	
BS III	April 2010*	Gasoline	2.30	0.20	-	0.15	-	
		Diesel	0.64	-	0.56	0.50	0.05	
BS IV	April 2017*	Gasoline	1.00	0.10	-	0.08	-	
		Diesel	0.50	-	0.30	0.25	0.025	
BS VI	01 April 2020	Gasoline	1.00	0.10+	-	0.06	0.0045++	6.0×1011++
		Diesel	0.50	-	0.17	0.08	0.0045	6.0×1011

For catalytic converter fitted vehicles; * Earlier introduction in selected NCR regions; + Additional NMHC = 0.068 g/km; ++ Applicable only to vehicles using direct injection engines.

Vehicular emission methods came under public scrutiny when on-board measurements with a Portable Emission Measurement Systems (PEMS) recorded emission values higher than those measured on a chassis dynamometer. This prompted the introduction of additional legislation to bridge the gap between measurements on a chassis dynamometer in an engine laboratory and on road emissions performance. The Real Driving Emissions (RDE) is derived from the emission data measured by the PEMS when the vehicle is driven along a pre-determined route. The current BS VI norms call for around ~90 per cent of conversion of the major pollutants CO, HC and NOx. The RDE is part of the BS VI notification and currently the testing agencies are in an on-going data acquisition phase. Once the results are collated the test agency will define a conformity factor *i.e.* the ratio of the levels of pollutants in the RDE study over the levels measured in the laboratory driving cycle. When this regulation comes into effect in 2023, India will be on par globally with the developed world.

The mass emission measurements are generally measured in an engine laboratory where a vehicle is mounted on chassis dynamometer (Figure 9.2). Before the measurement, the vehicle is stabilized by "soaking" it for 12 h at a controlled temperature of 25° C. The vehicle is subsequently driven according a prescribed driving cycle. The exhaust gas is collected in one or more expandable plastic bags. At the end of the test, the stored gas is analysed with various analytical methods. The most commonly used analytical methods are non-dispersive infrared analyser for CO and CO_2, flame ionization detector for hydrocarbons, chemiluminesence detector for NOx, gravimetric analysis for particulate matter and paramagnetic detector for oxygen. The units for measurement of emission levels are g/km.

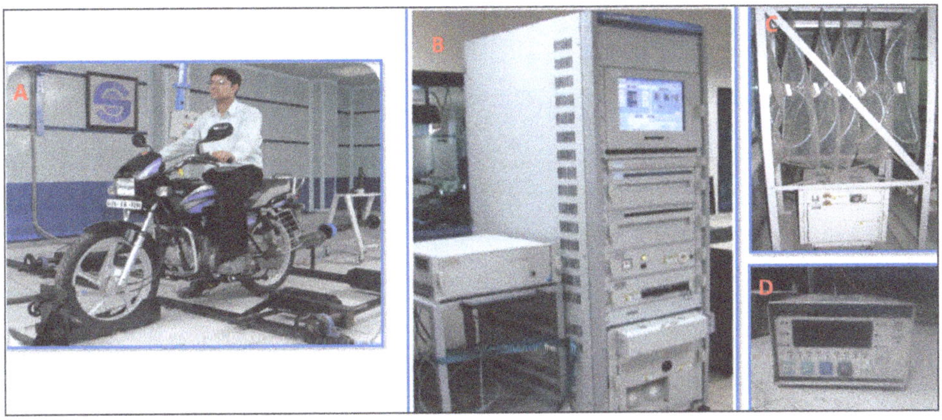

Figure 9.2: Typical Vehicle Testing Set-up and Instruments Used for the Measurement of Emission.

A) 2-W motorbike mounted on Chassis dyanometer, B) Horiba Mexa 7400D, C) Horiba CVS 7100 and D) Horiba Mexa 730 Lambda Analyzer.

The driving cycles used differ from segment to segment. A typical Worldwide Harmonized Motorcycle Test Cycle (WMTC) used for class 2.1 which is defined as engine capacity < 150 cm^3 and 100 km/h <Vmax< 115 km/h is shown in Figure 9.3 and the corresponding emission norms are shown in Table 9.2.

Table 9.2: Emission Standards for Two Wheeler Motorbike in India

Legis-lation	*Year*	*Vehicle class*		*CO*	*HC*	*NMHC*	*HC+ NOx*	*NOx*	*PM*
				g/km					
	1991	All Class	IDC	12-30	8-12		-	-	-
	1996			4.50	-		3.60		-
BS I	2000			2.00	-		2.00		-
BS II	2005.04			1.5	-		1.50		-
BS III	2010.04			1.00	-		1.00		-
BS IV	2016.04	Class 1 and 2-1	WMTC	1.40	-		0.79	0.39	-
		Class 2-2		1.97	-		0.67	0.34	
		Class 3-1 and 3-2		1.97	-		0.40	0.20	
BS VI	2020.04	All Class		1.00	0.10	0.068		0.06	0.0045*

* Gasoline direct injection only.

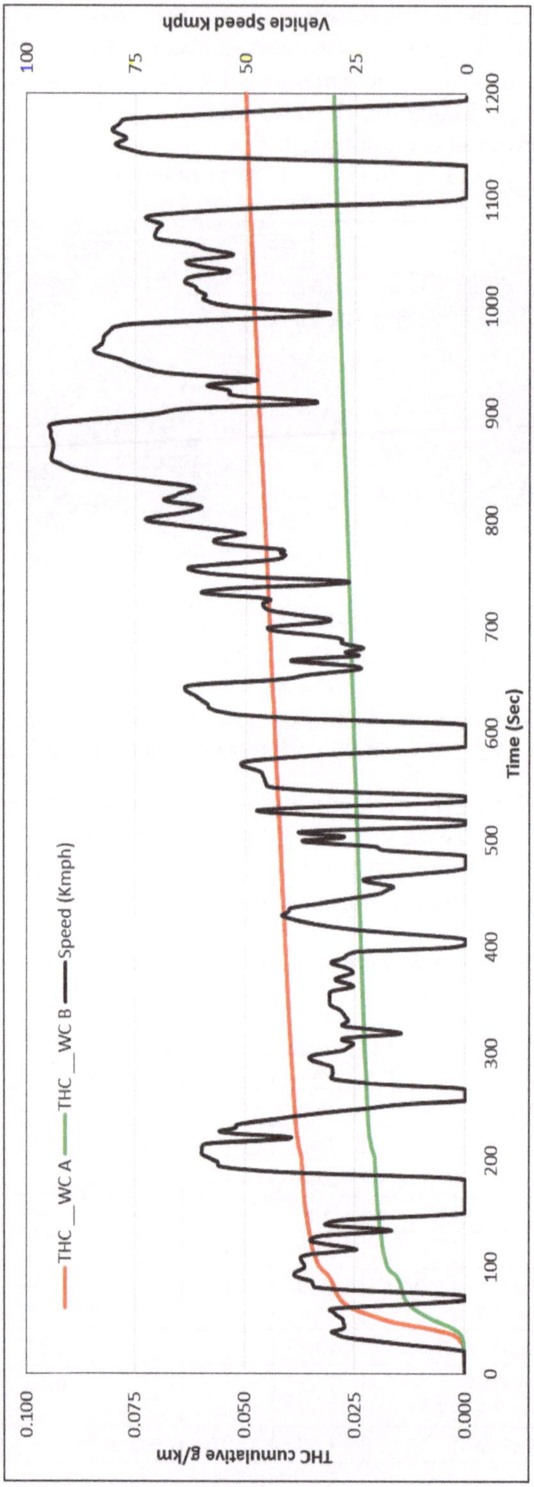

Figure 9.3: Worldwide Harmonized Motorcycle Emission Test Cycle (WMTC) with Cumulative HC as Function of Time.

9.3 Catalytic Converters for Emission Control

9.3.1 Monolithic Substrates

In early days of catalyst development of automotive emission control system, various catalysts and catalytic reactor designs were proposed. For example the bead shape catalytic reactor was commonly used in the USA until about 1993. However, high pressure drop, high attrition of catalyst and more complexity in the construction of catalytic convertor housing were the limitation of the bead shape catalytic reactors. As a result a completely different design of catalytic reactors were concomitantly developed, as in monolithic structures with various shape made from ceramics or metals.

Monolithic catalytic reactors have many small, parallel channels running axially through the part. The typical shape of the channel of ceramic substrates are square and triangular cells, while most metallic substrates have sinusoidal channels. The number of cells per unit frontal surface area, also usually measured as cells per square inch (cpsi), can vary between 10 and over 1200 cells. In the 1990s, ceramic substrates used in commercial applications had cell densities between 200 and 600 cpsi in which 400 cpsi was the most commonly used cell density in gasoline car applications. While increasingly higher cell densities even up to 1200 cpsi have been introduced in gasoline applications. Most substrates for diesel applications have cell densities in the range of 300-400 cpsi.

Figure 9.4: Image of Metallic and Ceramic Monolithic Substrates.

Metallic monolithic catalyst supports are made of thin metal foils that are composed of iron-chromium-aluminium alloys. This alloy materials provide good mechanical durability and resistance to thermal shock. The foils are corrugated to produce the honeycomb channels, typically resulting in a sinusoidal cell shape. To enhance turbulence in the flow of exhaust gas component within honeycomb channels, several structured monolithic substrates have been commonly used in commercial applications in gasoline as well as diesel applications. Major advantages of both ceramic and metallic monolithic supports include high geometric surface area per unit volume, large open frontal area (to reduce pressure drop), and excellent attrition resistance. Typical ceramic and metallic honeycomb substrates are shown in Figure 9.4.

The ceramic substrates most commonly used in the automotive emission control system is made of porous cordierite ($2MgO.\ 2Al_2O_3.\ 5SiO_2$), having approximate weight composition of 14 wt per cent MgO, 36 wt per cent Al_2O_3 and 50 wt per cent SiO_2. The porous cordierite substrate have large pores and very low specific surface areas of about 0.3 m^2/g, whereas foils used for metal substrates have no porosity. The channels of these substrates are coated with a thin layer of a mixture of inorganic materials which is commonly called as washcoat, providing a very high internal surface area to meet required conversion efficiency of exhaust gas components.

The washcoat increases surface area of monolithic substrates in the range of 50 to 200 m^2/g. Such high internal surface area is required for the adequate dispersion of the precious metals as well as for the adsorption of catalyst-poisoning elements present in the exhaust gas. The washcoat components not only support the catalytic functions of the precious metals but also play critical part in the catalytic reaction, *e.g.* in three-way catalyst, oxygen is stored in ceria-zirconia composite support materials when the exhaust is lean and released when the exhaust is rich, thus maintaining the gas composition closer to stoichiometry. For monolithic ceramic substrates, the typical washcoat layer has thickness of about 10 to 30 μm on the wall of the support, whereas the thickness in the corner of the support is about 100 to 150 μm (Figure 9.5).

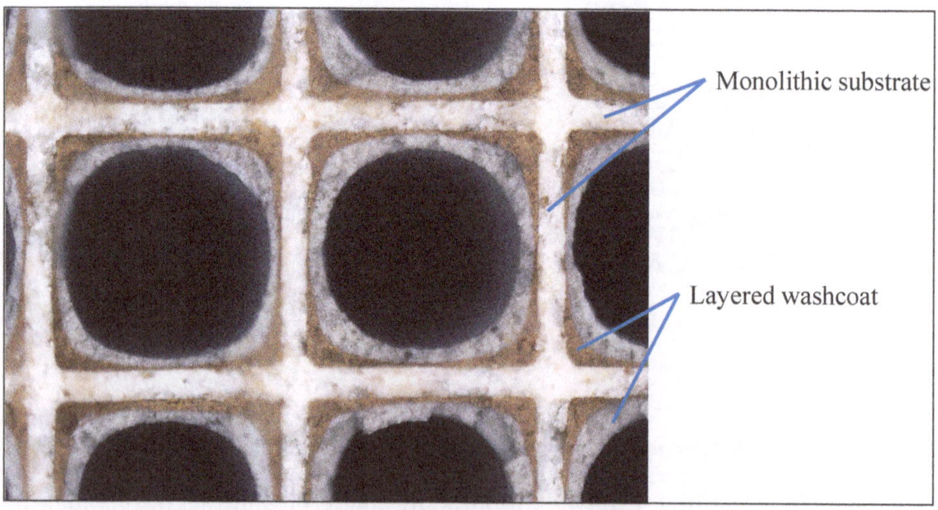

Figure 9.5: Washcoat on a Monolith Honeycomb.

9.4 Emission Control Technology

A. Three-Way Catalyst (TWC) for Gasoline Engines

In the last several decades, the concepts and the design of catalyst for gasoline engines have evolved to meet the increasingly stringent norms. The engine-out emissions of CO, hydrocarbons and NOx depends on the engine air-to-fuel ratio (A/F). For gasoline-powered engines, the A/F ratio at stoichiometric point is 14.7, where there is just enough air for the combustion of all hydrocarbons in the fuel. Nevertheless, the most common way to classify the engine-out exhaust gas composition is the lambda (λ) value,

$$\lambda = \frac{\text{Actual engine } \frac{A}{F}}{\text{Stoichiometric engine } \frac{A}{F}} \tag{1}$$

When the engine operates under excess fuel conditions, the exhaust gas will then contains more reducing agents such as CO and hydrocarbons than oxidizing agent such as NOx and oxygen. For such conditions the value of λ is less than 1. On the other hand, when the value of λ is more than 1, the engine operates under excess air conditions and the exhaust gas contains more oxidizing agent such as NOx and oxygen than reducing agents such as CO and hydrocarbons (See Figure 9.6).

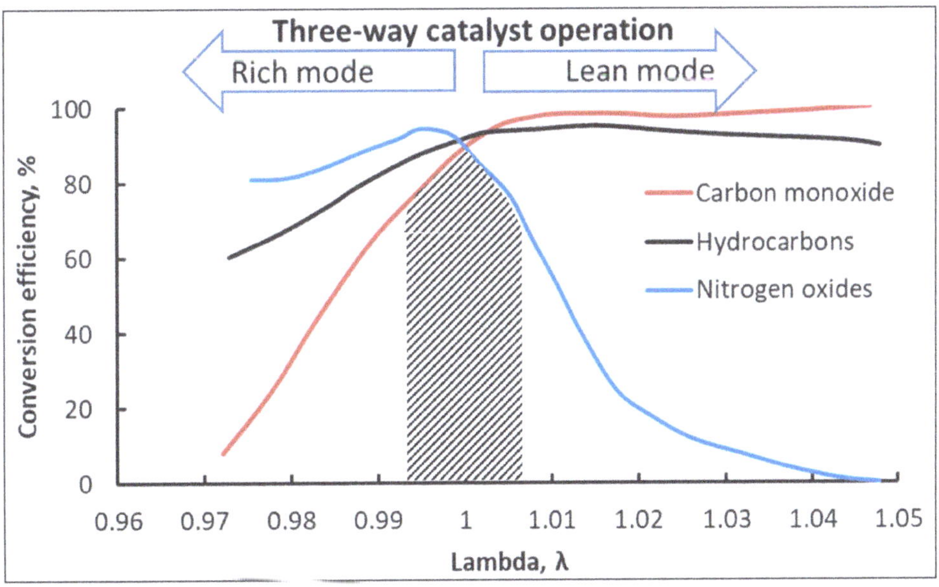

Figure 9.6: Basic Concepts for the After Treatment of the Exhaust from Spark-Ignition Internal Combustion Engines.

Depending on the operating conditions of the engine and the exhaust gas composition, the maximum efficiency of the three-way catalysts with conversion rates of the pollutants close to 100 per cent is achieved when the value of lambda is close to stoichiometric (lambda = 1) conditions. The main reactions that lead to the removal of the pollutants are CO and hydrocarbons oxidation reactions, NOx reduction reactions and water-gas and steam reforming reactions:

Oxidation Reactions

$$CO + 0.5O_2 \rightarrow CO_2 \tag{2}$$

$$C_3H_6 + 4.5O_2 \rightarrow 3CO_2 + 3H_2O \tag{3}$$

$$C_3H_8 + 5O_2 \rightarrow 3CO_2 + 4H_2O \tag{4}$$

$$H_2 + 0.5O_2 \rightarrow H_2O \tag{5}$$

Reduction Reactions

$$CO + NO \rightarrow CO_2 + 0.5N_2 \tag{6}$$

$$C_3H_6 + 9NO \rightarrow 3CO_2 + 3H_2O + 4.5N_2 \tag{7}$$

$$H_2 + NO \rightarrow H_2O + 0.5N_2 \tag{8}$$

Water-Gas and Steam Reforming Reactions

$$CO + H_2O \leftrightarrow CO_2 + H_2 \tag{9}$$

$$C_3H_6 + 3H_2O \leftrightarrow 3CO + 6H_2 \tag{10}$$

The extent to which each of the main reactions contribute to the removal of CO, hydrocarbons and NOx depends on the catalyst formulations and the catalyst operating conditions. Figure 9.3 illustratively shows the cumulative hydrocarbons emission along with driving cycle plotted for two different types of catalyst formations washcoat-A (WC-A) and washcoat-B (WC-B) with the same precious metal loading. Two important points can be observed from this plot:

1. More than 60 per cent hydrocarbon emission are eluted during the "cold start" in the first 100 second of driving cycle during which catalyst temperature is lower than its light-off temperature
2. Chemistry and architecture of the washcoats play significant role in controlling emission. Typical strategies adopted to optimize the washcoat are:
   a) Impregnate the precious metals on segregated wash coat components e.g Rh is impregnated on an alumina free wash coat to prevent adverse formation of rhodium aluminate
   b) Layering of the segregated washcoats where the Rh impregnated layer is placed in the upper layer so as to promote the reduction of NOx with CO before the CO can be oxidized to CO_2 on the bottom layer
   c) Zoning of the precious metal by loading higher levels at the front end helps in both lowering the light off temperature as well as reducing the precious metal inventory.

The chemical composition of the washcoat and its architecture forms part of the proprietary intellectual property of the manufacturers, but the most widely used materials are aluminium oxides, cerium-zirconium oxides as the major constituents.

Aluminium oxides provides high and stable surface area and thus also facilities a high dispersion of the precious metals. The typical range of surface area of various phases of γ-, δ-, and θ-aluminium oxides is 250 to 30 m^2/g^{-1}. The exothermicity of the oxidation reactions can raise the bed temperature in the catalysts to over 1000° C and therefore the thermal stability of the catalyst is important. Lanthanum is commonly added as dopant to the aluminium oxides. The addition of lanthanum/ barium remarkably improves the thermal stability, inhibiting sintering and phase transformations of the catalyst carriers. Similarly pore structure of support materials is also tailored to trade off the advantages of a high surface area with the need to limit intra-particle diffusion resistance. The Ceria-Zirconia mixed oxide performs several functions. The redox function of the ceria buffers the variability of the air/ fuel ratio by alternatively storing and releasing oxygen during the swing from lean to rich cycles of the engine. This enables the catalyst to perform within a narrower lambda window where the catalyst activity is optimum for both the oxidizing as well as the reducing function. The other functions of the ceria zirconia mixed oxide include:

a) Catalyse the WGS reaction for CO removal
b) Stabilizes the wash coat surface area against sintering at high temperature
c) Enhances the stability of the precious metal distribution because of the strong Metal Support Interaction (SMSI)

While cerium oxide is used for its redox properties, zirconia is intercalated into the ceria lattice to prevent sintering. The unique oxygen release and storage property of cerium-zirconium mixed oxide is further improved by select rare earth dopants such as Nd and Pr to more precisely suppress the oscillations of λ in exhaust gas within a narrow window and thus facilitating the catalyst to simultaneously perform both oxidation of CO and HC as well as reduction of NOx.

$$2CeO_2 \leftrightarrow Ce_2O_3 + 0.5O_2 \quad (11)$$

The oxygen storage and release ability widens the range of exhaust gas compositions at transient operation under which the closed-loop-controlled three-way catalyst has optimal conversion for CO, hydrocarbons and NOx simultaneously. The value of λ oscillates during operation with a typical frequency of 1 Hz and an amplitude of +/- 0.02. At values of λ, less than 1, the exhaust gas composition is marginally oxygen-deficient at which condition the cerium oxide partially reduces from Ce^{+4} to Ce^{+3} releasing oxygen to facilitate the conversion of CO and HCs. When the value of λ is more than 1, the exhaust gas composition becomes slightly oxygen-rich during which the partially reduced cerium oxide stores oxygen by reoxidising to Ce^{+4} and promotes the reduction of NOx with CO. The rate of oxygen storage and release allows the lambda ratio to remain near stoichiometric under which condition the catalyst performs optimally for both oxidation of CO and HC as well as reduction of NOx and significantly enhances the performance of TWC.

The commonly used precious metals in TWCs are platinum, palladium, and rhodium. The amount of precious metals and the mass ratio of platinum to palladium to rhodium depend on the specific application of the catalyst, and is also governed by factors such as the composition of engine-out emissions, the emissions targets to be reached, the catalyst operating conditions, and the properties of the fuels used.

In the past, most of the monolithic three-way catalysts contained platinum and rhodium, in mass ratios of about 5 to 20 : 1 Pt : Rh. The total precious metal loading is typically 0.9 to 2.2 g L^{-1} catalyst volume. With the innovative catalyst technology and the much-improved boundary conditions for the catalyst application – such as clean fuel with ultra-low sulphur content and advanced engine management systems – rhodium can be combined with either platinum or palladium to achieve a high performance catalyst. Some early designs of three-way catalysts used palladium together with platinum and rhodium, in which the palladium was a partial replacement of platinum. These catalysts are mainly used as light-off catalysts; these typically are small catalysts units located close to the engine outlet, and are used in combination with a main catalyst that contains rhodium together with platinum or palladium and which is located in the vehicle underbody downstream of the light-off catalyst.

The most common methods of incorporating PGM on the wash coat are:

- Equilibrium adsorption in excess solution
- Incipient wetness or pore volume filling
- Precipitation deposition

Typical emission recorded on commercial BS-VI 2W motorbike (class 2.1) with and without after treatment system are shown in Table 9.3.

Table 9.3: Typical Emission Recorded on Commercial BS-VI 2W Vehicle (class 2.1) with and without After Treatment System

Description	*Emission, g/km*		
Emission components	*CO*	*HC*	*NOx*
BS-6 limit	1.000	0.100	0.060
BS-6 limit (Deterioration factor)	0.769	0.077	0.046
Typical engine out emission of commercial 2W motorbike (~100cc engine)	1.508	0.346	0.256
Typical engine out emission with BS-6 catalyst on commercial 2W motorbike (~100cc engine)	0.178	0.033	0.007
Per cent of conversion efficiency	88.2	90.5	97.3

B. Diesel Catalysts

Compression-ignition engines differ from spark-ignition engines in terms of their air and fuel mixing, and ignition of the air/fuel mixture. Diesel engines provide better fuel economy and lower CO_2 emission. Typically these engines operate with lean burn combustion and contain high concentrations of oxygen in their exhaust gases at all operating conditions. Furthermore, the exhaust emission gas of diesel engines has a complex composition, as gaseous components are present together with liquid and even with solid components. The gaseous components of engine-out emissions are CO, hydrocarbons, NOx and SOx, whereas the typical solid component is a mix of soot and ash. The amount of particulate matters emitted is inversely proportional to the amount of NOx emitted.

CO and hydrocarbons from diesel exhaust gas emissions can be controlled with high efficiency by oxidation catalyst technologies. The hydrocarbons in the exhaust gas emissions may include organic fraction of diesel particulates or polynuclear aromatic hydrocarbons that condense on the soot particles and are measured as particulate matter (PM). The reduction of NOx from diesel exhaust gas emissions can commonly be controlled by either selective catalytic reduction using ammonia or lean NOx storage followed by rich reduction. Particulate matters are commonly captured on diesel particulate filters (DPF) and subsequently removed from the filter, continuously or periodically, through thermal or catalytic regeneration (Bhatt,2013; Daggolu, 2017). A generic schematic of diesel after treatment system is shown in Figure 9.7 and function of each catalyst is given in Table 9.4.

Table 9.4: Typical Catalysts Used for the After Treatment of the Exhaust from Compression-Ignition Internal Combustion Engines

Catalyst Technology	*Function*
Diesel Oxidation Catalyst	Oxidation of CO, hydrocarbons, oxidation NO to NO_2
Diesel Particulate Filter	Storage of particulate matter
Selective Catalytic Reduction	Reduction of NOx
Ammonia Oxidation Catalyst	Oxidation of excess ammonia

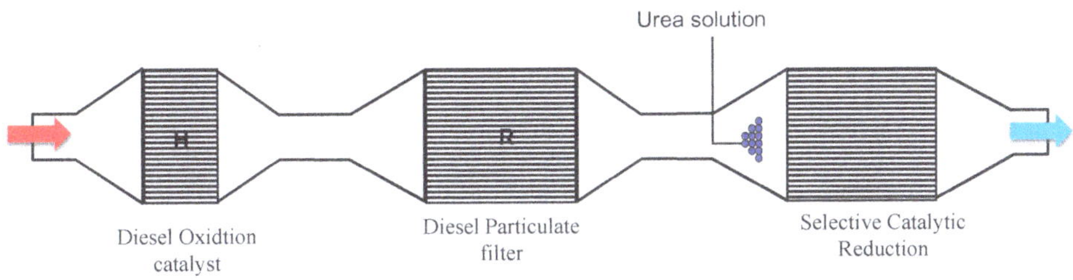

Figure 9.7: Basic Concepts for the After Treatment of the Exhaust from Compression-Ignition Internal Combustion Engines.

C. Diesel Oxidation Catalyst (DOC)

DOC promotes the oxidation of CO, hydrocarbons and organic fraction of particulate matters, commonly called soluble organic fraction (SOF). Moreover, several aldehydes and poly-aromatic hydrocarbons derived from diesel fuel via partial oxidation during engine combustion can also be oxidized along with a reduction or elimination of the odour of diesel exhaust. Following are the most common processes that take place over DOC:

$$CO + 0.5O_2 \rightarrow CO_2 \quad (12)$$

$$C_nH_{2m} + [n + m/2]O_2 \rightarrow nCO_2 + mH_2O \quad (13)$$

$$2SO_2 + O_2 \rightarrow 2SO_3 \quad (14)$$

$$SO_3 + H_2O \rightarrow H_2SO_4 \quad (15)$$

$$NO + 0.5O_2 \rightarrow NO_2 \quad (16)$$

$$C_{(s)} + NO_2 \rightarrow CO + NO \quad (17)$$

$$CO + NO_2 \rightarrow CO_2 + NO \quad (18)$$

The first two chemical reactions represents oxidation of gas phase CO and hydrocarbons, including SOF compounds. Due to oxidation capabilities of DOC catalysts, sulphur dioxide (SO_2) in diesel exhaust gas emission is oxidized to sulphur trioxide (SO_3). SO_3 then combines with water vapours in the exhaust stream to form sulphuric acid as shown in reaction 15. As the exhaust temperature decreases on the way to exit through tailpipe, the gaseous sulphuric acid combines with water to form liquid particles composed of hydrated sulphuric acid called sulphate particles. These particles condense on the soot and contribute in increasing the total PM emissions from diesel engines. Another important reaction in DOC applications is oxidation of NO to NO_2 which is known as excellent oxidizing agent. Nitrogen dioxide can efficiently oxidize carbon particles of PM emissions and hence commonly used for regeneration of DPFs. Designing DOCs with controlled NO_2 production are desirable for the operation of DOC/DPF systems.

The most basic DOC formulation was platinum on alumina washcoat (Pt/Al_2O_3). Nevertheless, the modern DOC commercial technologies are based on Pt, Pd, or their mixtures with certain ratio. Palladium based catalysts can be used for the oxidation of HCs and SOF; whereas, platinum-based formulations are used whenever NO_2 generation and/or the oxidation of CO and HC is required, particularly at low exhaust temperatures. Optimized Pt/Pd formulations are also commonly used in DOC catalysts for NO_2 generation in DPF and SCR systems.

D. Diesel Particulate Filter (DPF)

This is a device designed to physically capture particulate matter from the exhaust of a diesel engine. In addition to soot the particulates also contain approx. 20 per cent of the higher hydrocarbons and sulphates that condense as liquid on the soot particles, and are reported as PM. They are normally made of ceramics honeycombs with channels that are plugged at alternate ends as shown in the Figure 9.8. The exhaust gas flows through the walls, thereby providing a large filtration area. As the porous walls accumulate the soot and ash particles, the pressure drop across the filter increases and therefore the DPF is fitted with a monitoring mechanism that can sense the need for periodic regeneration. The soot particles are retained by a combination of depth filtration inside the filter pores as well as surface filtration along the channel walls. The soot load in the DPF increases the filter efficiency but increases the pressure drop.

Active Regeneration

This is carried out by increasing the temperature above 500 °C to burn of the soot with oxygen. As these temperatures are not available, use is made of a diesel burner to directly heat the exhaust entering the DPF or use the DOC to oxidise fuel over the catalyst. This process typically requires fuel dosing into the exhaust ahead of the DOC for approximately half an hour.

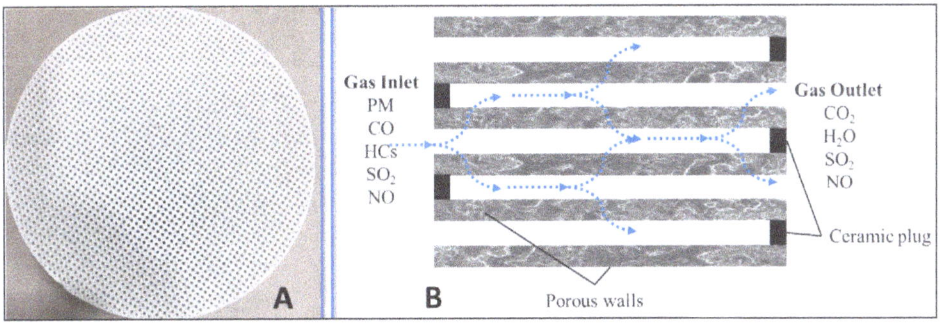

Figure 9.8: Ceramic Diesel Particulate Filter (DPF).
(A) Top view (alternate channels are blocked), (B) Schematic of functioning DPF.

Passive Regeneration

This strategy relies on the oxidation of soot with NO_2 which occurs at a much lower temperature of 250 to 400 °C. This requires a catalyst which can be loaded on the filter as in catalysed DPF or an upstream DOC that is designed to maximise NO conversion to NO_2.

Most systems are designed with a combination of active and passive regeneration and leave the driver with a choice depending on the temperatures reached during the driving condition. While the soot can be burned off the ash collected by the filter is not combustible and remains in the filter until it is physically blown away during periodic servicing.

E. NOx Reduction Technologies

Diesel engines when designed to operate at higher temperature give very low PM, however this engine modification leads to higher NOx emission. The two technologies that are adopted are:

a) Selective Catalytic Reduction
b) Lean NOx Trap (LNT)

Both approaches have been shown to be technically feasible to reduce the NOx level up to 90 per cent.

a) Selective Catalytic Reduction (SCR)

As the name implies this catalyst selectively reduces NOx by reacting it with ammonia in the presence of oxygen. The ammonia is sourced from a dilute solution of urea which is hydrolysed to ammonia over the catalyst.

The main reactions involved in selective catalytic reduction are as follows:

$$4NH_3 + 4NO + O_2 \rightarrow 4N_2 + 6H_2O \text{....Standard} \quad (19)$$

$$4NH_3 + 2NO + 2NO_2 \rightarrow 4N_2 + 6H_2O....\text{Fast} \quad (20)$$

$$8NH_3 + 6NO_2 \rightarrow 7N_2 + 12H_2O....\text{Slow} \quad (21)$$

$$4NH_3 + SO_2 \rightarrow 4NO + 6H_2O \quad (22)$$

$$4NH_3 + 4O_2 \rightarrow 2N_2O + 6H_2O \quad (23)$$

For an ammonia-SCR system, various commercial catalysts, such as Pt/metal oxide, base metal/zeolite and vanadium-tungsten-titanium (VWT) systems are available. Typical NOx conversion over VWT, Cu/ZSM-5 and Fe-SZM-5 is shown in Figure 9.9. The temperature window for the Pt-based catalysts is very narrow (175 to 250 °C). At less than 200 °C exhaust temperature, undesirable ammonium nitrate is formed whereas above 225 °C the catalyst loses its selectivity for reduction and oxidation reactions of NH_3 to NO_x and N_2O are dominant.

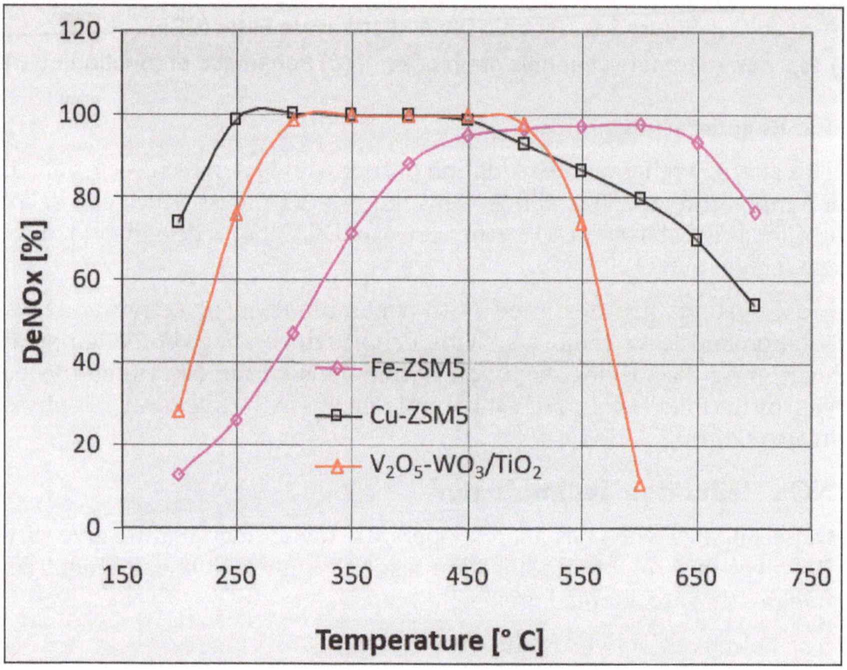

Figure 9.9: NOx Conversion over Various SCR Catalysts by Ammonia as Reducing Agent.

The VWT technology is recognized to be the most robust catalytic system for an ammonia-SCR. It is highly active at medium temperature range (230 to 450 °C) and has superior resistance against sulfur poisoning. V_2O_5 in a highly dispersed form on TiO_2 is the catalytically active system whereas WO_3 improves the thermal stability of TiO_2, increases the acidity of the catalysts and also suppresses the undesirable oxidation of SO_2. The catalyst temperature in the commercial system generally depends on factors such as, the distance of the catalyst from the engine, presence of an upstream particulate filter that could generate heat during the carbon burn out *etc.* VWT SCR technology is widely used where the NOx conversion requirement is lower than 80 per cent. Due to their resistance to sulfur poisoning, VWT SCR technology is preferred for diesel after treatment systems in emerging markets, particularly marine application.

Similarly, base metal (Cu or Fe) zeolite catalysts are commercially available. The Fe-based zeolite catalyst exhibits excellent activity in the high temperature regime, however the low temperature activity for the NOx conversion over Fe-zeolite is inferior (See Figure 9.9). Recently the Cu-based zeolites, particularly Cu-SSZ-13, has become more attractive due to its wide working temperature range and better durability. The main concerns of the base metal/zeolite catalysts for the SCR reaction were sulfur poisoning and thermal durability. The impact of sulfur is more severe, particularly on Cu-based catalysts than on Fe-based zeolite catalysts for the NOx activity. However, due to the availability of fuel that has less than 10 ppm sulfur for BS-VI application, the utilization of Cu-containing SCR catalysts becomes feasible for the after treatment systems. Typical catalytic performance for the conversion of NOx over Cu-SSZ-13 for fresh and hydrothermally aged monolithic catalysts is shown in Figure 9.10. It can be clearly seen that the conversion of NO_x over fresh and 700 °C hydrothermally aged Cu-SSZ-13 catalyst reaches more than 90 per cent conversion at 250 °C. Even after hydrothermally aging at 850 °C, the catalyst used in this study showed high NOx conversion rate and excellent stability.

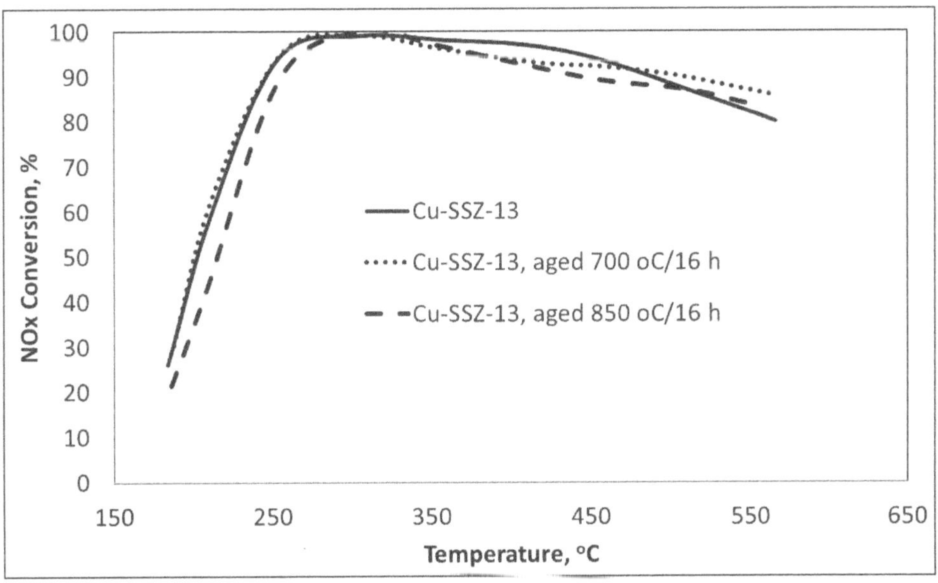

Figure 9.10: Typical Performance of Ammonia-SCR Cu-SSZ-13 Catalysts Measured on Model Gas Test Reactor with Monolithic Catalyst.

b) Lean NOx Trap

The SCR reaction requires an on-board tank to store urea as well as a mechanism for controlled injection of urea up stream of the catalyst. While this has been accepted for implementation on large trucks, space limitations in passenger cars has made the LNT a preferred option.

This system operates cyclically under oxidising and reducing atmosphere. During normal operation the engine operates in the lean mode where the NO is oxidised to NO_2 and is trapped by an alkali metal such as barium or strontium as nitrates. This has to be subsequently released during a brief regeneration cycle with an exhaust containing reducing gases such as CO, HC and H_2. This reducing atmosphere is generated by injection of fuel up stream of the diesel oxidation catalyst and involves a penalty of extra fuel consumption.

The catalyst design is typically formulated with both the alkali metal to trap the NO_2 as nitrate during lean operation and a precious group metal to convert the NOx to nitrogen when the fuel mixture is rich.

9.5 Conclusion

Since the introduction of legislation to control pollution from automobile exhausts the technology has come a long way and the polluting components have been reduced by a factor of more than 10x. A major step was adoption of BS VI in April 2020 bringing India on par with emission norms in rest of the world. However in view of the increasing density of vehicles a lot remains to be done. The contribution of the transport sector is still a matter of concern as well as a serious technical challenge.

Currently 22 people out of 1000 people own a car in India, compared to 980 persons and 850 per thousand individuals in the US and UK respectively (Narkhede, 2015). With a focus on urbanization the passenger car ownership is projected to grow to 175 per 1000 people by 2040.

The challenge for meeting the air quality standards will require coordinated efforts from a broad range of approaches that will include fuel quality, advanced engine design and management systems, transport system management and more efficient state of the art after treatment systems. While technical feasibility to achieve near zero emissions with catalytic converters has been demonstrated its economic viability requires in depth investigations into a complex web of interdependent parameters. The general efforts are directed towards reducing the precious metal inventory so as to ensure economic viability with a particular emphasis on achieving low light off temperature during the cold start which contributes near 60 per cent of total emissions as measured on the chassis dynamometer. To a large extent this is being achieved by developments in materials science of the washcoat in combination with innovative washcoat architecture and design of the substrate geometry.

The next stage of emission controls are the Real Driving Emissions which are scheduled to be adopted in April 2023 when the on road emissions as measured by the Portable Emission Measuring Systems will need to conform to BS VI norms. The opportunity lies in developing more intricate after treatment systems that can be suitably combined with better engine out emissions to significantly reduce the real emissions without an unacceptable cost escalation.

Acknowledgements

The authors gratefully acknowledge the support and encouragement of the management of Süd Chemie India Pvt. Ltd, and thank them for permission to present/publish this work.

REFERENCES

Lambert, C.K. (2019) Current state of the art and future needs for automotive exhaust catalysis. *Nat Catal* 2, 554–557. https: //doi.org/10.1038/s41929-019-0303-x.

Egbert S. J. Lox, (2008) "Automotive Exhaust Treatment", Handbook of Heterogeneous Catalysis 2nd Edition, pp. 2274-2345. https: //doi.org/10.1002/9783527610044.hetcat0120

NOAA (2020) National Centers for Environmental Information, State of the Climate: Global Climate Report for August 2020, published online September 2020, retrieved on October 4, 2020 from https: //www.ncdc.noaa.gov/sotc/global/202008 https: //en.wikipedia.org/wiki/Super_ultra-low_emission_vehicle.

Daggolu P.R., Gogia D.K., and Siddiquie T.A. (2017) Exhaust After-Treatment System for Diesel Locomotive Engines–A Review. In: Agarwal A., Dhar A., Gautam A., Pandey A. (eds) Locomotives and Rail Road Transportation. Springer, Singapore. https: //doi.org/10.1007/978-981-10-3788-7_8.

Bhatt, D., Waje, S., Babu, K., and Henn, J. (2013), "Achieving BS-IV Emission Targets for Diesel Small Commercial Vehicle (SCV) with Close Coupled Advanced EnviCat® DOC and Coated PM Metalit® After - Treatment System," SAE Technical Paper 2013-26-0054, https: //doi.org/10.4271/2013-26-0054.

Narkhede, V., Kumar, D., Cursetji, R., and Siddiquie, T (2013), "VWT - SCR Catalyst to Meet BS-IV Emission Standards and Beyond," SAE Technical Paper 2013-26-0056, 2013, https: //doi: 10.4271/2013-26-0056.

Narkhede, V., Kumar, D., Cursetji, R., and Siddiquie, T. (2015) "Low Temperature Selective Catalytic Reduction of NOx over Vanadium-Based Catalysts," SAE Int. J. Engines 8(1): 380-385, https: //doi.org/10.4271/2015-26-0108.

https: //auto.economictimes.indiatimes.com/news/passenger-vehicle/cars/india-has-22-cars-per-1000-individuals-amitabh-kant/67059021

Chapter 10

New Emerging Issue: Solid Waste Management from Hotel Industry in Vadodara City

Nisarg Bhatt and Deepa Gavali

Gujarat Ecology Society, Subhanpura, Vadodara – 390 023, Gujarat
e-mail: deepa.gavali@gmail.com

ABSTRACT

Hotel industry, is one of the largest sector in the hospitality section and is on increasing trend in recent years. Thereby, making them one of the major contributor to municipal solid waste generated. Today using the green strategies like recycling, waste minimization can provide a great benefit to hotel industry and environment overall. Therefore, the present study was undertaken to understand the quantum of waste collection and disposal practices of hotels in Vadodara city. Questionnaire based survey was undertaken to understand the quantum of the problem. The survey result indicated that hotels rely more on the civic bodies i.e., Vadodara Municipal Corporation (VMC) for daily waste collection. A major hurdle in the solid waste segregation was observed as 93 per cent of the hotels did not have separate staff to undertake the segregation and the waste is disposed directly into the collection vans. The concept of recycling is poor in the hotels with regards to amenities provided. The concerns towards the environmental issues especially contribution to climate change is very low.

Thus, the study brings out the lack of advantages in recycling of solid waste that would not only bring monetary gains but also help in reduction of waste dumped into the landfills. Proper solid waste management practices if followed by hotel industry can prevent adverse impact on environment and thereby reduce pollution. Also, the haphazard construction of hotels has led to encroachment on the banks of river Vishwamitri adding to the water logging issues in the city.

***Keywords**: Hotels, Solid waste management, Plastic, Vadodara, Tourism, Landfill site.*

10. 1 Introduction

Industrialization becomes very significant for developing countries like India with a large population. The process of industrialisation with the demand for cities and infrastructure facilities and amenities has led to boom in hotel industry. The Indian tourism and hospitality industry has emerged as one of the key drivers of growth among the services sector in India. The third-largest sub-segment of the services sector comprising trade, repair services, hotels and restaurants contributed nearly US$ 187.9 billion or 12.5 per cent to the Gross Domestic Product (GDP) in 2014-15, while growing fastest at 11.7 per cent Compound Annual Growth Rate (CAGR) over the period 2011-12 to 2014-15. Tourism is a potentially large employment generator besides being a significant source of foreign exchange for the country. The industry is expected to generate 13.45 million jobs (Omidiani and Hashemi Hezaveh, 2016). Hotel and restaurant industry is one of the major players in the growth of Tourism as a whole in India. Hotels are the massive producer of waste. The hotel sector has been identified as the largest consumer of durable and non-durable goods resulting in large amounts of waste generated (Matai 2019). The environmental impacts of hospitality services are different than those of manufacturing and other industry sectors and this necessitates an innovative approach to develop sustainable tourism (Silvestrelli, 2013; Sridhar, 2011). The expansion of the hotel industry today is suspected to have an impact on the environment, especially in terms of the amount of resources it consumes, and the amount of waste generated (Hsiao *et al.*, 2014). Typically, a hotel guest can produce 1 kg of waste per day (Matai 2019). Ferris (1995) reported that the variables like government regulation, storage capacity, beliefs of the organization, labor cost and others are responsible for the amount of wastage by hotel industry. Shanklin (1998) stated that effective strategy and practices can reduce the amount of waste. There should be an effective kind of disposal method for waste food. Todd and Hawkins (2001) explain that hotels that can reduce the amount of waste generated can reduce the costs associated with disposal, transportation and labor. Matai (2019) reported that hotels do not focus on waste management and have low level of sensitization towards environmental concerns. A focused sustainable waste management plan will decrease the pressure on landfills, reduce the risk of leaching and will benefit the management through cut in the waste disposal costs. The process of solid waste management at four and five star hotels in Bali showed a positive and significant influence to employee satisfaction and the hotel guest satisfaction. However, solid waste management in four-star and five-star hotels has no significant influence to the Hotel's financial benefits (Promon 2019). Radwan *et al.* (2010) discussed waste management practices in small hotels in UK and the role of Environmental Protection Act (EPA) 1990 to institutionalize the services of authorized waste carriers. Improvising upon the traditional hierarchical model of waste management, the authors have proposed composting as an important method where biotic wastes from hotels would flow directly to the composting facility.

Vadodara city is one of the most important urban centres in Gujarat with rapid expansion, spread over an area of 158 sq. km and comprises of 13 administrative wards under Municipal Corporation with total population of 21 Lakhs as per 2011 census. The Vadodara Municipal Corporation (VMC) manages 750 tonnes of solid waste of the generated every day. Household contribute 60 per cent of the waste, followed by commercial places like offices, markets and hotels (22.5 per cent) and the remaining 18 per cent includes the construction and industrial waste (Makwana, 2018) (Figure 10.1). The city has come to limelight in wake of its nearest proximity to Statue of Unity with road, air and railway with rest of India. This will benefit the city in terms of tourist inflow and the present study was undertaken to estimate the existing scenario of hotel waste management. The basic questions behind the study was - is the city prepared to handle the inflow of tourist and effective solid waste management?

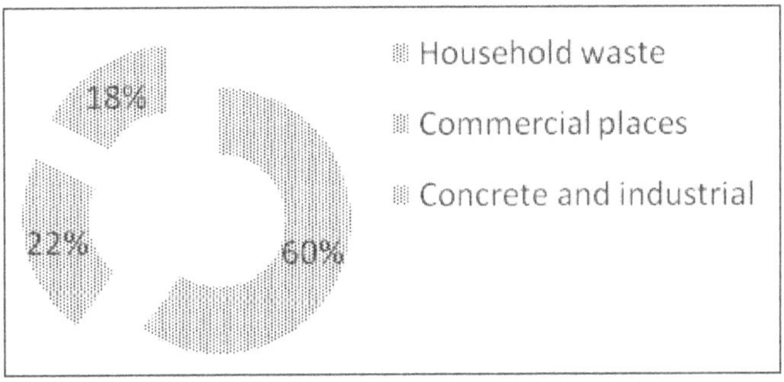

Figure 10.1: Different Categories of Solid Waste Generated in Vadodara City.

10.2 Material and Methods

There are more than 200 hotels in the city out of which only 135 are registered. However, this number excludes the eating outlets of the city. In order to access the solid waste management (SWM) practices in Vadodara city a questionnaire based survey in 74 hotels was carried out through one to one interaction with the owner or hotel manager. The questions enquired included the quantum of solid waste generated, the SWM practices followed, the hurdles for better SWM facilities within the city, practices followed by the hotel to reduce waste and options like recycling or composting practices. All data were collated and statistically analysed using SPSS version 14.

10.3 Results and Discussion

The survey results showed that solid waste generated is less than 100 kg/month/hotel in 60.8 per cent, while 29.7 per cent of hotels produced waste in the range of 100 to 500 kg/month/hotel (Figure 10.2). Only 8.1 per cent of hotels generated waste to the tune of 500 – 1000 kg/month/hotel and only one hotel had a waste generation of more than 1000 kg/month. The quantum of solid waste generated at present is low but tourist pressure is bound to increase due to tourism activity in Rajpipla. Vadodara is the nearest airport with connectivity from all over India and the City caters to the tourist inflow of 10 per cent of the state (Shukla 2014).

On type of solid waste generated in the hotels of Vadodara city, plastic represents 44 per cent followed by paper around 24 per cent, food (20 per cent)

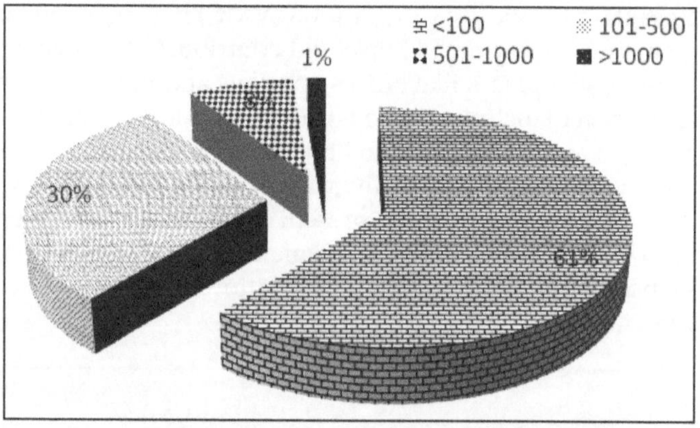

Figure 10.2: Quantum of Waste Generated in the Hotels Surveyed of Vadodara City.

and 6 per cent waste from packaging and green waste. Other waste included 3 per cent glass waste, 1 per cent cardboard and aluminum waste and milk-bag waste (Figure 10.3). Plastic waste is the most generated solid waste and 85.1 per cent of the hotels surveyed generated plastic waste. Plastic waste is major component in solid waste generated in hotels of other countries as well *viz.*, Vietnam (Trung and Kumar 2003), United Kingdom and Pennsylvania (Singh *et al.*, 2015). Irrespective of ban on plastic imposed by GOI, there is a long way to go before 100 per cent ban is achieved.

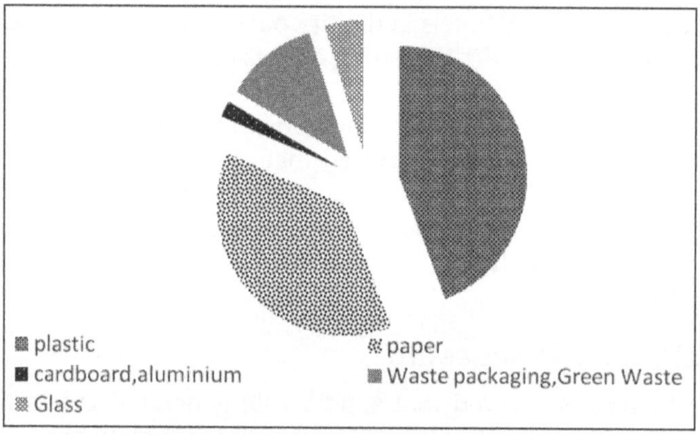

Figure 10.3: Type of Solid Waste Generated in the Hotels Surveyed in Vadodara City.

In terms of mechanism adopted for solid waste disposal surveye indicated that about 66.2 per cent of the hotels rely on the civic bodies for disposing their waste and 5.4 per cent hotels sell their recyclable waste to the concerned vendors (Figure 10.4). Whereas, 28.4 per cent hotels use both the modes of waste disposal *i.e.*, selling to recyclers and disposing through the civic authorities. The study clearly indicated the heavy dependency of hotel industry on civic authorities to dispose of the waste.

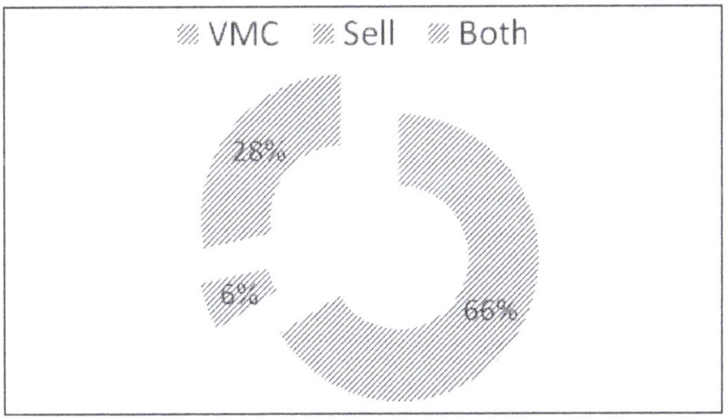

Figure 10.4: Mode of Disposal Adopted by the Hotels in Vadodara City.

The waste collected by municipal authorities is finally dumped into the land fill site, where there are issues of leaching risk into the groundwater. Another issue linked to this is the overloading of landfill sites and non-availability of space for creation of new landfill site for the urban centres. Thus, the best option available is to reduce the quantum of waste reaching the landfill site. Small steps taken by the hotel industry in reducing the waste generated will prove a long way in improving the environment.

Through the survey it is learnt that 89.2 per cent of the hotels do not pay any extra charges towards disposal of the waste into the landfill site. Once in the landfill site accidental fires can lead to major health hazard to the locals residing in the close vicinity. In 2019, Vadodara city reported two major fire incidences in the gap of 15 days and dangerous fumes were released into the atmosphere. There is need to avoid such incidences and protect the citizens from dangerous exposures.

Survey results indicted reluctance by 59 per cent of the hotels to switch over to eco-friendly mode of providing basic amenities in the form of refillable soap and shampoo dispensers. There exists lack of general awareness on environmental impact in the hotel industry and lack of awareness at large. There is need to bring in strict norms towards the use of refillable soap and shampoo dispensers for sustainable development.

The study revealed that there are severe lapses in the rule as the hotel industry comes under the purview of tourism department and not covered under the Department of Climate change and Environment. There is need to create awareness on the environment law to the tourism department and work in coherence to avoid future mishaps. Following are simple steps that can be followed to reduce the quantum of waste reaching the landfill site:

- Recycle and reuse of waste be promoted 100 per cent
- Promotion of energy saving measures.
- Usage of Copper and Clay Bottles.
- Composting of organic waste.
- Penalty can be imposed as imposed in different cities are listed below:

1.	EDMC (Delhi):	Rs. 5,000/- for dumping waste illegally.
2.	Chandigarh:	Dumping at any other public place Rs. 500/- fined
3.	Pune dumping waste in open spaces, riverbeds, canals and quarries.	Rs 25,000/-
4.	Chennai:	Rs. 2000/-
5.	Thiruvanthapuram:	Rs. 5000/- per M T
6.	Shillong:	Rs 5000/-
7.	Ghaziabad:	Rs 5,000/-
8.	Amravati:	Rs. 1000/-
9.	Vijayawada:	Rs 10,000/-
10.	Surat:	Rs. 100/- to 500/-

Other facts that were revealed in the study include that the staff or employee are not sent to attend environmental training. There is very poor enforcement of environmental laws and policy in the hotel industry and regular monitoring is absent.

10.4 Summary

Being one of the most populous country and fastest growing economy, India cannot afford to ineffective solid waste management. It seems that policy framework is available only on paper, but ground reality is alarming one. There is need to create awareness about the angle to the hotel industry of Vadodara and ensure better SWM practices. This will also ensure low garbage actually landing into the landfill site. With proper trained staff, the hotel can generate revenue and help in reducing the volume of solid waste reaching the landfill site

The study discussed in paper shows that waste generated in hotels of Vadodara city is on higher degree as compared to elsewhere. The quantum of aluminium waste and glass generated in the Vadodara hotels was low as compared to other cities. Awareness about composting was low among the hoteliers. Financial hurdles and lack of interest among hotels was identified as a major road block towards implementation of efficient solid waste management practices in the city. The recycling of solid waste can be promoted through campaigns that will encourage all the hotels to co-operate in waste separation and to purchase recycled products. The government should make separate rules for the Hotels as hotels are a major contributor to the GDP of the country as well as Solid waste. EMS certification and its popularity among the hotels should be attempted.

Hotels need to orient the customer regarding the best practices of disposal of waste generated during their stay. The government and non-governmental organisations (NGOs) should promote to reduce waste by awareness through promoting more programmes and campaigns continuously to all types of hoteliers, guests and also to the local communities. The most recent initiative is 'Swachchh Bharat Abhiyan', also known as 'Clean India Mission'. One of the objectives of this initiative is to aware citizens about the importance of proper waste management approach. The aim is to involve the citizens in the campaign of Clean India and to clean urban cities and also rural areas with public support. However, there is a need to ensure that all citizens are taken into confidence and involved morally to this Clean India Mission.

Acknowledgements

The authors are thankful to the respondents for sparing their valuable time during the process of documentation. Due acknowledgment goes to the Vadodara Municipal authorities for sharing the information on solid waste management.

REFERENCES

Dalem, A.A.G.R. (2002) Environmental Management System, Tri Hita Karana and its Implementation on Hotel. Study Group of Ecotourism, UNUD Bukit Jimbaran Bali and Postgraduate Program of Tourism Studies, Udayana University, Denpasar. http: //isjd.pdii.lipi.go.id/admin/Journal/81085862. pdf, downloaded on March 29, 2012.

Ferris, D.A. (1995). A comparison of methodologies used for waste characterization in foodservice operations, Dissertation. Kansas State University, Manhattan, KS.

Hsiao, T.Y., Chuang, C.M., Kuo, N.W. and Yu, S.M.F. (2014) 'Establishing attributes of an environmental management system for green hotel evaluation', *International Journal of Hospitality Management*, 36.

Makwana S. K. (2018), "Solid waste management of Vadodara city- feasibility study", *International Journal of Advance Research and Innovative Ideas in Education (IJARIIE)*, 4, (5), ISSN(O)-2395-4396.

Matai K (2019). Sustainable Tourism: Waste Management Issues. *Journal of Basic and Applied Engineering Research*. July 2019.

Omidiami A and S Hashemihezacech (2016), Waste Management in Hotel Industry in India: A Review, *International Journal of Scientific and Research Publications*, College of Architecture, BVP University, Pune, 6, (9): ISSN 2550-3153.

Pramon Jaya (2019). The Effect of the Solid Waste Management Toward Employee and Guest Satisfaction, also Financial Benefits at Four and Five Star Hotels in Bali. https: //www.researchgate.net/publication/339375176. 10 pp.

Radwan, H.R., Jones, E. and Minoli, D. (2010). Managing solid waste in small hotels. *Journal of Sustainable Tourism*, 18(2): 175–90. doi: 10.1080/09669580903373946

Shanklin, C.W. (1998), The role of dietitians in environmental protection. In Proceedings: The 2nd Asian Congress of Dietetics, pp. 104-109, Asian Dietetic Assoc., Seoul, Korea.

Singh, N., Sundari, S. K., and Nath, A. (2015). Contribution of Hotel and Restaurant Industry to Solid Waste: An Assessment of Its Impact on Environment. *Journal on Management*, 10(1).

Silvestrelli, P. (2013). Tourism development through heritage enhancement and hospitality innovation. *Int. J. of Globalisation and Small Business*, 5(1/2): 20–33.

Shukla P. K. (2014). "A study of tourism in Gujarat: a geographical perspective". Ph.D. Thesis Submitted to The Maharaja Sayajirao University of Baroda, Vadodara.

Todd, M. and Hawkins, R. (2001) Waste counts: a handbook for accommodation operators. Centre for Environmental Studies in the Hospitality Industry. Oxford: Oxford Brookes University.

Trung, D.N. and Kumar, S. (2005). Resource use and waste management in Vietnam hotel industry. *Journal of Cleaner Production*, 13 (2), 109-116.

Chapter 11

Bioremediation: An Efficient and Emerging Tool for Degrading Petroleum Hydrocarbons in Soil

Siddiqui Shagufta and Jadeja Rajendrasinh

Department of Environmental Studies, Faculty of Science, The Maharaja Sayajirao University of Baroda, Vadodara – 390 002, Gujarat
e-mail: siddiqui.shagufta-env@msubaroda.ac.in, rjadeja-chem@msubaroda.ac.in

ABSTRACT

The modern world is heavily dependent on crude oil as a primary source of energy. Due to advancement in technology, the use of petroleum is not just limited to energy. Today, most of the petroleum end products and by-product are used as industrial raw material or in construction. Ranging from paints, plastics, cosmetics to household cleaning products; petroleum has become indispensible part of human life. However, due to easy availability and extensive use; the problem of improper disposal has also arisen. The issue is addressed by means of various conventional methods at industrial scale but lately, the technique of bioremediation is gaining popularity across the globe because of eco-friendly approach towards the problem. Numerous studies have employed bioremediation to degrade petroleum and its components and those have achieved considerable success. The present chapter is a review on types of bioremediation and documentation of different methods and techniques that have potential to remediate petroleum hydrocarbon, focusing in detail upon various factors that generally affect the optimum bioremediation. Land is a vital resource and also most prone to oil contamination during processing, transport and disposal of petroleum and its products. Hence, the authors have made an effort to bring into perspective the holistic approach towards effective bioremediation.

Keywords: *Petroleum hydrocarbon, Bioremediation, Eco-friendly, Contamination, Sustainability.*

11.1 Introduction

Petroleum refining and transportation generally generates considerable amounts of a waste residue; known as oily sludge. One of the main issues faced by refineries and petrochemical industries is related to the safe disposal of this residue, since its destination and/or inappropriate treatment can cause serious impact to the environment and is a potential risk to general human health (Cameotra and Singh, 2008; Xu and Lu, 2010). Usually, this oily sludge contains water, sand, oils, grease, organic compounds, chemical elements, and metals. Among the organic compounds present, the most common are alkanes, straight chain alkenes, cyclic alkenes, benzene, toluene, ethyl-benzene, xylenes and polycyclic aromatic hydrocarbons (PAHs) and phenols (Kriipsalu *et al.*, 2008). The oil sludge has been classified by the United States Environmental Protection Agency (US EPA) as a hazardous organic complex (U.S. EPA, 1989, 2002, 2010). Polycyclic aromatic hydrocarbons (PAHs), in particular pose a serious environmental concern, as many of them are cytotoxic, mutagenic and potentially carcinogenic (Ubani *et al.*, 2013). The term "hydrocarbons" refers to compounds consisting of only carbon and hydrogen atoms. Chemically they are comprised of two or more benzene rings bonded in linear, cluster, or angular arrangements. The general characteristics of PAHs are high melting and boiling points (therefore they are solid), low vapor pressure, and very low aqueous solubility. The aqueous solubility tends to decrease with increasing molecular weight, on the contrary, resistance to oxidation and reduction increases.

Bioremediation of hydrocarbon-contaminated soils, which exploits the ability of microorganisms to degrade or detoxify organic contaminants, has been established as an efficient, economic, versatile and environmentally safe treatment (Singh and Lin, 2008). In the industrial plants that process petroleum, while treating residues which is mainly oily sludge, the land farming process is a frequently chosen method, due to its operational simplicity, low cost (when compared to other techniques), high potential for contaminant removal and high rate of residue application, varying from 83 to 252 m^3/ha/year which has been documented by many including Marín *et al.*, 2005 and Jacques *et al.*, 2007. However, the method still shows several limitations especially with regards to appropriate management and control, large areas requirements, the environmentally unfavorable conditions of microbial growth and inefficiency in the reduction of inorganic contaminants level (Hejazi *et al.*, 2003).

11.2 Total Petroleum Hydrocarbon (TPH)

They are a group of recalcitrant, bioaccumulative, semi-volatile organic pollutants that are widespread in the environment and can enter environmental components via natural sources like volcanoes and anthropogenic sources like industrial activities and agricultural activities (Shouse 2001; Quinn *et al.*, 2009; Dong 2012; Paulik 2016) (Figure 11.1). They are composed of two or more fused aromatics benzene rings and a system of hydrophobic and lipophilic double bonds throughout their hydrocarbon rings. These groups of compounds have potential harmful effects on ecosystems as well as human health because many of them have been shown to be carcinogenic, teratogenic and mutagenic (Wang *et al.*, 2010, 2011, 2012; Wick *et al.*, 2011; Zhou *et al.*, 2005, Zhu *et al.*, 2009, 2014).

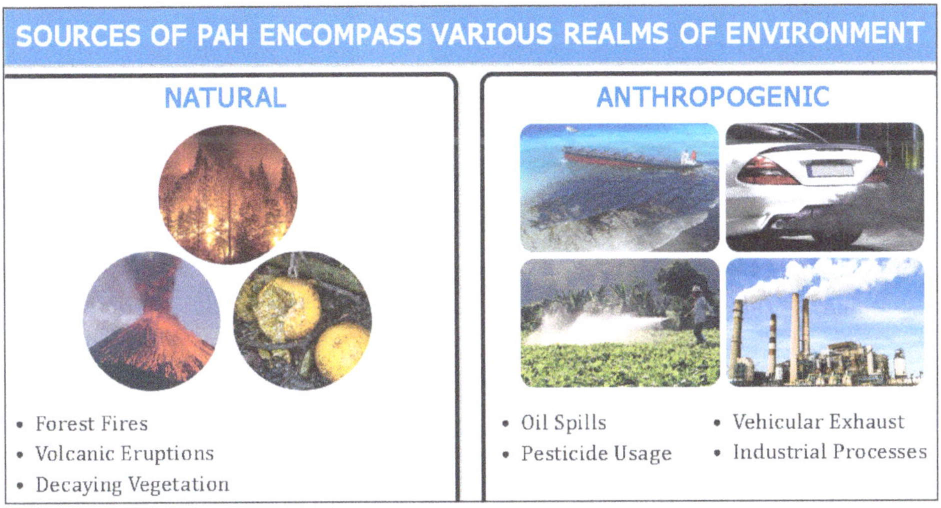

Figure 11.1: Natural and Anthropogenic Sources of Polyaromatic Hydrocarbons.

Several approaches and strategies including physical, chemical and biological strategies have been developed, optimized and utilized to amend petroleum contamination and reclaim polluted sites. Some of the available physical and chemical (conventional) techniques have been shown to have significant drawbacks such as technological complexity, high cost and a general lack of acceptance. Bioremediation, which is the use of biological systems to control pollution, has been shown to be a cost effective and environmentally friendly approach to remediate contaminated sites (Garbisu *et al.*, 2011).

Some of these compounds are mostly used as intermediates in pharmaceuticals, agricultural products, photographic products, thermosetting plastics, lubricating materials, and other chemical industries (Casarett and Doull's, 2008). However, the major source as a pollutant is the incomplete combustion of organic material such as coal, oil and wood. At each and every stage of the oil extraction lifecycle; right from well construction to extraction, operations, transportation and distribution, there is potential of air and water contamination. And workers employed or residents living near such operations are more susceptible to such exposures (Ellen *et al.*, 2018).

11.3 Bioremediation

Petroleum contamination caused during refining and exploration is treated by physical, chemical and biological approaches. The conventional methods like physical and chemical treatments have limitations such as high costs, required expertise and end products that can alter natural ecosystem. Today, biological treatment is considered as an interesting and wholesome approach towards removal of petroleum contamination (Sarkar *et al.*, 2005). Bioremediation is a technique in which biological systems such as microorganisms are applied to destroy or transform (degrade) harmful chemicals (Crawford *et al.*, 2006). Since a decade or so, employing hydrocarbon degrading bacteria to clean a petroleum contaminated soil has become very prevalent and is also economical as well as efficient method that leads to non toxic end products. Several bacterial (*Pseudomonas, Alcanivorax, Microbulbifer, Sphingomonas, Micrococcus, Cellulomonas, Dietzia, Gordonia, Marinobacter*) and fungal species (*Aspergillus* sp., *Trichocladium canadense*, and *Fusarium oxysporum*) capable of degrading PAH compounds have been isolated and characterized by different researchers (Silva 2009; Mirdamadian 2010; Silhavy 2010; Macek 2011; Llado 2012; Soleimani 2012; Cerqueira 2014; Roy 2018). Bioremediation technique has reduced unwanted negative effects on ecology as compared to traditionally conventional approaches. In addition, *in situ* treatment can be carried out without unnecessarily disturbing native ecosystems, compared to physical and chemical remediation methods (Ghoreshie *et al.*, 2017).

The combined action of plants and microbes has been suggested as a profitable bioremediation approach to deal with all possible problems caused by PAH contamination (Huguenot *et al.*, 2015).

Bioremediation involves harnessing the natural biodegradation processes of living organisms. This process can be further accelerated or encouraged via alteration of the contaminated media. Bioremediation of contaminated soils has been used worldwide, with many case studies taking place in India and other developing countries like China, Egypt, Africa and Iran. In remote areas, bioremediation may represent the most practical and cost-effective means of treating soils contaminated with petroleum hydrocarbons (Horel *et al.*, 2009). A flowchart showing the remediation technologies discussed in this chapter is given in Figure 11.2.

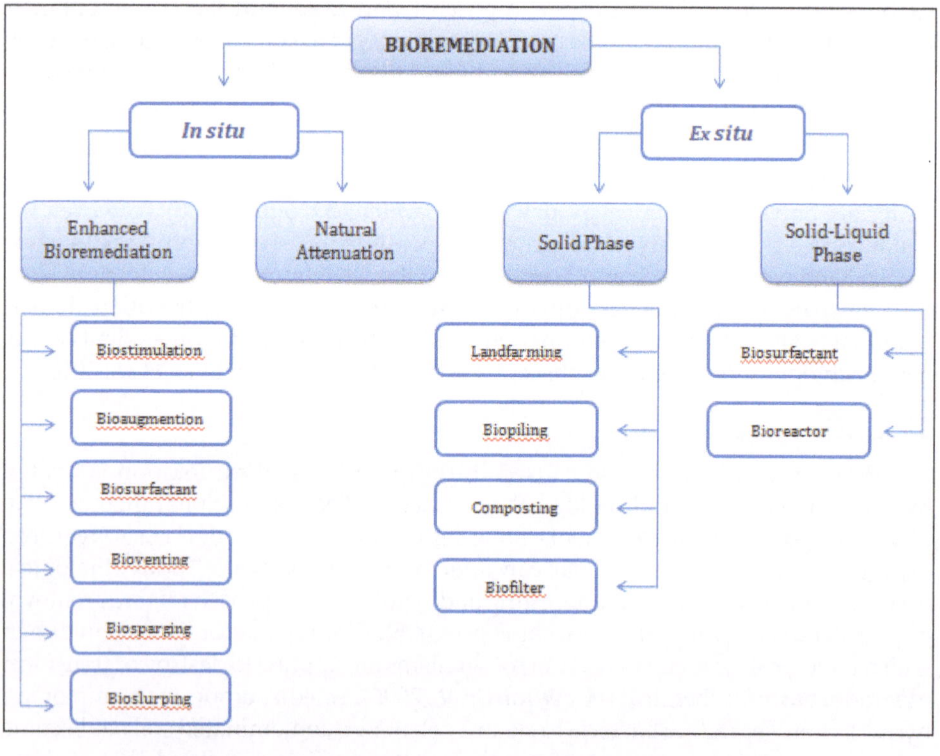

Figure 11.2: Techniques Involving Bioremediation in Soil.

Bioremediation can be broadly divided into *in situ* and *ex situ* types, which refers to treatment at the site of contamination itself and treatment by taking contaminated medium away from the site, respectively. During in situ treatment the natural course of intrinsic biodegradation where only monitoring is undertaken, is called natural attenuation. It is fairly cheaper way to undertake bioremediation without much input cost but is much slower compared to others. However, this process can be enhanced through engineered biodegradation by introducing certain amendments like biostimulation (altering the environment for stimulating the activity of indigenous microbial population), bioaugmentation (bioengineered microbes are introduced to the site of contamination), biosurfactant (use of surface active compounds having hydrophobic and hydrophilic moiety), bioventing (providing regulated stream of air/nutrient for enhanced biodegradation in unsaturated zone), biosparging (regulated stream of air/nutrients for enhanced biodegradation in saturated zone) and bioslurping (combines elements of bioventing and vacuum-enhanced pumping to recover by-products). Sometimes the contaminants cannot be treated on the site and have to be excavated and taken to another site for ex situ remediation. Bioreactors are used to treat such contaminations using technologies employing controlled and monitored biodegradation through anaerobic methods. Popular ex situ techniques like land farming, tilling, composting *etc.* are also well documented.

The success of bioremediation of any contaminated area depends on several factors; the most important ones are pollutant composition and concentration and its bioavailability to microbial attack, the presence of suitable microorganisms with appropriate catabolic capacity on the contaminated site or their introduction at the site, optimum environmental and nutritional conditions for microbial growth and metabolism. As far as presence of suitable microorganisms at the site is taken into account mainly two techniques have been proposed to increase the efficiency of bioremediation processes *i.e.* biostimulation and bioaugmentation. Biostimulation process mainly considers the introduction of nutrients like nitrogen and phosphorus in the form of organic and/or inorganic fertilizers, in the contaminated system to stimulate the growth and activity of already present microbial population. While on the other hand, bioaugmentation process aims at the introduction of microorganisms which have already been characterized as potential degraders of the targeted contaminant, in the given contaminated environment. The primary aim of bioaugmentation is to increase the microbial population and thereby increase the biodegradation rate at the site (Das and Chandran, 2011; Tyagi *et al.*, 2011; Taccari *et al.*, 2012).

As discussed in Table 11.1 and Figure 11.3 environmental factors play a very prominent role in making the pollutant components available for organisms to biodegrade. These factors range from moisture, pH, and temperature to even availability of gases. These environmental factors are also responsible for optimum physiological activities of microbes. Toxicity of the pollutants mainly decides its extent of degrability by the microbes. Thus, bioremediation is not a function of any one or two major factors but an optimum interaction between all the components and factors together.

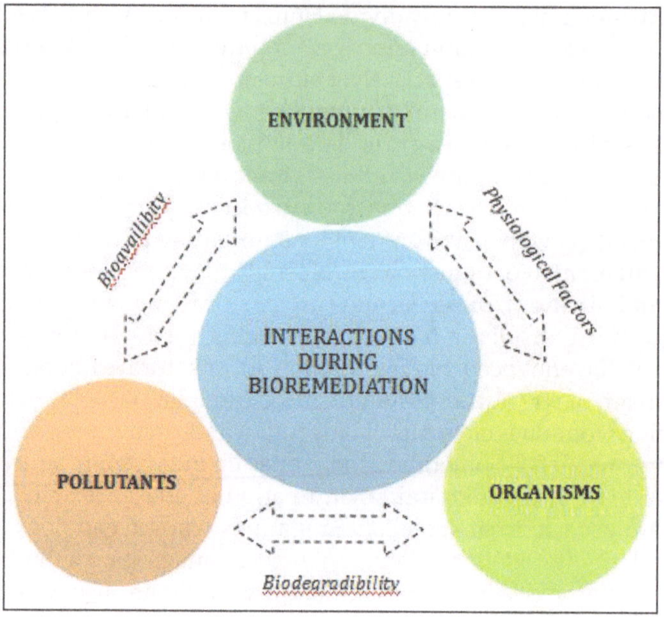

Figure 11.3: Interactions between Components Responsible for Optimum Bioremediation in Soil.

11.4 Factors Affecting the Bioremediation of Hydrocarbons in Soil

There are many researchers who have been successful in application of bioremediation and related technologies to polluted sites using approaches such as phytoremediation and microbe assisted remediation. Mostly, the studies have analyzed the efficiency of bioremediation under ideal laboratory conditions and on a pilot scale. However, it is to be noted that there are environmental factors such as soil texture, pH, nutrient availability and the bioavailability of the contaminant that are variable from site to site; which can greatly influence the process of bioremediation by inhibiting the growth of microorganisms helping in pollutant degradation. The success of any remediation eventually depends on the initial parameters of the soil which is being subjected to pollution.

Initial parameters of the soil are extremely important in deciding the ultimate success of any kind of remediation after an oil spill. Amongst the primary limiting factors are: soil texture, pH, temperature, water holding capacity and nutrient content. Maintenance of optimal temperature and pH, water, oxygen and inorganic nutrients are the prime requisites for supporting microbial cell growth and sustaining optimum biodegradation. The prominent environmental factors that can affect the feasibility and success of any bioremediation are summarized in the Table 11.1.

11.5 Conclusion

The residual petroleum compounds are classified as probable human

Table 11.1: Major Factors Affecting the Optimum Bioremediation of Aromatic Hydrocarbons

Factors	*Significance*	*Optimum Condition*	*Amendment*	*Reference*
pH	Extremely acidic or alkaline pH inhibits the microbial growth. The efficiency of microbes is much altered in suboptimal conditions.	Near neutral range (5.5-7.5)	The initial pH of the site can be adjusted by addition of lime or acids. Bioaugmentation can be induced with tolerant microbial strains.	Das and Chandran (2011)
Moisture Content	The role is statistically not very significant but it is crucial for the survival of the microbes.	More than 10 per cent is sufficient but higher saturation hampers gas exchange and restricts oxygen.	Frequent spraying of water.	Haghollahi *et al.* (2016)
Temperature	Extreme temperatures affect the survival of microbes. Very few studies have focussed on bioremediation by extremophiles. Higher temperature increases the solubility of HCs and in turn bioavailability but also reduces oxygen availability which acts as counter. Apart from this the seasonal variation of temperature also plays a role in efficiency of the remediation.	Mesophilic range(Moderate conditions, 20-45 °C)(Optimal conditions, 30-40 °C)	Planning for ex-situ remediation or optimum season wherever possible.	Lau *et al.* (2003)
Oxygen Availability	Very crucial for initiating oxidation of aromatic ring and helping in metabolic reactions of di-oxygenase enzymes. Survival of microbes in case of aerobic bioremediation.	Steady availability of oxygen along with optimising pH and controlling *in situ* heat and gas production.	Tilling or ploughing for surface contamination and steady supply of oxygen for subsurface contamination. Injecting H_2O_2 has also been documented as oxidising agent for subsurface contamination.	Grishchenkov *et al.* (2000); Petigara *et al.* (2002); Goi *et al.* (2009)
Texture of the medium	Factors like the matrix of the medium, pore space, bulk density, water holding capacity, capillary action, which decide the fate of degradation of HCs, are dependent on texture of the soil/sediment medium. Larger particle size like sand facilitates the process and finer medium like clay restrict bioavailability of the aromatic pollutants.	Bioremediation occurs most optimally in soil containing higher amount of sand compared to clayey particles.	Amendment of clayey medium by using organic inducers such as mulching agents, cotton stacks, rice husk and other agro-wastes. Application of surfactants to desorb hydrocarbons from clay matrix. Use of Fenton's treatment.	Haghollahi *et al.* (2016)
Nutrient Availability	Apart from readily available carbon source from pollutants, macro nutrients like N, P, and K are prime requisite for cellular metabolism; due to which it depletes quickly and need to be supplemented periodically.	C: N: P ratio in microbial biomass should range between 100: 15: 364 and 120: 10: 165 for optimum bioremediation.	Supplementation through nutrients containing Nitrogen and Posphorous in adequate amount.	Atagana *et al.* (2003)

Factors	*Significance*	*Optimum Condition*	*Amendment*	*Reference*
Bioavailability	Poly-aromatic Hydrocarbons (PAH) are compounds with low water solubility and are resistant to biological, chemical and photolytic breakdown. Thus, the bioavailability of a compound is inversely related to its persistence in environment. PAH adheres to the mineral surface of soil matrix (*e.g.* clay) and organic matters (*e.g.* humic and fulvic acids) rendering it unavailable for biodegradation	Presence of surface active compounds containing both, a hydrophobic and hydrophilic moiety; thus providing a 'bridge' between the hydrophobic PAH molecule and the hydrophilic microbial cell.	Enhancing desorption of PAH from soil matrix through techniques using surfactants or employing surfactant producing microbes like *Pseudomonas sp.*	Shukla *et al.* (2014); Suja *et al.* (2014); Sun *et al.* (2012)
Toxicity of the end-products	Most commercial bioremediations are more inclined towards monitoring extent of degradation of parent compound and seldom taking into consideration the toxic by-products or end metabolites. It has been observed that certain oxy-PAHs are more toxic than the parent PAHs, making it further crucial to ensure that the contaminated material is effectively detoxified towards the end of the treatment.	The breakdown of PAH and resulting metabolites need to be studied in detail for complete and optimum bioremediation success.	Employing Eco-toxicity and Bio-assay tests to monitoring the metabolites of bioremediation, specifically for toxic dead-end products, and assessing the toxicity of the material both before and after treatment	Ghosal *et al.* (2016); Adrion *et al.* (2016)

carcinogens. A major class of chemical carcinogens present in the environment is constituted by aromatic hydrocarbons. Several PAHs and their epoxides (organic compound whose molecule contains a three-membered ring involving an oxygen atom and two carbon atoms) are highly toxic, mutagenic and/or carcinogenic to microorganisms as well as to higher organisms including humans. Exposure to these always involves complex mixtures that may induce synergistic or antagonistic effects on the genotoxic properties making risk assessment more difficult. The degradation of polycyclic aromatic hydrocarbon compounds depends on several factors including: the environmental conditions, population and type of the microorganisms, nature and chemical structure of the chemical compound being degraded. They may undergo adsorption, volatilization, photolysis, and chemical degradation while converting to simpler compounds during this process of bioremediation.

Bioremediation is carried out through many natural agents like terrestrial and aquatic plants, earthworms and microorganisms. However, microbial degradation is the major degradation process and such PAH-degrading microorganism, include algae, bacteria, and fungi that help in breakdown of organic compounds through biotransformation as well as mineralization into less complex metabolites, and inorganic minerals like H_2O, CO_2 (aerobic pathway) or CH_4 (anaerobic pathway). Thus, bioremediation is the tool to transform such complex and harmful compounds to comparatively less hazardous or nonhazardous forms along with less input of chemicals, energy, and time. Due to the omnipresence of hydrocarbons in the environment, microorganisms that are able to biodegrade petroleum hydrocarbons are abundant in soil. Heterotrophic (mainly organisms that utilize organic carbon as a food source) bacteria and fungi are the predominant microorganisms biodegrading hydrocarbons. Many studies have found that microbial diversity is decreased in contaminated soil, likely due to the selection and subsequent dominance of microorganisms which are able to degrade petroleum hydrocarbons within the indigenous community, as well as the toxic effects of petroleum hydrocarbons on other organisms. These same studies have also shown a significant increase in the proportion of petroleum hydrocarbon-degrading bacteria in contaminated soils. A number of identified bacteria, fungi, and earthworms as discussed earlier are able to utilize various forms of hydrocarbons as a food source. However, nitrogen, phosphorous, sulfur and some other nutrients are necessary for microbial growth and activity. Therefore the contaminated soils should contain or supplied with ample amounts of these elements for a successful bioremediation process. As a thumb rule, the ratio of Carbon: Nitrogen: Phosphorous should be 100: 10: 1 to ensure a balanced nutrient medium for microbial growth.

Also, for optimum bioremediation to take place, the moisture content of soil should be at an optimum range. Low levels of moisture content decrease microbial activity, while excess water may create resistance to oxygen transfer and may also produce an unwanted leachate. During any bioremediation study, the moisture content is normally adjusted to a fraction of water holding capacity of the soil. The information in the literature on the optimum value of the moisture content of soil for the purpose of bioremediation is scarce, and more investigation is needed to address this case. The optimum value of moisture, however, is a function of soil type, pore size distribution, and soil texture. In general, optimum microbial activity is achieved by the maximum water content that does not restrict oxygen diffusion. Hence, the role of soil type and texture cannot be neglected for achieving best possible bioremediation. Fine grained soils like clay have low permeability and retard oxygen and nutrients transport in the soil and controlling the moisture content in fine grained soils is quite difficult due to smaller pore size and larger surface area. Clays can also catalyze humic acid formation and protect organic materials from decomposition within aggregates. However, here the methods like tilling, landfarming or even use of earthworms can come to rescue. They help not only in breakdown of soil aggregates but also improving the aeration of the soil matrix; thereby leading to optimum bioremediation without mechanical upturning and addition of chemical oxidizing agents.

The enhanced land farming treatment of oil sludge contaminated soil by adding cotton stalks or similar agricultural by-product like husk or hull is a useful bioremediation strategy. Compared with natural attenuation treatment, the advantages of land farming treatment for oil sludge contaminated soil are presented in the three aspects as follows: higher removals of TPH and petroleum fractions, improved soil properties, and higher soil microbial quantity and diversity.

Thus, Bioremediation as a wholesome approach towards removal of numerous synthetic contaminations from environment is now an established fact. Many studies have demonstrated removal of petroleum residue and hydrocarbons from soil by using different approaches of bioremediation. But factors like, soil type and texture, moisture content, pH, temperature, microbial population, compound toxicity, nutrient availability, and oxygen transport in soil; greatly affect the process of bioremediation. If these factors can be controlled or altered to optimum level then the success of bioremediation of petroleum hydrocarbons in soil is inevitable!

REFERENCES

Adrion AC, Singleton DR, Nakamura J, Shea D, and Aitken MD. (2016) Improving Polycyclic Aromatic Hydrocarbon Biodegradation in Contaminated Soil Through Low-Level Surfactant Addition After Conventional Bioremediation. Environ Eng Sci. 33(9): 659-670. https: //doi.org/10.1089/ees.2016.0128

Atagana HI, Haynes RJ, and Wallis FM. (2003) Optimization of soil physical and chemical conditions for the bioremediation of creosote-contaminated soil. Biodegradation 14: 297–307. https: //doi.org/10.1023/a: 1024730722751

Cameotra S.S, and Singh P. (2008) Bioremediation of oil sludge using crude biosurfactants. International Biodeterioration and Biodegradation; 62(3): 274–280. https: //doi.org/10.1016/j.ibiod.2007.11.009

Crawford RL. (2006) Bioremediation. In: The Prokaryotes, Volume1: Symbiotic Association, Biotechnology, Applied Microbiology. New York: Springer; 48-93. https: //doi.org/10.1007/0-387-30741-9_26

Cerqueira VS, Peralba MDCR, Camargo FAO, and Bento F.M. (2014) Comparison of bioremediation strategies for soil impacted with petrochemical oily sludge. International Biodeterioration and Biodegradation; 95: 338-345. http: //dx.doi.org/10.1016/j.ibiod.2014.08.015

Dong C, and Chen C. (2012) Determination of Polycyclic Aromatic Hydrocarbons in Industrial Harbor Sediments by GC-MS.*Int. J. Environ. Res. Public Health* 9(6): 2175-2188.https: //doi.org/10.3390/ijerph9062175

Ellen Webb, Julie Moon, LarysaDyrszka, Brian Rodriguez, Caroline Cox, Heather Patisaul, Sheila Bushkin and Eric London (2018) Neurodevelopmental and neurological effects of chemicalsassociated with unconventional oil and natural gas operations and their potential effects on infants and children. Rev Environ Health 33(1): 3–29.

Garbisu C, Allica J, Barrutia O, (2002) Phytoremediation: A Technology Using Green Plants to Remove Contaminants from Polluted Areas. Reviews on Environmental Health; 17(3): 173-188. https: //doi.org/10.1515/reveh.2002.17.3.173

Ghosal D, Ghosh S, Dutta TK, and Ahn Y. (2016) Current State of Knowledge in Microbial Degradation of Polycyclic Aromatic Hydrocarbons (PAHs): A Review. Front Microbiol. 7: 1837. https: //doi.org/10.3389/fmicb.2016.01369

Goi A, Trapido M, and Kulik N. (2009) Contaminated Soil Remediation with Hydrogen Peroxide Oxidation World Academy of Science, Engineering and Technology International Journal of Chemical and Molecular Engineering 3(4): 185-189

Golafarin Ghoreishi, Abbas Alemzadeh, Mohammad Mojarrad, Mohammad Djavaheri. (2017) Bioremediation capability and characterization of bacteria isolated from petroleum contaminated soils in Iran. Sustainable Environment Research Sustainable Environment Research 27: 195-202. https: //doi.org/10.1016/j.serj.2017.05.002

Grishchenkov VG, Townsend RT, McDonald TJ, Autenrieth RL, Bonner JS, Boronin AM. (2000) Degradation of petroleum hydrocarbons by facultative anaerobic bacteria under aerobic and anaerobic conditions. Process Biochemistry 35(9) : 889-896. https: //doi.org/10.1016/s0032-9592(99)00145-4

Haghollahi, Ali, Mohammad Hassan Fazaelipoor, and MahinSchaffie. (2016) "The Effect of Soil Type on the Bioremediation of Petroleum Contaminated Soils." Journal of Environmental Management180: 197–201. https: //doi.org/10.1016/j.jenvman.2016.05.038

Hejazi RF, Husain T, and Khan FI. (2003) Landfarming operation of oily sludge in arid region - Human health risk assessment. Journal of Hazardous Materials 99(3): 287–302. https: //doi.org/10.1016/S0304-3894(03)00062-1

Horel A, Schiewer S. (2009) Investigation of the physical and chemical parameters affecting biodegradation of diesel and synthetic diesel fuel contaminating Alaskan soils. Cold Regions Science and Technology 2009; 58: 113. https: //doi.org/10.1016/j.coldregions.2009.04.004

Huguenot D, Bois P, Cornu, JY, Jezequel K, Lollier M, Lebeau T. (2015) Remediation of sediment and water contaminated by copper in small-scaled constructed wetlands: effect of bioaugmentation and phytoextraction. Environ. Sci. Pollut. Res. 2015; 22: 721–732. https: //doi.org/10.1007/s11356-014-3406-6

Jacques, Rodrigo JosemarSeminoti Benedict, Fatima Menezes, Antoniolli, Zaida Ines and Camargo, FlávioAnastacio de Oliveira. Bioremediation of soils contaminated with polycyclic aromatic hydrocarbons. Rural Science 2007; 37 (4): 1192-1201.https: //doi.org/10.1590/s0103-84782007000400049

Kaminski NE, Faubert Kaplan BL, and Holsapple MP,(2008) In: Curtis D. Klaassen (Ed.), Casarett and Doull's Toxicology, the basic science of poisons, Mc-Graw Hill, Inc, 526, 2008

Kriipsalu M, Marques M, Nammari DR, and Hogland W. (2007) Bio-treatment of oily sludge: The contribution of amendment material to the content of target contaminants, and the biodegradation dynamics. Journal of Hazardous Materials. 148(3): 616–622. https: //doi.org/10.1016/j.jhazmat.2007.03.017

Lau KL, Tsang YY and Chiu SW. (2003) Use of spent mushroom compost to bioremediate PAH-contaminated samples. Chemosphere 2003; 52: 1539–1546. https: //doi.org/10.1016/s0045-6535(03)00493-4

Llado S, Solanas AM, de Lapuente J, Borras M, Vinas M. (2012) A diversified approach to evaluate biostimulation and bioaugmentation strategies for heavy-oil-contaminated soil. The Science of the Total Environment 435-6. https: // doi.org/10.1016/j.scitotenv.2012.07.032

Macek T, Francova K, Kochánková L, *et al.* (2011) Phytoremediation–Biological Cleaning of a Polluted Environment. Reviews on Environmental Health 19(1): 63-82.

Marin JA, Hernandez T, and Garcia C. (2005) Bioremediation of oil refinery sludge by landfarming in semiarid conditions: Influence on soil microbial activity. Environmental Research 2005; 98(2): 185–195.https: //doi.org/10.1016/j. envres.2004.06.005

Mirdamadian SM, Emtiazi G, Golabi MH, Ghanavati H. (2010) Biodegradation of petroleum and aromatic hydrocarbons by bacteria isolated from petroleum contaminated soil. J Petrol Environ Biotechnol 2010; 1: 1-5. https: //doi. org/10.4172/2157-7463.1000102

Das N. and Chandran P. (2011) Microbial Degradation of Petroleum Hydrocarbon Contaminants: An Overview. Biotechnology Research International 2011. 1-13pp. https: //doi.org/10.4061/2011/941810

Petigara BR, Blough NV, and Mignerey AC, (2002) Mechanisms of hydrogen peroxide decomposition in soils. Environ. Sci. Technol. 36: 639-645. https: // doi.org/10.1021/es001726y

Paulik LB, Donald CE, Smith BW, Tidwell LG, Hobbie KA, *et al.* (2016) Emissions of polycyclic aromatic hydrocarbons from natural gas extraction into air. Environ Sci Technol 50(14): 7921–9.https: //doi.org/10.1021/acs.est.6b02762

Quinn L, Pieters R, Nieuwoudt C, Borgen A, KylinH,Bouwman H. (2009) Distribution profiles of selected organic pollutants in soils and sediments of industrial, residential and agricultural areas of South Africa. J Environ Monit. 9: 1647–1657.https: //doi.org/10.1039/b905585a

Roy A, Dutta A, Pal S, Gupta A, Sarkar J, Chatterjee, A, *et al.* (2018) Biostimulation and bioaugmentation of native microbial community accelerated bioremediation of oil refinery sludge. Bioresource Technology 2018. doi: https: //doi.org/10.1016/j.biortech.2018.01.004.

Sarkar D, Ferguson M, Datta R, Birnbaum S. (2005) Bioremediation of petroleum hydrocarbons in contaminated soils: comparison of biosolids addition, carbon supplementation, and monitored natural attenuation. Environ Pollut 2005; 136: 187-195.

Silhavy TJ, Kahne D, Walker S. (2010) The bacterial cell envelope. Cold Spring Harbor Persp Biol. 2010. http: //dx.doi.org/10.1101/cshperspectives.org.

Silva IS, Grossman M, Durranta LR. (2009) Degradation of polycyclic aromatic hydrocarbons (2–7 rings) under microaerobic and very-low-oxygen conditions by soil fungi. IntBiodeteriorBiodegrad. 63: 224–229. https: //doi.org/10.1016/j.ibiod.2008.09.008

Singh C. and Lin J. (2008) Isolation and Characterization of Engine Oil Degrading Indigenous Microorganisms in Kwazulu-Natal, South Africa. African Journal of Biotechnology; 7: 1927-1932.https: //doi.org/10.5897/ajb07.728

Shouse B. (2001) The Gulf War's aftermath: Kuwait unveils plan to treat festering desert wound. Science 293.https: //doi.org/10.1126/science.293.5534.1410a

Soleimani M. Bioremediation: an environmental friendly solution for oil contaminated soils. J BioremedBiodegrad 2012; 3: 2155–6199. https: //doi.org/10.4172/2155-6199.1000e109

Shukla S, Mangwani N, Toleti, Subba Rao and Das S. (2014) Biofilm-Mediated Bioremediation of Polycyclic Aromatic Hydrocarbons. Microbial Biodegradation and Bioremediation; 203-232. https: //doi.org/10.1016/b978-0-12-800021-2.00008-x

Suja F, Rahim F, Taha MR, Hambali N, Rizal Razali M, *et al.* (2014) Effects of local microbial bioaugmentation and biostimulation on the bioremediation of total petroleum hydrocarbons (TPH) in crude oil contaminated soil based on laboratory and field observations. InternationalBiodeterioration and Biodegradation 90. https: //doi.org/10.1016/j.ibiod.2014.03.006

Sun GD, Xu Y, Jin JH, Zhong ZP, Liu Y, (2012) Pilot scale ex-situ bioremediation of heavily PAHs-contaminated soil by indigenous microorganisms and bioaugmentation by a PAHs-degrading and bioemulsifier-producing strain. Journal of Hazardous Materials; 233-4. https: //doi.org/10.1016/j.jhazmat.2012.06.060

Taccari M, Milanovic V, Comitini F, Casucci C, Ciani M. (2012) Effects of biostimulation and bioaugmentation on diesel removal and bacterial community. Int. Biodeterior. Biodegrad 66: 39-46. https: //doi.org/10.1016/j.ibiod.2011.09.012

Tyagi M, da Fonseca MMR, da Carvalho CCCR. (2011) Bioaugmentation and biostimulation strategies to improve the effectiveness of bioremediation processes. Biodegradation; 22: 231-241. https: //doi.org/10.1007/s10532-010-9394-4

Ubani O, Atagana HI, and Thantsha MS. (2013) Biological degradation of oil sludge: A review of the current state of development. African J. of Biotechnol; 12(47): 6544-6567. https: //doi.org/10.5897/AJB11.1139

USEPA (United States Environmental Protection Agency) (1989) Risk assessment guidance for super fund. Vol. I. USEPA, Washington (Human health evaluation manual (part A). EPA 540-1-89-002, Office of Emergency and Remedial Response.

USEPA (United States Environmental Protection Agency) (2002). Supplemental guidance for developing soil screening levels for superfund sites. OSWER; 2002.

USEPA 2010 (United States Environmental Protection Agency). Development of a relative potency factor (RPF) approach for polycyclic aromatic hydrocarbon (PAH) mixtures (External Review Draft). U.S. Environmental Protection Agency, Washington DC. EPA/635/R-08/012A; 2010

Wang W, Massey Simonich SL, Xue M, Zhao J, Zhang N, Wang R, Cao J, Tao S. (2010) Concentrations, sources and spatial distribution of polycyclic aromatic hydrocarbons in soils from Beijing, Tianjin and surrounding areas, North China. Environmental Pollution; 158(5): 1245–1251. https: //doi.org/10.1016/j.envpol.2010.01.021

Wang W, Huang MJ, Kang Y, Wang HS, Leung AOW, Cheung KC, Wang MH. (2011) Polycyclic aromatic hydrocarbons (PAHs) in urban surface dust of Guangzhou, China: status, sources and human health risk assessment. Science of the Total Environment; 409: 4519–4527.

Wang Y, Tian Z, Zhu H, Cheng Z, Kang M, Luo C, (2012) Polycyclic aromatic hydrocarbons (PAHs) in soils and vegetation near an e-waste recycling site in South China: concentration, distribution, source, and risk assessment. Science of the Total Environment; 439: 187–93. https: //doi.org/10.1016/j.scitotenv.2012.08.018

Wick AF, Haus NW, Sukkariyah BF, Hearing KC, Lee Daniels W. (2011)Remediation of PAH-contaminated soils and sediments: a literature review. Department of Crop and Soil Environmental Sciences Blacksburg, Virginia Polytechnic Institute and State University; 24061

Xu Y, and Lu M. (2010) Bioremediation of crude oil-contaminated soil: Comparison of different biostimulation and bioaugmentation treatments. Journal of Hazardous Materials 2010; 183(1–3): 395–401. https: //doi.org/10.1016/j.jhazmat.2010.07.038

Zhou JB, Wang TG, Huang YB, Mao T, Zhong NN. (2005) Size distribution of polycyclic aromatic hydrocarbons in urban and suburban sites of Beijing, China. Chemosphere; 61: 792–9. https: //doi.org/10.1016/j.chemosphere.2005.04.002

Zhu L, Lu H, Chen S, Amagai T, (2009). Pollution level, phase distribution and source analysis of polycyclic aromatic hydrocarbons in residential air in Hangzhou, China. Journal of Hazardous Material. 162(2–3): 1165–70. https: // doi.org/10.1016/j.jhazmat.2008.05.150

Zhu Y, Yang L, Yuan Q, Yan C, Dong C, Meng C, Sui X, Yao L, Yang F, Lu Y, Wenxing Wang W. (2014) Airborne particulate polycyclic aromatic hydrocarbon (PAH) pollution in a background site in the North China Plain: Concentration, size distribution, toxicity and sources. Science of the Total Environment 466–467: 357–368. https: //doi.org/10.1016/j.scitotenv.2013.07.030

Chapter 12

Environmental Impact of Arsenic Toxicity and its Remediation Approaches

Jyoti Mathur, Ritu Panwar, Pranci Dubey and Mamta Sharma

Bioscience and Biotechnology Department, Banasthali Vidyapith, Rajasthan – 304 022

e-mail: contact.srivastava@gmail.com

ABSTRACT

Arsenic pollution of soil and water is the major concern for the human health and environment throughout the world. Arsenic (As) is prevalent in nature and ambient pollutant in soil, air and water. Significant sources of As contamination are by geological, natural or anthropogenic activities. For several countries As contamination poses a threat. In the world As is the most talked-about metalloid, due to its negative effects on plants and animals. Threatening the health of millions of people due to its toxicity and carcinogenicity, the key routes for As exposure to human are either by drinking water or through soil. Biodegradation processes can remediate wide-ranging environmental pollutants. Plants and microbe like fungi and bacteria have the ability to reduce or detoxify harmful environmental contaminants into less toxic forms. This chapter will help to provide in depth knowledge on As occurrence, sources, speciation, impact in the environment, As toxicity and various methods of remediation approaches.

Keywords: *Arsenic toxicity, Health, Remediation approaches, Carcinogenicity, Pollution.*

12.1 Introduction

Pollution of soil, air and water with heavy metals is a worldwide environmental problem (Shahid *et al.*, 2015; Khalid *et al.*, 2017). There are 92 elements which have natural origin among them approximately 30 are metalloids and metals which are most likely lethal to the environment *i.e.*, lithium(Li), chromium(Cr), aluminium(Al), tin(Sn), manganese(Mn), nickel(Ni), lead(Pb), mercury(Hg), cadmium(Cd),copper(Cu), cobalt(Co) and arsenic(As) (Khlifi *et al.*, 2010; Jomova and Valko, 2011). Due to industrialization, the toxic metals concentration increases in water and soil (Joyce *et al.*, 2014; Mombo *et al.*, 2016). Contamination of heavy metal in soil and water is from industrial (like smelting, mining, refining) and agricultural activities (Shukla and Srivastava, 2017). Metals are categorized as essential (Zn, Co, Cu, Ni, *etc.*) and non-essential (Cd, As, Hg, Pb, *etc.*) based on their biota requirements (Nagajyoti *et al.*, 2010). In the environment even in small amount nonessential metals are very toxic while essential metals can also become toxic when exist in high concentration (Shukla *et al.*, 2019). Heavy metals toxicity poses severe threats, as metals can be biomagnified when enter into the food chain (Zhao *et al.*, 2010). Pb and As are widely toxic in nature (Khalid *et al.*, 2017).

Arsenic (As) is a lethal element which increases toxicity worldwide and affects the plants, animals and human health (Hughes *et al.*, 2011). It is an earth's crust element present in the soil and rock, atmosphere, organisms and water bodies (Khalid *et al.*, 2017). As is a metalloid *i.e.* an element that is not truly a metal but which also has some properties of metal. In the environment it is not present as free element, but mostly present in sulphur-containing compounds in which As present as metal arsenides (Singh *et al.*, 2007). As occurs in the environment in three allotropic forms which are α (yellow), β (black) and γ (grey) in metallic state and also found in numeral ionic forms (Orloff *et al.*, 2009). The common oxidation states of As are +3, +5 and –3 in which the element is able to form organic and inorganic compounds (Joshi *et al.*, 2009).

Absorption of As containing minerals is a key natural source, while industrial activities like paper, paint, fuel and fossil also contribute in As toxicity (Jaishankar *et al.*, 2014). The worldwide effect of As pollution in the environment is because of natural processes such as weathering reactions, volcanic emissions and through a range of anthropogenic activities (Jedynak *et al.*, 2012; Xiong *et al.*, 2016). As alone is insoluble in water but grouping with other elements (as salts) provides solubility depending upon the acidity and presence of other chemicals (Rajakovic *et al.*, 2018). The harmful impacts of As depend upon the chemical compounds and oxidation state. As that accumulated in the environment reach in human as it become an integral part of the food chain and cause health related problems such as skin cancer (Rai *et al.*, 2019). The Environmental Protection Agency (EPA) considered As, a famous carcinogen and one of the world's most harmful chemical (Shankar *et al.*, 2014).

12.2. Arsenic Occurrence in the Environment

Arsenic is the 12th trace element in the living system, 14th in the seawater, and 20th in the soil (Figure 12.1) (Mirza *et al.*, 2014; Hare *et al.*, 2018). It is highly poisonous in its inorganic form. Arsenic is broadly distributed as arsenic sulfide compounds, arsenide and metal arsenates (Singh *et al.*, 2007). In water, Arsenic is available as arsenate and when the water is oxygenated the oxidation state is V. It may be under reducing conditions, it is bound to be available as arsenite, with an oxidation territory of III (Ayala *et al.*, 2005). The inorganic forms of arsenic such as trivalent arsenite (As III) and pentavalent arsenate (As V) are more ubiquitous and toxic than the organic forms in terrestrial environment (Shankar *et al.*, 2014). Arsenic toxicity to the human beings relies upon their concentrations and natural compounds. Inorganic arsenic (iAs) species are commonly more lethal than natural As species. Arsenite is generally more toxic than Arsenate, and dimethyl arsinous corrosive (DMAAIII) and monomethyl arsinous acid (MMAAIII) are more harmful than their parent compound (Hussain *et al.*, 2019). As-V is the thermostable state in water, while As-III is prevalent in decreased redox responses to the environment (Smedley *et al.*, 2007). Aquatic organisms discharge As species privilege their bodies when exposed to the food and from different sources, for example water, soil and air (Rahman *et al.*, 2004).

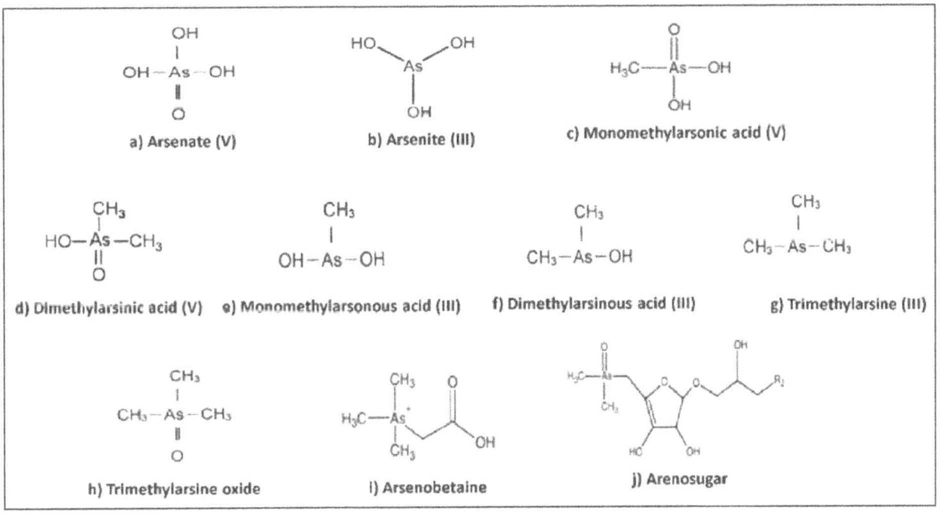

Figure 12.1: Various Structures and Forms of Arsenic Compounds.

12.3. Distribution and Sources of Arsenic Contamination

As mostly occurs in water, air, soil and plants. Arsenic trioxide (As_2O_3) is the most common inorganic As found in the air, whereas a multiple inorganic arsenites (AsO^{2-}) or arsenates (AsO_4^{3-}) are found in the soil, food and water (Chou *et al.*, 2016; Jomova *et al.*, 2011). It exists in more than 200 different minerals and the most common compound is called Arsenopyrite (Habashi *et al.*, 2001). In the earth's atmosphere about one third of the As is originated naturally. The other most important source is As containing vapors which are produced at low temperatures by liquid or solid forms of As salts (Xiong *et al.*, 2016). Geologically occurring inorganic As is present in groundwater which is used as drinking water in many countries, like India, Taiwan and Bangladesh (Smedley *et al.*, 2007). Organic As compounds which contains carbon are mainly found in marine animals, while some of these compounds are also found in the terrestrial animals (Popwich *et al.*,

2016). Volcanic emissions and woodland fires are regular sources that can discharge As in the atmosphere (Finnegan *et al.*, 2012). As is utilized in paints, drugs, colors, cleansers and semi-conductors. As is normally existing at elevated levels in the groundwater of various nations (Hu *et al.*, 2015). It is highly contaminated in its inorganic form (Calvo *et al.*, 2006).

12.3.1. Groundwater Pollution

Groundwater is a major source of drinking water and high level of As in water is having negative impact on human health (Calvo *et al.*, 2006). Elevated As concentrations mainly come from natural sources such as leaching and erosion or anthropogenic sources (Bhattacharya *et al.*, 2003). Furthermore, the use of As for industrial purposes, metal processing, mining, fertilizers and pesticides is another significant sources of As pollution (Shahid *et al.*, 2018).

12.3.2. Industrial Process

As mostly used in industrial practices to manufacture antifungal wood preservatives that contribute in the pollution of soil. Incineration of preserved wood products, As_2CrCuO_9-treated pressure was found to be a source of As pollution (Bhowmick *et al.*, 2013). As is utilized as alloying agent in the handling of glass, metal adhesion, pigments, textile, paper, wood ammunition and preservatives (Li *et al.*, 2014). In the glass and pharmaceutical industries, As is used in the manufacture of sheep dips, alloy, As-containing pigments, leather preservatives, poison baits and paints (Rahman *et al.*, 2004; Mishra and Bharagava, 2014). As compounds are also used in the optical and microelectronics industries. Inorganic As compounds, mostly sodium arsenite, were commonly used as weed killers and non-selective soil sterilant (Chung *et al.*, 2014).

12.3.3. Agricultural Sources

Arsenic contamination from ingested food usually comes from food crops grown in As polluted soil or irrigated with As polluted water (Chung *et al.*, 2014). The toxicity of As towards bacteria, insects and fungi directed to its use as a wood preservative (Rahman *et al.*, 2004). Usage of various agricultural support items, including pesticides like fungicides, herbicides and insecticides containing As (Hare *et al.*, 2019). Because of its germicidal capacity As has been commonly used in pesticides and insecticides. Insecticides such as lead hydrogen arsenate are standard trees insecticide (Arnold *et al.*, 2014), but contact with the compound has also resulted in brain damage among the workers. These arsenicals have been banned in agricultural activities except cotton farming (Mandal *et al.*, 2002). As is used in the cultivation of cotton as of pesticides, such as copper acetoarsenite, H_3AsO_4, Calcium arsenate, lead arsenate, cacodylic acid, monosodium methanearsonate (MSMA), copper-acetoarsenite and disodium methanearsonate (Arnold *et al.*, 2014).

12.3.4. Air Source

As is primarily bound to particulate matter in air and is existing as a mixture of arsenate and arsenite, with a small quantity of organic As species except in areas where As mediated biotic activity or pesticides are used (Argos *et al.*, 2012).

Methylated As is a negligible component in urban, suburban and industrial areas and that the main inorganic ratio of air is made up of pentavalent and trivalent compounds (Chung *et al.*, 2014). As concentrations in the atmosphere are generally low but as mentioned above, concentration is increased by the emissions from smelting, combustion of fossil-fuel, volcanic eruptions and other industrial activities (Hare *et al.*, 2019). Airborne As is transmitted by wet or dry deposition in water bodies and thus slightly enhance the concentration in aquatic system (Smedley *et al.*, 2007). In past when tobacco plants were treated with $PbHAsO_4$ an insecticide, the risk of As toxicity was elevated (Flanagan *et al.*, 2012). Plant can take up As present in the soil naturally (Chung *et al.*, 2014).

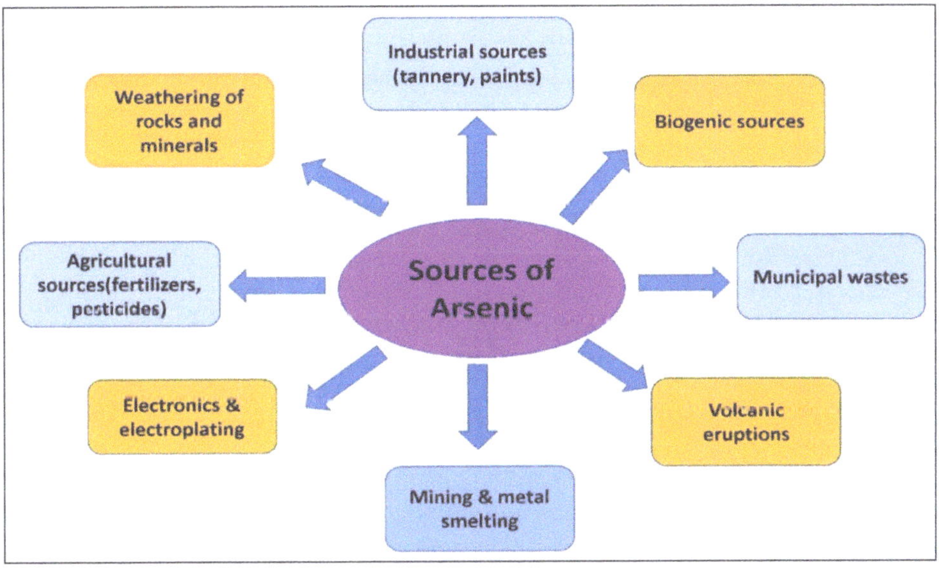

Figure 12.2: Major Sources of Arsenic.

12.3.5. Arsenic Pollution in Soil, water and air

The concentration of As is commonly more liable to sandy soil areas with high carbon content. Arsenite (AsIII) and Arsenate (AsV) dominate in aerobic and anaerobic soil (Mohan *et al.*, 2007). Soil have various concentrations of As; in most of the cases, the absorption in the soil is 5 to 10 mg kg^{-1}(Basu *et al.*, 2014). In the European soil, the concentration of As is 7 mg kg^{-1} (Karchzewska *et al.*, 2007; Stafilov *et al.*, 2010). The pollution of As increase in agricultural areas of the world (Mishra *et al.*, 2014). With the help of bacterial activities in the soil As cause problems, which effect microbes and human. With the help of soil these contaminants transfer to the plant roots (Pena-Fernandez, 2014). The high concentrations of As effects the plant and show symptoms like cell necrosis, inhibition of growth and chlorosis (Islam *et al.*, 2013). It is stated to be exists in higher concentrations in the soil than in rock (Flora *et al.*, 2015). The As concentrations in the earth's crust of various countries ranges around 0.1–40 mg kg^{-1}, 1–50 mg kg^{-1} and average 5 mg kg^{-1}, but vary considerably between geographical regions (Hare *et al.*, 2019).

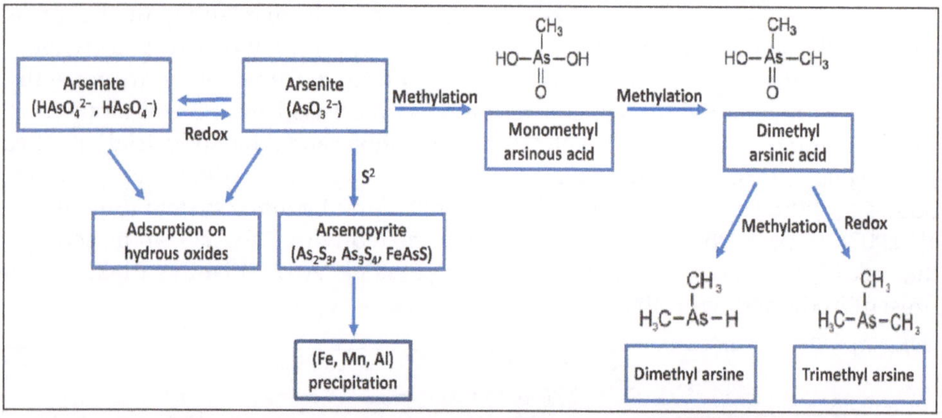

Figure 12.3: Mechanism of Uptake, Conversion and Detoxification of Arsenic.

The Environmental Protection Agency (EPA) described As concentration 10 μgL^{-1} in the drinking water (Sakurai *et al.*, 2004). Water sources in Argentina, India, Canada, Bangladesh, China, Laos, Mexico, Vietnam and US have been reported increasing arsenic concentration (Basu *et al.*, 2014). According to the WHO Provisional Drinking Water Guideline 10–50 ppb of As is consumed by around 45 million population mainly in Asian developing countries is at risk of the exposure of more than 50 ppb of As, which is the actual drinking water level in most of the Asian countries (Ravenscroft *et al.*, 2011). Many areas in India mainly northeastern states with concentrations of As greater than 0.05 mgl^{-1} indicates that billions of people are at risk of As poisoning (Singh *et al.*, 2004). According to a survey, As concentrations usually varies from 0.02 to 0.9mgl^{-1} while, it is 0.01 mgl^{-1} of WHO standard and 0.05 mgl^{-1} of Bureau of Indian Standard (BIS). While As poses negative impact on the aquatic environment, their groundwater association has the power to provide a simple means of As removal by adsorption and co-precipitation (Figure 12.3) (Kumar *et al.*, 2013).

The major source of As inhaled come from the burning of As containing fuels, manufacturing operations, cotton gins, manufacturing facilities for pesticides and tobacco smoke. In the workplaces, airborne As is typically present in the form of arsenic trioxide. Human exposure through air is normally found in low concentration from 0.4 to 30 ng/m^3 (Fowler *et al.*, 2015). The U.S. EPA has reported that humans usually inhale between 40 to 90 ng of As every day. Around 50 ng or less of As is inhaled every day in unpolluted areas (EPA, 2014). High level occupational exposures are unusual today, since the lowest acceptable amount of exposure is 0.01 mg/m^3 (Hare *et al.*, 2019). Smaller particles are more deeply incorporated into the respiratory tract (Mazumder *et al.*, 2011).

12.4. Environmental Hazards of Arsenic

12.4.1 Effect on Humans

Arsenic toxicity is a medical condition because of the high concentration of As in the body. Human are exposed to the heavy metals by the ingestion of polluted food or by consuming polluted water (Awasthi *et al.*, 2017). The signs of arsenic toxicity are abdominal pain, watery diarrhea and vomiting which holds blood. Long-term contact can result in thickening of the skin, diarrhea, abdominal pain, cancer and heart disease (Ratnaike *et al.*, 2003). The major regions in India are effected by the groundwater of Ganga and Bramhaputra river. West Bengal is the most effected state in India (Mukherjee *et al.*, 2015). There are several diseases caused by the arsenic toxicity (Figure 12.4):

1. **Arsenicosis and cancer:** The symptoms of cancer in humans showed when exposed to As in drinking water. The researcher's group of International Agency for Research on Cancer estimated and observed the statistics case studies from many countries (Hu *et al.*, 2015).
2. **Skin manifestation:** Pigmentation and keratosis are the skin diseases. To discover the frequency of pigmentation and keratosis with respect to arsenic exposure, the first survey based on the population was carried out on 7,683 members in West Bengal on As exposure data (Mukherjee *et al.*, 2015). Long-lasting As exposure leads to increase of skin cancer which changes the functions of organs with display of skin lesions through hyperkeratosis and pigmentation (Ratnaike *et al.*, 2003).
3. **Respiratory disease:** Twenty nine cases were studied in a hospital based on chronic exposure of As toxicity with lung disease and bronchiectasis in the people (Mazumder *et al.*, 2011).

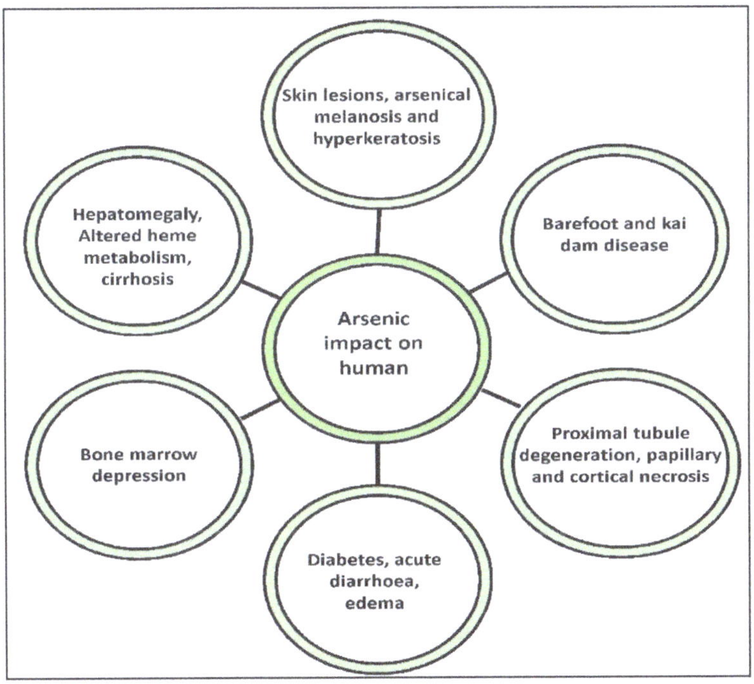

Figure 12.4: Effect of Arsenic Exposure in Humans.

4. **Liver disease:** Many workers were testified for liver harm after the treatment of patients with As in Fowler's solution. The first case in 1976 of non-cirrhotic portal fibrosis and hepatological effects, diagnosed in nine patients from Chandigarh, India had been chronically exposed to As through drinking water holding up to 549µg/L of As (Nandi *et al.*, 2005).
5. **Cardiovascular disease:** Blackfoot disease (BFD), acrocyanosis and Raynaud's syndrome, which is a form of vascular disease, testified among people in Taiwan were found to be one of the vital problems of As toxicity (Colognato *et al.*, 2007).

12.4.2. Effect of Arsenic on Plants

Plants are a major source of food, their adaptation to As-rich environments is important, as new studies on toxicity of arsenic and tolerance mechanisms in plants are indicated negative effects. Plants presentation towards As even at exceptionally low fixation, can caused numerous physiological, biochemical and morphological changes (Singh *et al.*, 2004). Various physiological procedures are affected to As toxicity. Cell layers become harmed in plants presented to As, causing electrolyte leakage (Singh *et al.*, 2006). Arsenic accessibility in soil can upset typical working of plant digestion, subsequently prompting low yield profitability (Shahid *et al.*, 2015). Decrease in leaf number, chlorosis, defoliation, decrease in shoot and root development, chlorophyll corruption, production of receptive oxygen species (ROS), prompting starch, protein and DNA fix also occur (Niazi *et al.*, 2017; Abbas *et al.*, 2018).

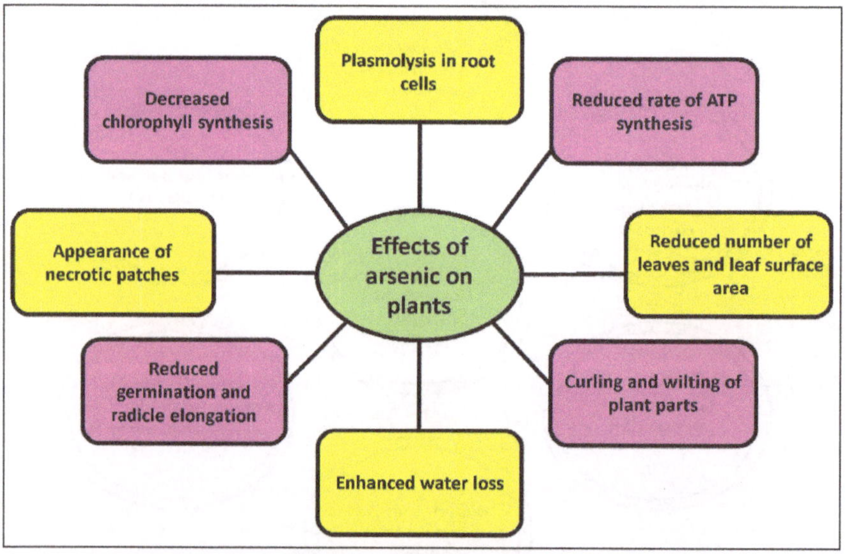

Figure 12.5: Impacts of Arsenic Toxicity on Plants.

Plants assimilate As by light collecting mechanical assembly can be influenced with a decrease in chlorophyll fixations and photosynthetic action II (Anjum *et al.*,

2012). Decrease in chlorophyll color amalgamation was accounted for because of lack of the versatile changes of photosystems-I and - II because of high As levels (Nandi *et al.*, 2005). Arsenic caused wounds in chloroplast layer and disordered elements of key photosynthetic procedure (Rasheed *et al.*, 2017). Arsenic has appeared to defiantly influence the photochemical productivity (Islam *et al.*, 2013). These outcomes are primarily reliable crosswise over examinations and recommend that As may cause an abatement in both root and leaf development, bringing about the presence of harmfulness indications, for example, wilting and violet colour of leaves (Rehamn *et al.*, 2001). Plant transpiration force can be diminished. Plant and animals exposure to As induce genotoxic responses (Francis *et al.*, 2017). Numerous studies have proposed that As genotoxicity is at first linked with the ROS during its transformation. In this way, the ROS produce protein and cause oxidative harm to the DNA (Rasheed *et al.*, 2017).

12.4.3. Effect on Aquatic Animals

In most marine warm-blooded animal, seabirds and ocean turtles, hazard to these aquatic animals might be somewhat low in spite of maintenance of high fixations in their tissues. Inorganic arsenic works as cancer-causing agent with the help of receptive oxygen species (Chakrabarti *et al.*, 2010). Fish are generally considered as a living being of decision for evaluating the impacts of natural contamination on biological systems (Caim *et al.*, 2015). Fishes are the good indicators of As toxicity that sentries the biomonitoring of marine environmental pollutants (Kunito *et al.*, 2008). Marine animals are model organism for the studies of toxic genomics owing to durable control of fish models as biomarkers towards the contact (Das *et al.*, 2012) (Figure 12.6).

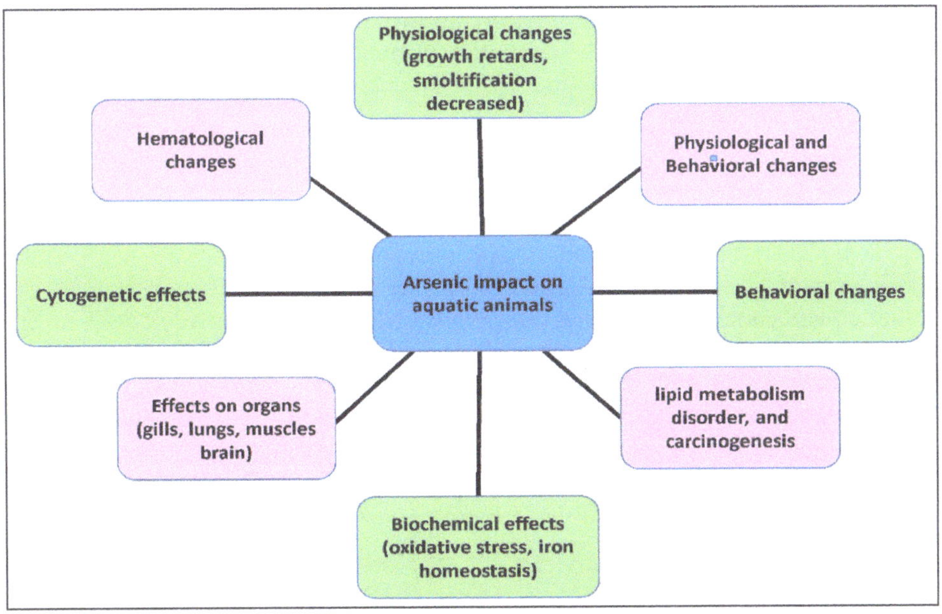

Figure 12.6: Effects of Arsenic Contaminated Water on Aquatic Animals.

Marine animals are evolved in various approaches for transformation of As to reduce contaminated forms and then willingly excret (Papry *et al.*, 2019). The high amount of As affects various physiological systems in an aquatic ecosystem such as ion regulation, growth, reproduction, enzyme activity, gene expression, smoltification and histopathology of organisms (Datta *et al.*, 2009; Xu *et al.*, 2013). Sublethal impacts including iron deficiency, gallbladder irritation, and liver disease, were seen at oceanic groupings of 9.64 mgL^{-1} and dietary fixations of 43.1–60 µg/g. White fish feed arsenate and showed histopathological changes like shedding of the epithelium, swelling and fibrosis in the sub-mucosal tissues in the gallbladders (Roy and Bhattacharya, 2006).

12.5. Mechanism of Arsenic Remediation

12.5.1. Plant-Mediated Arsenic Remediation

The term phytoremediation known as the plant-assisted remediation of contaminated air, water and soil, is an evolving economical technique, conventional and generally accepted form of pollutant removal or mitigation (Singh *et al.*, 2004). There are several procedures of phytoremediation, for example, phytoextraction, phytostabilization, rhizodegradation, and phytodegradation, (Kvesitadze *et al.*, 2015). Inorganic and organic contaminants accumulate in plants because of their natural ability. The organic pollutants are metabolized, and the rhizosphere facilitates bacterial degradation of organic contaminants (Abbas *et al.*, 2018). Phytoremediation convoluted the metal-tolerant hyperaccumulator plants that are capable to distillate the heavy metals in the plant tissues (Salido *et al.*, 2003). Then the generated plant biomass is harvested, collected, dehydrated, and either placed in dumping ground or supplemented to furnace feed (Kramer *et al.*, 2005). *Pteris vittata* have exceptional capacity to store and to develop an effective strategy for phytoremediation, while studies show the promising potential of *P. vittata* for As extraction (Indriolo *et al.*, 2010).

Chorella vulgaris a solitary cell phytoplankton help in the As remediation (Rehman *et al.*, 2001). There are various plants (*Pteris sort, P. longifolia, P. cretica, P. umbrosa, Pityrogramma calomelanos, Dryopteris juxtaposita* and *Nephrodium mole*) which can hype raccumulate As when growing soil include As (Hossain *et al.*, 2006; Huq *et al.*, 2004). Some other aquatic and terrestrial plant species are testified for the cleanup of As and its remediation from soil and water (Bharagava *et al.*, 2017). Aquatic plants such as *Apiumnodi florum* have help in the removal of As(V) from water, sediments and soil (Vlyssides *et al.*, 2005). Separation of As from the soil is very tough because it adhere to the soil, especially in oxidation state and low pH (Pickering *et al.*, 2000). Phytoremediation has various strategies to remediate the pollutants. Each method contains various mechanisms depending upon the collaboration among pollutants and plants (Figure 12.7). For the removal of As the important approaches are mentioned below:

A) Phytostabilisation

Phytostabilization uses contaminat-tolerant plants to stabilize mechanically degraded land to avoid bulk degradation, minimize airborne transportation and pollutant leaching. These plants only take up small quantities of Arsenic and another metals to avoid passing into the food web for wildlife (Fitz and Wenzel, 2002). It reduces the gathering of As in the plants and also decreases As leaching into the groundwater. Though this method is a short-term method, because only the transport of As is striped in the soil (Vangronsveld *et al.*, 2009).

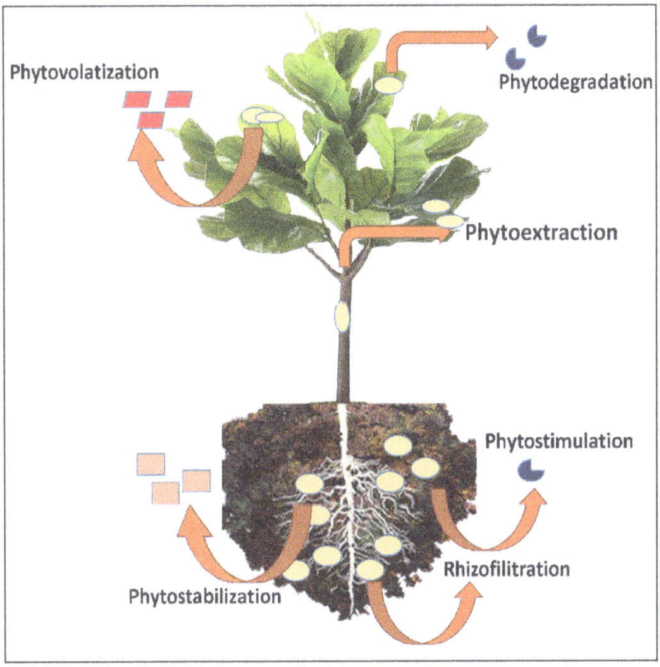

Figure 12.7: Mechanism of Various Phytoremediation Techniques and Metal Transport.

B) Phytoextraction

The uptake of contaminants from the substrate by the root system and their transport through stem are well-known as phytoextraction (Yoon *et al.*, 2006; Rafati *et al., 2011*). Capability of plants to store heavy metal in stems, depend upon their capacity of metal uptake and translocate it to the shoots preventing complex formation in the roots and metal is then seized into the vacuoles of shoot (Ali *et al.*, 2019). *P. vittata* is identified to have precise mechanisms that enhance the bioavailability of Arsenic uptake and passage to shoots. Arsenic make complex with reduced glutathione and phytochelatins, and restores most of the Arsenic as As^{3+} in vacuoles with the help of ACR3 (Arsenic Compounds Resistance 3) which is a specific transporter (Indriolo *et al.*, 2010). The left plant remains can be used in the manufacture of energy (Meers *et al.*, 2010).

C) Phytovolatilisation

Arsenic uptake from water and soil by the plants and its conversion into relatively reduced toxic volatile and organic forms, through the discharge of volatile heavy metal into air by the use of plant stems or leaf stomata is designated as phytovolatilization (Shukla *et al.*, 2019). This approach is achievable for those heavy metals that have unstable species, like Se, Hg and As. Between all the approaches, application of phytovolatilization is doubtful as it releases the metals in the air for further diffusion and is therefore hazardous for individuals employed in such region (Padmavathiamma *et al.*, 2007).

12.5.2. Microbial Remediation of Arsenic

Bioremediation by microorganisms is an important process of As removal with the technique's complexation, bio-methylation and redox reactions (Shrivastava *et al.*, 2015) (Figure 12.8). The main approaches of microbial removal of As-affected soil are: oxidation, reduction, biomethylation, solubilisation, biosorption and microbe-assisted remediation (Hughes *et al.*, 2002). Oxidation and reduction involves different ions of metal which have a dramatic impact on As movability. Microbes formed their biochemical paths to use as oxyanions, and as an acceptor of electrons. As $^{+5}$ for anaerobic respiration, or as donor of electrons promoting carbon fastening in the cell (Santini *et al.*, 2000; Rhine *et al.*, 2006). Even though bioremediation suffers sever limitation, due to their cost effectiveness it is gaining interest for the remediation of metal from the contaminated soil (Komarek *et al.*, 2013). Bioremediation technology is divided into two categories: engineered bioremediation and intrinsic bioremediation. Intrinsic bioremediation is usually referred to the deprivation of As by microorganisms and this method is more appropriate for the cleanup of soil (Azubuike *et al.*, 2016). Some of *Pseudomonas* species also help in bioremediation of arsenic in the environment like *P. putida* and *P. arsenitoxidans* (Shrivastava *et al.*, 2015). Engineered bioremediation often relies on involvement of human for enhancing the environmental conditions to support the proliferation and activity of microbes that lived in that area. As a result, the practice of engineered bioremediation is more favorable in the highly polluted areas (Mahimairaja *et al.*, 2005).

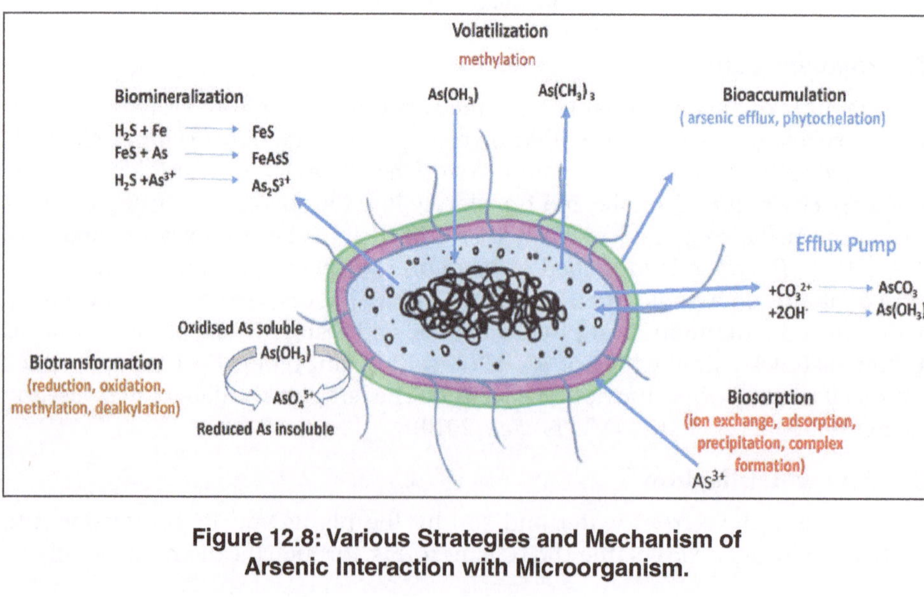

Figure 12.8: Various Strategies and Mechanism of Arsenic Interaction with Microorganism.

A) Oxidation/Reduction

Oxidation cycle extracts energy for microbial production. The enzyme Arsenite oxidase plays a crucial role in the oxidation of arsenites. Arsenite oxidase was described and placed in the family of DMSO reductase. Like other oxidases of arsenite, the one isolated from *Hydrogenophaga* sp. consists a heterodimer containing Fe and Mo as part of its catalytic assembly. Oxidation leads to As non-availability by immobilizing it, decreasing its toxicity in acidic water flow and neutralizing it with alkaline environment. The conversion of As (V) to As (III) in an anaerobic environment is intervened by mixed anaerobic communities such as fermentative, methanogens, SO_4^{2-} and Fe -reducing bacteria. Even when As is found bounded with $Fe(OH)_3$ microbial reduction of As (V) may occur (Langner and Inskeep, 2000). Zobrist *et al.* (2000) stated that As (V) reduction could result in its mobilization, without dissolving the sorbent process. In the presence of *Sulfurospirillum barnesii* the anaerobically incubating As (V) is co-precipitated with As hydroxide. In addition, the microbial reduction of As (V) to As (III) can occur relatively quickly under aerobic conditions in As-polluted soil, resulting in excessive mobilization and transportation from polluted soil to groundwater (Macur *et al.*, 2001). Arsenate (AsO_4^{3-}) reduction is a part of the respiration of anaerobic AsO_4^{3-} in certain bacteria *e.g. Shewanella*, where AsO_4^{3-} serve as a terminal electron acceptor.

B) Biomethylation

In the biological systems the oxidation state of Arsenic in the methylated and inorganic species is an important determinant of toxicity (Aposhian *et al.*, 2000). Inorganic As contains arsenate and arsenite and can be either methylated or dimethylated to form MMA (V) to DMA (V), (Hughes *et al.*, 2002). Binding of dihydrolipoamide Trivalent to the sulfydryl groups of inorganic arsenic hinders pyruvate dehydrogenase, consequently reduces the pyruvic acid to acetyl coenzyme A (CoA), while Krebs cycle and construction of ATP are reduced (Cullen *et al.*, 2014). It also prevents the uptake of glucose into the cells, fatty acid oxidation and the production of acetyl CoA (Miller *et al.*, 2002). The harmfulness of As (V) is due to its conversion to As(III), from which the noxious effects progress. At the significant level, pentavalent As compete with inorganic PO_4^{3-} and exchanges phosphate into the cellular and glycolytic respiration pathway (Hughes *et al.*, 2002). Methylation of the inorganic As has been considered as a decontamination mechanism (Cohen *et al.*, 2006).

C) Biosorption and Biofilm

Microorganisms bind and grow irreversibly on the surface and generate extracellular polymers that promote bonding and matrix development resulting in rate of growth and transcription of genes (Pandey *et al.*, 2018). There are many micro-organisms that can form microniches and biofilms which may coexist with variable physiological needs in a biofilm, allowing the coexistence of spatially different contradictory redox progressions at the same time (Labrenz *et al.*, 2000). In Arsenic polluted environment microbes stimulate oxidation and reduction, Arsenic methylation and redox transformation (Huang *et al.*, 2016). Arsenic-resistant halophilic bacteria *Bacillus vietnamensis* and *Kocuria flava* from the mangrove rhizosphere of Sundarban can endure 35 mM and 20 mM of As^{+5}, respectively.

In biosorption, retaining of metal on the cell surface occurs with the help of cationic elements (Gadd *et al.*, 2010). Amino, amide and Hydroxyl groups which are present in microbial cellwall and pH are responsible for the sorption of Arsenic (Giri *et al.*, 2013; Prasad *et al.*, 2013). Bacteria's like, *Bacillus subtilis, Bacillus cereus* (Hossain *et al.*, 2006), and many fungi like *Aspergillus niger, Penicillium purpurogenum* and *Penicillium chrysogenum* (Loukidou *et al.*, 2003; Pokharel *et al.*, 2006) showed biosorption activity with As(V), As(III) and MMA.

D) Bioaugmentation and Biostimulation

Bioaugmentation is the mechanism that takes advantage of metal-immobilizing microbes to turn highly toxic metal at the site itself into a less toxic state. Throughout biostimulation, the microbes already existing at the site are given a stimulus (Azubuike *et al.*, 2016). The stimulus might be a supplement, substrate for growth with or without electron acceptor or donor. It could be done through the use of biosurfactant foam technology to introduce nutrients known as biostimulators or microbial species called as bioaugmentors in the soil (Hug *et al.*, 2004). Biosurfactants improve metal ion availability [Fe^{3+}, As *etc.*] to microbes by decreasing interfacial stress and micelle formation.

E) Biomineralisation

During biomineralisation, the precipitation, *e.g.* As sulphide and $FeAsO_4$Å"$2H_2O$, keeps the As immobile. Around 300 As compounds are known to be present in different environments (Drahota *et al.*, 2009). The production of free Ca^{2+} limits the As movability by precipitation of $Ca_3(AsO_4)_2$ in As-contaminated suspended aquifer (Martinez-Villegas *et al.*, 2013).

12.5.3. Chemical Remediation

Compound remediation moves towards for the most part, including strategies, for example, adsorption by utilizing explicit media, immobilization, altered coagulation alongside filtration, precipitations, immobilizations, and complexation responses (Mahimairaja *et al.*, 2005). The coagulation alongside filtration technique for expelling arsenic from polluted sources is very economics however regularly show lower efficiencies (Duarte *et al.*, 2009). The development of stable stages, for instance, insoluble Iron (III) Arsenate and hydrous types of compound $FeAsO_4$Å"$2H_2O$, is valuable for the stabilization procedure (Komarek *et al.*, 2013). Remediation of Arsenic through Chemical methods are carried out by several approaches such as oxidation, flocculation, precipitation, coagulation, electrokinetics, ion exchange, adsorption *etc.*

A) Oxidation and Precipitation

The adding of chemical oxidants allows the lethal mobile As^{+3} to be converted to less toxic immobile As^{+5}. As^{+3} is oxidized by many organic oxidants like gaseous Cl, chloroamine, ClO_2, permanganate, H_2O_2, FeO_4^{2-} and Fenton's reagent (H_2O_2/Fe^{2+}) (Johnston *et al.*, 2001; Lee *et al.*, 2003; Vasudevan *et al.*, 2006; Sharma *et al.*,

2009; Mondal *et al.*, 2013). Photochemical oxidation uses UV rays for the oxidation of As^{+5} in the presence of catalyst like As sulfite, Fe_2O_3 or $C_6H_8O_7$ and oxidation of more lethal As^{+3} to less lethal As^{+5} and after oxidation As^{+5} is precipitated (Cantu *et al.*, 2016). The irradiation of UV rays photolysis $FeOH^{2+}$ helps to generate hydroxyl radicals in presence of oxygen (Yoon and Lee, 2005). Oxidation may also help to decrease Arsenic content in groundwater (Gupta *et al.*, 2009). The dissolved As can be remediate by forming compact minerals like $Ca_3(AsO_4)_2$, which has low solubility. These low-solubility solid minerals can be precipitated and removed by filtration and sedimentation (Duarte *et al.*, 2009). Coagulant such as ferric oxide, sulfate and alum are efficient in the removal of Arsenic from H_2O through the cycle of coagulation flocculation (Mondal *et al.*, 2007). Effluent water and iron in electrocoagulation produces loosely clumped mass of particle (Van Genuchten *et al.*, 2012). Electrocoagulation-chemical Arsenic Remediation (ECAR) model is based upon the principle of electrocoagulation (Amrose *et al.*, 2013).

B) Ion Exchange and Adsorption

Arsenic is having a strong affinity with minerals, like aluminum and iron hydroxide which adsorb As ions onto their surfaces (Sylvester *et al.*, 2007). After adsorption, pollutant can be refined by filtration and gravity flustering. Iron Based Sorbents (IBS) is a developing practice for the remediation of As. Hydroxyl groups are present in this technique as absorbent (Selvin *et al.*, 2000). Over the last decade this method has been modified to Zero valent iron technique because of non-toxic, cheap, abundant and easily available resource. In Zero valent Fe, oxygenated water comes in contact with Fe (II), Fe (III) and Z (VI) hydroxides to oxidize As and help to remediate it (Leupin *et al.*, 2005). In modern trend wide variety of absorbent like activated carbon fly ash, nanoparticles, aluminium loaded coral limestone, modified fly ash, or hydrous ferric oxide are used for remediation of As (Goswani *et al.*, 2000).

C) Immobilization

The use of organic chemicals which immobilize the contaminants and thus reduce the possible hazards triggered by the pollutants in the environment is referred as Immobilization (Palansooriya *et al.*, 2020). Because of its low economical cost and accessibility of process, attention has been paid towards the immobilization. Metal compounds (like Fe oxides), biochar and trash waste are generally used as immobilizers for As (Doherty *et al.*, 2017). The positive charged iron oxides formed complexes with negative charged As^{+5}, thus reducing the mobility of As. As sorption on the surface of iron oxides, precipitation of Fe with As^{+5} and oxidation of As^{+3} to As^{+5} could come up with immobilization. Fe helps to determine the behavior of Arsenic and has been used to develop a suitable technique for mobilization or immobilization (Violante *et al.*, 2012). In the course of remediation, the relationships between As and Fe, particularly their transformation of speciation and its impact on the mobility of Arsenic, require more investigations (Caporale, and Violante, 2016).

12.6 Conclusion

With the growth of industries arsenic pollution is a challenging task due of its toxicity toward the plants, humans and animals. As entering in the environment with different sources has become very toxic. For the remediation of such sites, phytoremediation and biological approaches seem to be the best idea. It is necessary to develop a combined use of the plants, microbes and agronomic practices for the remediation of As. The chapter discusses the literature available on the occurrence, behavior, processing, toxicity and removal of arsenic technologies used in the ecological media. Awareness and understanding of the origin, existence and transfer and chemistry of As in the ecological society is important due to its possible toxic effect towards the environment and human health, as well as for its effective removal of pollutants from the polluted soil and water. This chapter is pile up of the information from source to removal and provide a full knowledge of As in the environment. Knowledge of the toxicity and possible environmental impacts of arsenic exposure across various routes is very significant. A range of efficient, promising and environment sustainable remediation technologies are available for the management of soil and water polluted with As.

REFERENCES

Abbas, G., Murtaza, B., Bibi, I., Shahid, M., Niazi, N. K., Khan, M. I., and Hussain, M. (2018). Arsenic uptake, toxicity, detoxification, and speciation in plants: physiological, biochemical, and molecular aspects. International journal of environmental research and public health, 15(1), 59. https: //doi.org/10.3390/ ijerph15010059

Ali, W., Rasool, A., Junaid, M., and Zhang, H. (2019). A comprehensive review on current status, mechanism, and possible sources of arsenic contamination in groundwater: a global perspective with prominence of Pakistan scenario. Environmental geochemistry and health, 41(2), 737-760.

Amrose, S., Gadgil, A., Srinivasan, V., Kowolik, K., Muller, M., Huang, J., and Kostecki, R. (2013). Arsenic removal from groundwater using iron electrocoagulation: effect of charge dosage rate. Journal of Environmental Science and Health, Part A, 48(9), 1019-1030. https: //doi.org/10.1080/10934 529.2013.77321.

Anjum, N. A., Pereira, M. E., Ahmad, I., Duarte, A. C., Umar, S., and Khan, N. A. (2012). Phytotechnologies: remediation of environmental contaminants. CRC Press

Aposhian, H. V., Zheng, B., Aposhian, M. M., Le, X. C., Cebrian, M. E., Cullen, W., and Andrewes, P. (2000). DMPS–arsenic challenge test: II. Modulation of arsenic species, including monomethylarsonous acid (MMAIII), excreted in human urine. Toxicology and applied pharmacology, 165(1), 74-83. https: // doi.org/10.1006/taap.2000.8922

Argos, M., Ahsan, H., and Graziano, J. H. (2012). Arsenic and human health: epidemiologic progress and public health implications. Reviews on environmental health, 27(4), 191-195. https: //doi.org/10.1515/reveh-2012-0021

Arnold, L. L., Suzuki, S., Yokohira, M., Kakiuchi-Kiyota, S., Pennington, K. L., and Cohen, S. M. (2014). Time course of urothelial changes in rats and mice orally administered arsenite. Toxicologic pathology, 42(5), 855-862.

Awasthi, S., Chauhan, R., Srivastava, S., and Tripathi, R. D. (2017). The journey of arsenic from soil to grain in rice. Frontiers in Plant Science, 8, 1007.

Ayala-Fierro, F. (2005). Arsine. Encyclopedia of Toxicology, 173–176.

Azubuike, C. C., Chikere, C. B., and Okpokwasili, G. C. (2016). Bioremediation techniques–classification based on site of application: principles, advantages, limitations and prospects. World Journal of Microbiology and Biotechnology, 32(11): 11 https: //doi.org/10.1007/s11274-016-2137-x

Basu, A., Saha, D., Saha, R., Ghosh, T., and Saha, B. (2014). A review on sources, toxicity and remediation technologies for removing arsenic from drinking water. Research on Chemical Intermediates, 40(2), 447-485. https: //doi.org/10.1007/s11164-012-1000-4

Bharagava, R. N., Chowdhary, P., and Saxena, G. (2017). Bioremediation: an eco-sustainable green technology: its applications and limitations. In Environmental pollutants and their bioremediation approaches (pp. 1-22). CRC Press.

Bhattacharyya, R., Chatterjee, D., Nath, B., Jana, J., Jacks, G., and Vahter, M. (2003). High arsenic groundwater: mobilization, metabolism and mitigation–an overview in the Bengal Delta Plain. Molecular and Cellular Biochemistry, 253(1-2), 347-355. https: //doi.org/10.1023/A: 1026001024578

Bhowmick, S., Nath, B., Halder, D., Biswas, A., Majumder, S., Mondal, P., and Roman-Ross, G. (2013). Arsenic mobilization in the aquifers of three physiographic settings of West Bengal, India: understanding geogenic and anthropogenic influences. Journal of hazardous materials, 262, 915-923. https: //doi.org/10.1016/j.jhazmat.2012.07.014

Calvo, C., Bolado, S., Alvarez-Benedí, J., and Andrade, M. A. (2006). Arsenic uptake and accumulation in curly endives (Cichorium endivia L.) irrigated with contaminated water. Journal of Environmental Science and Health Part B, 41(4), 459-470. http: //dx.doi.org/10.1080/03601230600634778

Cantu, J., Gonzalez, L. E., Goodship, J., Contreras, M., Joseph, M., Garza, C., and Parsons, J. G. (2016). Removal of arsenic from water using synthetic Fe7S8 nanoparticles. Chemical Engineering Journal, 290, 428-437.

Caporale, A. G., and Violante, A. (2016). Chemical processes affecting the mobility of heavy metals and metalloids in soil environments. Current Pollution Reports, 2(1), 15-27.

Chakraborti, D., Das, B., Rahman, M. M., Chowdhury, U. K., Biswas, B., Goswami, A. B., and Hossain, A. (2009). Status of groundwater arsenic contamination in the state of West Bengal, India: A 20 year study report. Molecular nutrition and food research, 53(5), 542-551. https: //doi.org/10.1002/mnfr.200700517

Chou, Y. C., Lu, Z. Y. J., and Lin, C. J. (2018). Comparison of attitudes to the sexual health of men and women with intellectual disability among parents, professionals, and university students. Journal of Intellectual and Developmental Disability, 43(2), 164-173. https: //doi.org/10.3109/13668250.2016.1259465

Chung, J.-Y., Yu, S.-D., and Hong, Y.-S. (2014). Environmental Source of Arsenic Exposure. Journal of preventive medicine and public health, 47(5), 253-257. http: //doi.org/10.391/jpmph.14.036

Cohen, S. M., Arnold, L. L., Eldan, M., Lewis, A. S., and Beck, B. D. (2006). Methylated arsenicals: the implications of metabolism and carcinogenicity studies in rodents to human risk assessment. Critical reviews in toxicology, 36(2), 99-133. https: //doi.org/10.1080/10408440500534230

Colognato, R., Coppede, F., Ponti, J., Sabbioni, E., and Migliore, L. (2007). Genotoxicity induced by arsenic compounds in peripheral human lymphocytes analysed by cytokinesis-block micronucleus assay. Mutagenesis, 22(4), 255-261. https: //academic.oup.com/mutage/article/22/4/255/1077099.

Cullen, W. R. (2014). Chemical mechanism of arsenic biomethylation. Chemical research in toxicology, 27(4), 457-461. https: //pubs.acs.org/doi/abs/10.1021/tx400441h.

Das, S., Unni, B., Bhattacharjee, M., Wann, S. B., and Rao, P. G. (2012). Toxicological effects of arsenic exposure in a freshwater teleost fish, Channa punctatus. African journal of biotechnology, 11(19), 4447-4454. https: //doi.org/10.5897/AJB11.2808.

Datta, S., Ghosh, D., Saha, D. R., Bhattacharaya, S., and Mazumder, S. (2009). Chronic exposure to low concentration of arsenic is immunotoxic to fish: role of head kidney macrophages as biomarkers of arsenic toxicity to Clarias batrachus. Aquatic Toxicology, 92(2), 86-94. https: //doi.org/10.1016/j.aquatox.2009.01.002.

Doherty, S. J., Tighe, M. K., and Wilson, S. C. (2017). Evaluation of amendments to reduce arsenic and antimony leaching from co-contaminated soils. Chemosphere, 174, 208-217. https: //doi.org/10.1016/j.chemosphere.2017.01.100

Drahota, P., and Filippi, M. (2009). Secondary arsenic minerals in the environment: a review. Environment international, 35(8), 1243-1255. https: //doi.org/10.1016/j.envint.2009.07.004

Duarte, A. A., Cardoso, S. J., and Alçada, A. J. (2009). Emerging and innovative techniques for arsenic removal applied to a small water supply system. Sustainability, 1(4), 1288-1304. https: //doi.org/10.3390/su1041288

Environmental Protection Agency. Inorganic arsenic: toxicity and exposure assessment for children's health [cited 2014 Sep 16]. Available from: http: //www.epa.gov/teach/chem_summ/Arsenic_summary.pdf.

Finnegan, P., and Chen, W. (2012). Arsenic toxicity: the effects on plant metabolism. Frontiers in physiology, 3, 182. https: //doi.org/10.3389/fphys.2012.00182.

Fitz, W. J., and Wenzel, W. W. (2002). Arsenic transformations in the soil–rhizosphere–plant system: fundamentals and potential application to phytoremediation. Journal of biotechnology, 99(3), 259-278. https: //doi.org/10.1016/S0168-1656(02)00218-3.

Flanagan, S. V., Johnston, R. B., and Zheng, Y. (2012). Arsenic in tube well water in Bangladesh: health and economic impacts and implications for arsenic mitigation. Bulletin of the World Health Organization, 90, 839-846.

Flora, S. J. S. (2015). Handbook of arsenic toxicology. Academic Press 1–49. https: //doi.org/10.1016/b978-0-12-418688-0.00001-0.

Fowler, B. A., Selene, C.-H., Chou, Robert, J., Jones, Dexter, L., Sullivan Jr, W., and Chen, C.-J. (2015). Arsenic. Handbook on the Toxicology of Metals, 581–624. https: //doi.org/10.1016/b978-0-444-59453-2.00028-7.

Francis, O. O., Xiang, L., and Alepu, O. E. (2017). In-situ phytoremediation of arsenic from contaminated soil. Int J Waste Resour, 7(1), 1-5. https: //doi.org/10.4172/2252-5211.1000261

Gadd, G. M. (2010). Metals, minerals and microbes: geomicrobiology and bioremediation. Microbiology, 156(3), 609-643. https: //doi.org/10.1099/mic.0.037143-0

Gaim, K., Gebru, G., and Abba, S. (2015). The effect of arsenic on liver tissue of experimental animals (fishes and mice)–a review article. International Journal of Scientific and Research Publications, 5(5), 1-9.

Giri, A. K., Patel, R. K., Mahapatra, S. S., and Mishra, P. C. (2013). Biosorption of arsenic (III) from aqueous solution by living cells of Bacillus cereus. Environmental Science and Pollution Research, 20(3), 1281-1291. https: //doi.org/10.1007/s11356-012-1249-6

Goswani, D., and Das, A. K. (2000). Removal of arsenic from drinking water using modified fly ash bed, Int.J.Water, 1: 61–70.

Gupta, B. S., Chatterjee, S., Rott, U., Kauffman, H., Bandopadhyay, A., DeGroot, W., and Mukherjee, S. (2009). A simple chemical free arsenic removal method for community water supply–A case study from West Bengal, India. Environmental Pollution, 157(12), 3351-3353. https: //doi.org/10.1016/j.envpol.2009.09.014

Habashi, F. (2001). Arsenic, Antimony, and Bismuth Production. Encyclopedia of Materials: Science and Technology, 332–336. doi: 10.1016/b0-08-043152-6/00069-3

Hare, V., Chowdhary, P., Kumar, B., Sharma, D. C., and Baghel, V. S. (2019). Arsenic toxicity and its remediation strategies for fighting the environmental threat. In Emerging and eco-friendly approaches for waste management (pp. 143-170). Springer, Singapore. https: //doi.org/10.1007/978-981-10-8669-4_8

Hossain, M. F. (2006). Arsenic contamination in Bangladesh–an overview. Agriculture, ecosystems and environment, 113(1 4), 1-16. https: //doi.org/10.1016/j.agee.2005.08.034

Hu, Y., Wang, X., Dong, Z., and Liu, G. (2015). Determination of heavy metals in the groundwater of the Huaibei Plain, China, to characterize potential effects on human health. Analytical Letters, 48(2), 349-359. https: //doi.org/10.1080/00032719.2014.940530

Huang, Q., Luo, L., Alamdar, A., Zhang, J., Liu, L., Tian, M., and Shen, H. (2016). Integrated proteomics and metabolomics analysis of rat testis: mechanism of arsenic-induced male reproductive toxicity. Scientific reports, 6, 32518. doi: 10.1038/srep32518 (2016)

Hughes, M. F. (2002). Arsenic toxicity and potential mechanisms of action. Toxicology letters, 133(1), 1-16. https: //doi.org/10.1016/S0378-4274(02)00084-X

Hughes, M. F., Beck, B. D., Chen, Y., Lewis, A. S., and Thomas, D. J. (2011). Arsenic exposure and toxicology: a historical perspective. Toxicological Sciences, 123(2), 305-332. https: //doi.org/10.1093/toxsci/kfr184

Huq, S. I., Bulbul, A., Choudhury, M. S., Alam, S., and Kawai, S. (2004). Arsenic bioaccumulation in a green algae and its subsequent recycling in soil of Bangladesh. Natural Arsenic in Groundwater: Occurrence, Remediation and Management, 119-124.

Huq, S. I., Joardar, J. C., Parvin, S., Correll, R., and Naidu, R. (2006). Arsenic contamination in food-chain: transfer of arsenic into food materials through groundwater irrigation. Journal of health, population, and nutrition, 24(3): 305.

Hussain, M. M., Bibi, I., Shahid, M., Shaheen, S. M., Shakoor, M. B., Bashir, S., and Niazi, N. K. (2019). Biogeochemical cycling, speciation and transformation pathways of arsenic in aquatic environments with the emphasis on algae. In Comprehensive Analytical Chemistry 85: pp. 15-51).

Indriolo, E., Na, G., Ellis, D., Salt, D. E., and Banks, J. A. (2010). A vacuolar arsenite transporter necessary for arsenic tolerance in the arsenic hyperaccumulating fern *Pteris vittata* is missing in flowering plants. The Plant Cell, 22(6), 2045-2057.

Islam, A. B. M. R., Maity, J. P., Bundschuh, J., Chen, C. Y., Bhowmik, B. K., and Tazaki, K. (2013). Arsenic mineral dissolution and possible mobilization in mineral–microbe–groundwater environment. Journal of hazardous materials, 262, 989-996.

Jaishankar, M., Tseten, T., Anbalagan, N., Mathew, B. B., and Beeregowda, K. N. (2014). Toxicity, mechanism and health effects of some heavy metals. Interdisci Toxicology, 7(2), 60-72. https: //doi.org/10.2478/intox-2014-0009

Jedynak, L., Kowalska, J., and Leporowska, A. (2012). Arsenic uptake and phytochelatin synthesis by plants from two arsenic-contaminated sites in Poland. Pol J Environ Stud, 21, 1629-33.

Johnston, R., Heijnen, H., and Wurzel, P. (2001). Safe water technology. Chapter, 6, 390. In: M. Feroze Ahmed. *et al.*, 2001 (Eds). Technologies for Arsenic Removal from Drinking Water. A compilation of papers presented at the International Workshop on Technologies for Arsenic Removal from Drinking Water. Bangladesh University of Engineering and Technology, Dhaka, Bangladesh and the United Nations University, Tokyo.

Jomova, K., and Valko, M. (2011). Advances in metal-induced oxidative stress and human disease. Toxicology, 283(2-3), 65-87. https: //doi.org/10.1016/j.tox.2011.03.001

Joshi, N., Wang, X., Montgomery, L., Elfick, A., and French, C. E. (2009). Novel approaches to biosensors for detection of arsenic in drinking water. Desalination, 248(1-3), 517-523. https: //doi.org/10.1016/j.desal.2008.05.096

Joyce, T. S., Perez, V., Garry, M. R., and Alexander, D. D. (2014). Association of low-level arsenic exposure in drinking water with cardiovascular disease: a systematic review and risk assessment. Toxicology, 323, 78-94.

Karczewska, A., Bogda, A., and Krysiak, A. (2007). Arsenic in soils in the areas of former mining and mineral processing in Lower Silesia, southwestern Poland. Trace Metals and other Contaminants in the Environment, 9, 411-440.

Khalid, S., Shahid, M., Niazi, N. K., Rafiq, M., Bakhat, H. F., Imran, M., and Dumat, C. (2017). Arsenic behaviour in soil-plant system: Biogeochemical reactions and chemical speciation influences. In enhancing cleanup of environmental pollutants (pp. 97-140). Springer, Cham. https: //doi.org/10.1007/978-3-319-55423-5_4

Khlifi, R., and Hamza-Chaffai, A. (2010). Head and neck cancer due to heavy metal exposure via tobacco smoking and professional exposure: a review. Toxicology and applied pharmacology, 248(2), 71-88. https: //doi.org/10.1016/j.taap.2010.08.003

Komárek, M., Vanìk, A., and Ettler, V. (2013). Chemical stabilization of metals and arsenic in contaminated soils using oxides–a review. Environmental Pollution, 172, 9-22. https: //doi.org/10.1016/j.envpol.2012.07.045

Krämer, U. (2005). Phytoremediation: novel approaches to cleaning up polluted soils. Current opinion in biotechnology, 16(2), 133-141. https: //doi.org/10.1016/j.copbio.2005.02.006

Kumar, A., Kesari, V. P., and Khan, P. K. (2013). Fish micronucleus assay to assess genotoxic potential of arsenic at its guideline exposure in aquatic environment. Biometals, 26(2), 337-346. https: //doi.org/10.1007/s10534-013-9620-8

Kunito, T., Kubota, R., Fujihara, J., Agusa, T., and Tanabe, S. (2008). Arsenic in marine mammals, seabirds, and sea turtles. In Reviews of environmental contamination and toxicology (pp. 31-69). Springer, New York, NY. https: //doi.org/10.1007/978-0-387-77030-7_2.

Kvesitadze, G., Khatisashvili, G., Sadunishvili, T., and Kvesitadze, E. (2015). Plants for remediation: Uptake, translocation and transformation of organic pollutants. In Plants, Pollutants and Remediation (pp. 241-308). Springer, Dordrecht. https: //doi.org/10.1007/978-94-017-7194-8_12

Labrenz, M., Druschel, G. K., Thomsen-Ebert, T., Gilbert, B., Welch, S. A., Kemner, K. M., and Lai, B. (2000). Formation of sphalerite (ZnS) deposits in natural biofilms of sulfate-reducing bacteria. Science, 290(5497), 1744-1747.

Langner, H. W., and Inskeep, W. P. (2000). Microbial reduction of arsenate in the presence of ferrihydrite. Environmental Science and Technology, 34(15), 3131-3136.

Lee, Y., Um, I. H., and Yoon, J. (2003). Arsenic (III) oxidation by iron (VI) (ferrate) and subsequent removal of arsenic (V) by iron (III) coagulation. Environmental science and technology, 37(24), 5750-5756. https: //doi.org/10.1021/es034203+

Leupin, O. X., Hug, S. J., and Badruzzaman, A. B. M. (2005). Arsenic removal from Bangladesh tube well water with filter columns containing zerovalent iron filings and sand. Environmental Science and Technology, 39(20), 8032-8037.

Li, Y., Wang, H., Wang, H., Yin, F., Yang, X., and Hu, Y. (2014). Heavy metal pollution in vegetables grown in the vicinity of a multi-metal mining area in Gejiu, China: total concentrations, speciation analysis, and health risk. Environmental Science and Pollution Research, 21(21), 12569-12582. https: //doi.org/10.1007/s11356-014-3188-x.

Loukidou, M. X., Matis, K. A., Zouboulis, A. I., and Liakopoulou-Kyriakidou, M. (2003). Removal of As (V) from wastewaters by chemically modified fungal biomass. Water Research, 37(18), 4544-4552. https: //doi.org/10.1016/S0043-1354 (03)00415-9

Macur, R. E., Wheeler, J. T., McDermott, T. R., and Inskeep, W. P. (2001). Microbial populations associated with the reduction and enhanced mobilization of arsenic in mine tailings. Environmental science and technology, 35(18), 3676-3682. https: //doi.org/10.1021/es0105461

Mahimairaja, S., Bolan, N. S., Adriano, D. C., and Robinson, B. (2005). Arsenic contamination and its risk management in complex environmental settings. Advances in Agronomy, 86, 1-82. https: //doi.org/10.1016/S0065-2113(05)86001-8.

Martínez-Villegas, N., Briones-Gallardo, R., Ramos-Leal, J. A., Avalos-Borja, M., Castañón-Sandoval, A. D., Razo-Flores, E., and Villalobos, M. (2013). Arsenic mobility controlled by solid calcium arsenates: A case study in Mexico showcasing a potentially widespread environmental problem. Environmental Pollution, 176, 114-122. https: //doi.org/10.1016/j.envpol.2012.12.025.

Mazumder, D. G., and Dasgupta, U. B. (2011). Chronic arsenic toxicity: studies in West Bengal, India. The Kaohsiung journal of medical sciences, 27(9), 360-370.

Meers, E., Van Slycken, S., Adriaensen, K., Ruttens, A., Vangronsveld, J., Du Laing, G., and Tack, F. M. G. (2010). The use of bio-energy crops (Zea mays) for 'phytoattenuation'of heavy metals on moderately contaminated soils: a field experiment. Chemosphere, 78(1), 35-41.

Miller, W. H., Schipper, H. M., Lee, J. S., Singer, J., and Waxman, S. (2002). Mechanisms of action of arsenic trioxide. Cancer research, 62(14), 3893-3903.

Mirza, N., Mahmood, Q., Maroof Shah, M., Pervez, A., and Sultan, S. (2014). Plants as useful vectors to reduce environmental toxic arsenic content. The Scientific World Journal, 2014. https: //doi.org/10.1155/2014/921581.

Mishra, B. K., Dubey, C. S., Shukla, D. P., Bhattacharya, P., and Usham, A. L. (2014). Concentration of arsenic by selected vegetables cultivated in the Yamuna flood plains (YFP) of Delhi, India. Environmental earth sciences, 72(9), 3281-3291.

Mohan, D., and Pittman Jr, C. U. (2007). Arsenic removal from water/wastewater using adsorbents–a critical review. Journal of hazardous materials, 142(1-2), 1-53. https: //doi.org/10.1016/j.jhazmat.2007.01.006.

Mombo, S., Foucault, Y., Deola, F., Gaillard, I., Goix, S., Shahid, M., and Dumat, C. (2016). Management of human health risk in the context of kitchen gardens polluted by lead and cadmium near a lead recycling company. Journal of soils and sediments, 16(4), 1214-1224. https: //doi.org/10.1007/s11368-015-1069-7

Mondal, P., Balomajumder, C., and Mohanty, B. (2007). A laboratory study for the treatment of arsenic, iron, and manganese bearing groundwater using Fe3+ impregnated activated carbon: effects of shaking time, pH and temperature. Journal of hazardous materials, 144(1-2), 420-426. https: //doi.org/10.1016/j.jhazmat.2006.10.078

Mondal, P., Bhowmick, S., Chatterjee, D., Figoli, A., and Van der Bruggen, B. (2013). Remediation of inorganic arsenic in groundwater for safe water supply: a critical assessment of technological solutions. Chemosphere, 92(2), 157-170. https: //doi.org/10.1016/j.chemosphere.2013.01.097

Mukherjee, A., Saha, D., Harvey, C. F., Taylor, R. G., Ahmed, K. M., and Bhanja, S. N. (2015). Groundwater systems of the Indian sub-continent. Journal of Hydrology: Regional Studies, 4, 1-14. https: //doi.org/10.1016/j.ejrh.2015.03.005

Nagajyoti, P. C., Lee, K. D., and Sreekanth, T. V. M. (2010). Heavy metals, occurrence and toxicity for plants: a review. Environmental chemistry letters, 8(3), 199-216. https: //doi.org/10.1007/s10311-010-0297-8

Nandi, D., Patra, R. C., and Swarup, D. (2005). Effect of cysteine, methionine, ascorbic acid and thiamine on arsenic-induced oxidative stress and biochemical alterations in rats. Toxicology, 211(1-2), 26-35. https: //doi.org/10.1016/j.tox.2005.02.013

Niazi, N. K., Bibi, I., Fatimah, A., Shahid, M., Javed, M. T., Wang, H., and Shakoor, M. B. (2017). Phosphate-assisted phytoremediation of arsenic by Brassica napus and Brassica juncea: morphological and physiological response. International journal of phytoremediation, 19(7), 670-678. https: //doi.org/10.1080/15226514.2016.1278427

Orloff, K., Mistry, K., and Metcalf, S. (2009). Biomonitoring for environmental exposures to arsenic. J. Toxicol. Environ. Health B Crit. Rev. 12, 509–524.

Papry, R. I., Ishii, K., Al Mamun, M. A., Miah, S., Naito, K., Mashio, A. S. and Hasegawa, H. (2019). Arsenic biotransformation potential of six marine diatom species: effect of temperature and salinity. Scientific reports, 9(1): 1-16.

Padmavathiamma, P. K., and Li, L. Y. (2007). Phytoremediation technology: hyper-accumulation metals in plants. Water, Air, and Soil Pollution, 184(1-4), 105-126. https: //doi.org/10.1007/s11270-007-9401-5

Palansooriya, K. N., Shaheen, S. M., Chen, S. S., Tsang, D. C., Hashimoto, Y., Hou, D., and Ok, Y. S. (2020). Soil amendments for immobilization of potentially toxic elements in contaminated soils: a critical review. Environment international, 134, 105046. https: //doi.org/10.1016/j.envint.2019.105046

Pandey, N., Chandrakar, V., and Keshavkant, S. (2018). Mitigating Arsenic Toxicity in Plants: Role of Microbiota. In Mechanisms of Arsenic Toxicity and Tolerance in Plants pp. 191-218). Springer, Singapore.

Peña-Fernández, A., González-Muñoz, M. J., and Lobo-Bedmar, M. C. (2014). Establishing the importance of human health risk assessment for metals and metalloids in urban environments. Environment international, 72, 176-185.

Pickering, I. J., Prince, R. C., George, M. J., Smith, R. D., George, G. N., and Salt, D. E. (2000). Reduction and coordination of arsenic in Indian mustard. Plant Physiology, 122(4), 1171-1178.

Pokhrel, D., and Viraraghavan, T. (2006). Arsenic removal from an aqueous solution by a modified fungal biomass. Water Research, 40(3), 549-552. https: //doi.org/10.1016/j.watres.2005.11.040

Popowich, A., Zhang, Q., and Le, X. C. (2016). Arsenobetaine: the ongoing mystery. National Science Review, 3(4), 451-458. https: //doi.org/10.1093/nsr/nww061

Prasad, K. S., Ramanathan, A. L., Paul, J., Subramanian, V., and Prasad, R. (2013). Biosorption of arsenite (As+ 3) and arsenate (As+ 5) from aqueous solution by *Arthrobacter* sp. biomass. Environmental technology, 34(19), 2701-2708.

Rafati, M., Khorasani, N., Moattar, F., Shirvany, A., Moraghebi, F., and Hosseinzadeh, S. (2011). Phytoremediation potential of *Populus alba* and *Morus alba* for cadmium, chromuim and nickel absorption from polluted soil. International Journal of Environmental Research, 5(4), 961-970. https: //dx.doi.org/10.22059/ijer.2011.453

Rahman, F. A., Allan, D. L., Rosen, C. J., and Sadowsky, M. J. (2004). Arsenic availability from chromated copper arsenate (CCA)–treated wood. Journal of environmental quality, 33(1), 173-180. https: //doi.org/10.2134/jeq2004.1730

Rai, P. K., Lee, S. S., Zhang, M., Tsang, Y. F., and Kim, K. H. (2019). Heavy metals in food crops: Health risks, fate, mechanisms, and management. Environment International, 125, 365-385. https: //doi.org/10.1016/j.envint.2019.01.067

Rajakovic, L., and Rajakovic-Ognjanovic, V. (2018). Arsenic in Water: Determination and Removal. Arsenic: Analytical and Toxicological Studies, 9. https: //dx.doi.org/10.5772/intechopen.72058.

Rasheed, H., Kay, P., Slack, R., Gong, Y. Y., and Carter, A. (2017). Human exposure assessment of different arsenic species in household water sources in a high risk arsenic area. Science of the Total Environment, 584, 631-641.

Ratnaike, R. N. (2003). Acute and chronic arsenic toxicity. Postgraduate Medical Journal, 79(933), 391–396.

Ravenscroft, P., Brammer, H., and Richards, K. (2011). Arsenic pollution: a global synthesis (Vol. 94). John Wiley and Sons.

Rehman, A., and Shakoori, A. R. (2001). Heavy metal resistance Chlorella spp., isolated from tannery effluents, and their role in remediation of hexavalent chromium in industrial wastewater. Bulletin of environmental contamination and toxicology, 66(4), 542-547. https: //doi.org/10.1007/s001280041.

Rhine, E. D., Phelps, C. D., and Young, L. Y. (2006). Anaerobic arsenite oxidation by novel denitrifying isolates. Environmental Microbiology, 8(5), 899-908. https: //doi.org/10.1111/j.1462-2920.2005.00977.x.

Roy, S., and Bhattacharya, S. (2006). Arsenic-induced histopathology and synthesis of stress proteins in liver and kidney of *Channa punctatus*. Ecotoxicology and Environmental Safety, 65(2), 218-229. https: //doi.org/10.1016/j.ecoenv.2005.07.005.

Sakurai, T., Kojima, C., Ochiai, M., Ohta, T., and Fujiwara, K. (2004). Evaluation of in vivo acute immunotoxicity of a major organic arsenic compound arsenobetaine in seafood. International immunopharmacology, 4(2), 179-184. https: //doi.org/10.1016/j.intimp.2003.11.004

Salido, A. L., Hasty, K. L., Lim, J. M., and Butcher, D. J. (2003). Phytoremediation of arsenic and lead in contaminated soil using Chinese brake ferns (*Pteris vittata*) and Indian mustard (*Brassica juncea*). International Journal of Phytoremediation, 5(2), 89-103. https: //doi.org/10.1080/713610173.

Santini, J. M., Sly, L. I., Schnagl, R. D., and Macy, J. M. (2000). A new chemolithoautotrophic arsenite-oxidizing bacterium isolated from a gold mine: phylogenetic, physiological, and preliminary biochemical studies. Applied and environmental microbiology, 66(1), 92-97. https: //doi.org/10.1128/AEM.66.1.92-97.2000.

Selvin, N., Messham, G., Simms, J., Pearson, I., and Hall, J. (2000). The development of granular ferric media-arsenic removal and additional uses in water treatment. In Proceedings of Water Quality Technology Conference, Salt Lake City, UT (pp. 483-494).

Shahid, M., Dumat, C., Khan Niazi, N., and Khalid, S. (2018). Global scale arsenic pollution: increase the scientific knowledge to reduce human exposure. VertigO-la revue électronique en sciences de l'environnement, (Hors-série 31). https: //doi.org/10.4000/vertigo.21331

Shahid, M., Khalid, S., Abbas, G., Shahid, N., Nadeem, M., Sabir, M., and Dumat, C. (2015). Heavy metal stress and crop productivity. In Crop production and global environmental issues (pp. 1-25). Springer, Cham. https: //doi.org/10.1007/978-3-319-23162-4_1

Shankar, S., Shanker, U., and Shikha. (2014). Arsenic Contamination of Groundwater: A Review of Sources, Prevalence, Health Risks, and Strategies for Mitigation. The Scientific World Journal, 2014, 1–18. doi: 10.1155/2014/304524.

Sharma, V. K., and Sohn, M. (2009). Aquatic arsenic: toxicity, speciation, transformations, and remediation. Environment international, 35(4), 743-759. https: //doi.org/10.1016/j.envint.2009.01.005.

Shrivastava, A., Ghosh, D., Dash, A., and Bose, S. (2015). Arsenic contamination in soil and sediment in India: sources, effects, and remediation. Curr Pollution Rep 1: 35–46. https: //doi.org/10.1007/s40726-015-0004-2.

Shukla, A., and Srivastava, S. (2017). Emerging aspects of bioremediation of arsenic. In Green technologies and environmental sustainability (pp. 395-407). Springer, Cham. https: //doi.org/10.1007/978-3-319-50654-8_17.

Shukla, A., and Srivastava, S. (2019). A review of phytoremediation prospects for arsenic contaminated water and soil. In Phytomanagement of polluted sites (pp. 243-254). Elsevier. https: //doi.org/10.1016/b978-0-12-813912-7.00008-9.

Singh, A. K. (2004). Arsenic contamination in groundwater of North Eastern India. In Proceedings of 11th national symposium on hydrology with focal theme on water quality, National Institute of Hydrology, Roorkee (pp. 255-262).

Singh, H. P., Batish, D. R., Kohli, R. K., and Arora, K. (2007). Arsenic-induced root growth inhibition in mung bean (*Phaseolus aureus* Roxb.) is due to oxidative stress resulting from enhanced lipid peroxidation. Plant Growth Regulation, 53(1), 65-73. https: //doi.org/10.1007/s10725-007-9205-z.

Singh, N., Ma, L. Q., Srivastava, M., and Rathinasabapathi, B. (2006). Metabolic adaptations to arsenic-induced oxidative stress in Pteris vittata L and Pteris ensiformis L. Plant Science, 170(2), 274-282. https: //doi.org/10.1016/j.plantsci.2005.08.013.

Smedley, P. L., Knudsen, J., and Maiga, D. (2007). Arsenic in groundwater from mineralised Proterozoic basement rocks of Burkina Faso. Applied Geochemistry, 22(5), 1074-1092. https: //doi.org/10.1016/j.apgeochem.2007.01.001.

Stafilov, T., Aliu, M., and Sajn, R. (2010). Arsenic in surface soils affected by mining and metallurgical processing in K. Mitrovica Region, Kosovo. International journal of environmental research and public health, 7(11), 4050-4061. https: //doi.org/10.3390/ijerph7114050.

Sylvester, P., Westerhoff, P., Möller, T., Badruzzaman, M., and Boyd, O. (2007). A hybrid sorbent utilizing nanoparticles of hydrous iron oxide for arsenic removal from drinking water. Environmental Engineering Science, 24(1): 104-112.

Van Genuchten, C. M., Addy, S. E., PenÞa, J., and Gadgil, A. J. (2012). Removing arsenic from synthetic groundwater with iron electrocoagulation: An Fe and As K-edge EXAFS study. Environmental science and technology, 46(2): 986-994.

Vangronsveld, J., Herzig, R., Weyens, N., Boulet, J., Adriaensen, K., Ruttens, A., and van der Lelie, D. (2009). Phytoremediation of contaminated soils and groundwater: lessons from the field. Environmental Science and Pollution Research, 16(7), 765-794.

Vasudevan, S., Mohan, S., Sozhan, G., Raghavendran, N. S., and Murugan, C. V. (2006). Studies on the oxidation of As (III) to As (V) by in-situ-generated hypochlorite. Industrial and engineering chemistry research, 45(22), 7729-7732.

Violante, A., Pigna, M., Cozzolino, V., and Huang, P. M. (2012). Impact of soil physical, chemical and biological interactions on the transformation of metals and metalloids. Handbook of soil science resource of management and environmental impacts, Second Edition. CRC Press, Boca Raton, 1-30.

Vlyssides, A., Barampouti, E. M., and Mai, S. (2005). Heavy metal removal from water resources using the aquatic plant Apium nodiflorum. Communications in soil science and plant analysis, 36(7-8), 1075-1081.

Xiong, T., Dumat, C., Pierart, A., Shahid, M., Kang, Y., Li, N., and Laplanche, C. (2016). Measurement of metal bioaccessibility in vegetables to improve human exposure assessments: field study of soil–plant–atmosphere transfers in urban areas, South China. Environmental geochemistry and health, 38(6), 1283-1301.

Xu, H., Lam, S. H., Shen, Y., and Gong, Z. (2013). Genome-wide identification of molecular pathways and biomarkers in response to arsenic exposure in zebrafish liver. PloS one, 8(7): e68737.

Yoon, J., Cao, X., Zhou, Q., and Ma, L. Q. (2006). Accumulation of Pb, Cu, and Zn in native plants growing on a contaminated Florida site. Science of the total environment, 368(2-3), 456-464. https: / /doi.org/10.1016/j.scitotenv.2006.01.016

Zhao, F. J., McGrath, S. P., and Meharg, A. A. (2010). Arsenic as a food chain contaminant: mechanisms of plant uptake and metabolism and mitigation strategies. Annual review of plant biology, 61: 535-559.

Zobrist, J., Dowdle, P. R., Davis, J. A., and Oremland, R. S. (2000). Mobilization of arsenite by dissimilatory reduction of adsorbed arsenate. Environmental Science and Technology, 34(22): 4747-4753.

Chapter 13

Rapid Integrated Water Quality Evaluation of Thol Wetland Using Benthic Macroinvertebrates

Bhadrecha Mahendra[1]* *and Mankodi Pradeep*[2]

[1]*Scientific Officer, Gujarat Pollution Control Board, Sector 10-A, Gandhinagar, Gujarat*
[2]*Department of Zoology, Faculty of Science, The Maharaja Sayajirao University of Baroda, Vadodara, 390002 Gujarat*
**e-mail: pcmankodi@yahoo.com*

ABSTRACT

An integrated water quality assessment of Thol Wetland, in Mehsana District of Gujarat, India was done by studying physico-chemical parameters as well as Benthic Macroinvertebrates. Physico-chemical parameters analysed were pH, Colour, Turbidity, Conductivity, Total Solids, Total Dissolved Solids, Total Suspended Solids, Ammonical Nitrogen, Nitrate, Phosphate, Alkalinity, Total hardness, Sodium, Potassium, Calcium, Magnesium, per cent Sodium, Sodium Absorption Ratio, Dissolved Oxygen, Chemical Oxygen Demand, Biochemical Oxygen Demand, Chloride and Sulphate. The biological parameters calculated were BMWP (Bio Monitoring Working Party) Score, Saprobic Score and Sequential Comparison Index or Diversity Score. In total benthic invertebrates of 11 families were encountered. The findings indicate that the water quality of Thol Wetland is 'Moderately Polluted' with comparatively high organic content mainly of natural origin and suitable for irrigation purpose.

Keywords: *Benthic macroinvertebrates, BMWP Score, Diversity score, Biological water quality criteria, Central pollution control board (CPCB).*

13.1 Introduction

Wetlands are highly productive ecosystems that provide food and habitat to aquatic and amphibious organisms. It also provides multifarious ecosystem services that contribute to sustenance of human life on earth (Kamboj and Tatu 2017). It is thus necessary to periodically monitor the ecological health of wetlands especially the water quality. Mostly, water quality is assessed by studying various physico-chemical parameters such as pH, Biochemical Oxygen Demand (BOD), Chemical Oxygen Demand (COD), Dissolved Oxygen (DO) *etc*. However, for an overall assessment of an ecosystem, physico-chemical studies have to be supplemented with biological assessments (Cairns, and Pratt 1993; Zwart and Trivedi 1994). Central Pollution Control Board, Delhi has also established the role of biological parameters particularly Biomonitoring using Benthic Macroinvertebrates (CPCB 1995, 2002, 2016). They are indicators of water quality as these groups exhibit varying sensitivity to different pollution levels. This study is an attempt to integrate Bio monitoring and Physico chemical monitoring for effective assessment of overall water quality of Thol Wetland.

13.1.1 Study Area and General Habitat

Thol Bird Sanctuary is located about 22 km away from Kadi town of Mehsana district of Gujarat State, India. It has a total area of 699 sq. km. and the periphery is 5.62 km (Kamboj and Tatu 2017; PCCF Guj. Govt. 2016). There is a continuous earthen bund on its western, southern and eastern periphery, which helps in collection of water that flows into it during the monsoon from the catchment area (Figure 13.1). Thol water body also supports a canal based irrigation system. It remains covered with water in the rainy season. During winter it begins to dry and by summer the wetland is separated into water bodies of varying in size, the biggest being towards the western side. Study was carried out at sampling location no.1 (23° 07′ 53.9″ N, 72° 24′ 46.7″E) and Location no.2 (23° 08′ 01.8″ N, 72° 24′ 01.3″ E). Location 1 is characterized with almost no human intervention, profuse growth of water lilies in surrounding waters whereas location 2 is characterized by comparatively more human intervention with visitors camping and bird watching. Cattle wading, water pumping for irrigation was found in some spots of the Wetland. The surrounding land use mainly falls into arable, grazing and forest type. There was no discharge of water outside the wetland during the field visit.

13.2 Materials and Methods

Ample care was taken to ensure that all indicator families of Benthic Macroinvertebrates present are actually encountered. This was accomplished by sampling all different (micro) habitats in a sizeable stretch of the water body (CPCB 2016; Bhadrecha *et al.*, 2016). At each location; the water sampling point (GPS location) was fixed as reference point and keeping 250 m stretch on either side of it of the wetland bank; a total of 500 m X 2 m width of wetland stretch was covered for Benthic Macroinvertebrate collection. This 500 m stretch was divided into 5 sub stretches of 100 m length and in each sub stretch, Benthic Macroinvertebrate collection was carried out.

All possible micro habitats such as aquatic vegetation on edges of the wetland banks, algae, pebbles, wetland bed, detritus, submerged and floating vegetations *etc.* were explored for collection of Benthic Macroinvertebrates by using hand net,

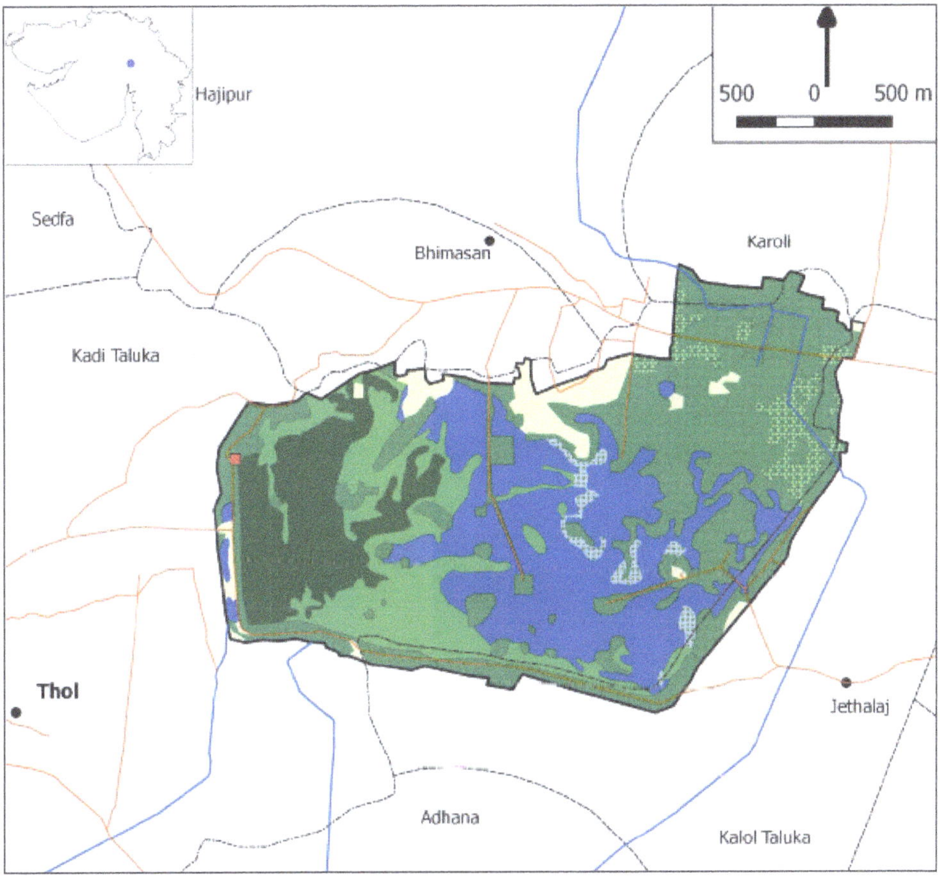

Figure 13.1: Location of Thol Wetland.

sieve and hand picking with forceps (CPCB 2002, 2016, Bhadrecha *et al.*, 2016) (Figure 13.2). Collection procedure was repeated five times in each sub stretch. Five grab samples of silt were picked from the wetland bed by the plastic scoop/shovel and washed by wetland water in a sieve with mesh size 0.6 mm. The animals were then picked by forceps from the sieve and transferred into the tray. Net was also moved all along the edges of grass/emergent aquatic vegetation all along the wetland bank and the animals were collected and transferred into the tray with forceps. The aquatic plants present in the sampling area were uprooted and washed directly into the net or into the white tray so as to detach the animals. All animals thus collected were preserved in 4 per cent formalin for further identification in the laboratory.

Grab water samples were collected from both the locations for the retrieval of physico-chemical parameters (Parikh and Mankodi, 2011) in the lab. Temperature was measured at the site. Dissolved Oxygen and COD were preserved at the sites itself using Winkler's reagent and Sulphuric acid respectively. The samples were then brought to the laboratory for further analysis. Benthic Macroinvertebrates were identified using 'Appendix: 6 Taxonomic Key for Biological Water Quality

Figure 13.2: Benthic Macroinvertebrate Sampling and Site Conditions.

Evaluation of the Manual on Integrated Water Quality Evaluation' (Zwart and Trivedi 1994). The abundance of each animal observed during sampling was also noted down.

The physico-chemical parameters of water samples were assessed as per standard Methods for Examination (APHA, 2005). Physico-chemical parameters retrieved were pH, Colour, Turbidity, Conductivity, Total Solids, Total dissolved Solids, Total Suspended Solids, Ammonical Nitrogen, Nitrate (NO_3-N), Phosphate, Alkalinity - as $CaCO_3$, Total Hardness as $CaCO_3$, Sodium, Potassium, Calcium, Magnesium, per cent Sodium, Sodium Absorption Ratio (SAR), Dissolved Oxygen, Chemical Oxygen Demand, Biochemical Oxygen Demand (3 days 27°C), Chloride as Cl^- and Sulphate. The biological parameters using Benthic Macroinvertebrates analyzed were BMWP (Bio Monitoring Working Party) Score or Saprobic Score and Sequential Comparison Index or Diversity Score (CPCB 2016).

13.2.1 Diversity Score (Sequential Comparison)

For calculation of Diversity Score, Benthic Macro-invertebrate specimens were randomly dispersed on a plastic tray marked with a grid of 1 inch x 1 inch (Figure 13.3) to facilitate the sequential comparison.

Figure 13.3: Biological Assessment of Benthic Macroinvertebrates in Laboratory.

A sequential comparison of organisms in each square of the gridded tray was carried out from left to right direction and starting from the uppermost extreme left grid and ending at lowermost extreme right grid. While comparing, the organism is assigned a mark (run) of 1 if it is different from the just previous organism and a mark of 0 if it is same. The Diversity Score is calculated as:

Diversity Score = Number of Runs/Total Number of Organisms.

13.2.2 Biological Monitoring Working Party (BMWP) Score

A quantitative inventory up to 'family' level of taxonomic precision was carried out for Benthic Macroinvertebrates collected from different microhabitats of the study area (Figure 13.4) by using BMWP score card designed by Central Pollution Control Board. All possible families having saprobic indicator value are classified on a score scale of 1 to 10. The families which are most sensitive to pollution are on the top of the list and are getting a score of 10 while the most pollution tolerant families are getting a score of 1 and 2. The other intermediately sensitive families are placed in between the scoring scale of 10 to 1. The saprobic scores of all the families were registered based on BMWP score. It was then averaged to produce Total BMWP Score. Abundance Scale of Families, A = Single (1 Individual), B =

Scarce (2-10 Individuals), C = Common (10-50 Individuals), D = Abundant (50-100 Individuals), E = Excessive (more than 100 individuals) is also noted based on field observation.

13.3 Results and Discussion

13.3.1 Biological Assessments

Total number of organisms collected at Location no. 1 and Location no. 2 were 201 and 163 respectively. Therefore the Diversity Score at Location no. 1 and Location no. 2 are found to be 105/201 = 0.52 and 81/163 = 0.49 respectively. At both Locations, 9 different families falling across 5 different taxonomic orders were found. However, in all 11 families falling across 5 taxonomic orders were encountered (Tables 13.1 and 13.2). The number of families found in each taxonomic order is depicted in the following graph (Figure 13.4).

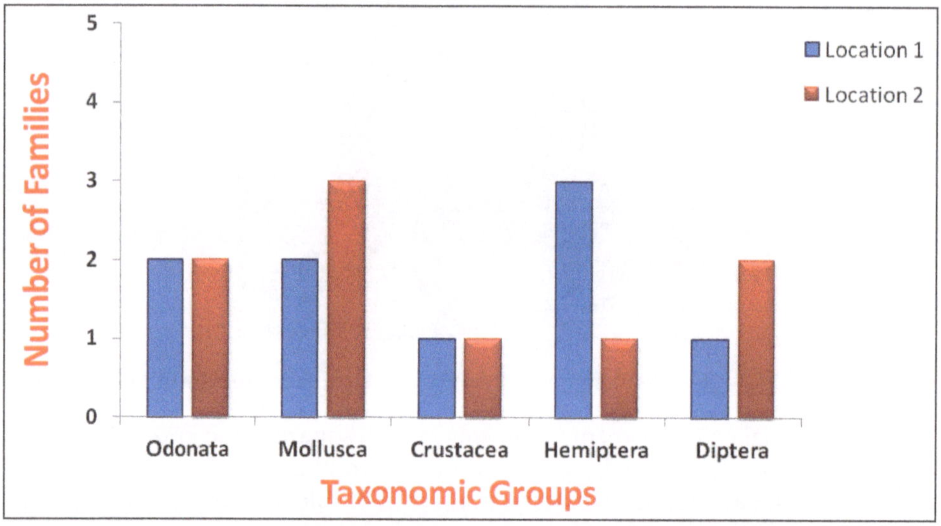

Figure 13.4: Location-wise Composition of Benthic Macroinvertebrates at Thol Wetland.

The BMWP Score was found to be 5.33 and 4.77 at location 1 and 2 respectively and the Diversity Score is found to be 0.52 and 0.49 at location 1 and 2 respectively. These values when compared with the BWQC (Biological Water Quality Criteria) developed by Central Pollution Control Board (CPCB 2002, 2016)(Table 3), collectively indicated that the water quality of Thol wetland at both the locations during the study was 'Moderately Polluted'.

13.3.2 Physico-chemical Assessment

The water samples were analyzed for pH, Colour, Conductivity, Total Solids, Total Suspended Solids, Total Dissolved Solids, Chlorides, Total Hardness, Calcium Hardness, Magnesium Hardness, Alkalinity, Turbidity, Ammoniacal Nitrogen, Chemical Oxygen Demand, Biochemical Oxygen Demand, Dissolved Oxygen,

Table 13.1: Calculation of BMWP Score (Saprobic Score) for Location No. 1

Taxonomic Group	*Taxonomic Families*	*Mark Encountered Families with Abundancy ASA, B, C, D, E*	*Total Families/ Species Encountered*	*BMWP Score*	*Multiplied Score*
Odonata	Lestidae	B			
	Gomphidae	C			
Total Families Encountered and Total Multiplied Score			2	X 8	16
Mollusca	Bithynidae	C			
Crustacea	Atydae	C			
Total Families Encountered and Total Multiplied Score			2	X 6	12
Hemiptera	Hydrometridae	C			
	Gerridae	D			
	Nepidae	B			
Total Families Encountered and Total Multiplied Score			3	X 5	15
Mollusca	Planorbidae	C			
Total Families Encountered and Total Multiplied Score			1	X 3	3
Diptera	Chironomidae	D			
Total Families Encountered and Total Multiplied Score			1	X 2	2
Grand Total Families Encountered and Grand Total Multiplied Score			**9**		**48**

Saprobic Score = Grand Total Multiplied Score ÷ Grand Total Number of Families Encountered = 48/9 = 5.33.

Table 13.2: Calculation of BMWP Score (Saprobic Score) for Location No. 2

Taxonomic Group	*Taxonomic Families*	*Mark Encountered Families with Abundancy ASA, B, C, D, E*	*Total Families/ Species Encountered*	*BMWP Score*	*Multiplied Score*
Odonata	Lestidae	B			
	Gomphidae	B			
Total Families Encountered and Total Multiplied Score			2	X 8	16
Mollusca	Bithynidae	C			
Crustacea	Atydae	B			
Total Families Encountered and Total Multiplied Score			2	X 6	12
Hemiptera	Hydrometridae	B			

Taxonomic Group	*Taxonomic Families*	*Mark Encountered Families with Abundancy ASA, B, C, D, E*	*Total Families/ Species Encountered*	*BMWP Score*	*Multiplied Score*
Total Families Encountered and Total Multiplied Score			1	X 5	5
Mollusca	Lymnaeidae	C			
	Planorbidae	C			
Total Families Encountered and Total Multiplied Score				·	
Diptera	Syrphidae	B			
	Chironomidae	D			
Total Families Encountered and Total Multiplied Score			2	X 2	4
Grand Total Families Encountered and Grand Total Multiplied Score			**9**		**43**

Saprobic Score = Grand Total Multiplied Score ÷ Grand Total Number of Families Encountered = 43/9 = 4.77

Table 13.3: Biological Water Quality Criteria (BWQC) Developed by CPCB

Sl. No.	*Range of BMWP Score*	*Range of Diversity Score*	*Water Quality Characteristics*	*Water Quality Class*	*Indicator Colour*
1	7 and more	0.2 - 1	Clean	A	Blue
2	6 - 7	0.5 - 1	Slight Pollution	B	Light Blue
3	3 - 6	0.3 - 0.9	Moderate Pollution	C	Green
4	2 - 5	0.4 and less	Heavy Pollution	D	Orange
5	0 - 2	0 - 0.2	Severe Pollution	E	Red

Sulphates and Nitrates. The values of physico-chemical parameters monitored are presented in Table 13.4.

The obtained average values of physico-chemical parameters are compared with IS: 10500 Drinking Water Specifications (IS: 2296: 1992; IS: 10500; ADSORBS/3/1978-79), and it is evident that the water quality is within the permissible limits. These obtained values are also within the usual range prescribed in FAO guidelines for irrigation water (Raychaudhuri 2014). However, the values of Chemical Oxygen Demand (COD) and Biochemical Oxygen Demand (BOD) are found to be comparatively high indicating accumulation of organic matter in wetland water (Table 13.4). This is mainly imparted by organic detritus in the form of dried leaves, twigs, flowers *etc.* falling from the surrounding trees and shrubs into the wetland water. The water column also had a greenish hue due to phytoplankton growth indicating organic content.

Table 13.4: Water Quality of Thol Wetland in Terms of Physicochemical Parameters

Physico-chemical Parameter	*Location 1*	*Location 2*	*Average Value*	*IS: 10500 Drinking Water Specification Acceptable and Permissible# Limit*	*FAO Guide-lines (1985) Usual Range in Irrigation water*
pH	7.92	9.39	8.66	6.5 to 8.5	6.0 -8.5
Colour (Hazen)	5	10	7.50	5 (15)	-
Turbidity (NTU)	2.27	2.21	2.24	1 (5)	-
Conductivity (μs/cm)	232	216.9	224.45	-	0 - 3000
Total Solids (mg/l)	166	196	181	-	
Total Dissolved Solids (mg/l)	134	156	145	500 (2000)	0 - 2000
Total Suspended Solids (mg/l)	32	40	36	-	-
Ammonical Nitrogen (mg/l)	0.29	0.42	0.36	0.5	-
Nitrate (NO_3-N) (mg/l)	0.17	0.048	0.109	45	0 - 10
Phosphate (mg/l)	0.016	0	0.008	-	0 - 2
Alkalinity - as $CaCO_3$ (mg/l)	100	80	90	200 (600)	200
Total Hardness as $CaCO_3$ (mg/l)	74.4	62.8	68.6	200 (600)	712*
Sodium (mg/l)	20.46	23.52	21.99	-	0 - 920
Potassium (mg/l)	3.1	2.5	2.80	-	0 - 2
Calcium (mg/l)	13.95	14.42	14.19	-	0 - 400
Magnesium (mg/l)	9.63	6.52	8.08	-	0 - 61
per cent Sodium	36.22	43.67	39.95	-	60^
SAR (milimole/l)	1.03	1.29	1.16	-	0 - 15
Dissolved Oxygen (mg/l)	4.55	13	8.78	-	-
COD (mg/l)	40	89	64.5	-	-
BOD (3 days 27°C) (mg/l)	10	26	18.0	-	-
Chloride as Cl^- (mg/l)	23	41	32	250 (1000)	0 - 1065
Sulphate (mg/l)	14	16	15	200 (400)	0 - 960

Permissible limits are shown in brackets.

* Values as suggested in Research Bulletin No.71, Directorate of Water Management, ICAR, 2014.

^ SAR maximum value as per BIS Standards (2002): Designated Use Class E (irrigation purpose)

Source: GEER Foundation [1].

13.4 Conclusion

The Integrated water quality of Thol wetland is observed to be 'Moderately Polluted' owing to comparatively high organic content. The results of Physico-chemical analysis are in consonance with the Biological Water Quality Criteria developed by Central Pollution Control Board. Further, the use of Biomonitoring for water quality assessment using Benthic Macroinvertebrates can be used as a complementary method along with the regular physico-chemical analysis for comprehensive water quality monitoring and thus turning out to be an effective tool for water quality management.

REFERENCES

Bhadrecha, M. H., Nitasha Khatri, and Sanjiv Tyagi(2016). 'Rapid integrated water quality evaluation of Mahisagar river using benthic macroinvertebrates.' Environmental monitoring and assessment 188: 254, 1-8.

Cairns, Jr. J. and J. R. Pratt (1993). A history of biological monitoring using benthic macroinvertebrates. 10-27 in D. M. Rosenberg and V. H. Resh (editors). Freshwater biomonitoring and benthic macroinvertebrates. Chapman and Hall, New York.

Central Pollution Control Board (September 1995), Parivesh News letter, Vol. II, Biomonitoring of Water.

Central Pollution Control Board (May 2002), Parivesh News letter, Biomonitoring of Water Bodies Part – II.

Central Pollution Control Board (July 2016). 'Benthic Macro invertebrates of River Ganga', Prepared and Published by PR Division, CPCB, Delhi- 110032.

D. de. Zwart (RIVM Bithoven, The Netherlands), R.C. Trivedi (CPCB, Delhi, India), (January 1994). Manual on Integrated Water Quality Evaluation; Appendix: 6 Taxonomic Key for Biological Water Quality Evaluation (and adopted by Gujarat Pollution Control Board, 2006 - as Manual on Biomonitoring of water bodies), Indo-Dutch Project Report 802023003.

IS: 2296: 1992 Water Quality Standards in India

IS: 10500 Drinking Water Specifications.

Kamboj R.D. and K. Tatu (2017): Important Wetland Destinations of Gujarat – A guide for ecotourists to explore some wetland jewels in Gujarat, Gujarat Ecological Education and Research (GEER) Foundation, Gandhinagar.

Principal Chief Conservator of Forest and Head of the Forest Force Govn. of Gujarat (2017) https: //forests.gujarat.gov.in, accessed on 21.12.2016 and 08.01.2018.

Parikh, A. N. and Mankodi, P.C. (2011). Water Quality Assessment of Harni Pond of Vadodara, Gujarat, Electronic Journal of Environmental Science, Vol. 4, 55-59.

Primary water quality criteria for various uses of fresh waters, Central Pollution Control Board, New Delhi, ADSORBS/3/1978-79.

Raychaudhuri, M., Raychaudhuri, S., Jena, S. K. Kumar, A. and Srivastava R.C. (2014). WQI to Monitor Quality for Irrigation and Potable Use. DWM Bulletin No. 71 Directorate of Water Management, Bhubaneshwar, India, 43p.

Water Environmental Federation, APH Association (2005). Standard methods for the examination of water and wastewater, 21st edition, APHA, AWWA, WEF.

Chapter 14

Role of Microbes in Bioremediation of Pollutants (Hydrocarbon and Pesticides) in Contaminated Soil and Water

Krupa Unadkat and Punita Parikh

Department of Botany, Faculty of Science, The Maharaja Sayajirao University of Baroda, Vadodara – 390 002, Gujarat
e-mail: unadkatkrupa@yahoo.co.in, punitaparikh@gmail.com

ABSTRACT

Environmental pollution is an anonymous process. Various anthropogenic activities, increase in population, globalization and urbanization, are major sources of environmental pollution. All these lead to addition of harmful pollutant in soil and in water. Major sources of soil and water pollutions, are pesticides and hydrocarbons. They have more negative impacts on ecological system when compared to its desired action. Pollutants remain in surface soil, leading to the toxicity of soil-water environment which are then taken up by organisms as food. When organisms in the higher food chain consume the organisms that may contain some toxins, they get biomagnified. Many conventional methods are used to remove contamination from soil and water but it have many disadvantages. Bioremediation is alternative solution to overcome these challenges. Bioremediation is invaluable technology for removal of various pollutants from environment. Bacteria, Archea and fungi are typical prime bioremediators. Bioremediation involves microorganisms for removal of many harmful pollutants through biodegradation. These microorganisms, in order to obtain energy and thus for production of biomass, utilize toxic pollutants of the environment. This novel technology of breaking down or transformation of toxic compounds into less toxic or nontoxic compounds through microbial bioremediation process is widely applicable technique of today. In this context, the present report is anticipated to provide basic perceptive of microbes useful in bioremediation process.

Keywords: *Microbes, Bioremediation, Pollutants, Pesticides, Soil, Water.*

14.1 Introduction

Soil contamination due to various anthropogenic activities has become a Global problem. A variety of xenobiotic compounds are released into soil due to intensification of Agriculture and inflation of industries. This poses severe ecological problems as they greatly affect wide range of organisms including humans resulting into imbalanced nature. Several chemical, thermal and physical processes such as excavation and polluted soil transportation are tried by the scientist all over the world. Researches in this area have proved that these methods are not only costly and labour intensive, but these also bring about undesirable and abrupt changes in important soil characteristics. To overcome this problem, certain alternative methods are available that have resulted in the development of bioremediation techniques. Microorganisms are widely distributed in the biosphere because of their metabolic ability and are very impressive as they can easily grow in a wide variety of environmental conditions.

The great significance of exploiting microorganisms for biodegradation of pollutants is the nutritional versatility. So these are used as bioremediations of various environmental pollutants like oils, Hydrocarbons, Heavy metals, dyes, pesticides *etc.* This is accomplished in enzymatic way through metabolic processes. Bacteria, archea and fungi are the examples of paramount bioremediators.

14.2 Pesticides and Hydrocarbons

Among diverse toxic compounds, Pesticides have made boundless contribution to the fight against pests and diseases caused by them. But simultaneously the one that are used to eradicate the pest create more negative impacts on ecological systems. Although they play a significant role in augmenting food production, their large scale and indiscriminate use in India has resulted in highly unstable ecosystem, development of resistance by insects, elimination of parasites, predators and pollinators, resurgence of minor pests and destruction of useful insects.

A hydrocarbon is any organic compound that consist of carbon and hydrogen atoms. Because of the unique covalent nature of carbon, there are thousands of hydrocarbons in the world. Gasoline, petroleum, kerosene, natural gas *etc.* are few forms of hydrocarbons. As they are used as fuel and also hvae applications in the manufacturing sector, there are chances of their spillage in the environment. It is believed that 60 per cent of air pollution is caused due to automobiles that run on these hydrocarbon-based fuels (Das and Adholeya 2012).

The major sources of hydrocarbon contaminants in wastewater effluents are oil spillage, automobile oils, pesticides, contaminated lands and urban storm water discharges. One of the main sources of hydrocarbon in wastewater is through oil

Table 14.1: Classification of Pesticides Based on their Chemical Nature

No	Chemical Group	Chemical names
1	Organochlorines	DDT,DDD, Dicofol, Eldrin, Dieldrin, Chlorobenziate, Lindane, BHC, Methoxychloro Aldrin, Chlordane, Heptaclor, Endosufan, Isodrin, Isobenzan, Toxaphene, Chloro propylate
2	Organophosphates	Dimefox, Mipafox, Methyl Parathion, Ronnel, enitrothion, Bidrin, Phorate, Fenthion, caumphos, Abate, Dichlorovas, Diptrex, Phosphomidon, Demetox, Oxydemeton-methyl, Malathion, Dimethoate, Trichlorofan
3	Carbamates	**Methyl** Carbaryl, Carbanolate, Prupoxur, Dimethan, Dimetilan, Isolan, Carbofuran, Pyrolan, Aminocarb, Aldicarb **Thio** Vernolate, Pebulate, Diallate, Monilate, Butylate, Cycloate, Trillate, Thiourea **Dithio** Methan, Thiram, Ferban, Amoban, Naban, Zineb, Maneb, Ziram Polyran, Dithane M- 45
4	Pyrethroids	Allethrin, Bonthrin, Dimethrin, Tetramethrin, Ptrethrin, Cyclethrin Furethrin, Fenevelerate, Alphamethrin, Decamethrin, Cypermethrin
5	Phenyl amides	**Carbanilates** Barban, Carbetamide, Chlororprofan, Prophan, Phenyl Urea, Fenuron, Monuron, Diuron,Flumeturon, Chloroxuron, Neburon, Bromuron **Acylanalide** Propanil, Solan, Dicryl, Karsil, Propachlor, Alachlor, Butachlor **Toluidines** Trifluralin, Dipropanil, Benefin, Oryzalin, Isopropanil, Nitralin **Acetamide** Diphenamid
6	Phenoxy alkonates	2,4-D(2,4 Dichloro phenoxy acetic acid) 2,4 5 T(2,4 5 Trichloro Phenoxy acetic acid) Dichloroprop, Mecoprop, Erbin, Sesone
7	Trazines	Atrazine, Simazine, Ametryn, Atratone, Chlorazine, Cynazine, Cyprazine, Metribuzin, Propazine, Turbutryn, Simetryn
8	Benzoic acid	Dicamba, Dichlorobenil, Chloroambin, Tricamba, Neptalan, Bromoxynil
9	Phtalimides	Captan, Diflotan, Folpet
10	Dipyrids	Paraquat, Diaquat
11	Others	Pentachlorophenol, Floroacetate, Phenyl mercuric acetate, Ethyl mercuric Phosphate,Methyl mercuric chloride, Sodium arsenate, Calcium arsenate, Lead arsenate, Cacodylic acid, Aluminium phosphide, Zinc phosphide

Source: Jayaraj, *et al.* (2016).

Table 14.2: Major Organochlorine Pesticides, their Chemical Structures, Uses and Toxicity

Sl. No.	*Chemical*	*LD_{50} Rat*	*Use*	*Persistence*	*WHO Classification Hazard Level*
1.	DDTC14H9Cl5	Oral 113-130 mg/kg Dermal 2510mg/kg	Acaricide Insecticide	Half life 2-15 years	Moderately hazardous
2.	DDD (p-chlorophynyl)ethane	4000mg/kg	Insecticide	5-10 years	Acute hazard unlikely
3.	DDE Dichlorodi-phenyldichlo-roethane	800-1240	Insecticide	10 years	Slightly
4	Dicofol $C_{14}H_9Cl_5O$	684-1495	Acaricide	60 Days	Moderately
5	Endrin $C_{12}H_8Cl_6O$	1.37g/kg	Avicide, Insecticide	1 day to 12 years	Highly
6	Dieldrin $C_{12}H_8Cl_6O$	46mg/kg	Insecticide	Highly persistent9 months	Highly
7	Methoxychlor	5to6g/kg	Insecticide	120 days	Acute Hazard unlikely

Sl. No.	Chemical	LD_{50} Rat	Use	Persistence	WHO Classification Hazard Level
8	Chlordane $C_{10}H_6Cl_8$	200 to 700mg/kg	Insecticide	High 10 years	Moderately
9	Heptachlor $C_{10}H_5Cl_2$	40-220 mg/kg	Insecticide	High 2 Years	High to Moderately
10	Lindane $C_6H_6Cl_6$	88-270mg/kg rat	Acaricide, insecticide, rodenticide	High persistence 15 month	Moderately
11	Endosulfan $C_9H_6Cl_6O_3S$	88-220mg/kg rat	insecticide,	Moderately persistence Alpha 35 days Beta 150 days	Highly Hazardous
12	Isodrin $C_{12}H_8Cl_6$	8.8mg/kg- rat	Insecticide	High persistence 0.5-6 years	Highly Hazardous
13	Aldrin $C_{12}H_8Cl_6$	39-60mg/kg	Insecticide	Moderate 4-7 years	Highly Hazardous
14	BHC $C_6H_6Cl_6$ Benzene Hexa chloride	10,000mg/kg	Insecticide Acaricide, rodenticide	High persistence 3-6 years	Acute Hazard is unlikely
15	Pentachlorophenol C_6Cl_5OH	27-211mg/kg	Fungicide insecticide, Herbicide	Moderate 45 days	Moderately hazardous

Source: Jayaraj *et al.* (2016).

spills. There may be accidental oil spills or may be through ruination of tankers, faults in the equipments *etc.* Oil spills can also be caused by nature and human activities, when large amount of oil is spilled from oil cargo ships or from ocean floors, as well as leaks that occur when petroleum products and other forms of oil are used on lands and later washed off into water bodies (Latimer *et al.*, 1990). It could be washed through runoffs from rain into the water bodies, thereby causing pollution (USEPA, 1996; Husaini *et al.*, 2008). Hydrocarbons are also released into environment due to discharges from urban storm water. In urban communities, roads and parking areas are continuously exposed to oil and gasoline from vehicles that get washed into drains and subsequently reach and pollute the environment (van Metre, *et al.*, 2000; FWR, 2008). The residual pesticides also get washed off into receiving water bodies during rainfall (Ward *et al.*, 1993). In addition to this, those lands that had carried out various industrial activities are also found to be contaminated. These lands not only get polluted by Hydrocarbons but also other chemicals of organic origin (FWR, 2008). A classification of pesticides is given in Table 14.1. Major organochlorine pesticides and their toxicity is shown in Table 14.2. Table 14.3 depicts their biochemical effects.

14.2.1 Adverse Effects

Chemicals are integral part of our daily life. Despite of their extensive and uncountable use and benefits, they have a huge influence on health and environment.

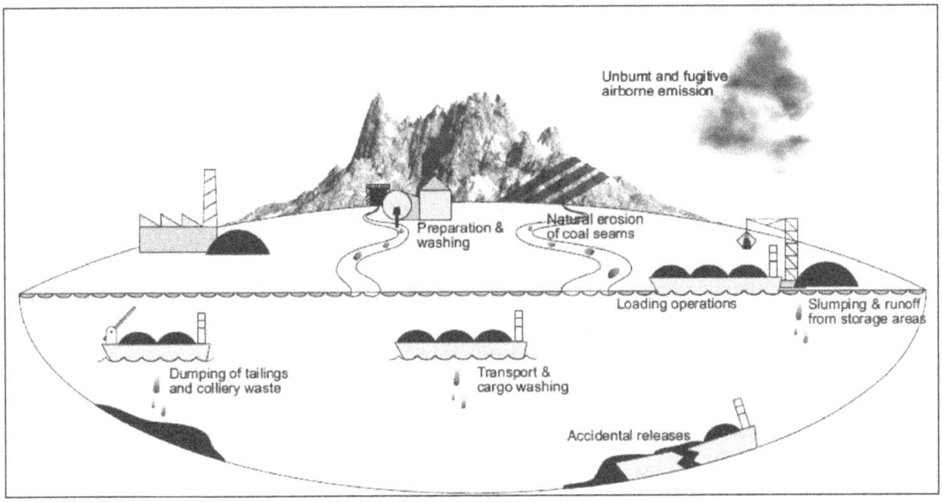

Figure 14.1: Coal Emission Sources to the Aquatic Environment (*Source*: Ahrens, and Morrisey, 2005).

In recent years, people have been exposed to several types of substances with broad spectrum due to the rapidly evolving technology. Simultaneously the pesticides have made a great impact on human health protection and preservation of foods fibre and other cash crops by controlling disease vectors and by keeping in check many species of unwanted insects and plants. The repeated use of pesticides and their environmental effects have prompted research into the environmental providence of these agents, which can be relocated from treated fields to air, land and water bodies. It is estimated that the financial cost of the damage to the environment and society is about $8.1 billion per year.

The natural biodegradation in polluted soils in the fields is often slow due to factors such as high hydrocarbons concentration, joint pollution with other pollutants (heavy metals). Other than this, limiting nutrient availability, inadequate supply of water or oxygen or poor bioavailability of pollutants. High concentration of hydrocarbons has led to stunted or reduced growth as well as premature death of agricultural crops. This in turn have led to alterations in normal morphological, chlorotic and necrotic foliar patterns that have resulted from major physiological disturbances in plant cells. Because of unequal distribution of photosynthate in plant metabolism, among various symptoms, foliar symptoms have proven unreliable indicators of oxidant effects on plant growth or yield. Pesticides like Hydrocarbons are known to reduce soil fertility, reduce nitrogen fixation, crop yield *etc.* Not only this, but they make deposition of silt in tanks and reservoirs. Figure 14.1 shows coal emission sources.

14.2.2 Biomagnifications

A trend of continuous rise in manufacturing and usage of various pesticides is observed in India. The problem becomes more serious because of lack of insufficient and ineffective waste treatment technology that involves physicochemical and

Table 14.3: Biochemical Effects of Major Organochlorine Pesticides

Sl.No	Chemical name	Organism	Biochemical effects	References
1	Aldrin and Dieldrin	Human	Neurotoxic, reproductive, developmental,immunological, genotoxic, tumerogenic effects, nausea, vomiting, muscle twitching and aplastic anemia	USEPA, 2003
		Mouse, rat, guniea pig, rabbit and dog	Convulsions, loss in body weight, depression, increased irritability, salivation, hyperexitability, prostration and death	
2	Chlordane	Human	Convulsions, tremor, mental confusion and incoordination	ATSDR, 1997
		Mice	Reduced fertility, liver cancer	
		Seals	Cancer, trauma, meningocephalitis	Kajiwara *et al.*, 2000
3	BHC/ DDE	Human	Cyst in hands, itching, psoriasis, eczema, leucoderma, skin rashes	Subramaniam & Solomon, 2006
4	DDT	Human	Prickling sensation of the mouth, nausea, dizziness, confusion, headache, lethargy, incoordination, vomiting, fatigue, tremors in the extremities, anorexia, anemia, muscular weakness, hyperexcitability, anxiety, and nervous tension	Klaassen *et al.*, 1996
		Mice	Liver tumors, liver changes including hepatocellular hypertrophy, margination and formation of lipospheres.	WHO, 1979
		Birds	Egg shell thinning	USEPA, 1975
		Fish	Affects membrane function and enzymes	
		Salmons	Impaired behavioral development	
5	Diazion	Rats	Neurotoxicity	USEPA, 2000 Peterson & Talcot, 2006
		Reptiles, fishes and Mammals	Lacrimation, salivation, anorexia, bradycardia, abdominal pain, hyperactivity, anxiety, depression and vomiting	
		Birds	Wing spasms, wing drop, hunched back, tenesmus, diarrhea, ptosis of eyelid, prostration, opisthotonos-like seizures or wing-beat convulsions.	
		Human	Dark or blurred vision, anxiety and restlessness, as well as psychiatric symptoms such as depression, memory loss, and confusion and acute pancreatitis.	Reigert & Roberts, 1999 Wagner, 1997 USEPA, 2000
6	Dicofol	Rats	Decrease in body weight and acute neurotoxicity	Phang *et al.*, 1996
		Dogs	Inhibition of ACTH (Adrenal cortical tropic hormone)	
7	Endosulfan	Human	Decreases the white blood cell count and macrophage migration, adverse effects on humoral and cell-mediated immune system. Affects semen quality, sperm count, spermatogonial cells, sperm morphology and other defects in male sex hormones DNA damage and mutation	Pandey *et al.*, 1990 Susan & Sania, 1999 Singh *et al.*, 2007
		Rats	Immunosuppression, neurological disorders, congenital birth defects, chromosomal abnormalities, mental retardation, impaired learning and memory loss and glomerulonephritis	Stockholm Convention, 2009
8	Lindane	Human	Damage human liver, kidney, neural and immune systems, and induces birth defects cancer, cause neurotoxicity, reproductive toxicity and hepatotoxicity	Sahoo *et al.*, 2008, Bano & Bhatt, 2010 Vijaya Padma *et al.*, 2011
		Rats	Alters gene expression of liver and hepatotoxicity	Sumida *et al.*, 2007 Videla *et al.*, 2004
9	Methoxychlor	Sea Urchins	Fertilization and early development of eggs	Pesando *et al.*, 2004
		Rats	Reduced fertility	Cummings & Gray, 1989
10	Polychlorinated Biphenyls (PCB)	Human	Neurological disorders and short term memory	Jacobson & Jacobson, 1996
		Fishes, rats, monkeys and mice	Cancer, Hodgkins lymphoma, decreased birth weight and decreased size of thymus gland	USEPA, 1996
11	Pentachlorophenol	Human	Inflammation of the upper respiratory tract and bronchitis, blood effects such as aplastic anemia, effects on the kidney and liver, immunological effects, and irritation of the eyes, nose, and skin	ATSDR, 1999
		Rats and Mice	Effects the cardiovascular system, blood, liver, immune system, and central nervous system (CNS)	

Source: Jayaraj, *et al.* (2016).

biological treatments. The waste generated after the production of pesticides has become the most challenging and serious environmental problem. In addition to this, certain residual compounds like phenols, Hydrocarbons and residues from paint and dye effluent are also present in wastewater. As a result, recalcitrant pesticide and hydrocarbons residues remain in surface soil, leading to the toxicity of

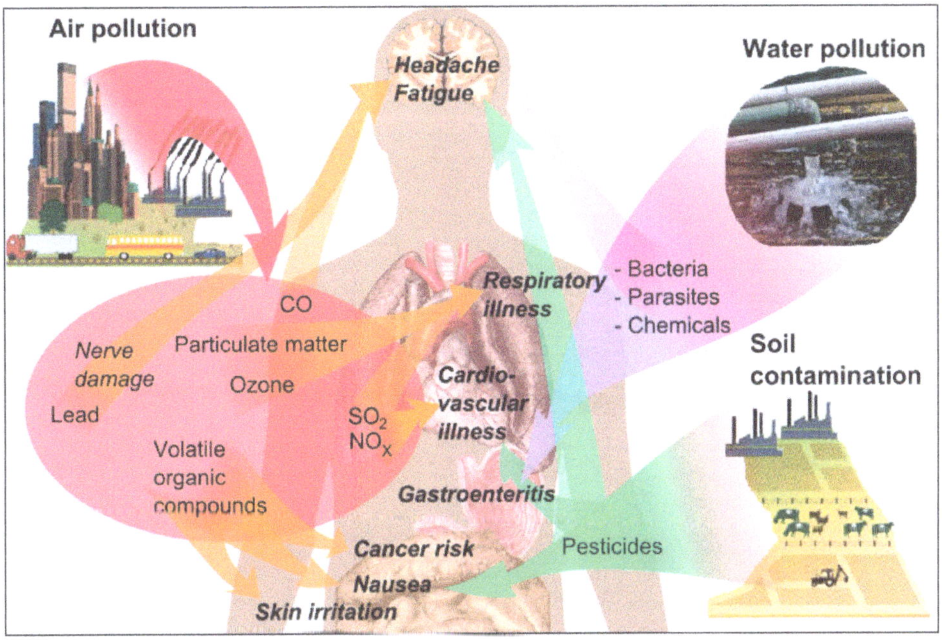

Figure 14.2: Impact of Pesticides on Human Health.

soil-water environment which are then taken up by organisms because of the food they consume or the intoxication of their environment. The hazardous substances then tend to build up inside the organism's cells. The toxins gradually become concentrated higher up in the food chain, when organisms of higher tropic level consume toxin containing organisms. As the process is a repetitive process in the ecosystem, the higher organisms are the ones that will concentrate most of the toxins (Figure 14.3). This increase of toxins in the food chains is known as biomagnifications. It stands for Biological Magnification. In short, In other words, Biomagnification is a process in which xenobiotic substances get transported from food to an organism and the concentration is much more as compared to its sources.

Thus, looking towards the environmental concerns and health hazards resulting due to the continuous use of these noxious xenobiotic pesticides and hydrocarbons, it is highly desirable to detoxify them. Figure 14.2 presents impact of pesticide on health.

14.2.3 Bioremediation: An Emerging Technology

There is an urgent need to remove chemo pollutants like pesticides and hydrocarbons, from the environment. Bioremediation including phytoremediation and rhizoremediation is expected to be a most effective and useful clean-up technology for soil contaminated with persistent organic pesticides (Matsumoto *et al.*, 2009). Conventional methods of pesticide detoxification are known to rely upon landfills and incineration. They not only generate secondary contamination problems but either simply dilute the pollutants or sequester and transfer them into another

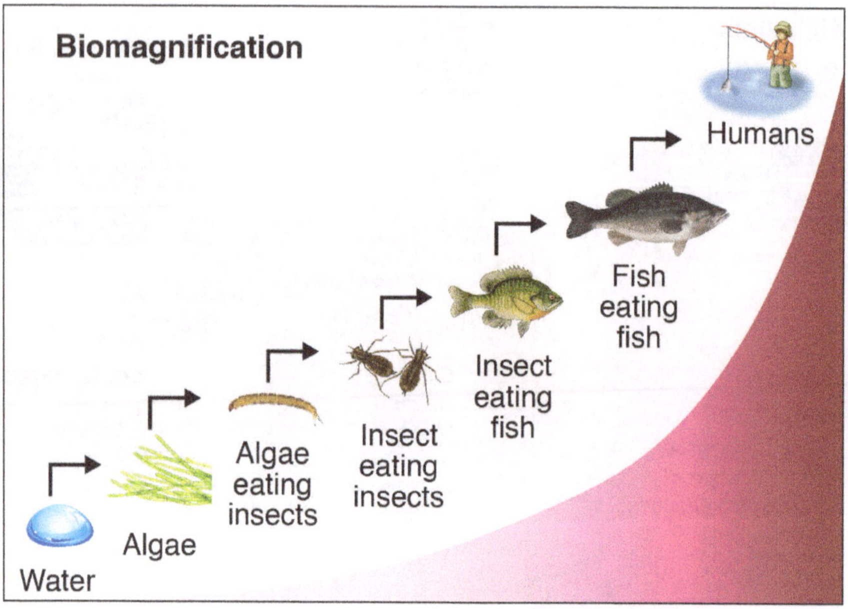

Figure 14.3: Process of Biomagnifications in Various Trophic Levels of Food Chains.

environment. Thus, they cause soil and groundwater contamination. Moreover, incineration, although approved by US EPA, is a very costly process requiring large amounts of energy (Chen and Mulchandani, 1998). From this point of view, biodegradation is a much safer and economical process for detoxifying pesticides. By contrast, bioremediation may be viewed as a more effective and environmentally benign clean-up technology, since it results in the partial or complete bioconversion of the organic pollutants, such as pesticides, to microbial biomass and stable non-toxic end-products (Bakers and Herson, 1994). It is a relatively recent addition to the battery of clean-up strategies currently employed to restore and rehabilitate contaminated sites. Bioremediation technologies have been broadly divided in to two categories, in situ and ex situ bioremediation techniques (Figure 14.4).

14.2.4 *In situ* Bioremediation Techniques

In situ bioremediation techniques involve enhancement of the biodegradation rate of organic contaminants within the affected area (soil, sediment and surface water or groundwater environments). In Latin, in situ means "in the origin place". It is the clean-up approach that directly involves the contact between microorganisms and the dissolved and sorbed contaminants for biotransformation (Alcalde *et al.*, 2006). The advantages of these technologies are that, they are cost effective, ecofriendly as well as reduce the release of quantum of volatile contaminants in the environment. The other Potential advantages are that they minimize site disturbances, less public exposure, and follow up simultaneous treatment of polluted soil and groundwater. The drawbacks of these techniques are that they are time consuming method as compared to other remedial methods.

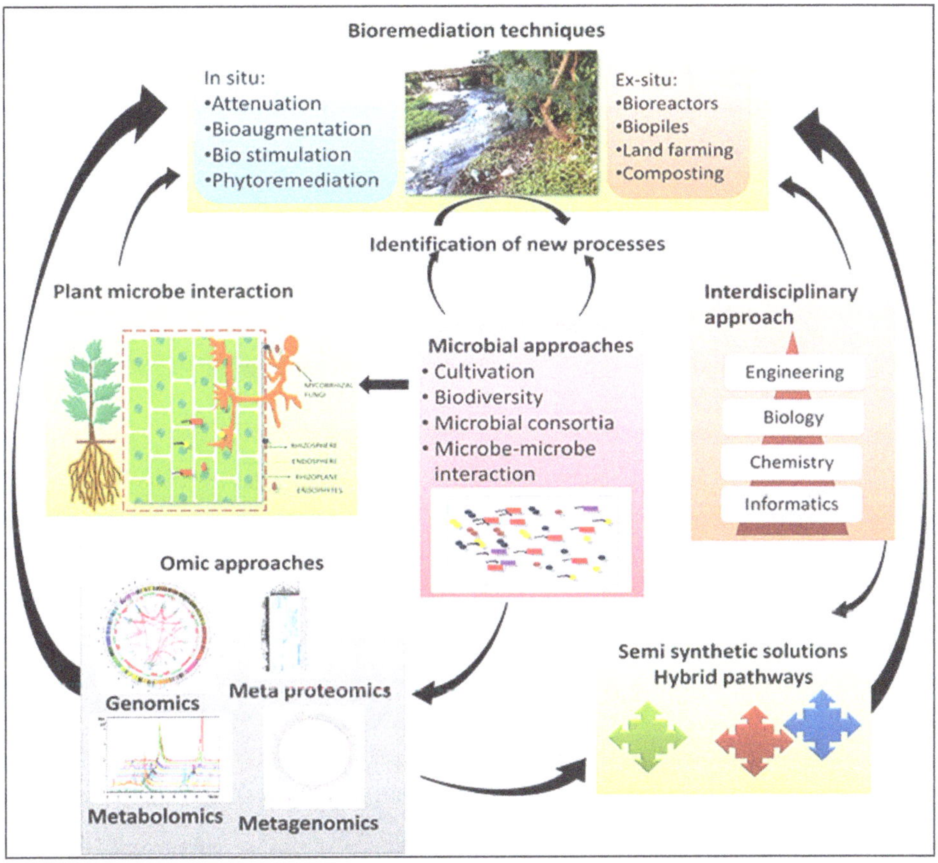

Figure 14.4: Novel Processes in Order to Improve Bioremediation of Polluted Environment (*Source*: Malla, *et al.*, 2018).

14.2.5 *Ex situ* Bioremediation Techniques

In *Ex situ* bioremediation techniques, contaminated soil or sediments are treated using solid or slurry phase systems. These mainly involve aerobic processes. It actually involves the removal of waste material from polluted area and their collection at a place distant from the contaminated site o facilitates microbial degradation. There are certain limitations and disadvantages of this technology besides being a costly process. Based on the physical nature of the contaminant under treatment, this technology is classified into two types: (i) Slurry phase bioremediation systems involving treatment of solid liquid suspensions in bioreactors and (ii) Solid phase bioremediation system including land farming (soil treatment units) and soil piles, compost heaps and engineered bio piles.

14.3 Microbial Diversity Used for Bioremediation

Microbes are well known bioremediators that are well utilized to clean up polluted sites. These include archaea, bacteria and fungi. They are the principal organisms that play a key role in bioremediation process (Figure 14.5, Table 14.4). They are universally accepted as tools to mitigate pollutants of various types. This is because of their advantages over other protocols for remediation procedure. Microorganisms are restoring the original natural surrounding and preventing further pollution (Demnerova *et al.*, 2005; Abatenh, 2017). Petroleum, its products and oils constitute hydrocarbons and if present in the environment causes pollution. Many of these microorganisms are able to build powerful surface active compounds known as surfactants. These compounds emulsify oil from water, thus enabling them to remove oil. This microbial emulsifier has proven to be nontoxic and are easily biodegradable. This is not observed in case with chemical surfactants. Pseudomonads, various *Corynebacteria*, *Mycobacteria* and some yeasts are the most promising agents that have capabilities to degrade petroleum.

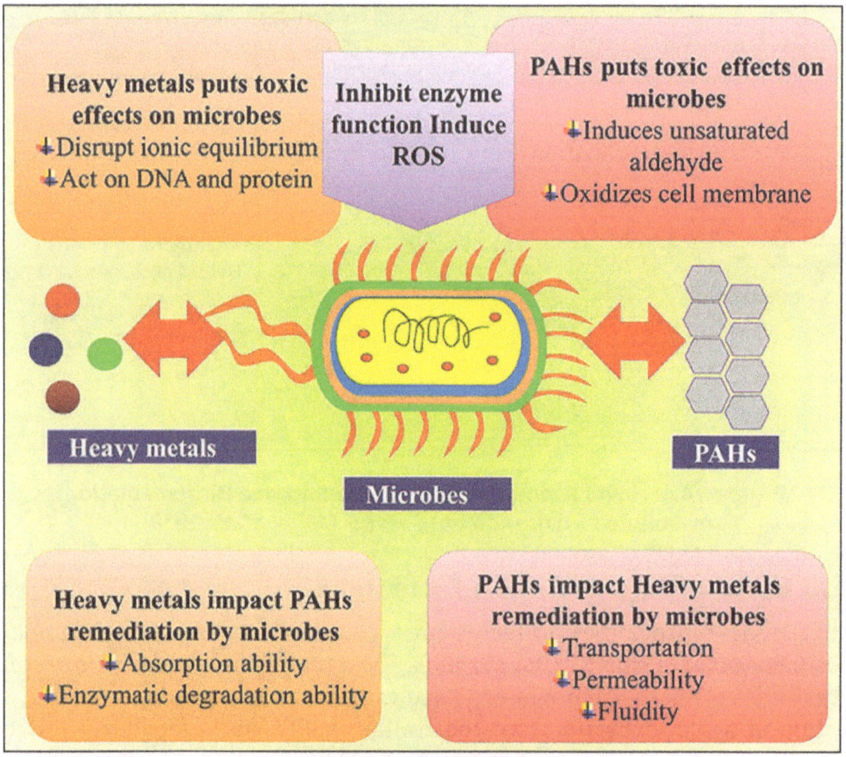

Figure 14.5: Microbial Based Bioremediation Process (Source: Malla *et al.*, 2018).

Not only the removal of hydrocarbons, microbes also have the ability to remove hazardous contaminants from industrial wastes. They lower down concentration of toxic cations of heavy metals (such as Selenium) and bring down the level to a much less toxic soluble form. Secretion from many algae and bacteria can attract toxic metals ions. Being bound to the secretions, it facilitates their removal. Dye degradation is also brought about by some anaerobic bacteria and fungi.

Table 14.4: Different Types of Microbes Used in the Bioremediation of Various Environmental Contaminants

Microorganism	Types of pollutant degraded	Significant outcomes	References
Brevibacterium epidermidis EZ-K02	Industrial wastes	*Brevibacterium epidermidis*, a bacterium that is capable of degrading waste waters contaminated with large scale dissolution of chemical compounds and nitrocellulose particles	Ziganshina et al., 2018
Bacillus sp. CDB3 *Lysinibacillus sphaericus*	Arsenic	*Bacillus* sp. CDB3, *Lysinibacillus sphaericus* Shows high resistance to arsenic contamination and aids in treating arsenic poisoning.	Rahman et al., 2016 Yang and Zhang, 2017
Mycobacterium dioxanotrophicus	Heterocyclic organic compounds (Dioxane)	*Mycobacterium dioxanotrophicus* is capable of remediating various heterocyclic organic compound contaminated environments by making the use of 1, 4-dioxane as a single source of energy and carbon and energy	He et al., 2017
Hyphomicrobium sp. Strain GJ21	Dichloromethane	*Hyphomicrobium* degrades halogenated contaminants by making the use of dichloromethane as a source of both energy and carbon	Bringel et al., 2017
Microbacterium oleivorans Strain A9	Radionuclides	*Microbacterium oleivorans* strain A9, a radionuclide-resistant actinobacterium capable of degrading uranium	Ortet et al., 2017
Plantibacter flavus Strain 251	Hydrocarbon-contaminated environments	*Plantibacter flavus* isolate 251 is known to possess novel biodegradation enzymes. The bacterium is anticipated to provide some of the novel insights into exploiting the hydrocarbon degrading pathways	Lumactud et al., 2017
Bacillus subtilis SR1	Polyaromatic hydrocarbon	*Bacillus subtilis* SR1 is a bacterium besides showing resistance to heavy-metals is also capable of degrading polyaromatic hydrocarbons	Kotoky et al., 2017
Alkaliphilus metalliredigens Strain QYMF,	Heavy metals	A metal-reducing bacterium capable of thriving in alkaline environments, a feature that is not commonly found in metal respiring microbes	Hwang et al., 2016
Pseudomonas veronii Strain 1YdBTEX2	Aromatic solvents *viz.*, Benzene, toluene, ethyl benzene, Xylene (BTEX)	*Pseudomonas veronii* bacterium contains genes that carry out the degradation of aromatic solvents via a catabolic pathway	Junca and Pieper, 2004 Morales et al., 2016 Moreno-Forero et al., 2016
Pseudomonas plecoglossicida TND35	Nicotine	*Pseudomonas plecoglossicida* TND35 besides being an effective nicotine-degrading bacterium also has genes responsible for the degradation of heavy metals, aromatic compounds, and biosynthesis of butanol	Raman et al., 2015
Microbacterium spp.,	Heavy metals	*Microbacterium* spp plays an important roe in phytoextraction and mobilization of heavy metals	Corretto et al., 2015
Arthrobacter sp. Strain SPG23	Hydrocarbon degradation	*Arthrobacter* is a hydrocarbonoclastic Gram-positive bacterium and is a potent bacterium for used for the remediation of the diesels fuels	Gkorezis et al., 2015
Raoultella ornithinolytica-TNT	Trinitrotoluene	*Raoultella ornithinolytica*-TNT is a Gram-negative bacterium. Strains of TNT make use of nitrate released from trinitrotoluene thereby making it less toxic. Hence is considered as a potent microbe in terms of bioremediation applications.	Thijs et al., 2014

Microorganism	Types of pollutant degraded	Significant outcomes	References
Pseudomonas taeanensis	Hydrocarbons (Petroleum compounds)	*Pseudomonas taeanensis* a bacterium that is able to degrade petroleum compounds like diesel, kerosene and gasoline	Lee et al., 2014
Caulobacter sp. Strain OR37	Heavy metals	Possesses tolerance to elevated concentrations of heavy metals *viz.*, cadmium, cobalt, uranium, nickel	Utturkar et al., 2013
Ochrobactrum pseudogrignonense	Arsenic pollutants	*Ochrobactrum pseudogrignonense*, a highly potent and efficient arsenate-resistant bacterium. The bacterium is involved in the degradation of the arsenic from arsenate-contaminated soils	Yang et al., 2013
Arthrobacter sp. Strain SJCon	2-Chloro-4-Nitrophenol	*Arthrobacter* sp. strains are useful in drafting the genetic pathways that are involved in the bioremediation of the aromatic compounds	Vikram et al., 2013
Brachybacterium sp. *Cytophaga sp.* *Sphingomonas sp.* *Pseudomonas sp.*	Oil spills	The bacteria are capable of remediating the oil-contaminated sites *via*, the processes of biostimulation and bioaugmentation	Angelim et al., 2013
Pseudomonas aeruginosa	Organic and Inorganic mercury	*Pseudomonas aeruginosa* bacterium is one of the potent agent that carries out the bioremediation of both organic and inorganic mercury in highly contaminated mercury sites	Dash and Das, 2012
Moraxella saccharolytica *Alteromonas Putrefaciens* *Klebsiella pneumonia* *Pseudomonas fragi*	Diesel hydrocarbon	*Moraxella saccharolytica, Alteromonas Putrefaciens, Klebsiella pneumonia, Pseudomonas fragi*, a complete bacterial consortium is proved to be one of the better and reliable choices for the rapid and complete remediation of the diesel hydrocarbon contaminated environments	Sharma and Rehman, 2009
Sphingomonadaceae/Sphingomonas	Degrades Hexachlorocyclohexae	Sphingomonads offer biostimulation of HCH polluted sites *via* addition of nutrients and aeration.	Dadhwal et al., 2009
Pseudomonas putida strains *P. putida* G7 *P.aeruginosa* PaK1 *P.putida* BS202 *P.* sp. strain U2 *Rhodococcus sp.* NCIMB12038 *Pseudomonas putida* OUS82 *Alcaligenes faecalis* AKF2 *Nocardiodes sp.* KP7	Degradation of PAHs Naphthalene and Phenanthrene	Isolation of the bacterial strains such as *Pseudomonas putida, P. putida* G7, *P.aeruginosa* PaK1, *P. putida* BS202, *P.* sp. strain U2, *Rhodococcus sp.* NCIMB12038 and *Pseudomonas* putida OUS82, *Alcaligenes faecalis* AKF2, *Nocardiodes sp.* KP7via the development of the modern genetic technologies provided a major breakthrough in the PAH remediation. The bacteria are capable of degrading substrates like phenanthrene, Pyrene and benzo-pyrene, fluoranthene, Naphthalene	Chauhan et al., 2008
Mycobacterium sp. PYR-1	Degradation of PAHs Pyrene	The strains of the *Mycobacterium* sp. although not producing the biosurfactants possess a strong ability to degrade even the low concentrations of aqueous-phase anthracene	Chauhan et al., 2008
Neisseria elongate Acinetobacter faecalis Staphylococcus. sp.	Crude petroleum oil	*Neisseria elongate, Acinetobacter faecalis, Staphylococcus. sp.*, are the potent bacterial isolates and carry upto 93, 94, and 95% of hydrocarbon degradation, respectively	Mukred et al., 2008

Sources: Malla *et al.*, 2018.

14.4 Microbial Potential for Degradation of Pesticides and Hydrocarbons (Figures 14.6 and 14.7)

Microorganisms provide a potential opulence in biodegradation process. The removal capacity of these organisms is directly linked to their adaptive nature because of persistence and prevalence of xenobiotics in the surroundings. In current times, genetic engineering is widely used to improve the performance of such microorganisms particularly those having essential properties required for biodegradation process (Schroll *et al.*, 2004). *Gordonia, Brevibacterium, Aeromicrobium, Dietzia, Burkholderia,* and *Mycobacterium* are the bacterial genera isolated from soil contaminated with petroleum has been proved to be the potential organisms for degradation of hydrocarbons. Various Fungal genera like *Amorphoteca, Neosartorya, Talaromyces,* and *Graphium* and yeast genera, for example *Candida, Yarrowia, Pichia, Aspergillus, Cephalosporium,* and *Pencillium* were also found to be the potential degrader of crude oil hydrocarbons. *Candida lipolytica, Rhodotorula mucilaginosa, Geotrichum* sp, *and Trichosporon mucoides* which are genera belonging to yeast are reported to degrade petroleum compounds.

14.5 Biochemical Mechanisms Involved in Microbial Degradation

Aerobic conditions is the most favorable condition that may lead to rapid and complete degradation of the majority of organic pollutants. The inceptive intracellular attack of organic pollutants is an oxidative process and its activation as well. It also incorporates enzymatic key reactions that are catalyzed by oxygenases and peroxidases. Peripheral degradation pathways lead to the conversion of organic pollutant step by step into intermediary metabolism, for instance, the tricaboxylic acid cycle. Central precursor metabolites serve as biosynthesis of cell biomass. Petroleum hydrocarbons degradation can be arbitrate by specific enzyme system. The Other mechanisms involves substrate attachment of microbial cells followed by production of biosurfactants. The mechanism of uptake which is linked to the attachment of cell to oil droplet is still unknown. Pesticides, polychlorinated biphenyls (PCBs), polyaromatic hydrocarbons (PAHs), heterocyclic compounds (such as pyridine or quinoline), pharmaceutical substances, radionuclides and metals are known to bioremediate by several microorganisms by this mechanism.

14.5.1 Microbial Degradation for Xenobiotic Compound

Xenobiotics are chemical substances found in the environment at unnaturally high concentration through anthropogenic activities. Many of these compounds not only reaches the environment but tend to accumulate as they get degraded very slowly and in extreme cases even become persistent. The degradation of xenobiotic compounds rely upon activity of microorganisms. Some example includes degradation of parathion. The degradation pathway of xenobiotic compound should be examine when single substrate is available there. In absence of oxygen there should be an alternative electron accepter like nitrate, sulphate, selenate, carbonate *etc.* There are no microbes or group of microbes that degrade all compounds. So there should be a group of organism, metabolically versatile that are applicable for the degradation of large number of compounds. They mineralize, transform or immobilize the pollutants. Examples of aerobic and anaerobic xenobiotics degradative bacteria are *Pseudomonas, Gordonia, Bacillus, Moraxella, Micrococcus, Escherichia, Sphingobium, Pandoraea, Rhodococcus,* and anaerobic xenobiotics degradative bacteria are *Pelatomaculum, Desulphovibrio, Methanospirillum, Methanosaeta desulfotomaculum, Syntrophobacter, Syntrophus etc.*

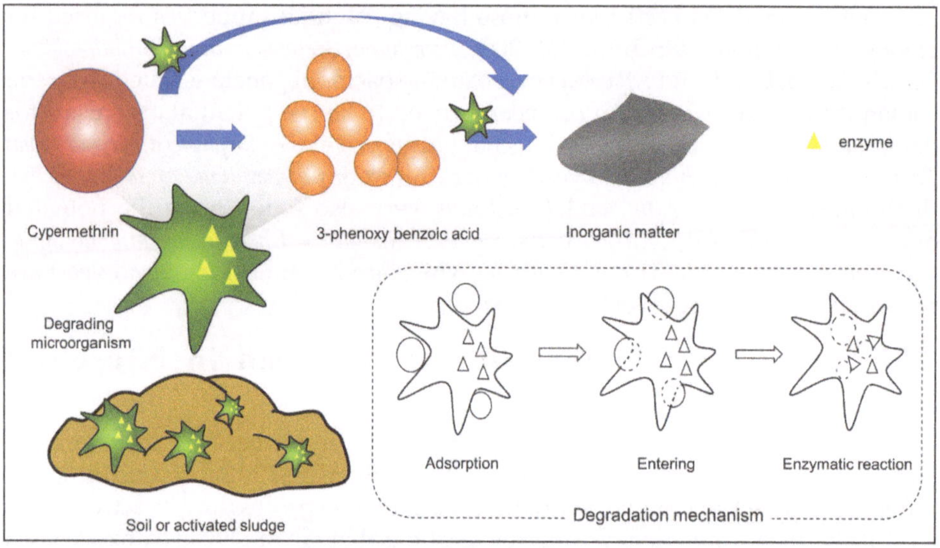

Figure 14.6: Mechanism Involved in Microbial Degradation of Pesticides.

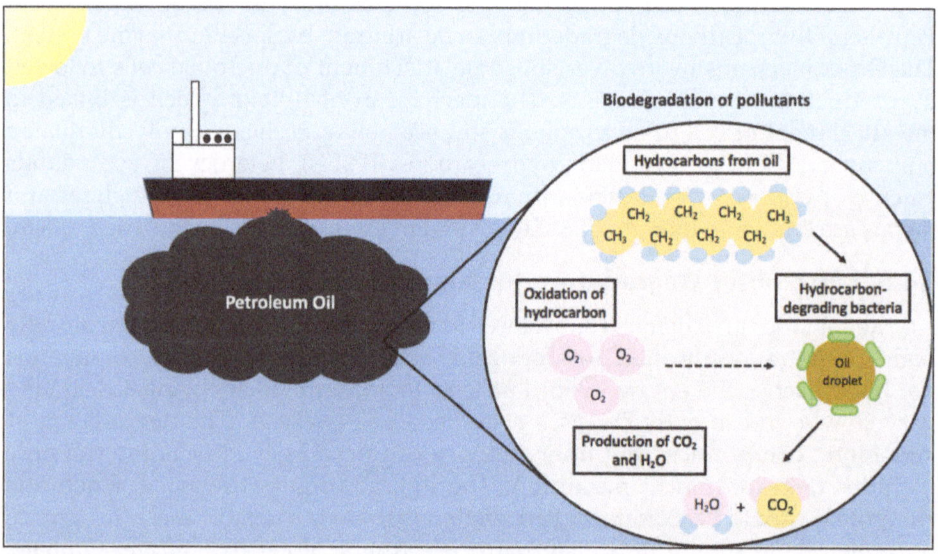

Figure 14.7: Mechanism Involved in Microbial Degradation of Hydrocarbons.

14.6 Advantages and Disadvantages of Bioremediation

Degradation by microorganisms though a natural process, is extensively explored as waste treatment system for all kinds of polluted environments. In the contaminated environment, their count increases manifold, which in turn enhances degradation. Their population gets declined when the material is degraded. The left out residues like cell biomass and carbon dioxide are harmless product which is the most positive side of microbial degradation process.

It requires a less effort and can often be carried out on site, often without causing a major disruption of normal activities. This also eliminates the need to transport large biomass of waste to far off sites and the potential threats to human health and the environment are avoided. The conventional technologies applied for removal of hazardous waste particularly oil-contaminated sites has manifested to be very costly (Montagnolli, *et al.*, 2015). On the other hand biodegradation process by microbe is cost effective. In degradation that involves microbes and use of other organisms, not only these are removed, but many hazardous compounds can be transformed to innocuous products. This unique feature decreases the chance of future accountability that is associated with treatment and waste disposal (Singh *et al.*, 2013). Moreover, it does not involve any harmful chemicals. The nutrients in the form of fertilizers are added to activate fast microbial growth. These fertilizers are commonly used in lawns and gardens, and are cost effective (Kumar *et al.*, 2011) and sustainable (Anno *et al.*, 2012).

Every technology has its own pros and cons and so the bioremediation has some limitations like it takes long duration and can harm the environmentalist. Biological processes are often highly specific. Important site factors required for success include the presence of metabolically capable microbial populations, suitable environmental growth conditions, and appropriate levels of nutrients.

14.7 Conclusion

Pollutants like Hydrocarbons and pesticides have a widely applicable consequence on every type of ecosystem. Tremendous use of fossil fuels and industrial revolution have made the problem more severe. In addition to population growth, frequent oil spills, divulges in pipelines, and uncontrolled use of pesticides contribute to substantial enhancement in pollution. These together are threatening the lives of animals and native microbiological population in land, air, and water surfaces and sub surfaces. Thus environmental remediation is the most important aspect of human survival. The chapter not only highlights the causes, but it also explains the techniques used in pollution rectifications.

REFERENCES

Abatenh, E., Gizaw, B., and Tsegaye, Z. (2017) Application of microorganisms in bioremediation-review *Jr. of Environ. Microbio.*1: 02-09.

Abudulai, M., and Shepard, B. M. (2001) Timing insecticide sprays for control of pod sucking bugs (Pentatomidae, Coreidae, and Alydidae) in cowpea (Vigna unguiculata L. Walpers). *Jr. of Agri. and Urban Entomology*. 18: 51– 60.

Achten, C., and Hofmann.T. (2009) Native polycyclic aromatic hydrocarbons (PAH) in coals – À hardly recognized source of environmental contamination. – *Science of the Total Environment*, 407, 2461–2473.

Anno, D., Beolchini, F., Rocchetti, L., Luna, G. M., and Danovaro, R. (2012) High bacterial biodiversity increases degradation performance of hydrocarbons during bioremediation of contaminated harbor marine sediments. *Environ Pollut* 167: 85–92.

Alcalde, M., Ferrer, M., Plou, F.J., and Ballesteros., A. (2006) Environmental biocatalysis: from remediation with enzymes to novel green processes. *Trends Biotech.*, 24(6): 281-287.

Ahrens, M.J. and Morrisey, D.J. 2005. Biological effects of unburnt coal on the marine environment. *Oceanogr. Mar. Biol. Annu. Rev.*, 43, 69-122.

Baker, K. H., and Herson D. S. (1994) Introduction and overview of bioremediation. In K.H. Baker and D.S. Herson, eds., Bioremediation.*McGraw-Hill*, New York, 1-7 pp.

Chen, W. and Mulchandani, A(1998) The use of live biocatalysts for pesticide detoxification. *Trends Biotech.*, 16(2): 71-76.

Das, M. and Adholeya A. (2012) Role of Microorganisms in Remediation of Contaminated Soil. Microorganisms in Environmental Management: *Microbes and Environment*. 81 – 105.

Demnerova,K., Mackova, M., Spevakova V., Beranova, K., Kochankova, L., Lovecka, P., Ryslava, E., and Macek, T. (2005) Two approaches to biological decontamination of groundwater and soil polluted by aromatics–characterization of microbial populations. *Int Microbiol* 8: 205–211.

FWR, (2008) What is pollution? Foundation for Water Research. http: //www. euwfd.com/html/source_of_pollution_-_overview.html.

Husaini, A., Roslan, H. Hii, K., and Ang, C.H. (2008) Biodegradation of aliphatic hydrocarbon by indigenous fungi isolated from used motor oil contaminated sites. *World Jr. Microbiol. Biotechnol.* 24: 2789-2797.

Jayaraj, R, Megha, P., and Shreedev, P. (2016) Organochlorine pesticides, their toxic effects on living organisms and their fate in the environment. *Interdiscip. Toxicol.* 9(3–4): 90–100.

Kumar, A., Bisht, B. S., Joshi, V. D., and Dhewa, T. (2011) Review on Bioremediation of Polluted Environment: A Management Tool. *Int. Jr. of environ. Sci.* 1: 1079-1093.

Latimer J., Hoffman E., Hoffman G., Fasching J. and Quinn, J. (1990) Sources of petroleum hydrocarbons in urban runoff. *Water Air Soil Pollut.* 52: 1-21.

Malla, M.A., Dubey, A. Yadav, S. Kumar, A., Hashem, A., and Fatehi, E. (2018) Understanding and designing the strategies for the microbe mediated remediation of Environmental Contaminants using omics approaches. *Frontiers in Microbiology*.9: 1-19.

Matsumoto, E., Kawanaka, Y., Yun, S.J., Oyaizu, H. 2009. Bioremediation of the Organochlorine pesticides, dieldrin and endrin, and their occurrence in the environment. *Appl Microbiol Biotechnol*. 84 (2): 205–216.

Montagnolli, R.N, Matos, Lopes, P.R., Bidoia, E. D. 2015. Assessing *Bacillus subtilis* biosurfactant effects on the biodegradation of petroleum products. *Environ. Monit. Assess* 187: 1-17.

Schroll, R., Brahushi, F., Dorfler, U., Kuhn, S., Fekete, J., and Munch, J.C. (2004) Biomineralisation of 1,2,4-trichlorobenzene in soils by an adapted microbial population. *Environ. Pollut*. 127, 395–401.

Sharma, S. (2012) Bioremediation: Features, Strategies and applications. *Asian Jr. of Pharm. and Life Science* 2: 202-213.

Sheela, S. C., and Papinazath, T. (2003) *Proc. 3 rd International conference on environment and Health* eds.,pp. 465, Chennai, India.

Singh, A, Kumar,V., and Srivastava, J.N. (2013) Assessment of Bioremediation of Oil and Phenol Contents in Refinery Wastewater via Bacterial Consortium. *Jr. Pet Environ Biotechnol* 4(3): 1-4.

Strong, P.J., and Burgess, J.E. (2008) Treatment methods for wine-related ad distillery wastewaters: A review. *Bioremediation Jr*., 12: 70-87.

USEPA, (1996) Indicators of the environmental impacts of transportation: Highway, rail, aviation and maritime transport. EPA 230-R-96-009, US Environmental Protection Agency, Washington, DC. http: / /ntl.bts.gov / lib / 6000 / 6300 / 6333 / indicall.pdf.

van Metre, P.C., Mahler, B.J., and Furlong, E.T. (2000) Urban sprawl leaves it's PAH signature. *Environ. Sci. Technol*., 34: 4064-4070.

Ward, C.W., R.G. Smith, S. Wachter, R.D. Lee, LE. Doxon, L.M. English, W.S. Cranshaw, K.N. Pinkston, G.W. Cuperus, W.P. Morrison, D.T. Langston, and Sorenson, C.E. (1993) Development and demonstration of an integrated pest management program for urban landscapes. New Mexico State Univ. Ext. Serv., USDA Project 92-EMP-1-0019. Annual Report 1993.

Chapter 15

Carbon Sequestration Potential of Major Forest Types of Central Himalaya

Tanuja Gahlot[1] and Arun Kumar Kushwaha[2]*

[1]Department of Botany, IPGCPG College, Haldwani, Nainital, Uttarakhand
[1]CSIR, National Botanical Research Institute, Lucknow – 226 001, Uttar Pradesh
**e-mail: tanugahlot8@gmail.com*

ABSTRACT

The rising CO_2 level in the atmosphere is a major concern for the 21st century. The increasing developmental activities are the main cause which effect environment directly or indirectly. For the mitigation of increasing CO_2 level in the atmosphere the nature has provided a big potential sink in the form of terrestrial biosphere. Carbon sequestration is a most natural way of CO_2 removal from the atmosphere. The terrestrial biosphere is a major platform where the green plants accumulate CO_2 in their biomass and sequester it for many years. The researchers are studying the role of trees and forests in carbon sequestration for almost two decades. The present paper is a critical review of past studies on carbon sequestration potential of major forests of Central Himalayan region. Every forest type has its own type of geography, aspect, soil and climatic conditions and hence the rate of sequestering atmospheric carbon in their biomass vary significantly. The Central Himalayan region has always been a forest dominated landscape. The rich forest cover of Central Himalaya is dominated with many types of major forest types. Therefore, to understand the potential of these forests in mitigation of climate change issue will be very interesting and it will certainly provide a baseline data to the policymakers of environment.

Keywords: *CO_2, Carbon sequestration, Central Himalaya, Forest.*

15.1 Introduction

Many dramatic changes in climate have been noticed in last century than any other century in the past. Some of the examples of these changes are: doubling of agricultural land, four-fold increase of human population, fourteen- fold increase in economy, fifteen-fold increase in energy use and seventeen-fold increase in carbon dioxide emission (Peterson *et al.*, 2003).

Forests are the major carbon pool in the global carbon cycle. The increasing concentration of atmospheric carbon dioxide is a result of increasing deforestation and degradation activities. The Reducing Emissions from Deforestation and Degradation (REDD) is now an important component of mitigating climate change (UNFCC, 2008). Enhancing rate of intense human activities directly or indirectly leave the impact on carbon sequestration ability of forests as a result of this the atmospheric carbon dioxide concentration keeps on increasing.

The atmospheric concentration of carbon dioxide has increased by over 30 per cent since pre-industrial levels and it is expected to exceed 500 ppm by 2100 (Janzen, 2004). About two-thirds of terrestrial C, exclusive of that sequestered in rocks and sediments, is sequestered in the standing forests, forest under storey plants, leaf and forest debris, and in forest soils (Sedjo *et al.*, 1998). Carbon sequestration can be described as the removal of CO_2 from the atmosphere (source) and storage into green plants (sink) where it can be stored for long-term (Watson *et al.*, 2000). The sinks can be above ground (trees) or living biomass below the ground in soil (roots and micro-organisms).

Sequestration can also be defined as storage of all forms of carbon, including storage in terrestrial, oceanic and geological ecosystems. Enhancing the natural processes that remove carbon dioxide from the atmosphere, is thought to be one of the most suitable methods of mitigating the carbon dioxide levels in the atmosphere.

Asian forests are recently estimated to be the net emitters of carbon dioxide, mostly due to forest degradation and deforestation (Dixon *et al.*, 1994). Restoration of degraded forests and soils has globally a potential for sequestering carbon in terrestrial forests corresponding to 50–75 per cent of the historic losses (Lal, 1999). The main reason being of high interest for forestry is the flexibility provided by increased carbon stocks in forests under the uncertainty regarding the impact of global warming (Solberg, 1997). By practices and technologies carbon sequestration seeks to quantify and boost the storage capacity of all potential sinks and expand the number and type of sinks in which storage of carbon is possible. If human activities are ultimately proved to cause global climate changes, increased stocks of carbon in biomass of wood will have an indefinite positive marginal value product for sinking the quantities of atmospheric carbon dioxide. Even if global warming is not as devastating as many fear, an increasing stock of timber will most likely have a positive marginal value for forest products, and other environmental gains (Hoen and Solberg, 1994).

The forests of Uttarakhand are located in the Central Himalayan region of India. Uttarakhand's forests are the repositories of great biological diversity and provide ecological stability, timber, fuel wood, fodder and green manure which are critical to the household economy of the local inhabitants (Rawat and Rawat, 2012). Eight of sixteen forest types existing in India are found in Uttarakhand (Champion and Seth, 1968). It means Uttarakhand is not only rich in forest cover but also in forest types. The variety of forests types can be noticed in few kilometres of area along the altitude. The tarai and bhabhar belt consists of forests of sal (*Shorea robusta*) and in the mid-altitude chir-pine occupied its area. However, at the higher altitude the forests of oaks can be noticed easily. These types of forests pattern can be seen from Kumaun Himalaya to Garhwal Himalaya. The rich forest diversity of Central Himalayan forests in Uttarakhand is the main reason of attraction for scholars, environmentalists and also tourists. Thus, the present study aims to document the rate of carbon sequestration through major forests types of Central Himalaya as to determine their role in climate change mitigation.

15.2 Major Forest Types in Central Himalaya

Central Himalaya is fortunately very rich in forest area. According to a forest report (FSI, 2011), forests cover approximately 41 per cent of geographical area in the Indian Himalayan Region (IHR), of which 16.9 per cent area is under very dense forest cover, 45.4 per cent under moderate forest cover and remaining 37.7 per cent under open forest cover. Himalayan states show that considerable forest area in these states is private (42 per cent), followed by area under community management (33 per cent) and revenue department (25 per cent). According to a report of the Task Force, Planning Commission (Anonymous, 2010) Uttarakhand is the only state where all types of ownerships exist, while Nagaland has mainly private forests. Even though the rest of the country seems to be witnessing a reduction in the dense forest cover, situation in the IHR appeared to be much better as per the FSI assessments of 2001 and 2011, and is generally believed to be a cumulative effort of many local and government bodies (*e.g.*, Van Panchayats).

Forests in Uttarakhand situated in IHR are widespread and diverse. The vegetation of forests differ significantly from tropical to temperate with respect to structure, function and phenology as well as in terms of ecosystem processes (Zobel and Singh, 1997). A variety of forests types present in Himalaya ranging from tropical dry deciduous forest in the foot hills to alpine scrubs near the timber line (Champion and Seth, 1968). Among these the sub-tropical forests (contain the dominating sal forests) situated at the altitude between 300 m to 1000 m and above this from 1100 m to 3000 m are characterized with extensive pure and mix coniferous-oak forests.

15.3 Carbon Sequestration in Central Himalaya

Understanding how the rate of carbon sequestration influence the global climate cycle is very important for forest conservators. Forests of Central Himalaya are rich in vegetation and biodiversity. The dominant tree species in the forests are generally represented by *Shorea robusta* Gaertn. f., *Pinus roxburghii* Sarg., *Quercus leucotrichophora* A. Camus, *Quercus floribunda* Lindl., *Quercus semecarpifolia* Smith. and *Cedrus deodara* Loud. from foothills to higher elevation.

The forests release 4.6×10^{12} g c yr^{-1} to the atmosphere. The carbon storage value for Central Himalayan forests ranged from an average of about 175 t c ha^{-1} for chir-pine forests to approximately 400 t c ha^{-1} for oak and sal dominated forests. Carbon sequestration rates of different tree species have been studied by many environmentalists (Singh *et al.*, 2004; Sah, 2007; Singh, 2008). The maximum carbon sequestration rates were observed for *Quercus floribunda* forests and minimum for *Cedrus deodara* forests. Singh and Singh (1986 and 1992) recorded 6.0-8.0 t c $ha^{-1}yr^{-1}$ carbon sequestration for different forests of Central Himalaya. Rana *et al.* (1989) also reported almost similar value for carbon sequestration for the seven forests of Central Himalaya (6.83-7.42 t $ha^{-1}yr^{-1}$). Under a project "Kyoto: Think Global, Act Local", carbon sequestration rates of three Van Panchayats have been studied which ranged between 3.0-4.5 t c ha^{-1} yr^{-1}. Jina *et al.* (2008) studied van panchayat forests of Central Himalaya (dominated by pine and oak) and reported 4-5 t c ha^{-1} yr^{-1} carbon sequestration rate. Pant and Tewari (2013) reported 3.1-6.07 t c ha^{-1} yr^{-1} carbon sequestration rate from the chir-pine forests of Central Himalaya.

Kaul (2010) studied the carbon sequestration rate in selected planted tree stands of Eucalyptus and Poplar and found the mean advancing range of 3-8.5 t c $ha^{-1}yr^{-1}$. However, Kaul reported a very less value of carbon sequestration in the forests of sal (1.2 t c $ha^{-1}yr^{-1}$) in the Central Himalaya. Overall, the potential rate of carbon sequestration is higher in Central Himalayan forests.

In oak forests of Central Himalaya, the carbon sequestration rate is generally reported to be very high. Lal and Lodhiyal (2016) reported carbon sequestration 7.99-9.96 t c $ha^{-1}yr^{-1}$ and Jina *et al.* (2008) found 5.5-6.2 t c $ha^{-1}yr^{-1}$ carbon sequestered by the oak dominated forests of Central Himalaya. Jina *et al.* (2008) also studied the degraded oak forest sites of Central Himalaya oak and recorded a very low rate of carbon sequestration (1.5-1.8 t c $ha^{-1}yr^{-1}$).

Degradation of forests is also prevalent at a very fast rate. Providing food and shelter to the huge population has become a threat to the biodiversity. In Uttarakhand and other Himalayan states most of the forest loss currently occurs as a result of forest degradation, in which forest stock drops without reduction in forest area. This represents a type of chronic disturbance (Singh 2002) in which the removal of biomass on a particular day is almost invisible, generally in the form of a few human loads of fodder, firewood and forest floor litter for manuring. The pressure of grazing results into seedling removal which results into clearance of species fate in the forest. The continuous lopping of branches of trees reduces seed settings and also reduces its regenerating foliage. The land eventually result into collection of stems with leafless branches (Singh, 2007).

On the basis of review of past studies, there appeared to be a high potential for enhancing the carbon sequestration in the vegetation and soils of the Himalayan region through improved management of degraded lands. With a cover, 2, 25, 557 km^2 the Indian Himalayan forests (about 38 per cent of geographical area) contain about 5.4 billion t carbon, and sequester nearly 65 million t carbon per year, equal to about 15 per cent CO_2 emissions from fossil fuels combustion from India around the year 2000.

Although less studies on carbon sequestration rate in the Central Himalayan forests have been reported, the results of most of the studies performed unto showed that there is a better capacity in the forests of Central Himalaya to sequester a large amount of carbon from the atmosphere in lesser time. All the species occurring in the Central Himalaya weather in the lower altitude or higher altitude are best suitable for their habitat and if proper management will be given to them then they can show their best capacity in sequestering carbon among other forests of Himalaya.

REFERENCES

Anonymous, (2010) Report of the Task Force- To look into problems of hill states and hill areas and to suggest ways to ensure that these states and areas do not suffer in any way because of their peculiarities. Planning Commission, Govt. of India, New Delhi.112pp.

Champion, H.G. and Seth, S.K. (1968). *A Revised Survey of the forest types of India.* Manager publication Delhi.

Dixon, R.K., Brown, S., Houghton, R.A., Solomon, A.M., Trexler, M.C., Wisniewski, J. (1994). Carbon pools and flux of global forest ecosystems. *Science* 263: 185–190.

Hoen, H.F., Solberg, B. (1994). Potential and economic efficiency of carbon sequestration in forest biomass through silvicultural management. *For. Sci.* 40 (3), 429–451.

Janzen, H.H. (2004). Carbon cycling in earth systems: a soil science perspective. *Agri. Ecosyst. Env.* 104: 399-417.

Jina, B.S., Sah, P., Bhatt M.D. and Rawat, Y.S. (2008). Estimating carbon sequestration rates and total carbon stockpile in degraded and non-degraded sites of oak and pine forest of Kumaun Central Himalaya. *Ecoprint*,15: 75-81.

Kaul, M., Mohren, G.M.J. and Dadhwal, V.K. (2010). Carbon storage and sequestration potential of selected tree species in India. *Mitigation and Adaptation Strategies for Global Change*, 15(5): 489-510.

Lal, R. (1999). World soils and greenhouse effect. *Global Change News Lett.* 37: 4–5.

Pant, H. and Tewari, A. (2013). Carbon Sequestration Potential of Chir Pine (Pinus roxburghii. Sarg) Forest on Two Contrasting Aspects in Kumaun Central Himalaya between 16501860 m Elevation. *Applied Ecology and Environmental Sciences*, 1(6): 110-112.

Peterson, G. D., Beard Jr. T. D., Beisner B. E., Bennett, E. M., Carpenter, S. R., Cumming, G. S., Dent, C.L., Havlicek, T.D. (2003). Assessing future ecosystem services: a case study of the Northern Highland Lake District, Wisconsin. *Conservation Ecology*, 7 (3): 1.

Rana, B.S., Singh, S.P. and Singh, R.P. (1989). Biomass and net primary productivity in Central Himalayan forest along an altitudinal gradient. *Forest Ecology and Management*, 27: 199–218.

Rawat, V.S. and Rawat, Y.S. (2012). Glimpse of Forestry Research in the Indian Himalayan Region. Pp. 97-101.Published for the ENVIS Centre on Himalayan

Ecology, G.B. Pant Institute of Himalayan Environment and Development, Almora by M/s Bishen Mahendra Pal Singh, Dehradun.

Sah, P. (2007). Plant species diversity assessment in undisturbed and disturbed oak forests of Central Himalaya. Ph.D. Thesis, Kumaun University, Nainital, India.

Sedjo, R.A., Sohngen, B. and Jagger, P. (1998). Carbon Sinks in the PostKyoto World. RFF Climate Issue Brief No. 13, Internet Edition.

Singh, G. (2008). Diversity of vascular plants in some parts of Kedarnath Wildlife Sanctuary (Western Himalaya). PhD. Thesis submitted to Kumaun University Nainital.

Singh, J.S. and Singh, S.P. (1992). Forest if Himalaya: Structure, Functioning and Impact of man. Gynodaya Prakashan, Nainital, India.

Singh, P., Dubey, P. and Jha, K. K. (2004) Biomass production and carbon storage at harvest age in superior *Dendrocalamus strictus* plantation in dry deciduous forest region in India. pp. 35. In: H. P. Singh and N.K. Daldani (eds.) Abstract Volume of VIIth World Bamboo Congress, 27th February-4 th March, New Delhi, India.

Singh, S.P. and Singh, J.S. (1986) Structure and function of the Central Himalayan oak forests. *Proceedings: Plant Sciences, 96*(3): 159-189.

Singh, S.P. (2002) Balancing the approaches of environmental conservation by considering ecosystem services as well as biodiversity. *Current Science, 82*(11): 1331-1335.

Singh, S.P. (2007) Himalayan forest ecosystem services. *Central Himalayan Environment Association, Nainital, Uttarakhand, India*, 53.

Solberg, B. (1997) Forest biomass as carbon sink–economic value and forest management/policy implications. Critical Rev. Env. Sci. Tech. 27, S323–S333, special issue.

United Nations Framework Convention on Climate Change (2008) Conference of the Parties 14, Poznan, December 2008.

Watson, R.T., I.R. Noble, B. Bolin, N.H. Ravindranathan and D.J. Verardo (eds.) (2000) Land Use, Land Use Change and Forestry. Special report of the intergovernmental panel on climate change. Cambridge University Press, Cambridge, UK.

Zobel, D.B. and Singh S.P. (1997) Himalayan forests and ecological generalizations. *BioScience*, 47: 735-745.

Part 3

Environmental Issues: Effects on Climate Change

Living Life

Life is not a race - but indeed a journey.
Be honest. Work hard. Be choosy.
Say "thank you", "I love you",
and "great job" to someone each day.
Go to church, take time for prayer.
The Lord giveth and the Lord taketh.
Let your handshake mean more than pen and paper.
Love your life and what you've been given, it is not accidental
– search for your purpose and do it as best you can.
Dreaming does matter. It allows you to become that which
you aspire to be. Laugh often.
Appreciate the little things in life and enjoy them.
Some of the best things really are free.
Do not worry, less wrinkles are more becoming.
Forgive, it frees the soul.
Take time for yourself –plan for longevity.
Recognize the special people you've been blessed to know.
Live for today, enjoy the moment.

Bonnie L. Mohr ©

Chapter 16

Climate Change and Marine Biodiversity

Samah M. Bassem and Fagr Kh. Abdel-Gawad

Centre of Excellence for Research and Applied Studies for Climate Change and Sustainable Development, (C3SD-NRC), Water Pollution Research Department, Environmental Research Division, Centre of Excellence for Advanced Science, National Research Center (NRC), Giza, Egypt

ABSTRACT

Climate change is a growing and pervasive universal menace to biodiversity and ecosystems. The rise in anthropogenic CO_2 emissions during the last century has altered oceanic conditions, impacting marine ecosystems as well as the goods and services that they present to society. The following chapter focuses on the recent literature representing how global marine biodiversity is related to climate change and also affected. It was noted that species response to climate change by alteration in morphology, behavior, also geographic range conveys; however, responses are not always regular across species. Then it is a must to study and understand the different impacts of climate change on marine biodiversity on our planet.

Keywords: *Climate change, Biodiversity, Range-shifts, Ocean acidification.*

16.1. Introduction

We live in a unique planet with a number of features, vast oceans with a variety of life forms therein. Biodiversity is usually defined as the variation among all living creatures (U.N., 1992). The ocean's biodiversity had been subjected to enormous changes during evolutionary times that have caused the variety of life through millions of years (Harnik *et al.*, 2012). Incorporated marine biodiversity management and conservation must consider maintaining and protecting the natural combination and functioning besides ensuring that seas submit the society needs (Elliott, 2011). This should include a lot of local actions and pressures, those emerging from inside the sea area being administered and also broader pressures, such as universal climate change, coming from outside (European Commission, 2008).

Nowadays, every country on every continent is being impacted by climate change affecting severely national economics and worldwide ecosystems. Weather patterns are variable, sea levels also rise, weather events are going more extreme, and the greenhouse gas emissions reached the maximum levels in history. Global warming may produce significant consequences to the marine environment, as temperature affects water column steadiness, availability of nutrients then primary production and thus composition, plenty trophic adequacy of zooplankton (Richardson 2008). The global climate change caused by greenhouse gas emissions resulted from different anthropogenic actions has affected both chemical and physical constituents of oceans (Stocker *et al.*, 2013). Many proofs showed significant aberration in temporal changes and spatial allocations of key environmental constituents in open-oceans and also coastal regions (Solomon *et al.*, 2007). Many changes affected the global ocean since 1950s such as; rising in sea surface temperature (SST), decreasing pH in addition to the rising in sea level. All of the previous changes are supposed to continue and grow during the 21st century with a surpassing rate of variation and reaching CO_2 concentrations which were not detected since the geologic era (Gattuso *et al.*, 2015).

Climate change may lead to some consequences in the whole biosphere affecting marine biodiversity and current human living models (IPCC 2014). Climate change impacts the growth, survival, reproduction and also distribution of species. Moreover, their effects can be represented in the level of communities or the whole ecosystems (IPCC 2014; Vieira *et al.*, 2015). Most of the aquatic species are very sensitive to any changes in water parameters that their temperature and the general biological performance changes to match the environment (Pörtner, 2002).

16.2 Biodiversity in the Context of Changing Climate

Recently, climate change and global warming gained a global concern, being a significant challenge for the society today and also becoming a focal subject, relying on that the frequency and intensity of extreme events are causing a global redistribution of life with essential modifications in food supply and also other products (Pecl *et al.*, 2017). The consequences of climate change are considered a global menace to the human life, involving the global economy, migration, public health, employment, universal peace and also security (United Nations Human Development Report 2008).

In the context of biodiversity conservation, improving life quality and sustainable use of the natural resources are topics that need to be harmonized with socioeconomic activities and also human growth of existing and coming generations, that all together with some global dynamics, comprising uncontrolled population increase, economic inequality and urbanization cause serious effects on ecosystems (United Nations Human Development Report 2008; IPCC 2014). Climate change involving extreme events have impacted our planet in different ways, represented in temperature increase, erratic and erosive precipitation causing soil and coastal areas erosion, impacting fishing communities and also great damage to ecosystems (IPCC 2014; Chiabai *et al.*, 2018).

16.3 Consequences of Changing Climate on Biodiversity

Climate change encompasses many modifications in the physical environment, resulting from human activities and emissions of atmospheric CO_2 beside some greenhouse gasses. In the marine environment, the primary keys of climate change that lead to shifts in biodiversity are rising in sea temperature, decrease in ocean pH, increasing rates and extent of extreme events (heat waves and storms). Also, sea-level rise. Climate change has already affected marine populations and communities as a result of high temperatures (Hoegh-Guldberg and Bruno 2010; Wernberg *et al.*, 2011). Any shifts in fundamental environmental properties such as; temperature (global climate warming), pH (ocean acidification) will cause direct physiological impacts on populations as seen in Figure 16.1.

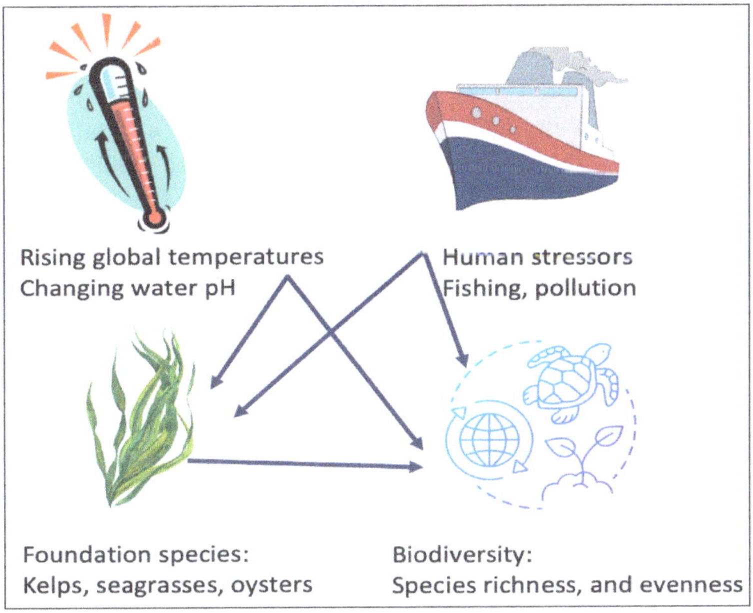

Figure 16.1: Impacts of Climate Change on Marine Environment.

Climate change causes physiological changes in individual species (blue and dark arrows), foundation species and also species living in a biogenic habitation. This lead to variation in species performance relying on their tolerance, some species will implement better (green plusses = winners) affected by the new conditions while others do worse (red minuses = losers), and this lead to balance shift of their interactions resulting in more direct impacts (light blue arrows). The presence of many species is dependent on existence of foundation species providing: food, protection and habitat so any effects on foundation species surely will have disproportionately big impacts on biodiversity and the whole ecosystem (big light blue arrow). Images and symbols used in the figure courtesy of the Integration and Application Network (ian.umces.edu/symbols/)

The temperature has a primary impact on biochemical reaction proportions, and all biological paths have an appropriate temperature above which the organism performance decreases, often swiftly (Kordas *et al.*, 2011). On the other hand, pH firstly impacts the availability of many forms of calcium carbonate, making it hard for calcifying marine organisms for maintaining growth rates as CO_2 concentrations rise (leading to ocean acidification). As a consequence, the direct effects of increasing water temperatures will be widespread onto all taxa and also biogeographical areas that impacts of ocean acidification will probably be more distinctive, unless where habitat formers like tropical corals are negatively influenced. Yet, impacts of ocean acidification exceed direct negative implications on calcifying aquatic organisms like; calcareous algae, molluscs and sea urchins. As an example, short small filamentous seaweeds, that can prohibit the induction of big habitat-forming seaweeds, may be positively impacted by acidification and modifications in carbon presence, especially beneath eutrophied properties (Russell *et al.*, 2009).

Moreover, ocean acidification effects may intensify the warming climate. Marine organisms exhibit a powerful link between their thermal tolerance borders and their spreading more than terrestrial organisms (Sunday *et al.*, 2012), and shifts in species movements have been one of the most vastly registered feedback to a warming climate (Wernberg *et al.*, 2011, 2013). Similar noted variations related to ocean acidification were not noticed, in spite of alterations in communities along natural pH ratios propose such variations are likely. It is probable that marine species are more resistant to ocean acidification than expected before (Hendriks *et al.*, 2010). Interestingly, most aquatic organisms possess several featured life stages that frequently have various tolerances to environmental properties. Likewise, many species in a community show various susceptibilities to climate change, and as they also possess several ecological missions, the optimum population and community-level effects of changing climate will represent that of the majority of sensitive species or life phases.

16.4 Impacts of Climate Change on Fish Functional Traits

Temperature has a direct impact on poikilothermic organisms such as fish; growth, physiology, reproduction and also general behavior. Environmental variations, particularly climate warming, may strongly affect the population and distribution of aquatic biota with species-specific physiological thresholds of temperature adaptation, or in responses to any other trophic fields (Sabates *et al.*, 2006, Rijnsdorp *et al.*, 2009). It was noted that variation in both temperature and ocean chemistry influenced physiology, growth and reproduction of marine biota (Sumaila *et al.*, 2011). Recently, it was proved that fish body size could be shrunk when exposed to climate change, mainly due to warming, decrease in oxygen and also resources availability as shown in Figure 16.2 (Sheridan and Bickford 2011, Cheung *et al.*, 2013).

By studying the meta-analysis of the impact of climate change on body size, Daufresne *et al.* (2009) declared that there is a significant increase in the ratio of small-sized fish and also young age groups and a reduction in size at an age that matched Bergmann's rule of temperature vs size. Oceans are subjected to be much warmer and also less oxygenated, according to the Intergovernmental Panel on Climate Change (IPCC 2014). So, the average maximum fish body size in the Mediterranean Sea is expected to decrease by 4 per cent to 49 per cent from 2000 to 2050 (Cheung *et al.*, 2013).

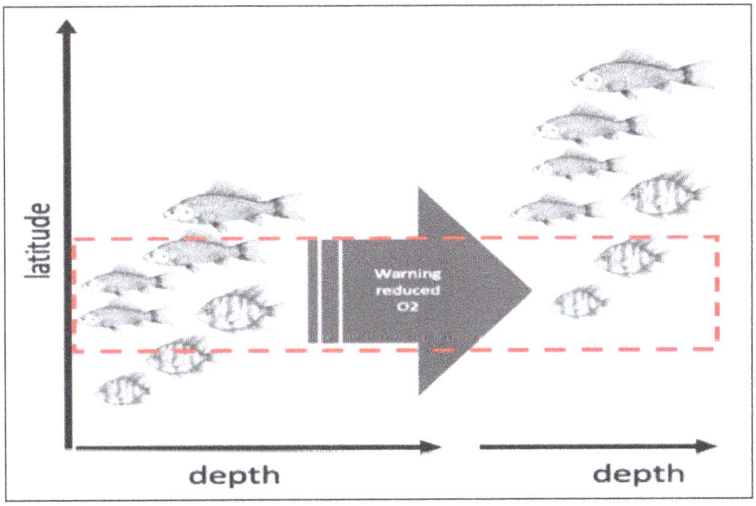

Figure 16.2: The Expected Modifications in Body Size at Individual and Assemblage Stages Referring to Cheung *et al.* (2013), due to the Invasion/Increased Abundance of other Smaller Species and Local Extinction/Decreased Abundance of Larger Species, the Average Maximum Body Weight is Expected to Shrink at the Assemblage Stage, with a Vertical Spreading Shift of Species.

In spite of the relative local heterogeneity, the estimated reduction in fish size is found to be the largest in the western Mediterranean (20 per cent -49 per cent). In contrary, the average body weight is projected to rise in the Gulf of Lion due to the migration of great exploited species. However, explaining international scale simulation outcomes at the regional level could be dangerous and require more precise scale regional researches. Moreover, variations in the whole body size structure proposed that climate and also ocean changes will give rise to dramatic variations in food web dynamics, the natural mortality ratio of fish species and size at organism maturity (Palomares and Pauly 1998, Cheung *et al.*, 2013). Fishes living in warmer waters are projected to have a decreasing body size and shorter size at first maturity with possible elevated natural mortality proportions (Sumaila *et al.*, 2011).

16.5 Adaptation to the Warming Climate

Aquatic species adapt themselves to the warming climate by shifting thermal niches and regulating niche widths (Pörtner 2010). Adaptation could be made within species (modifying allele frequencies in species population) or among species (modifying species recurrence in the community). Moreover, adaptation may encompass novel mutations or pick from standing genetic or species diversity. The last type is known as phenotypic plasticity. Consequently, based on environmental patterns, the genotype can regulate phenotypic properties of an organism referring to habitat requirements. Phenotypic plasticity comprises changes in life-history features, physiological and behavioral performance and is a crucial mechanism for organisms to conform rapidly fastly to variable environmental conditions. Developmental plasticity in behavioural and life history characteristics is a prevalent phenomenon in marine organisms, and climatic stimulant in early ontogeny may be a significant trigger to clarify plasticity (O'Connor *et al.*, 2007, Munday *et al.*, 2013).

The highest risk of extinction is found in species having prolonged generations cycles, limited population size and also overexploited species populations (Dulvy *et al.*, 2003). Particularly more complicated species need to "buy time" to live in a changing climate, that they require more time for permitting DNA fixed adaptive modification (Chevin *et al.*, 2010; Storch *et al.*, 2014). Munday *et al.* (2013) stated that even in case of long-lived species the genetic adaptation shouldn't be neglected that the adaptive choice within times of fast environmental variations, is depending on theoretical thinking of phenotypic plasticity and also genetic evolution (Chevin *et al.*, 2010). So, phenotypic plasticity, in addition to the organisms' evolutionary potential must be included when estimating the effects of climate change for marine living organisms.

Phenotypic plasticity, species sorting, genetic adaptation and trans-generational plasticity will all actively play a role in modifying biodiversity during the changing climate. Still, unfortunately, relative significance and schedule for all such different options are not readily addressed. To consider all the previous issues it requires a collection of experimental data on the main functional features combined with data on species movements with different temperature gradients to recognize thermal niches (Litchman *et al.*, 2012). In the same time, quantitative genetics, molecular approaches and evolution trials need to study temperature impacts on selection and mutation in isolation and also in integration with any other environmental stressors. The possibility of estimating species distribution and performance through merging ecological and evolutionary qualifications has already been expressed in some model approaches (Follows *et al.*, 2007; Thomas *et al.*, 2012).

REFERENCES

Cheung, W.W., Sarmiento, J.L., Dunne, J., Frölicher, T.L., Lam, V.W., Palomares, M.D., Watson, R. and Paul, Y. D. (2013) Shrinking of fishes exacerbates impacts of global ocean changes on marine ecosystems. Nat Clim Change 3: 254–258

Chevin, L.M., Lande, R. and Mace, G.M. (2010) Adaptation, plasticity, and extinction in a changing environment: towards a predictive theory. PLoS Biol 8: e1000357

Chiabai, A., Quiroga, S., Martinez-Juarez, P., Higgins, S. and Taylor, T. (2018) The nexus between climate change, ecosystem services and human health: towards a conceptual framework. Sci Total Environ 635: 1191–1204

Daufresne, M., Lengfellner, K. and Sommer, U. (2009) Global warming benefits the small in aquatic ecosystems. Proc Natl Acad Sci 106: 12788–12793

Dulvy, N.K., Sadovy, Y. and Reynolds, J.D. (2003) Extinction vulnerability in marine populations. Fish Fish 9: 261–285

Elliott, M. (2011) Marine science and management means tackling exogenic unmanaged pressures and endogenic managed pressures – a numbered guide. Mar. Pollut. Bull. 62: 651–655.

European Commission, (2008) Directive 2008/56/EC of the European Parliament and the Council establishing a framework for community action in the field of marine environmental policy (Marine Strategy Framework Directive). Off. J. Eur. Union L164, 19–40.

Follows MJ, Dutkiewicz S, Grant S, Chisholm SW. (2007) Emergent biogeography of microbial communities in a model ocean. Science 315: 1843–1846

Gattuso, J.P., Magnan, A., Billé, R., Cheung, W.W.L., Howes, E.L., Joos, F. *et al.* (2015) Contrasting futures for ocean and society from different anthropogenic CO_2 emissions scenarios, Science 349 (6243) (aac4722-1-10).

Harnik, P.G., Lotze, H.K., Anderson, S.C., Finkel, Z.V., Finnegan, S., Lindberg, D.R., Liow, L.H., Lockwood, R., McClain, C.R., McGuire, J.L. (2012) Extinctions in Ancient and Modern Seas Trends Ecol. Evol. 27: 608–617

Hendriks, *I.E.*, Duarte, C.M. and Alvarez, M. 2010. Vulnerability of marine biodiversity to ocean acidification: a meta-analysis. Estuar Coast Shelf Sci 86: 157–164

Hoegh-Guldberg, O. and Bruno, J.F. (2010) The impact of climate change on the world's marine ecosystems. Science 328: 1523–1528

IPCC, (2014) Climate Change, 2014. Impacts, adaptation, and vulnerability. Summaries, frequently asked questions, and cross-chapter boxes. In: Field CB, Barros VR, Dokken DJ, Mach KJ, Mastrandrea MD, Bilir TE, Chatterjee M, Ebi KL, Estrada YO, Genova RC, Girma B, Kissel ES, Levy AN, MacCracken S, Mastrandrea PR, White LL (eds) A contribution of working group II to the fifth assessment report of the Intergovernmental Panel on Climate Change. World Meteorological Organization, Geneva, 190 pp.

Kordas, R.L., Harley, C.D.G. and O'Connor, M.I. (2011) Community ecology in a warming world: the influence of temperature on interspecific interactions in marine systems. J Exp Mar Biol Ecol 400: 218–226

Litchman, E., Edwards, K.F., Klausmeier, C.A. and Thomas, M.K. (2012) Phytoplankton niches, traits and eco-evolutionary responses to global environmental change. Mar Ecol Prog Ser 470: 235–248

Munday, P., Warner, R.R., Monro, K., Pandolfi, J.M. and Marshall, D.J. (2013) Predicting evolutionary responses to climate change in the sea. Ecol Lett 16: 1488–1500

O'Connor, M.I., Bruno, J.F., Gaines, S.D., Halpern, B.S., Lester, S.E., Kinlan, B.P. *et al.* (2007)Temperature control of larval dispersal and the implications for marine ecology, evolution, and conservation. Proc Natl Acad Sci U S A 104: 1266–1271

Palomares, M.L.D. and Pauly, D. (1998) Predicting food consumption of fish populations as functions of mortality, food type, morphometrics, temperature and salinity. Mar Freshw Res 49: 447–453

Pörtner, H.O. (2002) Climate variations and the physiological basis of temperature dependent biogeography: systemic to molecular hierarchy of thermal tolerance in animals, Comp. Biochem. Physiol. A Mol. Integer. Physiol. 132: 739–761

Pecl GT, Araújo MB, Bell JD, Blanchard J, Bonebrake TC, Chen I-C, *et al.* (2017) Biodiversity redistribution under climate change: impacts on ecosystems and human well-being. Science 355: eaai9214

Pörtner, H.O. (2010) Oxygen- and capacity-limitation of thermal tolerance: a matrix for integrating climate-related stressor effects in marine ecosystems. J Exp Biol 213: 881–893

Richardson, A.J. (2008) In hot water: zooplankton and climate change. ICES J Mar Sci 65: 279–295

Rijnsdorp, A.D., Peck, M.A., Engelhard, G.H., Möllmann, C., Pinnegar, J.K. (2009) Resolving the effect of climate change on fish populations. ICES J Mar Sci J Cons: fsp056

Russell, B.D., Thompson, J-A.I., Falkenberg, L.J. and Connell, S.D. (2009) Synergistic effects of climate change and local stressors: CO_2 and nutrient-driven change in subtidal rocky habitats. Glob Chang Biol 15: 2153–2162

Russell, B.D., Harley, C.D.G., Wernberg, T., Mieszkowska, N., Widdicombe, S., Hall-Spencer, J.M. and Connell, S.D. (2012) Predicting ecosystem shifts requires new approaches that integrate the effects of climate change across entire systems. Biol Lett 8: 164–166

Sabates, A., Martín, P., Lloret, J., Raya, V. (2006) Sea warming and fish distribution: the case of the small pelagic fish, *Sardinella aurita*, in the western Mediterranean. Glob Change Biol 12: 2209–2219

Sheridan, J.A. and Bickford, D. (2011) Shrinking body size as an ecological response to climate change. Nat Clim Change 1: 401–406

Solomon, IPCC, S., Qin, D., Manning, M., Chen, Z., Marquis, M., Averyt, K.B., Tignor, M., Miller, H.L. (Eds.), (2007) Climate Change: the Physical Science Basis. Contribution of Working Group I to the Fourth Assessment Report of the Intergovernmental Panel on Climate Change, Cambridge University Press, Cambridge

Stocker, T.F., Qin, D., Plattner, G.K., Tignor, M., Allen, S.K., Boschung J., *et al.* (2013) Climate Change 2013: the Physical Science Basis. Contribution of Working Group I to the Fifth Assessment Report of the Intergovernmental Panel on Climate Change, Cambridge University Press, Cambridge

Storch, D., Menzel, L., Frickenhaus, S. and Pörtner, H.O. (2014) Climate sensitivity across marine domains of life: limits to evolutionary adaptation shape species interactions. Glob Chang Biol 20(10) : 3059–3067

Sumaila, U.R., Cheung, W.W., Lam, V.W., Pauly, D., Herrick, S. (2011) Climate change impacts on the biophysics and economics of world fisheries. Nat Clim Change, 1: 449–456.

Sunday, J.M., Bates,.AE. and Dulvy, N.K. (2012) Thermal tolerance and the global redistribution of animals. Nat Clim Change. doi: 10.1038/NCLIMATE1539

Thomas MK, Kremer CT, Klausmeier CA, Litchman E (2012) A global pattern of thermal adaptation in marine phytoplankton. Science 338: 1085–1088

U.N. (1992) Convention on Biological Diversity, United Nations, New York, 1992. http: //www.biodiv.org/.

United Nations (U.N.) (2008) Human development report 2007–2008: fighting climate change: human solidarity in a divided world. UNDP, New York. https: //doi.org/10.18356/3300d265-en

Vieira LR, Guilhermino L, Morgado F, (2015) Zooplankton structure and dynamics in two estuaries from the Atlantic coast in relation to multi-stressor exposure. Estuar Coast Shelf Sci 167: 347–367

Wernberg, T., Russell, B.D., Moore, P.J., Ling, S.D., Smale, D.A., Campbell, A., Coleman, M.A., Steinberg, P.D., Kendrick, G.A. and Connell, S.D. (2011) Impacts of climate change in a global hotspot for temperate marine biodiversity and ocean warming. J Exp Mar Biol Ecol 400: 7–16

Wernberg, T., Smale, D.A., Tuya, F., Thomsen, M.S., Langlois, T.J., de Bettignies, T., Bennett, S. and Rousseaux, C.S. 2013. An extreme climatic event alters marine ecosystem structure in a global biodiversity hotspot. Nat Clim Change 3: 78–82.

Chapter 17

Climate Change: Should I Care?

Sanskriti Suhas Mujumdar

Director, Institute of Climate Change Research, Civil Engineering Department, Faculty of Technology and Engineering, The Maharaja Sayajirao University of Baroda, Vadodara – 390 001, Gujarat

e-mail: ssmujumdar-ced2msubaroda.ac.in, directoralumni@msubaroda.ac.in

ABSTRACT

Of late there has been a lot of hue and cry about climate change. Terms like global warming, ozone layer depletion, green house effects etc. are now known to school children too. Quite often the press and media have also played an important role in making us aware about climate change. Earth's warming trend continued in 2019, making it the second-hottest year in NOAA's (National Oceanic and Atmospheric Administration, US Dept. of commerce)140-year climate record just behind 2016.The world's five warmest years have all occurred since 2015 with nine of the 10 warmest years occurring since 2005, according to scientists from NOAA's National Centres for Environmental Information (NCEI). It was also the 43rd consecutive year with global land and ocean temperatures, at least nominally, above average. The average temperature across the globe in 2019 was 1.71°F (0.95 of a °C) above the 20th-century average and just 0.07 of a °F (0.04 of a °C) cooler than the 2016 record.

So do we really care about it? Are we aware of the effects of climate change to each one of us as an individual? Can we really do something about it or is it just the government, NGOs and academicians who need to worry and can do something about it.

Keywords: *Climate change, Global warming, Causes, Threats, Challenges.*

17.1 Introduction

In recent years, global warming has been the subject of a great deal of political controversy. As scientific knowledge has grown, this debate is moving away from whether humans are causing warming and toward questions of how best to respond. The planet's average surface temperature has risen about 2.05°F (1.14 °C) since the late 19th century, a change driven largely by increased carbon dioxide and other human-made emissions into the atmosphere (www.ncdc.noaa.gov/monitoring-references/faq/indicators.php). Most of the warming occurred in the past 40 years, with the six warmest years on record taking place since 2014. Not only was 2016 the warmest year on record, but eight months out of that year – from January through September, with the exception of June – were the warmest on record for those respective months (www.giss.nasa.gov/research/news/20170118/).

Earth's climate has changed throughout history. Just in the last 650,000 years there have been seven cycles of glacial advance and retreat, with the abrupt end of the last ice age about 11,700 years ago marking the beginning of the modern climate era – and of human civilization. Most of these climate changes are attributed to very small variations in Earth's orbit that change the amount of solar energy our planet receives. The current warming trend is of particular significance because most of it is extremely likely (greater than 95 per cent probability) to be the result of human activity since the mid-20th century and proceeding at a rate that is unprecedented over decades to millennia. Earth-orbiting satellites and other technological advances have enabled scientists to see the big picture, collecting many different types of information about our planet and its climate on a global scale. This body of data, collected over many years, reveals the signals of a changing climate.

The heat-trapping nature of carbon dioxide and other gases was demonstrated in the mid-19th century. Their ability to affect the transfer of infrared energy through the atmosphere is the scientific basis of many instruments flown by NASA. There is no question that increased levels of greenhouse gases must cause Earth to warm in response.

Ice cores drawn from Greenland, Antarctica, and tropical mountain glaciers show that Earth's climate responds to changes in greenhouse gas levels. Ancient evidence can also be found in tree rings, ocean sediments, coral reefs, and layers of sedimentary rocks. This ancient, or palaeo climate, evidence reveals that current warming is occurring roughly ten times faster than the average rate of ice-age-recovery warming. Carbon dioxide from human activity is increasing more than 250 times faster than it did from natural sources after the last Ice Age.

In 1824, Joseph Fourier calculated that an Earth-sized planet, at our distance from the Sun, ought to be much colder. He suggested something in the atmosphere must be acting like an insulating blanket. In 1856, Eunice Foote discovered that blanket, showing that carbon dioxide and water vapour in Earth's atmosphere trap escaping infrared (heat) radiation. In the 1860s, physicist John Tyndall recognized Earth's natural greenhouse effect and suggested that slight changes in the atmospheric composition could bring about climatic variations. In 1896, a seminal paper by Swedish scientist Svante Arrhenius first predicted that changes in atmospheric carbon dioxide levels could substantially alter the surface temperature through the greenhouse effect. In 1938, Guy Callendar connected carbon dioxide increases in Earth's atmosphere to global warming. In 1941, Milutin Milankovic linked ice ages to Earth's orbital characteristics. Gilbert Plass formulated the Carbon Dioxide Theory of Climate Change in 1956.

Signs that the Earth is warming are recorded all over the globe. The easiest way to see increasing temperatures is through the thermometer records kept over the past one and a half century. Around the world, the Earth's average temperature has risen by few °F (0.8°C) over the last century and is continuously rising every year.

The concepts of climate and weather are often confused. Weather is the state of the atmosphere at any given time and place. Weather patterns vary greatly from year to year and from region to region. Familiar aspects of weather include temperature, precipitation, clouds, and wind that people experience throughout the course of a day. Severe weather conditions include hurricanes, tornadoes, blizzards, and droughts.

Climate is the average weather conditions that persist over multiple decades or longer. While the weather can change in minutes or hours, identifying a change in climate has required observations over a time period of decades to centuries or longer. Climate change encompasses both increases and decreases in temperature as well as shifts in precipitation, changing risks of certain types of severe weather events, and changes to other features of the climate system.

Climate scientists overwhelmingly agree that humans are causing recent global warming. The consensus position is articulated by the Intergovernmental Panel on Climate Change (IPCC) statement that 'human influence has been the dominant cause of the observed warming since the mid-20th century' (Qin *et al.*, 2014). The National Academies of Science from 80 countries have issued statements endorsing the consensus position. Nevertheless, the existence of the consensus continues to be questioned. Despite the diversity of sampling techniques and approaches, a consistent picture of an overwhelming consensus among experts on anthropogenic climate change has emerged from these studies. Another recurring finding is that higher scientific agreement is associated with higher levels of expertise in climate science (Oreskes 2004, Verheggen *et al.*, 2014).

We are already experiencing changes in the frequency, severity, and even the location of some weather and climate phenomena, including extreme temperatures, heavy rains and droughts, and some other kinds of severe weather, and these changes are projected to continue. This means that areas already experiencing health-threatening weather and climate phenomena, such as severe heat or hurricanes, are likely to experience worsening impacts, such as even higher temperatures and increased storm intensity, rainfall rates, and storm surge. It also means that some areas will experience new climate-related health threats.

A close observation of the rainfall patterns across the district, state and country will give a proof of this phenomenon. Areas which were known for abundant rainfall have suddenly registering below average values and areas otherwise known for scarce rainfall is experiencing heavy bouts.

17.2 Let us First Understand the Broad Effects of Climate Change

- ✰ A warmer atmosphere causes the planet's snowpack, glaciers and sea and freshwater ice to melt rapidly. Melting glaciers and polar ice sheets contribute to unprecedented sea level rise. Melting sea ice exposes darker ocean waters, which absorb more sunlight than ice – heating the ocean more and triggering a relentless cycle of melting and heating. This phenomenon leads to inward regress of sea into land.
- ✰ The Weather is getting more extreme. Heatwaves are more frequent worldwide. The increased evaporation of water is like fuel for storms, extreme events such as hurricanes. Rising sea levels make storm surges capable of much greater damage. The areal extent of naturally arid areas, droughts and wildfires intensify. Average temperature has increased and is increasing every year. Average precipitation has increased since 1900, but some areas have experienced increases greater than the national average, and some areas have experienced decreases. Heavy downpours are increasing nationally, especially over the last two decades.
- ✰ The oceans are getting hotter, expanding and becoming more acidic. They are getting hotter because they absorb 90 percent of the extra heat in the climate. This shift causes the oceans to expand, contributing to higher sea levels. Meanwhile, nearly a third of carbon dioxide emissions end up in the oceans, triggering a chemistry change that makes the water more acidic, dissolving the shells of sea creatures. The ocean is almost 40 percent more acidic than it used to be.
- ✰ Climate change is a major threat to agriculture. Where, how and when we grow food is vitally connected to our climate's normal patterns. Worldwide, farmers are struggling to keep up with shifting weather patterns and increasingly unpredictable water supplies. Farms are more likely to face attacks from weeds, diseases and pests, which reduce yield.
- ✰ Infrastructure and transportation are at risk. Hot weather, flooding and other extreme weather events damage infrastructure, put heavy burdens on electrical supplies and disrupt how we travel and commute. Also capacity of existing stormwater drains will be less to carry water for heavy downpours resulting in flooding of the urban areas.
- ✰ Warmer, polluted air affects our health. A warmer atmosphere increases the formation of ground-level ozone – also known as smog – in polluted regions. Smog irritates lungs and triggers asthma attacks. Smoke from wildfires further degrades the air. Extreme summer heat means more deaths during heatwaves. Warmer freshwater makes it easier for disease-causing agents (such as bacteria) to grow and contaminate drinking water.
- ✰ Weather and climate affect the survival, distribution, and behavior of mosquitoes, ticks, and rodents that carry diseases.

- ☆ Oil and natural gas infrastructure is vulnerable to the effects of climate change and the increased risk of disasters such as storm, cyclones, flooding and long-term increases in sea level. Minimising these risks by building in less disaster prone areas, can be expensive and impossible in countries with coastal locations or island states.

All thermal power stations depend on water to cool them. Not only is there increased demand for fresh water, but climate change can increase the likelihood of drought and fresh water shortages. Another impact for thermal power plants, is that increasing the temperatures in which they operate reduces their efficiency and hence their output. The source of oil often comes from areas prone to high natural disaster risks; such as tropical storms, hurricanes, cyclones, and floods.

Changes in the amount of river flow will correlate with the amount of energy produced by a dam. Lower river flows because of drought, climate change, or upstream dams and diversions, will reduce the amount of live storage in a reservoir; therefore reducing the amount of water that can be used for hydroelectricity. The result of diminished river flow can be a power shortage in areas that depend heavily on hydroelectric power.

- ☆ Global warming, together with resultant changes in food and water supplies, can indirectly cause increases in a range of adverse health outcomes, including malnutrition, diarrhoea, injuries, cardiovascular and respiratory diseases, and water-borne and insect-transmitted diseases.
- ☆ Due to extreme temperatures and atmospheric pollution, we prefer the use air conditioners. This has increased our consumption of electricity drastically. Electricity bills have doubled for most of the houses in the last decade. Even lower middle class people now have air conditioners at their homes.

Similarly people staying in colder regions have increased use of heaters. Centrally air-conditioned offices, homes, shopping complexes and malls are big culprits for electricity consumption.

Higher consumption of electricity in turn plays a role in increasing the atmospheric heat and thus adding to climate change.

Polar sea ice coverage continued its downward trend in 2019. Northern Hemisphere snow cover was close to average in 2019; at 9.57 million square miles Both the Arctic and Antarctic oceans recorded their second-smallest average annual sea-ice coverage during the 1979–2019 period of record. The month was in fact Earth's second-hottest December 2019 on record, logging an average temperature 1.89°F (1.05°C) above the 20th-century average. Only December 2015 was warmer. Parts of central Europe, Asia, Australia, southern Africa (including the island of Madagascar), New Zealand, Alaska, Mexico and eastern South America had record-high average land temperatures in 2019. The annual globally averaged sea surface temperature was the second highest on record at 1.39°F (0.77 of a °C) above the 20th-century average and just behind 2016 (www.ncdc.noaa.gov/sotc).

Total anthropogenic GHG emissions have continued to increase over 1970 to 2010 with larger absolute increases between 2000 and 2010, despite a growing number of climate change mitigation policies. Anthropogenic GHG emissions in 2010 have reached 49 ± 4.5 $GtCO_2$-eq/yr[3]. Emissions of CO_2 from fossil fuel combustion and industrial processes contributed about 78 per cent of the total GHG emissions increase from 1970 to 2010, with a similar percentage contribution for the increase during the period 2000 to 2010. Globally, economic and population growth continued to be the most important drivers of increases in CO_2 emissions from fossil fuel combustion. The contribution of population growth between 2000 and 2010 remained roughly identical to the previous three decades, while the contribution of economic growth has risen sharply. Increased use of coal has reversed the long-standing trend of gradual decarbonisation (*i.e.*, reducing the carbon intensity of energy) of the world's energy supply. Various cropping pattern in different parts of the world will change due to change in climate. Table 17.1 shows various health risks associated with climate change.

Table 17.1: Health Risks Associated With Climate Change

Health Conditions	*Possible Influences of Climate Change*
Alzheimer's Disease	Persons with cognitive impairments are vulnerable to extreme weather events that require evacuation or other emergency responses.
Asthma	Asthma is exacerbated by changes in pollen season and allergies and exposures to air pollutants affected by changes in temperature, humidity, and wind.
Chronic Obstructive Pulmonary Disease (COPD)	COPD patients are more sensitive than the general population to changes in ambient air quality associated with climate change.
Diabetes	Diabetes increases sensitivity to heat stress; medication and dietary needs may increase vulnerability during and after extreme weather events.
Cardio-vascular disease	Cardiovascular disease increases sensitivity to heat stress.
Mental Illness	Mental illness may impair responses to extreme events; certain medications increase sensitivity to heat stress.
Obesity	Obesity increases sensitivity to high ambient temperatures.
Disability	Persons with disabilities may find it hard to respond when evacuation is required and when there is no available means of transportation or easy exit from residence.

17.3 What can be done as an individual?

☆ Plant more Trees

Trees not only attract rain but also stabilize the soil and thus prevent excess surface run off thus reduce floods to a large extent.

Trees absorb carbon dioxide and release oxygen into the air thus purifying it.

☆ Use Water Judiciously

Water is one of the five essential elements necessary for survival of life. However, the available quantity and quality of water for human consumption is decreasing with passage of time. Running taps, long showers, leaking taps, *etc.* are responsible for unaccountable wastage of water.

Besides the direct use of water for our daily requirements, human beings also use water indirectly for production of many items like paper, leather, clothes, *etc.* Thus, less wastage of other commodities will also result in less wastage of water indirectly.

Water conservation, harvesting and recharge can play a big role in climate change parameters besides increasing the availability of water for human use and improving its quality.

Reduce, Reuse, Recycle and Recharge are the four essential Rs to be implemented by each one of us.

☆ Reduce Electricity Consumption

Consumption of electricity in general can be reduced easily. This can reduce the burden on water resources too. Generation of electricity depends on hydropower/ thermal resources in most of the cities. Consumption of coal plays a big role in harming the environment and hydropower projects require large investments and space.

Judicious use of electricity plays a big role in climate conservation. Proper use of home appliances like air conditions, refrigerators and washing machines can reduce not just the consumption but also our bills. Use of washing machine on full load can reduce use of water as well as electricity. Putting things in organised manner and removing unnecessary storage can reduce the load of the refrigerator. Managing proper ventilation and designing houses keeping in mind the wind directions can reduce the temperatures within the houses by 1-2°. Use of reflecting material on terraces can reduce transfer of heat in summers. This can reduce the load on air conditioners finally reducing the consumption of electricity. Use of solar water heaters instead of electric water heaters should be preferred.

We can opt for alternate sources of energy like solar and wind for generation of electricity. Rechargeable batteries can be used for storing the energy generated by solar and wind resources. This can be used for lights, water heating, cooking, *etc.* Unused space of the terrace can be used for installation of solar panels and wind mills.

☆ Reduce Consumption of Fossil Fuels

Increase in the spending capacity of consumers and lack of proper public transport system has resulted in more vehicles on roads. Along with higher consumption of fuels, the emissions from these vehicles are also responsible for pollution and atmospheric heat.

Earlier people used to have about 2-3 two wheelers in each family. Today most of them have two wheelers as well as four wheelers and prefer to use four wheelers over two wheelers. Also each member of the family uses individual vehicles. Vehicle pooling can be easily managed by proper planning and communication amongst family and friends. Short distances can be travelled by walking/cycles. CNG or Electric battery driven vehicles can be preferred over fuel driven vehicles.

All this can reduce the consumption of fossil fuels and thus reduce emission of gases which lead to atmospheric heating.

17.4 Conclusions

Climate change is real. There will always be uncertainty in understanding a system as complex as the world's climate. However there is now strong evidence that significant global warming is occurring. The evidence comes from direct measurements of rising surface air temperatures and subsurface ocean temperatures and from phenomena such as increases in average global sea levels, retreating glaciers, and changes to many physical and biological systems. It is likely that most of the warming in recent decades can be attributed to human activities (IPCC 2001).

In answer to the question, can we do something? Consumption of electricity can be reduced easily. Proper use of home appliances like air conditions, refrigerators and washing machines can reduce not just the consumption but also our bills. Managing proper ventilation and designing houses keeping in mind the wind directions can reduce the temperatures within the houses by 1-2°. Use of reflecting material on terraces can reduce transfer of heat in summers. This can reduce the load on air conditioners. Vehicle pooling can be easily managed by proper planning and communication amongst family and friends. Short distances can be travelled by walking/cycles. Use of CNG or electric driven vehicles should be promoted. All these can reduce emission of gases which may lead to atmospheric heating. Each small step of ours can contribute a lot.

To conclude the discussion, each and every one of us can play a small but significantly important role in controlling climate change. Many small drops collectively form an ocean and when it comes to climate change every drop matters.

Small changes/precautions by each and every individual will collectively lead to a big change including spreading awareness amongst others. Authority and administration are responsible for big infrastructural changes but it is the individual habits that has lead to the disaster we are facing today ... so individuals have a bigger role to play as compared to the authorities.

At this juncture Swami Vivekanand's words "**ARISE, AWAKE AND STOP NOT TILL THE GOAL IS ACHIEVED**" seem the most applicable for all of us.

REFERENCES

https: //climate.nasa.gov/faq/17/do-scientists-agree-on-climate-change/

http: //www.globalcarbonatlas.org/en/CO_2-emissions

http: //bit.ly/Global201912

http: //www.ncdc.noaa.gov/sotc

https: //www.ncdc.noaa.gov/monitoring-references/faq/indicators.php

http: //data.giss.nasa.gov/gistemp

https: //www.giss.nasa.gov/research/news/20170118/

Oreskes N. (2004) Beyond the ivory tower. The scientific consensus on climate change *Science* **306** 1686

Qin D, Plattner G K, Tignor M, Allen S K, Boschung J, Nauels A, Xia Y, Bex V and Midgley P M (2014) *Climate Change* 2013: *The Physical Science Basis* (ed T Stocker) Cambridge University Press

Verheggen B, Strengers B, Cook J, van Dorland R, Vringer K, Peters J, Visser H and Meyer L (2014) Scientists' views about attribution of global warming *Environ. Sci. Technol.* **48** 8963–71

Chapter 18

Medicinal Plants in India: Impact on Biodiversity, Pharmaceutical Industries and Production Chain Due to Climate Change

*Akankhya Guru, Payel Saha, Vivek Kumar, Akash Hidangmayum and Padmanabh Dwivedi**

Department of Plant Physiology, Institute of Agricultural Sciences, Banaras Hindu University, Varanasi – 221 005, Uttar Pradesh
**e-mail: pdwivedi25@rediffmail.com*

ABSTRACT

India is home for rich biodiversity and a major contributing nation towards the 3000 years old Indian system of healing known as Ayurveda. Global climate change has posed a serious threat to the biodiversity of medicinal plants. Lack of rainfall, increase in temperature, elevation in CO_2 level, changes in soil properties, etc. have caused a severe change in the biodiversity of plants. Many of the drugs derived from such medicinal plants are in practice in Ayurveda since ancient times to modern practice as 'tradition to trend'. For better therapeutic leads, the potentials of Ayurveda drugs and practices should be explored further with the validation of modern scientific approaches. The multicentric and multi-disciplinary approach in the field of medical research by ICMR has validated several traditional treatments. A study has shown that among the young generation of both developing and developed countries, exploration for forest-based medicinal plant products having new pharmaceuticals as well as the demand for well-known medicinal herbs is increasing drastically with growing consciousness towards health. This chapter highlights some issues regarding medicinal plants status and their conservation owing to climate change which has affected their production-market chain, supply and demand, challenging the pharmaceutical industries. Various states in India have set up Medicinal Plant Boards to improvise the status of medicinal plants present in their areas by either ex-situ or in *situ conservation practices. Climate change has resulted in phenological shifts in flowering plants and insect pollinators resulting in mismatches between plant and pollinator populations, leading to the extinction of the plant as well as the pollinator with projected consequences on the architecture of plant-pollinator systems. India serves as home to an anomalous diversity of climatic zones, ranging from tropical in the south to temperate and alpine in the Himalayan north, where regions at the higher elevations receive incessant snowfall during winter. These tremendous climatic variations may instigate a divergence in the phytochemical composition of the plant species. This chapter has reviewed some of the intervention which could successfully implement to conserve and commercialize medicinal phytochemicals for therapeutic and pharmaceutical products. Since climate change is real there is a need to conserve and reduce its impact on medicinal plants through technological intervention.*

Keywords: *Ayurveda, Biodiversity, Conservation, Climate change, Health, Medicinal plants, Phytochemicals*

18.1 Introduction

"Look closely at nature. Every species is a masterpiece, exquisitely adapted to the particular environment in which it has survived. Who are we to destroy or even diminish biodiversity?"

— E.O.Wilson

India, the land of unity in diversity, is one of the 17 mega-biodiversity nations, accounting for about 2.4 per cent of land area and 7 per cent of biodiversity across the globe. The key factor of this much variation in life is known to be the geographical location of this country along with 15 agro-climatic zones. India has a vast variety of landscapes, starting from the great Himalayan mountain ranges at the north to the coastal areas and tropical forests of the south, from the barren deserts of the west to the rain forests and river valleys of the east (Fact sheet, National Medicinal Plant Board). The nation is the natural habitat of about 45,000 plant species (both flowering and non-flowering) belonging to various ecosystems due to a wide variety of environmental conditions and climate variations. Among these, around 15000 species are known to have medicinal importance and 7000-7500 herbal species are used by different communities of this country to cure various diseases (Parasuraman *et al.*, 2014). These plants can be found in different ecosystems, from the mountains of Himalayas to the coasts of the oceans, or from the deserts to the tropical rain forests. The country is so rich in biodiversity that 4 of the total 34 global biodiversity hotspots can be found in India, which are: Nicobar Islands,Western Ghats,NorthEastern rain forests and the Himalayas (India, 2015).This richness in biodiversity is the major contributing factor towards the 3000 years old Indian system of healing known as Ayurveda, where different types of medicinal plants have been listed along with the medicinal properties and uses. Along with the Siddha and Unani system of medicines, Ayurveda has greater than 90 per cent of plant-based formulations, which not only has herbal formulations for general health and wellness, but also curative and preventive properties against different types of diseases (Ravi, and Bharadvaja, 2019). According to the survey conducted by WHO in 2019, around 80-99 per cent of the world population depends on the indigenous system of healing and traditional medicines across the world (WHO 2019, n.d.). For the year 2014-15, the domestic demand for medicinal plants in India was estimated at 1,95,000 MT. On the other hand, the overall consumption of raw herbal drugs in India was estimated at 5,12,000 MT in the same fiscal year. According to a study conducted by the NMPB, around 22 per cent of the production of these drugs were sourced through cultivation. It has also been studied that about 1178 medicinal plant species have been used in trade. Among these species, 242 were traded in an excess of100MT/year. India's export demand for medicinal herbs was estimated at approximately 1,34,500 MT, with an export value of 3211Crore for the period of 2014-15 (Fact sheet, NMPB).

When people refer to medicinal herbs and the age-old traditional system of healing in India, the first thing that comes in our mind is the system of Ayurveda, which is considered to be the oldest healing science in th eworld by many scholars and researchers. According to modern researchers, Ayurveda dates back its origin around 6000 BC (Ernst *et al.*, 2007) which roots from the ancient Vedic period, when people used to be taught these practices in an oral tradition, from an accomplished master, known as"Guru", to their disciples, known as "Shishya". The entire system of healing was based on finding the balance between the body and different types of energies which constitute the very essence of life. Ayurvedic practices suggest the use of different medicinal plants or plant parts alone or in combination to cope up with different diseases. These plants are rich in active compounds, known as secondary metabolites, which are known to have medicinal properties. With the advancement of modern researches and technologies, these active compounds are further investigated in present days to find the actual chemical components which are responsible for their medicinal properties. In the present age, advanced medical science has been combined with many indigenous medicines and traditional practices from different fields, in order to achieve wellbeing to a greater extent for mankind. People are also growing more consciousness towards their health so that they can cope up with their busy stressful routine and harness a better living. In this scenario, Ayurveda is gaining more interest among the people, not only in India but also across the globe. Studies show that people have started to incline more towards the medicinal herbs and natural remedies after the growing popularity of Ayurveda in the present day. The current market for plant-based drugs is projected at about 40 billion, and in the next 3-4 years it is expected to grow by 16 per cent. (Semwal *et al.*, 2019). Plants like turmeric, tulsi, ashwagandha, giloy, *etc.* are used for their immense medicinal properties for thousands of years and many of such plants are exclusively found in the Indian subcontinent. For this obvious reason, many national as well as multinational pharmaceutical companies, have jumped into this field to formulate the herbal drugs from these plants in order to meet the growing demand for those medicines. These companies are exploiting the rich biodiversity of India for the procurement of many such medicinal plants from their natural habitats. Most of these plants are collected from the natural biodiversity hotspots such as Himalayan regions or tropical rain forests, and very few of those species are cultivated for commercial purposes. Not only that, these companies are also gaining a sheer amount of profit from their billion-dollar businesses, by capturing the domestic and international markets (Lange, 2004). But, due to this over-exploitation of natural resources, many medicinal herbs along with other plants are gradually disappearing from their wild habitats. Most of these species have become endangered and even extinct in recent years. It's not a surprise that global warming, climate change and deforestation are also playing crucial roles in this case. Lack of rainfall, increase in temperature, elevation in CO_2 level, changes in soil properties, *etc.* have caused severe changes in the biodiversity of plants (Harish *et al.*, 2012). In the coming years, most Ayurvedic medicines could be adulterated to gain enormous benefit, and it will undoubtedly cause more side effects rather than therapeutic effects (Semwal *et al.*, 2019).

Secondary Metabolites obtained from different plant species are not only important for the plant itself for it's defense mechanism, but also it is very useful for us to treat various ailments. These are considered to have various pharmacological activities that can be described on the basis of their chemical structure and functional groups.Terpenoids, flavonoids, phenolics, alkaloids and glycosides are important secondary metabolites. In both nutraceuticals and modern medicines, they serve as a valuable source of single bioactive ingredients. Medicinal plant-based secondary metabolites often have very strong antioxidant properties that can be used in nutraceuticals as an efficient source of natural antioxidants. As they specifically interact with the signalling receptors, nucleic acids and cell membranes present in the body, several secondary metabolites have a wide range of therapeutic activities. (Velu *et al.*, 2018).

Since medicinal plants are extremely important to human livelihoods, a number of studies have been conducted on the potential effects on medicinal plants for climate change (both beneficial and adverse effects). There is a lot of evidence that climate change can have noticeable effects on life cycles as well as the distribution of medicinal plant species throughout the globe. The effects of climate change on the quality and quantity of the production of secondary metabolites, however, is still uncertain. (Harish *et al.*, 2012). Elevated CO_2 levels have been shown to have beneficial effects on the secondary metabolite synthesis in a few medicinal plants. A study has shown that ultra-high CO_2 levels (3,000 µl CO_2/litre) have increased the weight of fresh leaf, shoot and root of *Thymus vulgaris* L. (thyme), *Mentha spicata* L. (spearmint) and *Mentha aquatica* L. (water mint) as compared to ambient air and high levels of CO_2 (Tisserat, 2002). On the contrary, another study shows the adverse effect of elevation in atmospheric tempreratur resulted by climate change on the habitats of several medicinal plantspecies. Researchers have found that many medicinal plant species which are adapted to cold temperature, inhabiting in the alpine-environment, begun to move upwards gradually to avoid this sudden rise in atmospheric temperature. In many cases, these plants keep migrating upwards until and unless there are no high-altitude areas to inhabit. At this point these plants may get extinct from the face of this world (Keutgen *et al.*, 1997).

18.2 Status of Medicinal Plants in India: Significance in Modern Drug Industries on Commercial Basis

Development of Ayurveda

The origin of Ayurveda can be traced from the ancient knowledge of Atharvaveda and Rigveda. It depicts the scientific tradition of harmonious living with nature. From the ancient age, Ayurveda is one of the traditional systems of medicine and healthcare in India. For centuries, it has formulated several medicines most of which are from medicinal herbs. The promise of Ayurvedic drugs and practises should be further explored with the validation of modern scientific methods for better clinical outcomes. This ancient medical system deals with different classical formulations, such as arka, aristas, asavas, taila, churna, bhasma, gutika, vati, *etc*. Different lead molecules from Ayurvedic herbs, which are known to have various important therapeutic properties, have been invented and developed.Chemo-profiling of these Ayurvedic drugs is important not only for product quality determination, but also for the quantification of bioactive compounds, determination of spurious and related drugs, standardisation, stability and chromatographic fingerprinting of Ayurvedic products. By identifying, purifying, providing protection, drug material, physical and biological properties, *etc.*, the therapeutic efficacy of Ayurvedic herbal medicines can be dramatically increased. Plants based medicines are important for their validation to be explored using modern scientific approaches. Many scholars have made several attempts to study and highlight the important aspects of Ayurveda for the promotion and production of herbal medicines (Mukherjee *et al.*, 2016).

Present Status: Modern Medical Science

There are many initiatives taken by ICMR in India towards the research on medicinal herbs available in this country and thoroughly mentioned in Ayurveda. The multicentric and multi-disciplinary approach in the field of medical research by ICMR have validated several traditional treatments *e.g.* Vijayasar (heartwood of *Pterocarpus marsupium* Roxb.) to cure diabetes mellitus, Kshaarasootra (medicated Ayurvedic thread) to cure anal fistula, encouraging micro as well as macrofilaricidal activity of Shakotak (bark from the stem of *Streblus asper* Lour.) in experimental studies an iridoid glycosides fraction isolated from root/rhizomes of *Picrorhiza kurroa* Royle ex Benth. (Picroliv) for curing viral hepatitis, *etc.* (Tandon and Yadav, 2017). Other developmental as well as compilation of researches on Indian medicinal herbs have resulted in publications of 4 volumes of 120 phytochemical reference standards; 17 volumes of quality standards, consisting of 580 Indian medicinal herbs; 22 volumes of review monographs on around 1420 medicinal herb genus; one publication each on perspectives of Indian medicinal plants for management of lymphatic filariasis, liver disorders and diabetes mellitus and one publication on safety reviews on 75 selected Indian medicinal plants (*ICMR*).

18.2.1 Status of Medicinal Plants in India

In India, individuals from all over the country use various types of medicinal plants in their everyday lives. More than 7500 species of medicinal herbs are estimated to be used by various ethnic groups in India annually. In this current scenario, in the modern Indian economy, medicinal herbs play a vital role. Around 70 per cent of India's forest product exports are accounted for by NTFPs (Non-timber forest products) and demand for phytochemicals is expected to emerge as a new trade frontier in the near future. For thousands of years, India has been recognised to have the richest and most diverse cultural practises in the use of medicinal herbs.

The Table 18.1 shows the representative species as well as species diversity of several biogeographical zones in India proposed by Ved *et al.* (2001).

Table 18.1: Medicinal Plants: Representative Species as well as Species Diversity of Several Biogeographical Zones in India (Ved *et al.*, 2001)

Biogeographical Region of India	*Estimated Number of Medicinal Herbs*	*Examples of some Representative Medicinal Species*
Himalayan region	2500	*Ferula jaeshkeana*Vatke, *Aconitum heterophyllum*Wall. ex Royle and *Saussurea costus* (Balc). Lipsch., *Nardostachys grandiflora* D.C. *Rhododendron anthopogon* D.Dun, *Taxus wallichiana* Zucc and *Panax pseudoginseng* Wall.
Trans Himalayan region	700	*Hippophae rhamnoides* L., *Ephedra geradiana* Wall., *Arnebia euchroma* (Royle) John
Gangetic Plain	1000	*Mallotus philippensis* (Lam.) Muell –Arg., *Holarrhenaq pubescens* (Buch-Ham.) Wall. ex DC., *Peganum harmala* L. and *Pluchea lanceolata* C.B. Clarke.
North-East India	2000	*Smilax glabra* Roxb., *Aquilaria malaccensis* Lam., *Hydnocarpus hurzii* (King) Warb. and *Ambroma augusts* (L.) L.f.
Western Ghats	2000	*Garcinia indica* (Thou.) Choisy, *Myristica malabarica* Lam., *Vateria indica* L. and *Utleria salicifolia* Bedd.
Deccan Peninsula	3000	*Shorea tumbuggaia* Roxb., *Terminalia pallida* Brandis, *Decalepis hamiltonii* Wigh and Arn, and *Pterocarpus santalinus* L.f.
Semi-Arid	1000	*Caesalpinia bonduc* (L.) Roxb, *Commiphora wightii* (Arn.) Bhandari, *Tribulus rajasthanensis* Bhandari and Sharma.
Desert	500	*Convolvulus microphyllus* Seib ex Spreng., *Citrulus colocynthis* (L.),Schrader, *Tecomella undulate* (Sm.) Seem. and *Cressa crertica* L.
Islands	1000	*Claophyllum inophyllum* L., *Aisandra butyracea* (Roxb.), Baehni, *Barringtonia asiatica* (L.), Kurz and *Adnanthera pavonina* L.
Coasts	500	*Acanthus ilicifolius* L., *Rhizophora mucronata* Lam., *Sonneratia caseolaris* (L.) engl. and *Avicennia marina* Vierth

A study has shown that exploration for forest-based medicinal plant products with new pharmaceuticals as well as demand for well-known medicinal herbs has grown dramatically among the young generation of both developing and developed countries with growing consciousness towards health (Farnsworth and Soejarto, 1991). It's unfortunate that most of the traded herbal materials in bulk are still from their wild habitats and very few of some well-known species (like turmeric, cinnamon, *etc.*) are cultivated to meet the demand. According to studies conducted by the International Trade Centre, Geneva, India was ranked second among the countries exporting medicinal plants with an annual export volume of 326000 tonnes, with a market value of Rs 45.95 million (approx. US$ 1.4 million) during 1992-95. The country which was ranked first is China. A further increase in trade with a new and growing herbal cosmetic industry that plays a very important role in meeting demand for medicinal plants worldwide is indicated by current trends. There is a significant rise in domestic trade in medicinal herbs in India, in addition to international trade and marketing. Another estimation (Ved 1997; Ved *et al.*, 2001) had shown that, in the year 2000, the turnover of this booming industry in India was projected as Rs 4000 million (approx. US$ 88 million). This increasing trade in medicinal herbs, however, has serious consequences for the survival of numerous plant species, many of which are threatened and even at risk of extinction. Unfortunately, due to the rapid loss of natural ecosystems as well as the overexploitation of these herbs from the wild, this rich biodiversity of medicinal herbs in India faces an enormous challenge and difficulty in survival. To meet the rising demands of the Indian herbal industries, which is projected to have an annual turnover of US$ 300 million, medicinal herbs are harvested and exploited from nearly 165,000 ha of forests in India every year (FRLHT, 1997).

Not just that, most of India's natural forests and some of the forest plantations are facing risks from a rising human population. In order to promote agricultural operations, forests are rapidly losing their species diversity mainly due to illegal harvesting and clearing (Luoma Aho, A. T., Pacific, A., and International Plant Genetic Resources Institute. 2004). Following are few species of medicinal herbs in India which have not only been considered as endangered but also threatened over a decade: *Alpinia galanga, Acorus calamus, Commiphora wightii, Dendrobium pauciflorum, Dendrobium nobile, Diplomeris hirsuta, Dioscorea deltoidea, Gentiana kurroo, Paphiopedilum druryi, Nelumbo nucifera, Podophyllum hexandrum, Santalum album, Saussurea lappa* and *Rauvolfia serpentina*. However, other than these, many species of medicinal herbs can be added to this list, *e.g. Berberis aristata, Saraca asoca, Picrorrhiza kurroa, Gloriosa superba, Costus speciosus, etc.* (Ayensu, 1986).

In September 1996, The Medicinal Plant Specialist Group met in Nairobi for the identification of "Top 50 medicinal plant species" of India which require immediate attention for conservation. Both the global as well as regional priority species were identified by this Group in five steps which have been listed by them. The following are the species which were identified in January 1998 by the Indian Subcontinent Plant Specialist Group for the purpose of detailed study and protection: *Centella asiatica, Streblus asper, Adhatoda vasica, Abrus precatorius, Costus speciosus, Tribulus terrestris, Rauvolfia serpentina, Gloriosa superba, Withania somnifera* and *Saraca asoca.*

Fifty-four forest genebank sites known as Medicinal Plant Conservation Areas (MPCA) have been established by respective State Forest Departments (SFDs) of Tamil Nadu, Kerala, Maharashtra, Andhra Pradesh and Karnataka; in association with UNDP (United Nations Development Programme) and DANIDA (Danish International Development Policy); and consulted by Foundation for Revitalisation of Local Health Traditions (FRLHT). Since 1993, a network of 54 MPCAs, with an area around 200- 500 ha each, has been established gradually. These MPCA represent each and every forest type with major bio-climatic as well as soil regime variation. These gene banks harbour about 45 percent of documented flowering and medicinal herb populations from Peninsular India, along with 70 percent of red-listed plant specimens. In order to satisfy the increasing demand of the herbal drug industry, these germplasms maintained by the MPCA network can be used to provide genuine, high-quality planting materials for commercial cultivation purposes. Another great thing about MPCA is that it constitutes study sites for the recovery researches on threatened species. For equitable benefit sharing and sustainable use, MPCAs have proven their key roles in capacity building for forestry personnel, local communities and researchers in order to protect medicinal plants of India in their natural habitats (State of Plant Genetic Resources for Food and Agriculture in India 1996- 2006: A Country Report, 2007). This experience helped a lot in the implementation of plans as well as programs under the Biological Diversity Act 2002, Medicinal Plants Board, and National Biodiversity Strategy and Action Plan (NBSAP). Medicinal Plant Boards have been set up by various states in India to improvise the status of medicinal plants present in their regions by *ex situ* or *in situ* conservation practises (State of Forest Genetic Resources Conservation and Management in India, n.d.).

18.2.2 Market Demand and Supply of Medicinal Plants

NMPB made the first significant and successful attempt to assess the demand for and availability of medicinal products at national level in the year 2001-02 when it commissioned a survey conducted by CERPA (Centre for Research, Planning and Action, New Delhi) in order to understand the level of annual trading of 162 medicinal plants, specifically selected in India (Ved *et al.*, 2008). In the year 2006-07, NMPB commissioned another study at the national level on the demand as well as supply of medicinal plants. This study was conducted by FRLHT, which, for the first time properly addressed a number of intricacies in the herbal drug market, contributing to the awareness of the subject related to the diversity of raw drug trading entities, their botanical correlation, the annual amount of trade as well as the sources of supply. NMPB has extensively surveyed the market of medicinal herbs of this country in collaboration with ICFRE (Indian Council of Forestry Research and Education, Dehradun), to assess the demand and supply scenario of medicinal herbs and related drugs. For the year 2014-15, the estimated consolidated commercial demand for raw, plant-based drugs was 5,12,000 MT. On the other hand, estimated Exports of plant-based raw drugs along with herbal extracts was estimated as 1,34,500 MT for the same fiscal year. Consumption of the domestic market by Herbal Industries was estimated at 1,95,000 MT. It was also estimated that approximately 1,67,500 MT of plant-based raw drugs were consumed annually by rural households. Around 1178 medicinal herbs documented in the trading practises, of which approximately 242 plant species were reported to be used annually with a quantity of more than 100 MT (Demand and Supply Position of Medicinal Plants | NMPB | Gov. of India. n.d.).

18.2.3 Significance in Modern Drug Industries on Commercial Basis

It is unfortunate that inspite of the establishment of various Councils and Boards by the Government, many major constraints are still faced by the medicinal plant sector and herbal drug industries due to lack of coordination among different stakeholders in India. For the promotion and development of this sector, NMPB has come up with remarkable strategies and policies in order to conserve, properly harvest, low-cost cultivation, research and development, processing, storage, and marketing of herbal raw materials for the promotion and development of this sector. Pharmaceutical Export Promotion Council (PHAREXCIL) serves the purpose of handling various pharmaceutical items such as intermediates and bulk drugs, herbal concoction formulations, nutraceutical products, biotechnologically as well as biologically derived products, clinical trials, contract manufacturing and consultancy services. Shellac and Forest Products Export Promotion Council (SHEFEXCIL) supervises the export of lac and minor forest products such as resins, gums, natural dyeing materials (myrobalan fruits) and herbal extracts, whereas Chemicals, Pharmaceuticals and Cosmetics Export Promotion Council (CHEMEXCIL) promotes and supervises the export of basic inorganic and organic chemical substances, important essential oils and castor oil. The application of biotechnological approaches was very effective for the cultivation and propagation of medicinal plants in order to obtain more yield of plant organs with a greater amount of phytoconstituents (Tiwari *et al.*, 2014). The following are a few examples of the same:

- ***Chlorophytum borivilianum*** Sant. *et* Fernand. is a commercially important medicinal plant, which is valued for its aphrodisiac activity. Due to overexploitation for several decades, it reached a rare status in nature. However, the species gained the attention of researchers and farmers due to its increasing demand in several institutions. Researchers then experimented to heighten the production of the root by controlling sucrose concentrations and CO_2 levels. This resulted in an increase in the overall production level (Joshi *et al.*, 2008).
- Another experiment was carried out by Baker for the improvement of seed yield and L-Dopa content of ***Mucuna pruriens*** during the year 1996–2000 through a conventional breeding approach, where 17 parental lines of *M. pruriens* from different regions of India were evaluated for characters associated with seeds, along with L-Dopa content. After that, those were subjected to conventional breeding programs. The parental lines with non-itching and itching characteristics were taken under consideration. The breeding resulted in the formation of a stable line, named "*Zandu Kauncha*" which showed significantly high yield and L-Dopa content than the previous lines, and pods were developed with the absence of trichomes on them (Nishteswar, 2014).

Another disappointing fact is that some of the most popular formulations of Ayurveda like *Dashamoolarishta, Ashokarishta,Chyavanaprash, etc.* are not genuinely prepared and marketed by the commercial industries as per classical references of Ayurveda. It has been found that during the preparation of classical formulations, the manufacturer replaces or deletes or adds the adulterated drugs instead of the original product. Herbs are also necessary for most of the minerals and metals for their purification as well as incineration processes. The rapid disappearance of forest areas and plants with medicinal importance finally resulted in the loss of those classical prescriptions from the clinical practices and industry is being subjected to a critical situation to process even their proprietary medicines. Banning mercurial preparations is another threat to the medical system of Ayurveda. Hence, it's very important to conserve the biodiversity of plants with medical importance in order to keep this ancient system of medicinal science alive. A proper audit should be carried out on consumption of raw and unprocessed plant-based drugs being exploited annually by individual pharmaceutical companies which would suggest them to cultivate their highly used plant material (instead of procuring them from the wild), either through buyback policy with farmers or in their own areas. In present situation, due to improper harvesting methods or post-harvesting, and primary manufacturing techniques, pharmaceutical companies use raw plant-based products of poor quality. It has been studied that in the name of superior Ayurvedic medicines, most of the medications are actually sold with low quality raw material, manufactured and marketed at extraordinarily high prices. Whereas, drugs with optimum potentialities alone are considered to be useful in treatment of various ailments.

It is a matter of concern that classical Ayurvedic formulations are put into cold storage for their long-term preservation and commercial interests. Ayurvedic drugs are gradually becoming costlier than allopathic drugs without showing any of the promised therapeutic effects. It has been observed that many pharmaceutical companies are marketing herbal drugs for the treatment of diseases which are caused by modern faulty lifestyle habits like hypertension, diabetes, bronchial asthma, *etc.* For the prevention of complications and effective management of these diseases, life long usage of medicines with improvement in life style is necessary. Thus, it is important for herbal drug manufacturers to formulate medicines with optimum therapeutic potentialities (Nishteswar, 2014).

18.3 Climate Change Problems and their Agents

From contemporary times, climate change is a highly concerned subject. Increased temperature and glacier melting are severe threats for humankind. It has a significant impact on water resources, food security and health. Climate change may be defined as the change in the weather pattern including temperature, humidity, wind and hydrological cycle. Globally, the average temperature has increased by 0.85°C in the last three decades and the average air temperature was expected to rise by 1.4-5.8°C during the 21st century (Kovacevic, 2017). The temperature increses, because of the increase in the atmospheric CO_2 concentration with a rate of 20 ppm per decade (Wigley, 2004). The probable effects of global climate change includes more intense heatwaves, frequent wildfires, longer drought, unpredicted storms, a rise in sea level and other extreme climatic events. It is projected that global sea level will rise by 40-63 cm during the 21st century due to the melting of glaciers and polar icecaps (https: //www.un.org/en/sections/issues-depth/climate-change/). Farmers are most vulnerable to extreme weather fluctuations because of climate change. Cultivated crops including cereals, horticultural crops, spices, ornamental and medicinal plants perform critically due to the hazardous effect of the changing climate (Shahrajabian *et al.*, 2020). It limits the availability of water, light and nutrient which affects plant's vital processes such as photosynthesis, respiration, biomass accumulation, pollination, insect, pest and disease invasion (Melillo *et al.*, 1990; Applequist *et al.*, 2019). At the initial stage, photosynthesis will increase but in the longer-term, it will reduce due to less availability of nutrients. The population of plant species and biodiversity distribution is threatened by the changing climate

(Kaky and Gilbert, 2019). The occurrence of extreme meteorological events and unpredictable weather patterns affects production and change the biochemical properties and quality components of medicinal crops, which reduce the profit of produce (Applequist *et al.*, 2019). Farm operations are also affected by adverse climate due to changes in soil texture and insect attacks. Fertilizer, nutrient and water use efficiency decline due to the stress conditions and pollution in soil and water bodies. Increased atmospheric CO_2 concentration has a significant impact on plant-microbe interaction, thus climate change can affect nitrogen fixation (Melillo *et al.*, 1990).

The magnitude of climate change depends on the emission of GHGs and amount of radiative forcing in the next few decades (https: //climate.nasa.gov/effects). Both natural and anthropogenic emissions are responsible for global climate change. But the anthropic factors are the primary cause of the alterations in the global climate in the 20th century. Agents produced by human activity include greenhouse gases responsible for aerosols production and altering of cloud properties and increase of the radiative forcing (Melillo *et al.*, 1990). Man-made emissions account for 64 per cent, 17 per cent, and 6 per cent of the total amount of CO_2, CH_4, and N_2O, respectively in the atmosphere today (Sadatshojaie and Rahimpour, 2020). Fuel combustion produces CO_2 and N_2O, while methane is mainly produced by livestock and rice cultivation. Nitrogenous fertilizers increase the emission of nitrous oxide. The deposition of nitrogen and other nutrients increase pollution in the atmosphere. The warming effects of fluorinated gases are 23000 times greater than CO_2 (Barla, 2016). Hydroxyl radical (OH^-) has the potential to react with CH_4 and CFCs and manage their concentration in the atmosphere (Lelieveld *et al.*, 1998). An increase in radiative forcing due to changes in ozone concentration is influenced by solar and infra-red radiation (Fergusson, 2001).

To mitigate the adverse impacts of climate change, various sustainable approaches are being applied by scientists. Conservation of natural resources, afforestation, cultivation of climate resilience crop, green-house plantation and modern irrigation technologies are crucial for maintaining sustainability in earth climate (Reyer *et al.*, 2009; Maheswari *et al.*, 2019). Replacing the traditional energy sources with renewable and clean energy sources, and adoption of progressive technology for energy production are the needs of the day for fighting climate change. Adoption of innovative ideas to manage carbon emission, application of clean development mechanism, funding to ecosystem services, reduction of food waste and promotion of sustainable food production will help in mitigation of the climate crisis (Munang *et al.*, 2011).

18.4 Climate Change and Biodiversity

Undoubtedly, there will be serious consequences of climate change on biodiversity as it has been estimated that by the year 2100, the temperature of 1.4 °C will be further increased to 5.8 °C. Climate change is already having an influence on biodiversity by compelling it for the adoption of an unstable habitat, varying life cycles, and the establishment of novel physical characteristics. Species having the capability to stretch their ranges swiftly or the species that can endure a broad range of climatic conditions might be favored by an abruptly changing climate; many invasive plant species share these two traits. At the most primitive level of biodiversity, the genetic diversity of populations can be diminished by climate change as a consequence of direction-dependent selection and sudden migration, which meanwhile could affect the functioning of ecosystem and resilience (Botkin *et al.*, 2007). Climate change has resulted in phenological shifts in flowering species and insect pollinators that create incompatibilities between plant and pollinator populations, leading to the extinction of the plant as well as the pollinator with projected outcomes on the architecture of plant-pollinator systems (Kiers *et al.*, 2010; Rafferty and Ives, 2010). At a higher biodiversity level, large fluctuations in vegetation communities can be induced by virtue of climate change which can alter the integrity of the biome.

The expected shift in the type of vegetation may cause significant forest degradation and biodiversity loss particularly in the switch over between forest types, for example, increment in temperature, at a higher altitude, will bring about transitions of lower altitude temperate and sub-temperate forest, to higher altitude alpine and sub-alpine forests following the extinction of several temperate flora types. Since climate change is expected to be more wide-ranging for the regions having higher elevations, the mountainous forests including the Himalayan dry temperate forest, the Himalayan moist temperate forest and sub-alpine and alpine forest are prone to the negative consequences of climate change (Gopalakrisnan *et al.*, 2011).

18.5 Impact of Climate Change on the Diversity of Medicinal Plants

Climate change has turned out to be one of the biggest obstacles to medicinal plants dwelling on Earth. The occurrence of climate change is due to variations in seasonal patterns, weather circumstances, temperature scales and other associated phenomena across the globe. It has been warned by scientists that the intensity and frequency of detrimental effects of climate change will rise in the future, particularly as long as environmentally devastating anthropogenic activities continue uninterrupted (Walther *et al.*, 2002). There exists concern over its widespread impact affecting medicinal value as well as secondary metabolites of numerous medicinal plants which are greatly important on both economic and commercial fronts. Although it is unknown for many scientists whether climate change threatens medicinal plant species more prominently or immediately in comparison to other risks, it has the potency to exert rising thrust upon the species and populations of medicinal plants in the future. The medicinal plants may be affected significantly because of their value as medicines in the traditional systems and as economically beneficial plants. Though in the future, climate change effects are uncertain to a great extent, existing evidence indicates that there is an impact of these phenomena on medicinal plants, and also there are more or less promising threats in line for consideration and discussion. In the 21st century, according to the current assumptions, there will be a rapid climate change affecting heterogeneous levels of biodiversity. As per the estimation of the Intergovernmental Panel on

Climate Change (IPCC), an overall temperature rise of 2–3 °C amplifies the risk of extinction for 20–30 per cent of species (Warren *et al.*, 2013; Stocker *et al.*, 2014). By 2080, it is assumed that the niche of about 58 per cent of plants will be lost depending on their immediate dispersal rates (Warren *et al.*, 2013).

Western Ghats, one of the rich biodiversity regions of India, is threatened by climate change. Priti *et al.* (2016) reported that in practical terms, elevating temperature and fluctuation of rainfall patterns can exhibit serious impacts on the feasible allocation, range shifts of numerous species in addition to a widespread deterioration in the well-suited habitats in the Western Ghats. Very few attempts have been made yet to study the distribution of medicinally important species in the western ghats (Chandran *et al.*, 2010; Priti *et al.*, 2016), but under future climate change, the distribution of *Garcinia indica*, indigenous medicinal species in India, belonging to Maharashtra and Goa of the Western Ghats region, is the first attempt for rehabilitation and reintroduction (Pramanik *et al.*, 2018). The seed oil is specifically used for the preparation of 'Red Mango' (ayurvedic medicine) that treats skin disorders (Vidhate and Singal, 2013). Other health benefits include pain relief, inflammation suppression, reduction in allergic reactions, immunity boost-up, prevention against chronic illness, heat strokes, diarrhea, and dysentery treatment (Mathew *et al.*, 2001). A study suggested that bio-climatic variables such as temperature seasonality, isothermality, annual precipitation, precipitation seasonality, *etc.* are anticipated to display substantial changes, impelling extensive region of the ongoing distribution to turn into minimal potential by 2050, that steadily widens these low potential locations by 2070. The findings denote that predicted climate can affect the diminution of the species' fitness, especially under the greenhouse gas scenario in which the gases will surge continuously at the latest rate till the end of the 21st century (Pramanik *et al.*, 2018). The western ghat is the center of origin for *Piper nigrum*, a high-value crop that possesses anti-inflammatory, anticonvulsant, analgesic and neuroprotective properties. Utilizing ecological niche models, forthcoming climate scenarios expected a decline in the habitat of *P. nigrum* in Southern Western Ghats, harboring several wild cultivars of *P. nigrum*. Moreover, a directional shift was observed in the niche centroid of this particular species (Sen *et al.*, 2016).

Our country, India serves as a homeland to an anomalous diversity of climatic zones, extending from tropical in the south to temperate and alpine in the Himalayan north, where areas at the higher elevations receive incessant snowfall during winter. These tremendous climatic variations may instigate a divergence in the phytochemical composition of the plant species. *Aloe barbadensis*, commonly known as *Aloe vera* is an abundant source of secondary metabolites having manifold attributes *i.e.*, antibacterial, anti-inflammatory, sunburn relief, immune-boosting, anticancer, antioxidant, anti-aging and antidiabetic potentials (Wintola *et al.*, 2011). Phytochemical screening in *Aloe vera* samples collected from six diverse agro-climatic zones *viz.* highland, semi-arid, arid, humid subtropical, tropical wet and dry and tropical wet respectively, of India revealed the presence of most of the bio-active compounds such as glycosides, alkaloids, reducing sugars, phenolic compounds, steroids and terpenoids, flavonoids, tannins, and saponin glycosides, though quantities varied in different accessions. Accessions from highland (low temperature) showed the maximum antioxidant potential while the extracts of tropical zones possessed the least anti-oxidant activity. Lower temperature led to the production of a high amount of phenolics (Kumar *et al.*, 2017).

In India, the Himalayan region is regarded as a great repository of medicinal plants. Because of their importance in traditional health care and trading system, high-altitude medicinal plant species are of huge concern across the Himalayan region. Important species of medicinal plants, for example, species of *Aconitum, Taxus, Ephedra, Dactylorhiza, Fritillaria, Polygonatum, Podophyllum, Picrorrhiza, Nardostachys, etc.* and many other high altitude plants of Himalayan origin have considerable market potential but excessive exploitation and climate change factors have resulted in their restriction to either remote areas or the shielded environments (Kala, 2008). These plants are useful for the treatment of many chronic diseases in humans (Kala, 2009). In the Indian Himalayan region, particularly in the alpine ecosystem, medicinal plants have experienced changes in phenology as well as distribution in relation to the variability of temperature and snow pattern (Khanduri *et al.*, 2008). However, investigations on the effects of climate change on secondary metabolites are limited in the context of the Indian Himalayan region.

18.6 Implications of Climate Change on Secondary Metabolites of Plants

Due to massive chemical diversity, fewer side effects and economic values, plant secondary metabolites are used in many aspects including drug development. Climatic shifts can affect plant secondary metabolites as well as the growth of plants because of changes in the metabolic pathways that modulate physiology, signaling and defense mechanisms. Elevation of atmospheric CO_2 concentration has become a serious issue for plant life since the industrialization era. CO_2 emission resulting from anthropogenic activities has shown a drastic increase in past years (Field *et al.*, 2014). Medicinal plants possess the capability to acclimatize to variable environmental conditions. Secondary metabolites are known to contribute to the elasticity in the metabolic pathways of medicinal plants, but sometimes, this elasticity may affect the synthesis of active constituents which forms the base for their pharmacological activities (Mishra, 2016). *Catharanthus roseus*, native to Madagaskar island of the Indian ocean, possesses anticancer, astringent and antibacterial properties, used to cure cancer (blood, lung), diabetes, Alzheimer's diseases (Kaushik *et al.*, 2017). Exposure of *C. roseus* to the elevated atmospheric CO_2 concentrations (560 ± 25 ppm) significantly increased the total alkaloid and total phenolic content at different levels of nitrogen supply (Table 18.2). The enhancement of phenolic compounds may be incurred as a consequence of the stimulatory effect of elevated CO_2 on the activity of phenylalanine ammonia-lyase (PAL), a chief enzyme of the phenylpropanoid pathway resulting in the production of the increased phenolics (Singh *et al.*, 2015). *Ginkgo biloba*, cultivated in Indian gardens as an ornamental, has been reported to be used for treating Alzheimer's disease and vascular or mixed dementia (Weinmann *et al.*, 2010). When *Ginkgo biloba* is exposed to high concentrations of CO_2 and O_3, the terpenoid profile gets altered in conjunction with a 15 per cent elevation in

Table 18.2: Effects of Climate Change Factors on Secondary Metabolites of Medicinal Plants in India

Sl. No.	*Plants*	*Geographical Region*	*Medicinal Properties*	*Climate Change Factor*	*Changes in Secondary Metabolites*	*References*
1.	*Catharanthus roseus*	Indian ocean island	Anticancer, astringent, and antibacterial	Elevated CO_2	Increase in total alkaloid and total phenolic content	Singh *et al.* (2015)
2.	*Picrorhiza kurroa*	Himalaya	Treats liver, upper respiratory tract disorders	Elevated CO_2	Increased picroside-I and picroside-II contents in rhizomes	Kumar *et al.* (2017)
3.	*Zingiber officinale*	South west and North east India	Cold, nausea, arthritis, migraines, and hypertension treatments	Elevated CO_2	Higher accumulation of phenolics and flavonoids	Ghasemzadeh *et al.* (2011)
4.	*Hypericum perforatum*	Western Himalayas	Anti-oxidant, anti-bacterial, and neuroprotective	Elevated CO_2	A decline in hypericin content	Sharma *et al.* (2020)
5.	*Ginkgo biloba*	Central Himalayan mountains	Treats asthma, bronchitis, Alzheimer's disease	Elevated CO_2 and ozone	Elevated quercetin aglycone and reduced bilobalide levels	Huang *et al.* (2010)
6.	*Sida cordifolia*	Tropical and sub-tropical regions	Anti-inflammator, analgesic, hepato-protective	Ozone	Enhanced synthesis of stigmasterol, β caryophyllen, alkanox, and caryophyllene oxide	Ansari *et al.* (2020)
7.	*Salvia officinalis*	North Western Himalaya	Anti-inflammatory, anti-allergic, antifungal, and antiseptic	Ozone	Increased rosmarinic acid, gallicacid and caffeic acid levels	Pellegrini *et al.* (2015)
8.	*Withania somnifera*	Semi-arid, Deccan Peninsula	Sedative, hypnotic, anti-inflammatory	Cold stress	Increased production of withanolides	Mir *et al.* (2015)
9.	*Ajuga bracteosa*	Sub-Himalayan tract, plains of Punjab and the upper Gangetic plain	Aids to cure cancer, diabetes, gastrointestinal disorders, urinary disorders tuberculosis	Low temperature	Higher concentrations of phenylpropanoids	Rani *et al.* (2017)
10.	*Mentha piperita*	Indo-Gangetic plain, Himalayas	Cures indigestion, antispasmodic, diuretic,antiemetic, mild sedative, relieves irritation	Elevated temperature	Reduced flavonoids, phenolics; increased tannin, terpenoid contents	Alhaithloul *et al.* (2020)
11.	*Coleus* species	Western Himalayas, Nilgiri Hills	Treats digestive disorders, dysentery	Salinity	Maximum terpene content, increased levels of anthraquinones,cardiac glycoside, saponinsetc	Kotagiri *et al.* (2017)
12.	*Centella asiatica*	Eastern Himalaya	Used to treat bronchitis, leprosy, fever, asthma	Salinity	Elevated levels of phenolics, flavonoids, terpenoids, alkaloids	Ibrahim *et al.* (2018)

quercetin aglycone and a decline in the levels of isorhamnetin, kaempferol aglycone and bilobalide, respectively (Huang *et al.*, 2010). *Picrorhiza kurroa* is an endangered medicinal plant of the Himalayan region (3000–5000 m) possessing tremendous medicinal values attributable to the presence of iridoid glycosides (Kumar *et al.*, 2017). This plant species showed enhanced accumulation of secondary metabolites, principally picroside-I and picroside-II in rhizomes in addition to a reduction in picroside-I and augmentation in picroside-II content in leaves at 550 ± 5 μmol mol^{-1} CO_2. Biochemical traits such as total phenolics, flavonoids content were altered in a tissue-specific manner in response to elevated CO_2 and temperature (Kumar *et al.*, 2020). *Zingiber officinale* that aids to heal motion sickness, nausea and vomiting is cultivated in Indian states like Kerala, Andhra Pradesh, Uttar Pradesh, West Bengal and Maharashtra. Atmospheric CO_2 enrichment (800 $\mu molmol^{-1}$) increased total phenolics and flavonoids in younger parts of *Z. officinale*, especially in leaves and rhizomes as evaluated against ambient CO_2 condition (Ghasemzadeh *et al.*, 2011). The plant, *Hypericum perforatum*, found in temperate western Himalayas (2000-3000 m) is well characterized by various secondary metabolites including hyperforin, hypericin, pseudohypericin (Barnes *et al.*, 2001). It is utilized for the treatment of cancer, depression, anxiety, and also as an anti-oxidant, anti-bacterial and neuroprotective agent (Klemow *et al.*, 2011). An experiment carried out by Sharma *et al.* (2020) on *Hypericum perforatum* highlighted higher hypericin content at elevated temperatures (2.5 ± 0.5 °C above ambient) and lower (reduced by 22 per cent) at elevated CO_2 concentration (550 ± 50 μmol mol^{-1}) than control plants. An increase in CO_2 level *i.e.* up to 700 μmol mol^{-1} stimulated the internal CO_2 concentration, leading to a decrease in reduction capacity ($NADPH+H^+$: $NADP^+$) which directly caused a decline in the rate of biosynthesis of highly reduced secondary metabolites (Sharma *et al.*, 2020).

Another factor contributing to climate change, 'ozone', an allotrope of oxygen, is produced in the troposphere layer of earth's atmosphere under the availability of high sunlight through multiplex chemical reactions between carbon monoxide, oxides of nitrogen, methane and non-methane volatile organic compounds (Kleanthous *et al.*, 2014; Harmens *et al.*, 2019). Since the past few decades, the tropospheric O_3 level is uninterruptedly increasing as a consequence of exacerbated human activities as well as industrialization (Sharma *et al.*, 2016; Yadav *et al.*, 2019; Harmens *et al.*, 2019). An elevation in O_3 found in the troposphere also puts forward a threat to crops, including medicinal plants. For example, *Sida cordifolia* has been cultivated predominantly in tropical and subtropical parts of India, where a high O_3 concentration is realized. This is a medicinally important plant exhibiting anti-inflammatory, analgesic, hepato-protective and hypoglycemic activities (Silva *et al.*, 2006; Swathy *et al.*, 2010) owing to the presence of certain biologically active components. Ansari *et al.* (2020) noticed that some important metabolites like stigmasterol, β-caryophyllene, alkanox and caryophyllene oxide displayed enhancement in their activities in elevated ozone exposed leaves of *Sida cordifolia* compared with control plants while variable responses of terpenes were observed in leaves and roots. The concentrations of secondary metabolites might be influenced due to atmospheric ozone via inhibition of carbohydrate accumulation or elicitation of compounds involved in the defense signaling pathway. When

Salvia officinalis plants were exposed to a high ozone concentration of 120 ± 13 ppb for 90 days consecutively, an elevation in phenolics content, 122 per cent increase in rosmarinic acid concentration along with the increase in gallic acid and caffeic acid levels up to 2-fold and 8-fold, respectively was reported (Pellegrini *et al.*, 2015). In a study, Sun *et al.* (2012) examined the effect of ozone on the production of secondary metabolites in the cell suspension culture of *Pueraria thomsnii.* On exposure of suspension cultures to 90 nL L^{-1} ozone for 20 h, the concentration of puerarin, an isoflavonoid showed no elevation. However, after 35 h of exposure duration, the highest amount (2.6-fold increase) of puerarin was found. Further, this study highlighted the function of abscisic acid (ABA) in mediating ozone-induced puerarin production. Interestingly, results revealed that the inhibition of ozone-triggered puerarin content by fluridone could be reversed by exogenous application of low ABA dose (0.2 mg L^{-1}) in the suspension cells and thus, indicating that a low concentration of ABA is required to mediate the ozone-induced puerarin synthesis despite its insufficiency to stimulate puerarin accumulation in the cells. Similarly, in cell cultures of *Hypericum perforatum*, the effect of different doses of O_3 on hypericin production at multiple phases of culture was studied (Xu *et al.*, 2011). Cultures treated with 60 to 180 nL L^{-1} ozone showed a significantly higher concentration of hypericin in comparison to cell cultures without ozone exposure, suggesting that ozone contributes to the stimulation of hypericin synthesis. Moreover, comparing to lag and stationary phases of cell culture, ozone-induced hypericin production was higher at the exponential phase, which indicates that the exponential phase was more responsive to ozone effects while the cell cultures at lag and stationary phases responded comparatively less. Additionally, at the exponential phase, the culture obtained the highest (4-fold of the control) hypericin production while exposing to 90 nL L^{-1} ozone for three hours (Xu *et al.*, 2011).

Fluctuation in temperature is likely to affect the formulation of bioactive compounds found in medicinal plants. Low temperature is a critical environmental factor that can optimize the production of secondary metabolites in medicinal plants such as *Withania somnifera* (Mir *et al.*, 2015). Exposure of *W. Somnifera* to cold temperature showed increased accumulation of withanolides in leaves and roots which could act as a mechanism for reactive oxygen species (ROS) scavenging. On the first day of cold treatment, the plant roots displayed a lower amount of withanolide-A but recovered thereafter until day 7. Cold temperature exposure elicited the accumulation of withanone content in leaves; on the other hand, in roots withanone content increased on the first day of cold treatment and then decreased up to the fifth day, and then again it reached the recovery phase. Furthermore, withaferin A content was significantly and constantly increased for 7 days of cold treatment in leaves, signifying its protective action against cold stress. The authors reported that this increase in withanolides content might be due to the enhanced accumulation as well as the expression of antioxidant enzymes. Besides, the regulation of genes involved in withanolide metabolic pathway might play a potential role in improving withanolides accumulation in the plant species during cold stress (Mir *et al.*, 2015). Likewise, in *Ajuga bracteosa*, a high value endangered medicinal plant found in the sub-Himalayan tract, plains of Punjab and the upper Gangetic plains of India, a low temperature (15°C) treatment resulted in stress-induced accumulation of pharmaceutically active compounds. The highest levels (12.9 mg gallic acid/g dry weight and 9.3 mg quercetin/g dry weight) of total phenolic and flavonoid contents were obtained at 15°C temperature. Also, the study revealed the detection of phenylpropanoids such as cinnamic acid, caffeic acid and quercitol at higher concentrations in samples exposed to the temperature of 15°C and thereby activating plant antioxidant defense system to combat stress stimulus (Rani *et al.*, 2017).

By contrast, the range of higher temperatures suppressed the production of ginsenoside in *Panax quinquefolius* cell cultures (Jochum *et al.*, 2007). Shohael *et al.* (2006) investigated the effects of temperature on the enhancement of eleutheroside B, E, E1, total phenolics, flavonoids and chlorogenic acid accumulation in the somatic embryos of *Eleutherococcus senticosus*. The results showed that lower (12°C and 18°C) and higher (30°C) temperature ranges reduced the accumulation of total phenolics, flavonoids and total eleutheroside but the lower range of temperature enhanced the production of eleutheroside E in somatic embryos. Exposing *Mentha piperita* and *Catharanthus roseus* plants to high temperatures (35°C) for seven and fourteen days caused a reduction in the levels of secondary metabolites including total phenolics, flavonoids and saponins. However, other active compounds such as tannins, alkaloids and terpenoid contents were significantly increased at the same temperature (Alhaithloul *et al.*, 2020). These compounds act as key signaling components, hence their accumulation modulates the response mechanism of plants. Increased production of metabolites helps plants to protect the structural and functional features of cells and membranes under stressed conditions.

Increasing natural and anthropogenic activities has caused the salinity problem that induces changes in plant physiology and metabolism and thus, affecting overall growth and development. Cultivated soils are becoming more saline because of excessive fertilizer usage, irrigation with water of low quality and desertification (Ramadoss *et al.*, 2013). In addition to growth and productivity, a high concentration of salts also affects the composition of secondary metabolites in medicinal plants (Gil *et al.*, 2002). The medicinal plant, *Coleus* has gained its importance in India for exhibiting several therapeutic properties including the treatment of digestive disorders and dysentery (De Souza *et al.*, 1983). For the investigation of salinity effects on the accumulation of bioactive compounds, quantitative estimation of flavonoids, phenols, tannins, lignins, alkaloids, steroids, cardiac glycosides, anthraquinones, saponins and terpenes was performed in leaves, stem and roots of *Coleus* species subjected to salinity stress (Kotagiri *et al.*, 2017). Mild (100 mM), moderate (200 mM) and severe (300 mM) salinity treatments resulted in maximum accumulation of plant secondary metabolites in leaves followed by stem and roots. However, at higher concentrations *i.e.* above 300 mM salt treatment, a decline in the levels of secondary metabolites was noted in *Coleus* species. Among various secondary metabolites studies, terpene content was found to be maximum (123-315 mg g^{-1}) in salt-stressed *Coleus* plants. The authors stated that the enhancement in secondary metabolites accumulation might be due to the induction of enzymatic activity during salinity stress (Kotagiri *et al.*, 2017). Under *in vitro* culture conditions, an increase in withanolide content was greatest at 50 mM NaCl treatment than other treatments in

Withania somnifera (Sabir *et al.*, 2012). Ibrahim *et al.* (2018) experimented in *Centella asiatica*, grown in marshy places all over India up to an altitude of 200 m. While exposing *C. asiatica* to four different salinity treatments *viz.* 0, 50, 100 and 150 mM, the presence of higher amounts of phytochemicals *viz.* phenolics, flavonoids, terpenoids, alkaloids and saponins were noticed in comparison to the control. When the salinity level increased to 50 mM, phenolics, terpenoids, and tannins were detected while increasing the level of salinity to 100 mM, and 150 mM salt concentration increased the accumulation of all the bioactive compounds in *C. asiatica*. The production of phytochemicals may be utilized by plants as a mechanism for regulating osmotic balance together with the maintenance of membrane fluidity under salinity stress conditions (Ibrahim *et al.*, 2018).

18.7 Conclusion

There is a high diversity among species of medicinal plants producing secondary metabolites which are very useful for the treatment of acute diseases including cancer, tuberculosis, HIV, ulcers, *etc.* in human beings. Changing climatic background has a serious impact on medicinal plants growing in variable Indian geographical locations such as the Himalayas, Western ghats. In particular, the endemic medicinal plants are addressed as more vulnerable to the effects of climate change and may face a strong risk of extinction as a result of their limited geographic distribution. Changes in temperature, CO_2 level, O_3 concentration, abiotic stressors such as salinity stress have led to an alteration in the normal functioning and physiology of medicinal plants, which ultimately affects the production of biologically active compounds like phenolics, terpenoids, alkaloids, saponins, *etc.* There is a need to comprehend the behavior of medicinal plants at the molecular level in response to climate change due to multiple factors. Furthermore, research on the habitat range and secondary metabolites production efficacy of threatened medicinal plants under climate change scenario is necessary to formulate conservation strategies and agrotechnologies for cultivation.

REFERENCES

AICEP. (1994). Summary report of the All India coordinated ethnobiological project (AICP).

Alhaithloul, H. A., Soliman, M. H., Ameta, K. L., El-Esawi, M. A., and Elkelish, A. (2020). Changes in Ecophysiology, Osmolytes, and Secondary Metabolites of the Medicinal Plants of *Mentha piperita* and *Catharanthus roseus* subjected to Drought and Heat Stress. *Biomolecules*, 10(1), 43.

Ansari, N., Agrawal, M., and Agrawal, S. B. (2020). An assessment of growth, floral morphology, and metabolites of a medicinal plant *Sida cordifolia* L. under the influence of elevated ozone. *Environmental Science and Pollution Research*, 1-14.

Anthropological survey of India (1994). People of India Project Report for 1994.

Applequist, W. L., Brinckmann, J. A., Cunningham, A. B., Hart, R. E., Heinrich, M., Katerere, D. R., and Van Andel, T. (2020). Scientists¼ warning on climate change and medicinal plants. *Planta Medica*, 86(1), 10-18.

Ayensu, E.S. (1986). World Medicinal Resources. In: Chopra, V.L. and T.N. Khoshoo (eds.) *Conservation for Productive Agriculture*. ICAR, New Delhi.

Barla, M. (2016). Impacts on climate change on tribal economy: a study of Jharkhand state of India. In Proceedings of the International Conference on Poverty and Sustainable Development, Vol. 3, No. 1, pp. 25-36.

Barnes, J., Anderson, L.A., Phillipson, J.D. (2001). St. John's wort (*Hypericumperforatum* L.): a review of its chemistry, pharmacology and clinical properties. *Journal of Pharmacy and Pharmacology*, 53, 583-600.

Botkin, D.B., Saxe, H., Araujo, M.B., Betts, R., Bradshaw, R.H.W., Cedhagen, T. *et al.* (2007). Forecasting the effects of global warming on biodiversity. *Bioscience*, 57, 227–236.

de Souza, N.J., Dohadwalla, A.N., Reden, J. (1983). Forskolin: a labdane diterpenoid with antihypertensive, positive inotropic, platelet aggregation inhibitory, and adenylate cyclase-activating properties.*Med Res Rev. Apr-Jun;* 3(2),201-19.

Demand and Supply Position of Medicinal Plants | National Medicinal Plants Board | Government of India. (n.d.). www.nmpb.nic.in. Retrieved September 22, 2020, from https: / /www.nmpb.nic.in/content/demand-and-supply-position-medicinal-plants

Ernst, Edzard; Pittler, Max H.; Wider, Barbara, (2007). *Complementary Therapies for Pain Management: An Evidence-based Approach*. Elsevier Health Sciences. p. 108. ISBN 9780723434009.

Farnsworth, N., and Soejarto, D. (1991). Global Importance of Medicinal Plants. In O. Akerele, V. Heywood, and H. Synge (Eds.), *Conservation of Medicinal Plants* (pp. 25-52). Cambridge: Cambridge University Press. doi: 10.1017/ CBO9780511753312.005

Fergusson, A. (2001). Ozone Depletion and Climate Change. Meteorological Service of Canada.

Field, C.B., Barros, V.R., Dokken, D.J., Mach, K.J., Mastrandrea, M.D., Bilir, T.E., *et al.* (2014). IPCC, 2014: Climate Change 2014: Impacts, Adaptation, and Vulnerability. Part A: Global and Sectoral Aspects. Contribution of Working Group II to the Fifth Assessment Report of the Intergovernmental Panel on Climate Change. Cambridge University Press, Cambridge and New York.

Forest genetic resources conservation and management/ : proceedings of the Asia Pacific Forest Genetic Resources Programme (APFORGEN) Inception Workshop, Kepong, Kuala Lumpur, Malaysia, 15-18 July, 2003. International Plant Genetic Resources Institute.

FRLHT. (1997). *Trade in Medicinal plant. Amruth* 1997: 2-8. FRLHT, Bangalore, India.

Ghasemzadeh, A., and Jaafar, H. Z. (2011). Effect of CO_2 enrichment on synthesis of some primary and secondary metabolites in ginger (*Zingiber officinale* Roscoe). *International Journal of Molecular Sciences*, 12(2), 1101-1114.

Gil, A., De La Fuente, E.B., Lenardis, A.E., Loopez Pereira, M., Suaorez, S.A., Bandoni, A., Van Baren, C., Di Leo Lira, P., Ghersa, C.M. (2002). Coriander essential oil composition from two genotypes grown in different environmental conditions. *Journal of Agricultural Food Chemistry*. 50, 2870-2877.

Gopalakrisnan, R., Mathangi, J.R., Bala, G., Ravindranath, N.H. (2011). Climate change and Indian Forest. *Curr Sci*, 101(3), 25-29.

Harish, B., Dandin, S., Umesha, K., and Sasanur, A. (2012). Impact of climate change on medicinal plants - A review. *Ancient Science of Life*, 32(5), 23. https: //doi. org/10.4103/0257-7941.111986

https: //climate.nasa.gov/effects/

(https: //cloud1.arc.nasa.gov/solveII/outreach/highschool-ozonedepletion.pdf)

https: //www.un.org/en/sections/issues-depth/climate-change/

Huang, W., He, X.Y., Liu, C.B., Li, D.W. (2010). Effects of elevated carbon dioxide and ozone on foliar flavonoids of Ginkgo biloba. *Adv. Mater. Res.*, 113, 165169. Trans Tech Publications.

Ibrahim, M.H., Shibli, N.I., Izad, A.A., and Zain, N.A.M. (2018). Growth, Chlorophyll Fluorescence, Leaf Gas Exchange and Phytochemicals of *Centellaasiatica* exposed to Salinity Stress. *Annual Research and Review in Biology*, 1 13.

India. (2015, October20) IUCN. https: //www.iucn.org/asia/countries/india#: ~: text=India per cent 2C per cent 20a per cent 20megadiverse per cent 20country per cent 20with

Indian Medicinal Plants - Fact Sheet | National Medicinal Plants Board | Government of India. (n.d.). www.nmpb.nic.in. Retrieved September 21, 2020, from https: // www.nmpb.nic.in/content/medicinal-plants-fact-sheet

Jochum, G.M., Mudge, K.W., and Thomas, R.B. (2007). Elevated temperatures increase leaf senescence and root secondary metabolite concentration in the understory herb *Panax quinquefolius* (Araliaceae). *Am. J. Bot.*, 94, 819–826.

Joshi, N., Dave, A., Vyas, S., and Purohit, S. (2008). Growth and shoot proliferation in ChlorophytumborivilianumSant. et Fernand. In vitro under different carbon dioxide environment. *Indian J Biotechnol.*, *8*, 323–327.

Kaky, E., and Gilbert, F. (2020). Allowing for human socioeconomic impacts in the conservation of plants under climate change. *Plant Biosystems*-An International Journal Dealing with all Aspects of Plant Biology, 154(3), 295-305.

Kala, C.P. (2008). High altitude medicinal plants: A promising resource for developing herbal sector. *Hima- Paryavaran*, 20(2), 7-9.

Kala, C.P. (2009). Medicinal plants conservation and enterprise development. *Med. Plants*, 1(2), 79-95.

Kaushik, S. H. U. C. H. I., Tomar, R. S., Gupta, M., and Mishra, R. K. (2017). An overview of *Catharanthus roseus* and medicinal properties of their metabolites against important diseases. *Eur Acad Res*, 2, 1237-1247.

Keutgen, N., Chen, K., and Lenz, F. (1997). Responses of strawberry leaf photosynthesis, chlorophyll fluorescence and macronutrient contents to elevated CO_2. *Journal of Plant Physiology, 150*(4), 395–400. https: //doi.org/10.1016/s0176-1617(97)80088-0

Khanduri, V.K., Sharma, C.M., Singh, S.P. (2008). The effects of climate change on plant phenology. *Environmentalist*, 28, 143-147.

Kiers, E.T., Palmer, T.M., Ives, A.R., Bruno, J.F., and Bronstein, J.L. (2010). Mutualisms in a changing world: an evolutionary perspective. *Ecol. Lett.*, 13, 1459–1474.

Klemow, K.M., Bartlow, A., Crawford, J., Kocher, N., Shah, J., Ritsick, M., Benzie, I.F.F., Wachtel-Galor, S. (2011). (Eds.), *Herbal Medicine: Biomolecular and Clinical Aspects*. Chapter 11, Medical Attributes of St. John's Wort (*Hypericumperforatum*), CRC Press/Taylor and Francis 11, 142-241.

Kotagiri, D., Beebi, S. K., and Chaitanya, K. V. (2017). Secondary metabolites and the antimicrobial potential of five different Coleus species in response to salinity stress. *bioRxiv*, 220368.

Kovacevic, V., Antunovic, M., Varga, I., Iljkic, D., andJoviæ, J. (2017). Response of soybean and barley to Fertdolomite application on acid soil. Columella: *Journal of Agricultural Environmental Sciences*, 4(2), 7-15.

Kumar, R., Joshi, R., Kumari, M., Thakur, R., Kumar, D., and Kumar, S. (2020). Elevated CO_2 and temperature influence key proteins and metabolites associated with photosynthesis, antioxidant and carbon metabolism in *Picrorhiza kurroa*. *Journal of Proteomics*, 103755.

Kumar, S., Yadav, A., Yadav, M., and Yadav, J. P. (2017). Effect of climate change on phytochemical diversity, total phenolic content and in vitro antioxidant activity of Aloe vera (L.) Burm. f. *BMC Research notes*, 10(1), 60.

Kumar, V., Bansal, A., Chauhan, R.S. (2017). Modular design of picroside-II biosynthesis deciphered through NSG transcriptomes and metabolic intermediates analysis in naturally variant chemotypes of a medicinal herb, *Picrorhizakurroa, Front. Plant Sci.*, 8,564.

Lange, D. (2004). Medicinal and aromatic plants: Trade, production and managemen of botanical resources. *ActaHorticulturae, 629*, 177–197. https: //doi.org/10.17660/actahortic.2004.629.25

Lelieveld, J. O. S., Crutzen, P. J., and Dentener, F. J. (1998). Changing concentration, lifetime and climate forcing of atmospheric methane. Tellus B, 50(2), 128-150.

Luoma-Aho, A. T., Pacific, A., and International Plant Genetic Resources Institute. (2004).

Maheswari, M., Sarkar, B., Vanaja, M., Rao, M. S., Prasad, J. V. N. S., Prabhakar, M., and Bhaskar, S. (2019). Climate resilient crop varieties for sustainable food production under aberrant weather conditions. ICAR-Central Research Institute for Dryland Agriculture, Hyderabad. P, 64.

Melillo, J. M., Callaghan, T. V., Woodward, F. I., Salati, E., and Sinha, S. (1990). Effects on ecosystems. Climate change: The IPCC scientific assessment, 283-310.

Mir, B. A., Mir, S. A., Khazir, J., Tonfack, L. B., Cowan, D. A., Vyas, D., and Koul, S. (2015). Cold stress affects antioxidative response and accumulation of medicinally important withanolides in *Withania somnifera* (L.) Dunal. *Industrial Crops and Products*, 74, 1008-1016.

Mishra, T. (2016). Climate change and production of secondary metabolites in medicinal plants: a review. *Int. J. Herb. Med.*, 4 (4), 27-30.

Mukherjee, P. K., Harwansh, R. K., Bahadur, S., Banerjee, S., Kar, A., Chanda, J., Biswas, S., Ahmmed, S. M., and Katiyar, C. K. (2017). Development of Ayurveda - Tradition to trend. *Journal of Ethnopharmacology, 197*, 10–24. https: //doi. org/10.1016/j.jep.2016.09.024

Munang, R. T., Thiaw, I., and Rivington, M. (2011). Ecosystem management: Tomorrow's approach to enhancing food security under a changing climate. *Sustainability*, 3(7), 937-954.

Nishteswar, K. (2014). Depleting medicinal plant resources: A threat for survival of Ayurveda. *Ayu*, *35*(4), 349–350. https: //doi.org/10.4103/0974-8520.158972

Nishteswar, K. (2014). Depleting medicinal plant resources: A threat for the survival of Ayurveda. *Ayu*, *35*(4), 349–350. https: //doi.org/10.4103/0974-8520.158972

Parasuraman, S., Thing, G. S., and Dhanaraj, S. A. (2014). Polyherbal formulation: Concept of ayurveda. *Pharmacognosy reviews*, *8*(16), 73–80. https: //doi. org/10.4103/0973-7847.134229

Pellegrini, E., Francini, A., Lorenzini, G., Nali, C. (2015). Ecophysiological and antioxidant traits of *Salvia officinalis* under ozone stress. *Environ. Sci. Pollut. Res. Int.*, 22 (17), 1308313093.

Pramanik, M., Paudel, U., Mondal, B., Chakraborti, S., and Deb, P. (2018). Predicting climate change impacts on the distribution of the threatened Garcinia indica in the Western Ghats, India. *Climate Risk Management*, 19, 94-105.

Priced Publications | Indian Council of Medical Research | Government of India. (n.d.). Main.Icmr.Nic.In. Retrieved October 7, 2020, from https: //main.icmr.nic.in/ priced-publications/1352

Rafferty, N.E., and Ives, A.R. (2010). Effects of experimental shifts in flowering phenology on plant–pollinator interactions. *Ecol. Lett.*, 14, 69–74.

Ramadoss, D., Lakkineni, V.K., Bose, P., Ali, S., Annapurna, K. 2013. Mitigation of salt stress in wheat seedlings by halo tolerant bacteria isolated from saline habitats. Springer Plus. 2, 1-7.

Rani, R., Khan, M. A., Kayani, W. K., Ullah, S., Naeem, I., and Mirza, B. (2017). Metabolic signatures altered by *in vitro* temperature stress in *Ajugabracteosa* Wall. ex. Benth. *Acta Physiologiae Plantarum*, 39(4), 97.

Ravi, S., and Bharadvaja, N. (2019). Market Analysis of Medicinal Plants in India. *Current Pharmaceutical Biotechnology*, *20*(14), 1172–1180. https: //doi.org/10.2 174/1389201020666190819154516

Reyer, C., Guericke, M., and Ibisch, P. L. (2009). Climate change mitigation via afforestation, reforestation and deforestation avoidance: and what about adaptation to environmental change?. *New Forests*, 38(1), 15-34.

Sabir, F., Sangwan, R. S., Kumar, R., and Sangwan, N. S. (2012). Salt stress-induced responses in growth and metabolism in callus cultures and differentiating in vitro shoots of Indian ginseng (*Withania somnifera* Dunal). *Journal of Plant growth Regulation*, 31(4), 537-548

Sadatshojaie, A., and Rahimpour, M. R. (2020). CO_2 emission and air pollution (volatile organic compounds, *etc.*)–related problems causing climate change. *In Current Trends and Future Developments on (Bio-) Membranes* (pp. 1-30). Elsevier.

Semwal, D. K., Chauhan, A., Kumar, A., Aswal, S., Semwal, R. B., and Kumar, A. (2019). Status of Indian medicinal plants in the International Union for Conservation of Nature and the future of Ayurvedic drugs: Shouldn't think about Ayurvedic fundamentals? *Journal of Integrative Medicine*, 17(4), 238–243. https: //doi.org/10.1016/j.joim.2019.04.008

Sen, S., Gode, A., Ramanujam, S., Ravikanth, G., and Aravind, N. A. (2016). Modeling the impact of climate change on wild *Piper nigrum* (Black Pepper) in Western Ghats, India using ecological niche models. *Journal of Plant Research*, 129(6), 1033-1040.

Shahrajabian, M. H., Sun, W., and Cheng, Q. (2020). Climate Change, Acupuncture and Traditional Chinese Herbal Medicines. *Pharmacognosy Communications*, 10(1).

Sharma, S., Walia, S., Rathore, S., Kumar, P., and Kumar, R. (2020). Combined effect of elevated CO_2 and temperature on growth, biomass and secondary metabolite of *Hypericum perforatum* L. in a western Himalayan region. *Journal of Applied Research on Medicinal and Aromatic Plants*, 16, 100239.

Shohael, A.M., Ali, M.B., Yu, K.-W. *et al.* (2006). Effect of temperature on secondary metabolites production and antioxidant enzyme activities in *Eleutherococcus senticosus* somatic embryos. Plant Cell Tissue Org. Cult. 85, 219–228.

Singh, A., and Agrawal, M. (2015). Effects of ambient and elevated CO 2 on growth, chlorophyll fluorescence, photosynthetic pigments, antioxidants, and secondary metabolites of *Catharanthusroseus* (L.) G Don. grown under three different soil N levels. *Environmental Science and Pollution Research*, 22(5), 3936-3946.

State of Forest Genetic Resources Conservation and Management in India. (n.d.). www.fao.org. Retrieved September 22, 2020, from http: //www.fao.org/3/ad871e/ad871e09.htm

State of Plant Genetic Resources for Food and Agriculture in India (1996- 2006): A Country Report. (2007). National Bureau of Plant Genetic Resources, (Indian Council of Agricultural Research, New Delhi. 70p

Stocker, T. F., Qin, D., Plattner, G. K., Tignor, M. M., Allen, S. K., Boschung, J.,. and Midgley, P. M. (2014). Climate Change 2013: The physical science basis. contribution of working group I to the fifth assessment report of IPCC the intergovernmental panel on climate change.

Sun, L., Su, H., Zhu, Y., Xu, M. (2012). Involvement of abscisic acid in ozone-induced puerarin production of *Pueraria thomsnii* Benth. suspension cell cultures. *Plant Cell Rep.* 31 (1), 179185.

Tandon, N., and Yadav, S. S. (2017). Contributions of Indian Council of Medical Research (ICMR) in the area of Medicinal plants/Traditional medicine. *Journal of Ethnopharmacology*, *197*, 39–45. https: //doi.org/10.1016/j.jep.2016.07.064

Tisserat, B. (2002). Influence of Ultra-High Carbon Dioxide Levels on Growth and Morphogenesis of Lamiaceae species in Soil. *Journal of Herbs, Spices and Medicinal Plants*, *9*(1), 81–89. https: //doi.org/10.1300/j044v09n01_09

Tiwari, L., Rai, N., Sharma, R. K., Dhar, R., and Mangal, A. (2014). Conservation and sustainability of trade potential medicinal plants: Conservation, cultivation and exploration of therapeutic potential of medicinal plants. *New Delhi: CCRAS, Dept. of AYUSH, Govt. of India*, 61.

Ved Prakash. (2001). Indian Medicinal Plants; Current Status.InSamant SS, Dhar U, Palni, LMS (eds), *Himala-yan Medicinal Plants: Potential and Prospects.*Nainital: Gyanodaya Prakashan;: 45–65

Ved, D. K., Goraya, G. S., India, and National, India. (2008). *Demand and supply of medicinal plants in India*. Bishen Singh Mahendra Pal Singh. https: //books.google.co.in/books/about/Demand_and_Supply_of_Medicinal_Plants_in.html?id=CnftAAAAMAAJ and redir_esc=y.

Ved, D.K. (1997). Trade in medicinal plans: the state of our ignorance. Amruth 1(10): 2-8.

Velu, G., Palanichamy, V., and Rajan, A. P. (2018). Phytochemical and Pharmacological Importance of Plant Secondary Metabolites in Modern Medicine. *Bioorganic Phase in Natural Food: An Overview*, 135–156. https: //doi.org/10.1007/978-3-319-74210-6_8

Walther, G. R., Post, E., and Convey, P. (2002). Ecological responses to recent climate change. *Nature*, 416, 389–95.

Warren, R., Van Der Wal, J., Price, J., Welbergen, J. A., Atkinson, I., Ramirez-Villegas, J., and Lowe, J. (2013). Quantifying the benefit of early climate change mitigation in avoiding biodiversity loss. *Nature Climate Change*, 3(7), 678-682.

Weinmann, S., Roll, S., Schwarzbach, C., Vauth, C., Willich, S.N. (2010). Effects of Ginkgo biloba in dementia: systematic review and meta-analysis. *BMC Geriatr*. 10, 14.

WHO GLOBAL REPORT ON TRADITIONAL AND COMPLEMENTARY MEDICINE 2019. (n.d.) https: //www.who.int/traditional-complementary-integrative medicine/Who Global Report On Traditional And Complementary Medicine 2019.pdf?ua=1

Wigley, T. M. (2004). Modeling climate change under no-policy and policy emissions pathways (pp. 221-248). OECD Publications.

Wintola, O.A., Afolayan, A.J. (2011). Phytochemical constituents and antioxidant activities of the whole leaf extract of Aloe ferox Mill. Pharmaco Mag.,7(28),325–33.

Chapter 19

Climate Change as an Influencing Factor of Global Agriculture and Food Security

Raman Kumar Ravi and R.Y. Hiranmai

School of Environment and Sustainable Development, Central University of Gujarat, Gandhinagar – 382 030, Gujarat
e-mail: smartramanravi@gmail.com

ABSTRACT

The climate change is now a reality and is already affecting the environment. The climate change is a change in the long-term weather patterns that characterize the regions of the world. It has put adverse impact on biogeochemical pathways. According to general consensus, the global climate is changing, which may also affect agricultural and livestock production. This time potential impact of climate change on food production is widely debated. Climate change induced by increasing greenhouse gases is likely to impact directly on food production across the world, and that can vary from region to region. In the long run, the climatic change could affect agriculture in several ways such as quantity and quality of crops in terms of productivity, growth rates, photosynthesis and transpiration rates, moisture availability etc. Increase in the mean seasonal temperature can reduce the duration of many crops and hence reduce the yield. The areas where temperatures are already close to the optimum physiological maxima for crops, warming will impact yields more immediately. The alterations in atmospheric composition can also influence food production directly by its impacts on plant physiology. The chapter explore the influence of climate change on agricultural productivity directly related to food security and their respective measures to combat with.

Keywords: *Climate change, Environment, Agriculture, Food security.*

19.1 Introduction

Climate change refers to any significant long term change in the expected patterns of average weather of region or the entire earth over a long period of time. These changes may take decades, hundreds or perhaps millions of years. However, the increase in human activities, such as industrialization, urbanization, deforestation, agricultural practices, and changes in land use methods, has led to the emission of greenhouse gases, so the rate of climate change is much faster (Gillett *et al.*, 2012; Stott *et al.*, 2016). The issue of climate change is increasingly receiving attention from scientists, public, and policy makers. The United Nations Climate Change Convention defines climate change as "a change, which is directly or indirectly attributed to human activities, changes the composition of the global atmosphere, and in addition to natural climatic conditions, a comparable period of time." Furthermore, the convention defines adverse effects of climate change as changes that have deleterious effects on ecosystems, socio-economic systems, and human and animal welfare (United Nations, 1994).

The greenhouse effect is a natural process and plays an important role in shaping the earth's climate. It has created a relatively warm and friendly environment near the surface of the earth, and humans and other organisms have been able to develop and prosper. However, greenhouse gases (GHGs) such as carbon dioxide (CO_2), water vapor (H_2O), methane (CH_4), nitrous oxide (N_2O), hydrofluorocarbons (HFCs), perfluorocarbons (PFCs) and hexafluoro Sulphur (SF6) levels have increased due to human activities that have led to an overall increase in the earth 's temperature, leading to global warming (Tripathi *et al.*, 2016;Stocker and Intergovernmental Panel on Climate Change, 2013).

There are various sources of greenhouse emission. The energy supply by different sources emits maximum (25.9 per cent) of greenhouse followed by industry (19.4 per cent), forestry (17.4 per cent), agriculture (13.5 per cent), transport (13.1 per cent), residential and commercial buildings (7.9 per cent) and waste and wastewater (2.8 per cent) (Figure 19.2). Due to greenhouse gas emissions, the surface atmosphere is heated, thereby raising the global atmospheric temperature for a long time. This long-term surface temperature change adversely impacts on rainfall is called climate change.

The average global surface temperature (14.6°C) has increased by 0.74°C since the late 19th Century and is expected to increase by 1.4°C - 5.8°C by 2100 AD with significant regional variations (Tripathi *et al.*, 2016; IPCC, 2007; Stocker and Intergovernmental Panel on Climate Change, 2013).The concentration of atmospheric CO_2 has increased from 280 ppm to 395 ppm, whereas that of CH_4 from 715 ppb to 1882 ppb and N_2O from 227 ppb to 323 ppb from the year 1750 and 2012. The Global Warming Potential (GWP) of these gases *i.e.*, CO_2, CH_4 and N_2O is 1, 25 and 310 respectively (Stocker and Intergovernmental Panel on Climate Change, 2013). The projected global warming scenario indicates that by 2100, the global average surface temperature may rise by 1.4 to 5.8 ° C (Figure 19.1). The expected rate of warming in the previous year was unprecedented in 10,000 years.

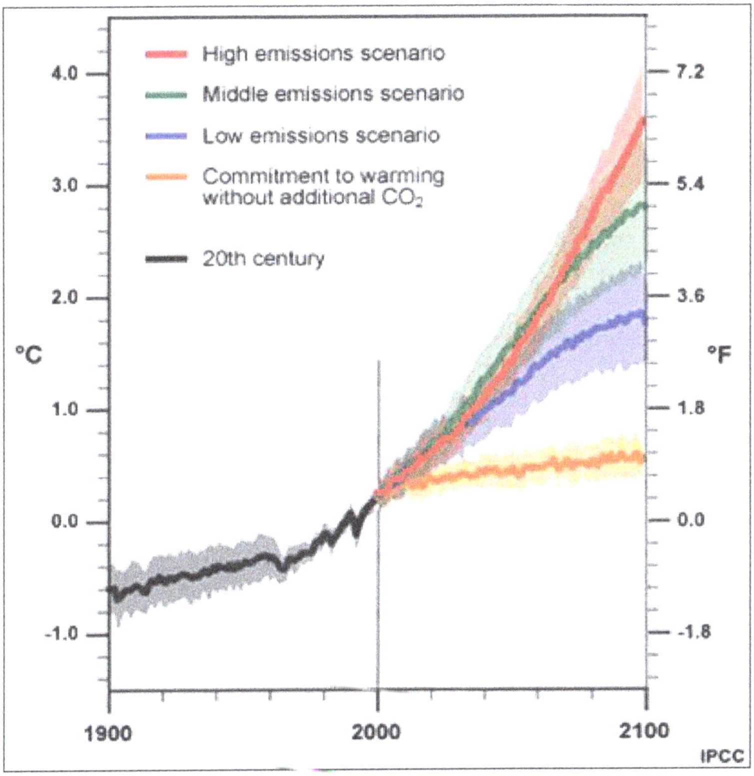

Figure 19.1: Increase in Temperature with Various Emission Scenarios (*Source*: IPCC-2007).

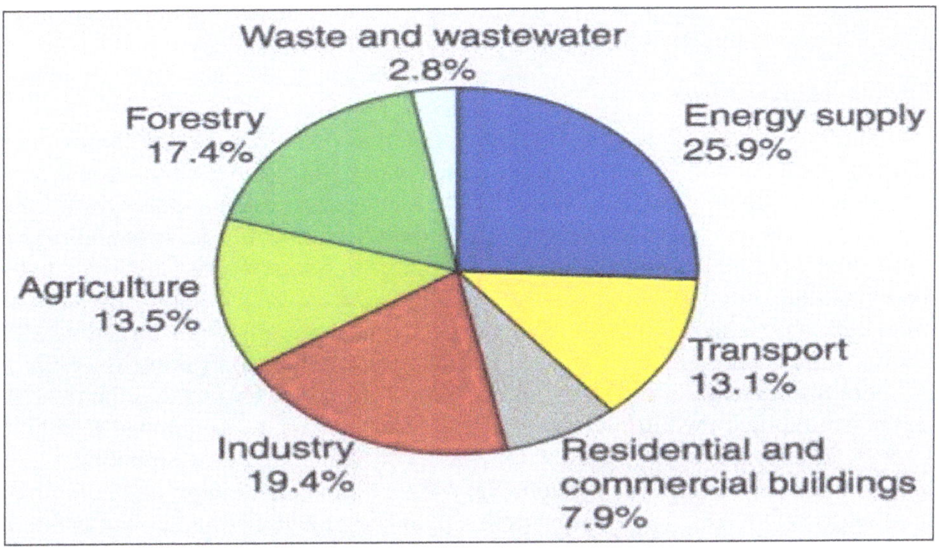

Figure 19.2: Contribution of different Greenhouse Producing Sources.

The climate change implies adverse impacts on different sectors including agriculture, terrestrial and marine ecosystems, health, infrastructure and pose severe societal challenges (IPCC, 2012). About 90 per cent of the disasters are caused by weather related hazards, such as floods, storms, extreme temperatures, and droughts (UNISDR, 2015). The climate change is considered as one of the most important environmental threats to human health, as it is expected that its impacts will be larger in the future as warming progresses (IPCC, 2013; Gasparrini *et al.*, 2017). It can affect the human health through various pathways, including indirect (*i.e.*, spread of disease vectors, increase in food insecurity and migration) and direct effects associated to the increase in frequency and intensity of extreme climatic conditions such as floods, droughts, hurricanes, and heat waves (Patz *et al.*, 2005; McMichael *et al.*, 2012). Six of the ten countries most vulnerable to climate change are in the Asia-Pacific region. Bangladesh ranked first, followed by India, Nepal, Philippines, Afghanistan and Myanmar. For example, in Bangladesh, where the loss of farmland due to sea levels rise by 1.5 m, approximately one fifth of the country's population will be displaced. The land area of the Maldives Islands in the Indian Ocean will be submerged by half while the sea level rises by 2 m.

The current climate of India is highly diverse, ranging from the subfreezing Himalayan winters to the tropical climate of the south. The states of Assam and West Bengal experience extremely damp, rainy, and humid conditions, while the regions of Rajasthan and Gujarat make up part of the arid Great Indian Desert. Based on precipitation and temperature, India can be divided into six climatic regions: the Himalayas, Assam and West Bengal, the Indo-Gangetic Plain, the Western Ghats and coast, the Deccan (the interior of the Peninsula south of the Narmada River), and the Eastern Ghats and coast.

19.2 Climate Change: A Decisive Factor of Global Changes

The climate change scenarios include rising temperatures, changes in precipitation, drought, flood and increased CO_2 concentration in the atmosphere.

19.2.1 Temperature

The temperature fluctuation may directly affect growth rate crops, occurrence of pests, appetite and livestock growth, and water supply in reservoirs and soil. According to Solomon and co-workers (2007) the temperature rise was seen during the 21st century is expected to be highest over land and at high latitudes in the Northern hemisphere during the winter period. The change in temperature is expected to be lowest at the coast and to increase going land inwards. In geographically similar areas, warming is typically larger in arid than in moist regions. Near-term (by 2050) warming projections are little affected by different scenario assumptions. The highest temperature increase is projected at high Northern latitudes and over land, with less warming over the Southern oceans and North Atlantic (Solomon *et al.*, 2007). In the next 50 years, it is very likely, *i.e.* with more than 90 per cent probability, that temperatures, averaged over all habitable continents and over many sub-continental land regions, will rise at a greater extent than the global average rate and by an amount substantially in excess of natural variability. It is also very likely that hot extremes and heat waves will continue to become more frequent and long lasting

(Bernstein *et al.*, 2007; Solomon *et al.*, 2007). In the mid and high latitudes, almost everywhere a decrease in frost days and an increase in growing season length are projected to occur. According to Solomon and co-workers (2007), "Equilibrium climate sensitivity is likely to be in the range 2-4.5° C, with a best estimate value of about 3° C, based upon multiple observational and modeling constraints. It is very unlikely to be less than 1.5° C." As for the global average surface air warming for the end of the 21st century (2090–2099), the range of best estimate and the likely range for six Special-Report-on-Emission-Scenarios (SRES) emission marker scenarios are 1.8–4° C and 1.1–2.9°C and 2.4–6.4°C, respectively (Solomon *et al.*, 2007).

19.2.2 Precipitation

According to IPCC (2007), the amount of precipitation is very likely to increase at high latitudes, while in most subtropical land regions (especially at the margins of the subtropic towards the poles) decreases in precipitation are likely, as a consequence of a general intensification of the global hydrological cycle (Solomon *et al.*, 2007). It is very likely that the annual precipitation in European and African regions close to the Mediterranean will greatly decrease as well as the winter rain in South-Western Australia. Extremes of daily precipitation are considered very likely to increase in Northern Europe, Southern and Eastern Asia, Australia, and New Zealand, as well as in many other regions (Solomon *et al.*, 2007).

19.2.3 Drought

The projected tendency for summer drying of the mid-continental areas during summer will imply a greater risk of drought in those regions (Solomon *et al.*, 2007).

19.2.4 Flood

Excess rainfall is caused when the atmosphere becomes "stuck" in one pattern of circulation, giving wet weather for an extended period in some regions. In this situation, soils and groundwater reservoirs can become saturated, which can leads to flooding as additional rain runs off the land surface.

19.2.5 Atmospheric Carbon Dioxide

Atmospheric carbon dioxide (CO_2) concentration has seen a continue increasing from the pre-industrial time and is now almost 100 ppm above. Due to the future global warming the capacity of land and ocean to absorb anthropogenic CO_2 will be reduced. Therefore a higher amount of anthropogenic CO_2 will stay in the atmosphere under a warmer climate (Solomon *et al.*, 2007). In the 21st century, due to the increase of the atmospheric CO_2, the buffer capacity of the marine sea will be altered with the net result of ocean increase acidity; a reduction ranging between 0.14 and 0.35 pH units, in the average pH of global surface ocean, will occur. The increased atmospheric CO_2 concentrations have a positive effect on crop yields and may offset some of the negative effects related to climate change (Tubiello *et al.*, 2007a). However, the magnitude of this effect is highly uncertain and the subject of ongoing research (Ainsworth *et al.*, 2008; Long *et al.*, 2006; Parry *et al.*, 2004; Rosenzweig *et al.*, 2013; Tubiello *et al.*, 2007b). The CO_2 effect on plant growth and water use efficiency varies considerably between regions as it depends on crop type and interactions with climate, soil properties, nutrient availability, pests and diseases, air pollutants, and other factors. Hence, beneficial CO_2 effects may not be experienced equally in all regions across the globe (Deryng *et al.*, 2016; McGrath and Lobell, 2013; Tubiello *et al.*, 2007a).

19.3 Climate Change and Agriculture

Climate change poses major risks for world agriculture in the 21st century, with implications for the livelihoods of farmers and the food security of communities across the globe. The impact of climate on agriculture is related to changes in local climate rather than in global climate patterns. Since 1880, the average surface temperature of the earth has risen by 1.5 °F (0.83 °C). Consequently, agronomists consider any assessment has to be individually consider each local area. Agricultural yields and production are significantly influenced by climatic conditions during the growing season. The climate change impacts on agroecological variables and ultimately on the cropping and crop yields (Schmidhuber and Tubiello, 2007). The plant growth processes are driven by different climatic variables like temperature, precipitation, solar radiation. The soil physico-chemical parameters such as soil moisture, temperature, acidity, bulk density and nutrient concentrations, and also on atmospheric carbon dioxide and ozone concentrations, and microbial processes (Schauberger *et al.*, 2016), which are all directly or indirectly affected by climate change. The climate change may directly affect global food production. The increase in average seasonal temperature will reduce the duration of many crops, thereby reducing the final yield. In areas where the temperature is close to the physiological maxima of crops, warming will more directly affect yield (IPCC, 2007). Due to global warming, the world's agriculture is facing a severe decline in this century. It is expected that, by 2080, agricultural productivity across the world is expected to fall between 3 per cent and 16 per cent. In developing countries, many of which have average temperatures close to or exceeding crop tolerance levels, it is expected that agricultural productivity will decline by an average of 10 per cent to 25 per cent by the 2080s. Rich countries with lower average temperatures will have a milder or even lower average impact, rising from 8 per cent of productivity to a 6 per cent reduction. Individual developing countries are facing a greater decline. For example, India may decline by 30 per cent to 40 per cent.

19.3.1 Crop Productivity

Despite technological developments such as improved varieties, genetically modified organisms and irrigation systems, weather is still an important factor in crop productivity. Climate change has a direct impact on global food production. The increase in average seasonal temperature will reduce the duration of many crops, thereby reducing the final yield.

19.3.2 Indian Scenario of Agriculture

Since ancient times, India's agriculture has dependent on the monsoon. Any change in monsoon trends will seriously affect agriculture. Even rising temperatures are affecting agriculture in India. In the Ganges Plain of India, these pre-monsoon changes will mainly affect wheat (>0.5°C increase in time slice 2010-2039; IPCC 2007). In the state like Jharkhand, Orissa and Chhattisgarh alone, during severe droughts (about one year in five), rice production reduces at the average of 40 per cent of total production, with an estimated value of $ 800 million (Pandey, 2007).

Increasing carbon dioxide to 550 ppm can increase rice, wheat, beans and oilseed production by 10-20 per cent. For every 1°C increase in temperature, the yields of wheat, soybeans, mustard, peanuts and potatoes will decrease by 3-7 per cent. The loss is much higher at higher temperatures. By 2020, due to rising temperatures, volatile rainfall and reduced irrigation water, the productivity of most crops has only declined slightly, but by 2100 it will decline by 10-40 per cent. The main impact of climate change will be rain-fed or unirrigated crops, of which nearly 60 per cent of the arable land is planted with such crops. It is expected that the winter temperature increase in India by 0.5°C will reduce the yield of rain-fed wheat per hectare by 0.45 tons. The production chickpeas, babies corn, sorghum, millet and coconut may have increased in West Coast. Due to reduced frost damage, production of potatoes, mustard and vegetables in northwestern India is reduced. Increased drought and flooding may increase production variability (Tables 19.1 and 19.2).

Table 19.1: Projected Changes in Climate in India (2070-2099)

Region	*Jan- March*	*April-June*	*July-Sept.*	*Oct.-Dec.*
Temperature Change (°C)				
Northeast	4.95	4.11	2.88	4.05
Northwest	4.53	4.25	2.96	4.16
Southeast	4.16	3.21	2.53	3.29
Southwest	3.74	3.07	2.52	3.04
Precipitation Change (per cent)				
Northeast	-9.3	20.3	21.0	7.5
Northwest	7.2	7.1	27.2	57
Southeast	-32.9	29.7	10.9	0.7
Southwest	22.3	32.3	8.8	8.5

Source: Cline (2007).

The latest research from the Indian Agricultural Research Institute shows that for every 1°C increase in temperature throughout the growing season, wheat production may lose 4-5 million tons. If the temperature is increased by 2°C, rice production is expected to decrease by nearly a ton/ha. In Rajasthan, a 2°C increase in temperature will reduce pearl millet production by 10-15 per cent. If the maximum and minimum temperatures increase by 3°C and 3.5°C, respectively, then the soybean yield in Madhya Pradesh will decrease by 5 per cent compared to 1998. Agriculture in the coastal areas of Gujarat and Maharashtra will be the most severely affected, as fertile areas are susceptible to flooding and salinization.

Table 19.2: Effect of High Temperature and CO_2 on Crop Productivity

Crops	Effect of Increasing Temp. °C (Per cent reduction in yield)				CO_2 Fert. Effect (per cent)	Threshold Temp. for Net Loss
	+ 1 °C	+2 °C	+3 °C	+4 °C	560 ppm	
Rice	5	10	15	20	15	3.0
Wheat	10	20	30	40	15	1.5
Chickpea	5	10	15	20	22	4.5
Groundnut	8	16	24	32	20	2.5
Green gram	6	12	18	24	20	3.0
Mustard	4	8	12	16	20	5.0
Potato	6	12	18	24	20	3.5

19.4 Impact of Climate Change on Food Security

Climate change is one of the significant and growing threat to food security and it has already affecting vulnerable populations in many developing countries and expected to affect an ever increasing number of people across the world in the future if immediate actions are not taken.

Food security is directly or indirectly related to climate change. Any changes in climatic parameters that control crop growth (temperature, humidity) will directly affect the amount of food production. It can change the suitability of the land, which leads to an increase in suitable cultivated land in higher latitudes and hinders potential cultivated land in low latitudes. A slight increase in ambient temperature (from 1°C to 3°C) is expected to benefit crop yields in temperate regions, but it will have adverse effects in seasonally dry and tropical regions, especially cereal crops. However, as climate warming exceeds 3°C, all regions are expected to face negative impacts (Stocker and Intergovernmental Panel on Climate Change, 2013; Tripathi *et al.*, 2016).

19.5 Assessment of Liability of the Food Security Systems to Climate Change

To deal with emerging food safety risks caused by climate change, a large variety of disciplines are needed ranging from natural to social science and, as a consequence, a holistic approach is advisable to adequately tackle the complexity encountered.

Typically the impact assessment on different aspects of human health, including the safety of our food, refers as maximum to 20 years, therefore allowing the use of an array of methodologies "at near or mid period". The impact assessment of the vulnerability of the food safety systems to climate change implies as a minimum up to 50 years' time scale, therefore needing impact assessment based on scenario formulation. In this last respect, impact assessment under transient scenario is preferred.

An important issue is "how much likely" climate change will have an impact on each of the specific topics related to food safety. In other words, when assessing the specific emerging risks, an issue is to discern among "speculative" and "knowledge-based" risks, also in view of assigning priority to adaptation measures. Qualitative or quantitative correlations with climate (change) indicators have been reported for many food safety issues. In some cases, correlations are supported by experimental data, while some of them are not based on validated models, or are given as hypothesis, even though on an extremely reasonable basis. These uncertainties should be added to those of the expected climate change, as treated by Solomon and co-workers (2007), thereby enhancing the overall uncertainty of the assessment of the vulnerability of food safety to climate change.

The approach for the impact assessment implies (i) the qualitative identification of climate-related determinants of the vulnerability for each topic; (ii) the evaluation of the degree of knowledge on the correlation between each topic and one or more climatic determinants; and (iii) the comprehensive analysis linking the identified determinants of predicted climate change with each topic on a regional basis. In this respect, an emblematic case is represented by mycotoxins, for which knowledge-based quantitative correlations with temperature, humidity, and insect attack pattern are available, whilst correlations with wind and climate-related characteristic of the soil have been hypothesized, but not quantitatively nor semi-quantitative stated. Similarly, each climate-related topic of food safety has its own pattern of "knowledge based" and "speculative" previsions. Likewise, the prediction of and warning systems for the occurrence of harmful algal blooms, which may be prone to phycotoxin production and contamination of shellfish, are in a well-developed state. Also the recently observed movement of zoonotic agents caused by altered dissemination of their vectors, such as bluetongue by midges, may also provide a useful resource of data. In order to prioritize adaptation measure, also an evaluation of the societal health concerns linked to the food safety risks induced by climate change is needed. In this respect, the analysis should be based both on the severity of the hazard and on the likely of the risk.

19.6. Possible Measures to Combat Influence of Climate Change

19.6.1 Adaptation and Mitigation

Adaptation strategies is very helpful to mitigate climate change impact on agriculture. Adaptation is defined as "reductions in risk and vulnerability through the actions of adjusting practices, processes, and capital in response to the actuality or threat of climate change" (Porter *et al.*, 2014), while mitigation measures include measures that first reduce greenhouse gas emissions that cause climate change, for example by converting to renewable energy sources, such as solar, wind or nuclear energy, rather than burning fossil fuels in thermal power plants.

Adaptation is context-specific and depends on local conditions, including regional historical climate and future climate projections, biophysical constraints, traditional agricultural practices, socioeconomic conditions, as well as political and institutional settings (Belliveau *et al.*, 2006). In the context of climate change, adaptation includes measures taken to reduce the adverse effects of climate change, such as relocating the communities living close to the sea shore, for instance, to cope with the rising sea level or switching to crops that can withstand higher temperatures.

Case Study

This case study about is of an African country Ethiopia, which is among those countries most vulnerable to climate risks in Africa (USAID, 2011), and it is the historically prone country to extreme weather events, which was a reason for droughts that cause famine (World Bank, 2010). In Ethiopia, climate change is expected to intensify the already high hydrological variability and frequency of extreme events. More than changes in mean annual rainfall, these two factors may have a significant negative effect on the development of the agricultural sector and on the Ethiopian economy as a whole (You *et al.*, 2010). Over the last decades, the temperature in Ethiopia increased at about 0.2° C per decade. Similarly, the increase in minimum temperatures is more pronounced with roughly 0.4° C per decade. Precipitation, on the other hand, remained fairly stable over the last 50 years when averaged over the country. However, the spatial and temporal variability of precipitation is high, thus large-scale trends do not necessarily reflect local conditions (Bread For All, 2009).

Somali regional state where 86.1 percent of the total population is rural (FDRE Population Census Commission, 2008) is known by occurrences of drought lead famine resulted from frequent weather changes (IIED, 2006). Thus, as part of the region the study area also shares the problem; as in Geladi districts, the growth of *Birkeds* has meant the disruption of the previous patterns of wet and dry season grazing as a result of climate change (Smali Regional State, 2011; Jama Sugule and Robert, 1998). However, the percentage of pastoralist proportion of the study area has different figure; *i.e.* 95 per cent (DPPC, 2004), from the region's profile. The livelihood of the Ethiopian Somali depends mainly on livestock, complemented by rain fed and irrigated crop production(FDRE Population Census Commission, 2008). Whereas, the area has mono-agriculture, hence experiences a single land use type, *i.e.* range land. Furthermore, the economic activities practiced in the study area include pastoralism (90 -95 per cent) and small trade activities (5 -10 per cent) (DPPC, 2004).

Pastoralism is the finely-honed symbiotic relationship between local ecology, domesticated livestock and people in resource-scarce, climatically marginal and highly variable conditions. It represents a complex form of natural resource management, involving a continuous ecological balance between pastures, livestock and people (Oxfam, 2008). It is a livelihood system and way of life practiced by people dwelling in arid and semi-arid environments, using mainly traditional knowledge to optimize the interaction between humans, the environment and livestock (PFE, IIRR and Development Fund, 2010). Pastoralism is therefore an economic and social system well adapted to dry land conditions and characterized by a complex set of practices and knowledge that has permitted the maintenance of a sustainable equilibrium among pastures, livestock and people Pasture/grazing land, pasturage, hay field (Koocheki and Gliessman, 2005).

Therefore, although livelihoods have constantly adapted to change, the impacts of climate change may push people beyond their capacity to cope and adapt, because of an increasing magnitude, frequency or rate, particularly of weather-related disasters such as drought, storms and floods (Osbahr, 2007). As a result, the researcher has selected some but which can easily identified and understood by the pastoralists based on his judgment to differentiate which one of mitigation measures are practicing or not by the pastoralists. Mitigation measures at the same time are different from place to place as a result of local variances. Accordingly, among the practices the most likely practiced ones are herd mobility, traditional system of wet and dry grazing areas, establishment of land closures, and resettlement programs, with less confident to report the use of renewable resources.

Therefore, the effects of the changing climate are imposing impacts up on the social, economic and environmental conditions of the pastoralists affecting their livelihoods causing poor living, housing, health, conditions *etc.*, by reducing the number at the same time production of livestock, where the life of the pastoralists is highly dependent. Similarly, these reduction in the assets/properties of the pastoralists is as a result of lowering the carrying capacity of their environment where the animals depend for forage, *i.e.* the pasture land resulted from the effects of the changing climate.

As a result, the way the pastoralists are adapting to the changing climate raises other question and which demand further and detail study. This is because, the pastoralist have poor economic, vulnerable environment and sensitive livelihood to climate change, but better social structure. However beyond this, when we see to their adaption practices, they are poorly accomplished, especially those related to livestock and the pasture lands and relatively some good to the impacts close to the pastoralists; and the same thing to their mitigation measures. However, the impacts of the changing climate, for the poor are significantly higher, because of their less adaptive capacity as a result of lower economic development. Therefore, both the practices (adaptation and mitigation) are not a matter of choice but a matter of survival posing higher burden up on them.

19.7 Conclusion

Climate is the main determinant of agricultural productivity that directly affects global food production. The agricultural sector is the most sensitive to climate change because the climate of a region/country determines the nature and characteristics of vegetation and crops. The increase in average seasonal temperature will reduce the duration of many crops, thereby reducing the final yield. Food production systems are extremely sensitive to climate changes such as changes in temperature and precipitation, which can lead to outbreaks of pests and diseases, thereby reducing harvests and ultimately affecting the country's food security. The net impact of food security will depend on the exposure to global environmental changes and the ability to respond to and recover from global environmental changes. In order to deal with the impact of climate change on agriculture, it will require careful management of resources such as soil, water and biodiversity. To address the impact of climate change on agriculture and food production, India will need to take action at the global, regional, national and local levels. To this end, there is an urgent need for coordinated efforts to strengthen research to assess the impact of climate change on agriculture and other sectors such as forestry, animal husbandry, aquatic life and other organisms.

REFERENCES

Ainsworth, E. A. and Long, S. P. 2005 What have we learned from 15 years of free-air CO_2 enrichment (FACE)? A meta-analytic review of the responses of photosynthesis, canopy. New Phytol. 165, 351 –371.

Belliveau, S., Bradshaw, B., Smit, B., Ramsey, D., Tarleton, M., and B. Sawyer. (2006). Farm-level adaptation to multiple risks: Climate change and other concerns (Occasional Paper No. 27). Guelph, Ontario, Canada: University of Guelph, Department of Geography.

Bernstein, L., Bosch, P., Canziani, O., Chen, Z., Christ, R., Davidson, O., Hare, W., Huq, S., Karoly, D., Kattsov, V., Kundzewicz, Z., Liu, J., Lohmann, U., Manning, M., Matsuno, T., Menne, B., Metz, B., Mirza, M., Nicholls, N., Nurse, L., Pachauri, R., Palutikof, J., Parry, M., Qin, D., Ravindranath, N., Reisinger, A., Ren, J., Riahi, K., Rosenzweig, C., Rusticucci, M., Schneider, S., Sokona, Y., Solomon, S., Stott, P., Stouffer, R., Sugiyama, T., Swart, R., Tirpak, D., Vogel, C., Yohe, G., 2007. IPCC, 2007: Climate Change 2007: Synthesis Report. Contribution of Working Groups I, II and III to the Fourth Assessment Report of the Intergovernmental Panel on Climate Change. Intergovernmental Panel on Climate Change, Geneva. <http: //www.ipcc.ch/ipccreports/ar4-syr.htm>

Bread For All, 2009. Climate Risks and Development Projects: By Marius Keller. Assessment Report for a Community-Level Project in Guduru Oromiya, Ethiopia.

Cline William R. Global Warming and Agriculture: New Country Estimates Show Developing Countries Face Declines in Agricultural Productivity, Center for Global Development. 2007, 1-4.

Deryng, D., Elliott, J., Folberth, C., Mu¨ller, C., Pugh, T.A.M., Boote, K.J., *et al.*, 2016. Regional disparities in the beneficial effects of rising CO_2 concentrations on crop water productivity. Nat. Clim. Change 6 (8), 786790. Available from: https: //doi.org/10.1038/nclimate2995.

DPPC, 2004. Qorshaha Waxaadobka Shan Sano EE (2003-2007 E.C). By Ahmed Farah, Xafiska DPPC Dagmada Galaadi; (unpublished).

FDRE Population Census Commission, 2008. Summary and Statistical Report of the 2007 Population and Housing Census. United Nations Population Fund (UNFPA), Addis Ababa.

Gasparrini, A., Guo, Y., Sera, F., Vicedo-Cabrera, A.M., Huber, V., Tong, S., de Sousa Zanotti Stagliorio Coelho, M., Nascimento Saldiva, P.H., Lavigne, E., Matus Correa, P., Valdes Ortega, N., Kan, H., Osorio, S., Kyselý, J., Urban, A., Jaakkola, J.J.K., Ryti, N.R.I., Pascal, M., Goodman, P.G., Zeka, A., Michelozzi, P., Scortichini, M., Hashizume, M., Honda, Y., Hurtado-Diaz, M., Cesar Cruz, J.,

Seposo, X., Kim, H., Tobias, A., Iñiguez, C., Forsberg, B., Åström, D.O., Ragettli, M.S., Guo, Y.L., Wu, C.-F., Zanobetti, A., Schwartz, J., Bell, M.L., Dang, T.N., Van, D.D., Heaviside, C., Vardoulakis, S., Hajat, S., Haines, A., Armstrong, B., 2017. Projections of temperature-related excess mortality under climate change scenarios. Lancet Planet. Health 1, e360–e367. https: / /doi.org/10.1016/ S2542-5196(17)30156-0.

Gillett, N. P., Arora, V. K., Flato, G. M., Scinocca, J. F., and von Salzen, K. (2012). Improved constraints on 21st-century warming derived using 160 years of temperature observations: Observationally-constrained projections. Geophysical Research Letters, 39(1), 344356

IIED, 2006. Sustainable Development Opinion; Climate Change and Pastoralists: Investing in People to Respond To Adversity. By Ced Hesse And Lorenzo Cotula, 3 Endsleigh Street, London, UK.

IPCC (2007) Summary for Policy makers, Climate Change 2007: Mitigation. Contribution of Working Group III to the Fourth Assessment Report of the IPCC. Cambridge University Press, Cambridge, United.

IPCC, 2012. Managing the Risks of Extreme Events and Disasters to Advance Climate Change Adaptation. A Special Report of Working Groups I and II of the Intergovernmental Panel on Climate Change. Cambridge University Press, Cambridge, UK, and New York, NY, USA, 582 pp.

IPCC, 2013. Climate change 2013: the physical science basis. In: Stocker, T.F., Qin, D., Plattner, G.-K., Tignor, M., Allen, S.K., Boschung, J., Midgley, P.M. (Eds.), Contribution of Working Group I to the Fifth Assessment Report of the Intergovern -Mental Panel on Climate Change. Cambridge University Press.

Jama, S., and Robert, W., 1998. Changing Pastoralism in the Ethiopian Somali National Regional State, south East Range Land Project (SERP); United Nations Development Program Emergencies group for Ethiopia.

Koocheki, A., and Gliessman, R. (2005). Pastoral Nomadism: A Sustainable System for Grazing Land Management in Arid Areas. In Journal Of Development Agriculture, Vol. 25, Issue 4.

Long, S.P., Ainsworth, E.A., Leakey, A.D., No¨sberger, J., Ort, D.R., 2006. Food for thought: lower-than-expected crop yield stimulation with rising CO_2 concentrations. Science 312 (5782), 19181921.

McGrath, J.M., Lobell, D.B., 2013. Regional disparities in the CO_2 fertilization effect and implications for crop yields. Environ. Res. Lett. 8 (1), 014054. Available from: https: / /doi.org/10.1088/1748-9326/8/1/014054.

McMichael, T., Montgomery, H., Costello, A., 2012. Health risks, present and future, from global climate change. BMJ 344, https: / /doi.org/10.1136/bmj.e1359.

Osbahr, H., 2007."Building resilience: Adaptation mechanisms and mainstreaming for the poor,"Human Development Occasional Papers (1992-2007) HDOCPA-2007-10, Human Development Report Office (HDRO), United Nations Development Programme (UNDP).

Oxfam, 2008. Survival of the Fittest; Briefing Paper 116.

Pandey, S. 2007. Economic costs of drought and rice farmers' coping mechanisms. International Rice Research News 32: 5-11

Parry, M.L., Rosenzweig, C., Iglesias, A., Livermore, M., Fischer, G., 2004. Effects of climate change on global food production under SRES emissions and socio-economic scenarios. Glob. Environ. Change A 14, 5367. Available from: https: //doi.org/10.1016/j.gloenvcha.2003.10.008.

Patz, J.A., Campbell-Lendrum, D., Holloway, T., Foley, J.A., 2005. Impact of regional climate change on human health. Nature 438, 310–317. https: //doi. org/10.1038/nature04188.

PFE, IIRR and DF, 2010. Pastoralism and Land: Land Tenure, Administration and Use in Pastoral Areas of Ethiopia.

Porter, J.R., L. Xie, A.J. Challinor, K. Cochrane, S.M. Howden, M.M. Iqbal, D.B. Lobell, and M.I. Travasso, 2014: Food security and food production systems. In: Climate Change 2014: Impacts, Adaptation, and Vulnerability. Part A: Global and Sectoral Aspects. Contribution of Working Group II to the Fifth Assessment Report of the Intergovernmental Panel on Climate Change [Field, C.B., V.R. Barros, D.J. Dokken, K.J. Mach, M.D. Mastrandrea, T.E. Bilir, M. Chatterjee, K.L. Ebi, Y.O. Estrada, R.C. Genova, B. Girma, E.S. Kissel, A.N. Levy, S. MacCracken, P.R. Mastrandrea, and L.L.White (eds.)]. Cambridge University Press, Cambridge, United Kingdom and New York, NY, USA, pp. 485-533.

Rosenzweig, C., Elliott, J., Deryng, D., Ruane, A.C., Mu¨ller, C., Arneth, A., *et al.*, 2013. Assessing agricultural risks of climate change in the 21st century in a global gridded crop model intercomparison. Proc. Natl. Acad. Sci. Available from: https: //doi.org/10.1073/pnas.1222463110201222463.

Schauberger, B., Rolinski, S., Mu¨ller, C., 2016. A network-based approach for semi-quantitative knowledge mining and its application to yield variability. Environ. Res. Lett. 11 (12), 123001. Available from: https: //doi.org/10.1088/1748-9326/11/12/123001.

Schmidhuber, J., Tubiello, F.N., 2007. Global food security under climate change. Proc. Natl. Acad. Sci. 104 (50), 1970319708.

Solomon, S., Qin, D., Manning, M., Alley, R.B., Berntsen, T., Bindoff, N.L., Chen, Z., Chidthaisong, A., Gregory, J.M., Hegerl, G.C., Heimann, M., Hewitson, B., Hoskins, B.J., Joos, F., Jouzel, J., Kattsov, V., Lohmann, U., Matsuno, T., Molina, M., Nicholls, N., Overpeck, J., Raga, G., Ramaswamy, V., Ren, J., Rusticucci, M., Somerville, R., Stocker, T.F., Whetton, P., Wood, R.A., Wratt, D., 2007. Technical summary. In: Solomon, S., Qin, D., Manning, M., Chen, Z., Marquis, M., Averyt, K.B., Tignor, M., Miller, H.L. (Eds), Climate Change 2007: The Physical Science Basis. Contribution of Working Group I to the Fourth Assessment Report of the Intergovernmental Panel on Climate Change. Cambridge University Press, Cambridge and New York. http: //www.ipcc.ch/pdf/assessment-report/ar4/ wg1/ar4-wg1-ts.pdf

Somali Regional State, 2011. Climate Change: Impacts, Vulnerabilities and Adaptation Strategies in Somali Region, Regional Program of Plan to Adapt to Climate Chang; Environmental Protection and Energy and Mines Resources Development Agency. Jigjiga, Ethiopia.

Stocker, T. F., and Intergovernmental Panel on Climate Change (2013). Climate change 2013. In T. F. Stocker, and Intergovernmental Panel on Climate Change (Eds.), The physical science basis; summary for policymakers, a report of Working Group I of the IPCC, technical summary, a report accepted by Working Group I of the IPCC but not approved in detail and frequently asked questions; part of the Working Group I contribution to the fifth assessment report of the Intergovernmental Panel on Climate Change. Intergovernmental Panel on Climate Change. New York: Cambridge University Press.

Stott, P. A., Christidis, N., Otto, F. E. L., Sun, Y., Vanderlinden, J.-P., Zwiers, F. W. (2016). Attribution of extreme weather and climate-related events: Attribution of extreme weather and climate-related events. Wiley Interdisciplinary Reviews: Climate Change, 7(1), 2341.

Tripathi, A., Tripathi, D. K., Chauhan, D. K., Kumar, N., and Singh, G. S. (2016). Paradigms of climate change impacts on some major food sources of the world: A review on current knowledge and future prospects. Agriculture, Ecosystems and Environment, 216, 356373.

Tubiello, F.N., Amthor, J.S., Boote, K.J., Donatelli, M., Easterling, W., Fischer, G., *et al.*, 2007b. Crop response to elevated CO_2 and world food supply: a comment on "Food for Thought." by Long *et al.*, Science 312: 19181921, 2006. Eur. J. Agron. 26 (3), 215223.

Tubiello, F.N., Soussana, J.-F., Howden, S.M., 2007a. Crop and pasture response to climate change. Proc. Natl. Acad. Sci. 104 (50), 1968619690.

UNISDR, 2015. The Human Cost of Weather-Related Disasters 1995–2015. United Nations Initiative for Disaster Reduction. Available at: https: //www.unisdr. org/files/48588_ unisdrannualreport2015evs.pdf.

United Nations, 1994. United Nations Framework Convention on Climate Change. United Nations, New York. <http: //www.unfccc.int/essential_background/ convention/items/2627.php>.

USAID, 2011. Climate Change and Conflict in Pastoralist Regions of Ethiopia; Mounting Challenges, Emerging Responses. By Jeffrey Stark, Katsuaki Terasawa, and Mersie Ejigu; USAID CMM Discussion paper no. 4

World Bank, 2010. Economics of Adaptation to Climate Change, Ethiopia. World Bank Group 1818 H Street, NW Washington, DC 20433.

You, G. J. Y., and Ringler, C., 2010. Hydro-Economic Modeling of Climate Change Impacts in Ethiopia. IFPRI Discussion Paper No. 960. International Food Policy Research Institute, 2010, Washington, DC.

Chapter 20

Conservation Importance of Trees as Urban Green Spaces: A Case Study of Birds in Pratapgunj Residential Area, Vadodara

Tanaya Ambadkar[1], Geeta Padate[1] and Arun Arya[2]*

[1]*Department of Zoology,* [2]*Department of Botany,*
Faculty of Science, The Maharaja Sayajirao University of Baroda, Vadodara – 390 002, Gujara
**e-mail: tanayaa@gmail.com*

ABSTRACT

The green spaces in urban areas that provide an opportunity for people to connect with nature are also essential for urban biodiversity and their conservation. Trees in urban areas are the keystone habitats for Urban wildlife. Tree diversity and density impact quality of habitat in an urbanised landscape wherein birds are ideal models for evaluating the impact of urbanisation. Birds are known to effectively indicate quality of habitat in an area. Present study aims to highlight the bird diversity of Pratapgunj – an important residential area in centre of Vadodara city. It is an area of comparatively low disturbance and high tree count, hence a good refuge for birds. In this study, diversity of bird species and the supporting tree diversity are explored. Avian survey techniques and citizen science/ citizen survey technique were used for data collection.

A total of 54 species of birds belonging to 32 families have been observed that are using 28 species of trees present in 29.4 acres residential area. The area is dotted with several large trees like Ashoka (Asopalav or Mast tree- Polyalthia longifolia Benth and Hk.f.), Neem (margosa- Azadirachta indica

---

* Independent Researcher, Ishan Flats, Pratapgunj, Vadodara – 390 002, Gujarat, Corresponding author.

A. Juss.) and Peepal (Ficus religiosa L.). Several young Saptaparni or Devil's tree (Alstonia scholaris (L.) R. Br.) adds to the green spaces. The implications of this fast-growing species on the bird diversity of Pratapgunj are discussed.

Keywords: *Urban green space, Biodiversity conservation, Bird diversity, Tree diversity.*

20.1 Introduction

Urban Green Spaces (UGS) and waterways are important public spaces, that act as areas of respite from the effects of rapid urbanization and provide substantial public health as well as environmental health benefits (Gerrish and Watkins, 2018; Huang *et al.*, 2017; Wolch *et al.*, 2014; Hartig *et al.*, 2014). The conservation of UGS has gained recognition amongst scientists and environmentalists in recent years, however, it is yet to be incorporated into the city planning and development in India.

India is a mega-biodiverse nation, harbouring approximately 7 to 8 per cent of the world's recorded species in less than 2.4 per cent of the total land area (MOEFCC, 2014). India is also the world's second most populous country, with particularly high concentrations of its citizens living in cities (UN DESA, 2018). The expanding human settlements are always shrinking green spaces resulting in a constant battle for space between humans and wildlife. Therefore, UGS play a crucial role in supporting biodiversity conservation (Threlfall, *et al.*, 2017) as well as creating a connect between urban land and nature (Bilgili and Gökyer 2012). In recent years the urban green spaces that used to be natural or semi-natural ecosystems are being converted into urban concrete spaces for need and greed (Bilgili and Gökyer 2012).

UGS include all trees in the urban areas, inclusive of individual street trees, cluster of park trees and peri-urban forests that extend to the outer metropolitan area. Urban forests, parks and gardens, campus areas, backyard gardens/kitchen gardens, and any other green spaces such as riparian corridors, vacant plots, rooftops, nurseries, *etc.* are parts of UGS. These trees provide millions of dollars' worth of ecosystem services to cities globally (Endreny, 2018). Ecosystem services from trees can be cultural (*e.g.*, spiritual, recreational), utilitarian (*e.g.*, fruits, water), environmental (*e.g.*, urban heat island reduction, flood control), and supporting wildlife habitat, and carbon storage. In addition, improved mental health, lower rates of cardiovascular morbidity and mortality, reduced obesity and diabetes, improved pregnancy outcomes are some of the positive health effects of physical activity and socialization in available urban green spaces (Sugiyama *et al.*, 2008).

Bird diversity is a good indicator of habitat quality (Xi *et al.*, 2018). Birds are the most commonly studied group in assessing urban species diversity or impact of urbanisation on species (McKinney, 2008). A wide range of behaviour is exhibited birds *i.e.* they can be locally abundant, easily observed with appropriate taxonomic identification, make them important group of species for ecological studies. Positive and negative impacts of urbanisation on birds have been documented and amply reported, making birds' ideal models for comparative studies (Xu *et al.*, 2015). In this study the tree diversity and supported bird diversity are explored for Pratapgunj, a residential area in Vadodara.

Pratapgunj shares its boundaries on two sides with the Maharaja Sayajirao University campus and on one side with the Sayajibaug Public Garden. These areas extend the green habitat for birds. Pervious observations in 1990s (Unpublished) documented high bird diversity in this area –wherein over 125 species have been listed. The second author has observed a reduction in species diversity as the density of human settlement has increased over the last 30 years.

20.2 Materials and Methods

20.2.1 Study Area

In early years of Baroda State, Pratapgunj was on the outskirts of city and full of large native trees and land for grazing cattle. Though a part of it was a cemetery the remaining area had single large residential bungalows surrounded by gardens (History of Vadodara). After the establishment of The Baroda College in 1881 the area started getting urbanized with development of railway station and state transport bus station just outside the area (Figure 20.1b). Thus, the age of Pratapgunj residential area can be estimated to be approximately 120 years. It is named after the son of Mahraja Sayajirao Gaekwad III, Maharaja Pratapsinghrao Gaekwad. The M.S. University of Baroda was established during his regime. It spreads in about 24.5 acres Long 22.3072°N, Lat 73.1807°E, which included 52 bungalows in private plots, one society garden and a railway officer's colony. Many of these plots have been developed now as multi-storeyed apartment buildings to provide residence to small urban families. However, UGS are still maintained in the area (Figure 20.1a).

Figure 20.1a: Aerial View of Study Area.

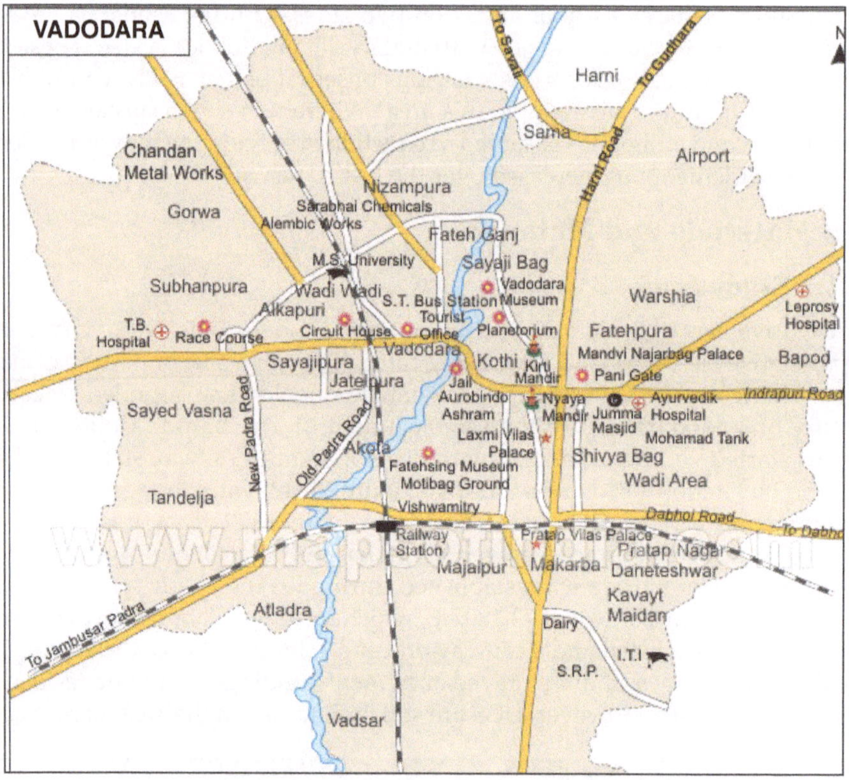

Figure 20.1b: Map of Vadodara Showing The M.S. University, Pratapgunj.

20.2.2 Methodology

The survey of bird species was carried out through random sampling and point count methods. Surveys were continuously conducted from January 2019 to December 2019. Field binoculars (8X40) and a Canon Digital Single Lens Reflex Camera (DSLR) were used. Bird species were included only if they were seen to be inhabiting/utilizing the area. Birds observed to fly over the area are not included in the list. Observations were carried out during morning and evening hours' *viz.* 06: 00 to 10: 00 and 16: 00 to 19: 00 h respectively.

Tree species survey was conducted on foot in all accessible areas of Pratapgunj. In order to collect data from private residential plots citizen-science method was utilized. All owners were contacted and asked to self-report the species inside their property, photographs were requested to confirm tree identity. Few plants are included in Table 20.1. On occasions when respondents were unaware, their plots were visited with permission in order to eliminate error by omission.

20.3 Results

Pratapgunj residential area is a mix of old bungalows with gardens, new residential apartments and a society garden. The western boundary of the area is marked by a Bhukhi Nala, a tributary of the Vishwamitri River. The residential area being old has still retained several large trees along its internal roads – this makes the area green and a good habitat for birds. A total of 28 species of trees (Table 20.1) could be identified in the area. The most remarkable feature of the area is its stretch of 75 Ashoka (Asopalav) trees, all of which are more then 60 years old. The 28 species include 16 fruiting species. Neem (*Azadiracta indica*), Ashoka *(Saraca asoka)*, Bottlebrush *(Callistemon viminalis)*, Flame of forest *(Butea monosperma)*, Banyan (*Ficus benghalensis*), Kadayo gum *(Sterculia foetida)*, *etc.* are some examples of the large trees present in Pratapgunj. In recent years Saptaparni *(Alstonia scholaris)* or the Indian Devil has been planted along road side in large numbers. Saptaparni is a medium-sized evergreen tree that has characteristic bunches of seven (sapta) leaves (parni).

Table 20.1: Tree Species Present in Pratapgunj Area, Vadodara

Sl.No.	Common Name	Scientific Name
1	Sita Ashoka	*Saraca asoka*
2	Bottle brush	*Callistemon viminalis*
3	Kadayo gum	*Sterculia foetida*
4	Custard apple (Sitaphal)	*Annona squamosa*
5	Drumstick (Saragvo)	*Moringa oleifera*
6	Ashoka (Ashopalav)	*Polyalthia longifolia*
7	Fig	*Ficus*
8	Flame of the forest (Kesudo)	*Butea Monosperma*
9	Goras amli	*Pithecellobium dulce*
10	Guava	*Psidium guajava*
11	Wild Almond (Badam)	*Terminalia cattapa*
12	Indian Bael (Bili patra)	*Aegle marmelos*
13	Indian Banyan (Vad)	*Ficus benghalensis*
14	Indian Gooseberry (Amla)	*Phyllanthus emblica*
15	Indian laburnum (Garmalo)	*Cassia fistula*
16	Jamun/Jambu	*Syzygium cumini*
17	Mango	*Mangifera indica*
18	Mulberry	*Morus nigra*
19	Neem	*Azadirachta indica*
20	Peepal	*Ficus religiosa*
21	Poon/Kadayo gum	*Sterculia foetida*
22	Prosopis/Gando Baval	*Prosopis juliflora*
23	Sapodilla (Chickoo)	*Manilkara zapota*
24	Saptaparni/Indian devil tree	*Alstonia scholaris*
25	Silk cotton	*Bombax ceiba*
26	Spanish cherry (Borsali)	*Mimusops elengi*
27	Tamrind (Imli)	*Tamarindus indica*
28	Elephant apple (Kavith)	*Limonia acidissima*

During this study, 54 species of birds have been observed in Pratapgunj, belonging to 33 different families (Figure 20.2).

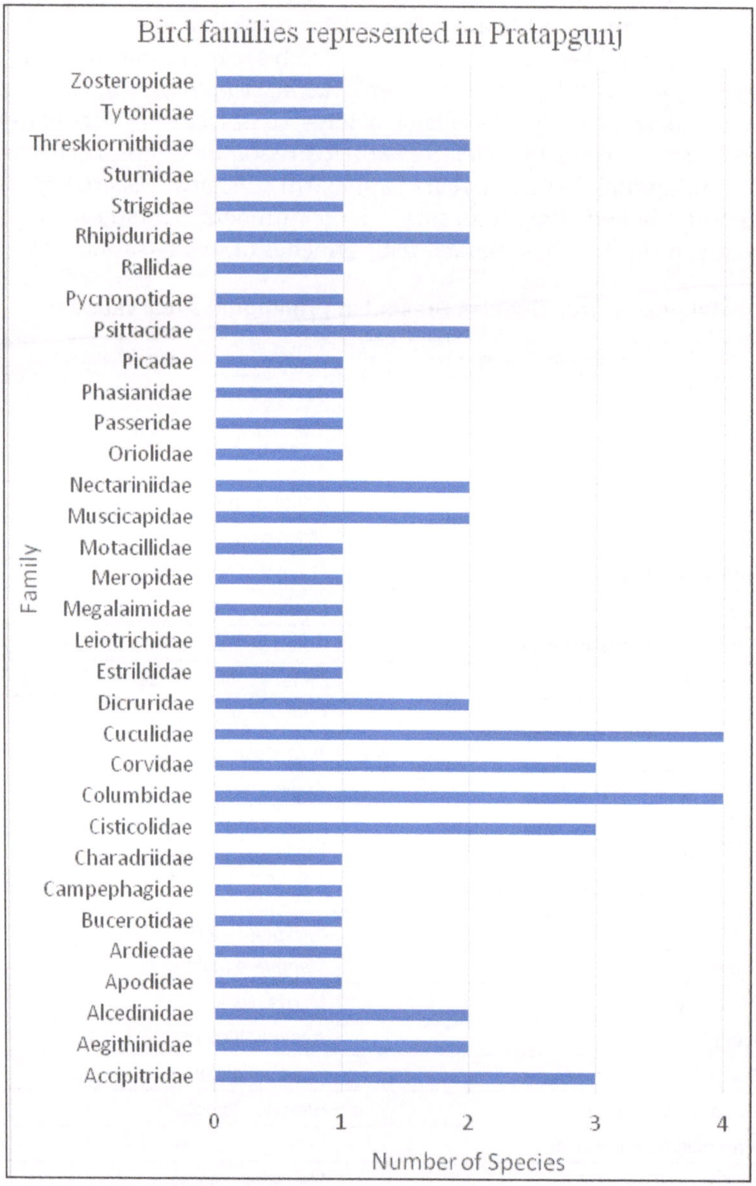

Figure 20.2: Bird Families found in Pratapgunj.

Most of the resident bird species (species that do not migrate) found in Pratapgunj are observed to breed successfully in the area. Two migratory species – Rosy Starling *(Pastor roseus)* and White Wagtail *(Motacilla alba)*, are observed in good numbers. Two species observed are classified as 'Near Threatened' in

IUCN red lists, Alexandrine Parakeet (*Psittacula eupatria)* and Black-headed Ibis (*Threskiornis melanocephalus*). Further, as per the same list which indicates global population trends of species, in Pratapgunj area 7 species with decreasing trend, 9 with increasing trend, 28 having stable global population and 10 species whose trend is not known (Figure 20.3) have also been observed.

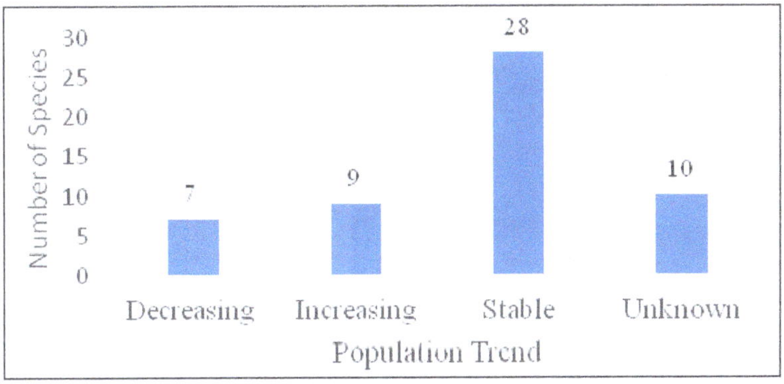

Figure 20.3: Population Trend of Bird Species found in Pratapgunj.

Three stripped palm squirrel *(Funambulus palmarum)*, Gray Langur (*Semnopithecus entellus)*, Indian Grey Mongoose *(Herpestes edwardsi)*, Indian flying fox *(Pteropus medius)* and Asian house shrew (*Suncus murinus)* are commonly found mammals in the area. Reptiles like – Oriental garden lizard *(Calotes versicolor)*, Spectacled Cobra (*Naja Naja)*, Indian rat snake *(Ptyas mucosa)* are also reported.

During surveys, it is observed that the residents of Pratapgunj are benefited by the green spaces. Children, adults and senior citizens can be seen walking, exercising and playing inside the park or along the internal roads.

20.4 Discussion

The conservation value of large trees is high as they provide varieties of ecosystem services (Mahmoud *et al.*, 2015). Presence of large structure trees in green space greatly influences biodiversity of an area (Stagoll *et al.*, 2012). Large trees are considered keystone structures because they provide "resources, shelter, or 'goods and services' crucial for other species" (Tews *et al.*, 2004; Manning *et al.*, 2006; Fischer *et al.*, 2010). They provide critical habitat for a range of taxa, including bats (Fischer *et al.*, 2010), ground-dwelling, and arboreal mammals (Lindenmayer *et al.*, 2011), birds (Seymour and Dean 2010; Stagoll *et al.*, 2010), and invertebrates (Carpaneto *et al.*, 2010). Large trees are important for the production of coarse woody debris (Killey *et al.*, 2010) and provide a distinct microclimate, with increased soil nutrients, plant species richness and structural complexity (Manning *et al.*, 2006). Furthermore, large trees fulfil a range of landscape-scale ecological functions, including increasing habitat connectivity, which may facilitate species' range expansions and thus capacity to adapt to climate change (Manning *et al.*, 2009). In addition to these biological functions, large trees are also important socially, culturally, and aesthetically (Jim 2004).

Plate 20.1: Birds Occurring in Pratapgunj

(a) Black-rumped Flameback (*Dinopium benghalense*); (b) Grey Hornbill (*Ocyceros birostris*); (c) White-throated Kingfisher (*Halcyon smyrnensis*); (d) Green Bee-eater (*Merops orientalis*); (e) Shikra (*Accipiter badius*); (f) Rufous Treepie (*Dendrocitta vagabunda*).

Within India about 1349 species of birds are recorded (Lepage, Avibase - Bird Checklists of the World - India, 2016), of these 613 species are reported from Gujarat (Lepage, 2018). In an area of 29.4 acres 54 species have been noted over a period of 1 year. Birds of all feeding habits – carnivores, insectivores, omnivores, nectarivores, frugivores and granivores are observed in the Pratapgunj residential area, indicating availability of good and diverse food sources in the area. High number and diversity of fruiting trees supports a good diversity of small birds (Carlo, *et al.*, 2004).

The study area has presence of several large trees both inside private plots and along the internal roads. Large trees are beneficial in urban areas as the ecosystem services derived from them are higher (Yotapakdee, *et al.*, 2018). Large trees have consistent positive relationship with bird diversity, larger the tree more positive is its effect on birds (Stagoll, *et al.*, 2012). Presence of continuous trees creates a miniature green space network in the society. This is important because it allows even small birds to move within the area, since smaller birds avoid flying over open space as a survival strategy.

Saptaparni (*Alstonia scholaris*) is growing in large numbers in the area. It is a hardy and adaptable species that has high tolerance for pollution. This capability allows it to grow in the conditions that may not support other species. Further, it is an evergreen tree with a dense crown. Its flowers blooms between October to December. The sweet smellig flowers attract bees, butterflies and very few species of birds like the purple sunbird and purple rumped sunbird. However, the fast spread of Saptaparni is not favourable in the long term. It is important that the diveristy of tree species be maintained when planting new trees. A mix of tree species is more favourable for supporting biodiversity (Wang, *et al.*, 2019). Further, Saptaparni has been noted as an allergen for many people in urban areas (Hussain, *et al.*, 2014). Thus, there is a stigma attached to the species. If Saptaparni numbers are to rise, they would cause increase in severe seasonal allergies – this may anatagonise the residents into felling the trees, reducing green space in the area. It is important to note that residents of Pratapgunj have noticed more Saptaparni growing in recent years. Few residents also wished to replace all Saptaparni with other species of trees.

Pratapgunj Residential Area currently has good tree density and bird diveristy. However, it is imperative to take steps towards active management of the green spaces in Paratapgunj. For many years, unregulated, unmonitored development in the area has led to felling of trees, concretization of open spaces, introducion of ornamental or exotic plants, *etc.* Going further, in order to prevent loss of biodiveristy, the locals and the adminstrators need to work in an ecologically sensitive manner. This applies to all urban areas.

Introduction of rules such as no felling of old/large structure trees, introduction of ecosensitive architecture that avoids felling whereever possible, ban on introduction of exotic species, ban on unscientific and unseasonal trimming of trees, introduction of tree and shrub plantation programs *etc.* should be strictly implemented to protect and imporve the biodiveristy of any area.

Imporvement in green space and biodiversity also result in imporved human wellbeing, decrease pollution and Senior citizens and children will be most benifited as they are most impacted by pollution in residential areas (UNEP, 2018). Imporved or well maintained green spaces are expected to lead to reduction in pollution levels and imporved quality of recreational area leading to improved health for all. The urban green spaces are not only important for human health but provide various ecosystem benefits including support to biodiversity.

20.5 Conclusion

Pratapgunj a residential area near railway station is a mix of old bungalows with gardens, and new residential apartments. The residential area being old has still retained several large trees along its internal roads. More than 28 species of trees including 16 fruiting species can be located in the area. The most remarkable feature of the area is its stretch of Asopalav and Saptparni trees. During the present study, 54 species of birds have been sighted belonging to 33 different families. Pigeon is most common invasive species present. There is also presence of other diversity like three stripped palm squirrel, Gray Langur, Indian Grey Mongoose, mouse and Asian house shrew are commonly found mammals in the area. Reptiles like – Oriental garden lizard, and during rainy season snakes are reported. During present surveys, the residents narrated their experiences and benefits of the green spaces. Residents are aware of the importance of greenery and except the removal of a small park near Vikram baug gate, no major cutting of trees have been noticed.

Ackowledgements

We are grateful for the prompt and unconditional support from people who participated in the study and helped us with their anecdotal knowledge of the study area.

REFERENCES

Bilgili, C., and Gökyer, E. (2012). Urban green space system planning. In D. Ozyavuz (Ed.), *Landscape Planning*. InTech.

Endreny, T. A. (2018). Strategically growing the urban forest will improve our world. *Nature Communications, 9* (1160 (2018).

Gerrish, E., and Watkins, S. L. (2018). The relationship between urban forests and income: A meta-analysis. *170*, 293-308.

Hartig, T., Mitchell, R., Vries, S. d., and Frumkin, H. (2014). Nature and Health. *Annual review of Public Health, 35*, 207-228.

History of Vadodara. (n.d.). *Nawabs of Vadodara*. Retrieved from History of Vadodara: https: / /historyofvadodara.in/nawabs-of-baroda/#

Huang, C., Yang, J., Lu, H., Huang, H., and Yu, L. (2017). Green Spaces as an Indicator of Urban Health: Evaluating Its Changes in 28 Mega-Cities. *Remote Sens., 9*, 1266.

Hussain, M., Mandal, J., and Bhattacharya, K. (2014). Aerobiological, clinical, and immunobiochemical studies on Alstonia scholaris pollen from Eastern India. *Environ Monit Assess., 186* (1), 457-67.

IUCN. (2020). *The IUCN Red List of Threatened Species. Version 2020-2.*

Lepage, D. (2018). *Avibase - Bird Checklist of the World - Guajarat*. Retrieved 01 11, 2020, from Avibase - The World Bird Database: https: //avibase.bsc-eoc.org/ checklist.jsp?region=INnwgj and list=howardmoore

Lepage, D. (2016). *Avibase - Bird Checklists of the World - India*. Retrieved 01 11, 2020, from Avibase - The World Bird Database: https: //avibase.bsc-eoc.org/ checklist.jsp?lang=EN and p2=1 and list=ioc and synlang= and region=IN and version=text and lifelist= and highlight=0

Mahmoud, T., Gairola, S., and El-Keblawy, A. (2015). Large old trees need more conservation attention: A case of Tamarix aphylla in the arid deserts of the United Arab Emirates. *Journal of Asia-Pacific Biodiversity, 8* (2), 183-185.

McKinney, M. (2008). Effects of urbanization on species richness: A review of plants and animals. *11*, 161-176.

MOEFCC, M. (2014). *India's Fifth National Report to the Convention on Biological Diversity*. Government of India.

Stagoll, K., Lindenmayer, D., Knight, E., Fischer, J., and Manning, A. D. (2012). Large trees are keystone structures in urban parks. *Conservation Letters, 5* (2).

Sugiyama, T., Leslie, E. R., Giles-Corti, B., and Owen, N. (2008). Associations of neighbourhood greenness with physical and mental health: do walking, social coherence and local social interaction explain the relationships? *62* (9).

Threlfall, C. G., Mata, L., Mackie, J. A., Hahs, A. K., Stork, N. E., Williams, N. S., *et al.* (2017). Increasing biodiversity in urban green spaces through simple vegetation interventions. *Journal of Applied Ecology, 54*, 1874-1883.

UN DESA, P. (2018). *Revision of the World Urbanization Prospects*. Department of Economic and Social Affairs of the United Nations.

UNEP. (2018, Oct 18). *Blogpost: Young and old, air pollution affects the most vulnerable*. Retrieved from United Nations Environment Program: https: //www. unenvironment.org/news-and-stories/blogpost/young-and-old-air-pollution-affects-most-vulnerable

Wang, X., Hua, F., Wang, L., Wilcove, D., and Yu, D. (2019). The biodiversity benefit of native forests and mixed species plantations over monoculture plantations. *Divers Distrib., 25*, 1721-1735.

Wolch, J., Byrne, J., and Newell, J. (2014). Urban green space, public health and environmental justice: the challenge of making cities 'just green enough'. *Landscape and Urban planning, 125*, 234-244.

Xi, X., Yujing, X., Ke, Q., Zukui, L., and Xiangrong, W. (2018). Detecting the response of bird communities and biodiversity to habitat loss and fragmentation due to urbanization. *624*, 1561-1576.

Xu, H., Cao, M., Wu, J., Cai, L., Ding, H., and *et al.* (2015). Determinants of Mammal and Bird Species Richness in China Based on Habitat Groups. *10* (12).

Yotapakdee, T., Asanok, L., Torlarp, K., Norsangsri, M., Karnasuta, N., Navakam, S., *et al.* (2018). Benefits and Value of Big Trees in Urban Areas: A study in Bang Kachao Green Space, Thailand. *Environment and Natural Resources Journal, 17* (1), 33-43.

Part 4

General Topics on Environment: Including Changes in Lifestyle during COVID

अमन की गंगा

गंगा की लहरें हों,
हो बहता हुआ समंदर,
ये जुनून रहे तुम्हारा
है ये मेरा सम्बोधन॥

हो अमन की पूजा घर-घर
न घमासान दुनिया में
हो कश्मीर में शांति
डल झील का निर्मल पानी॥

ना बंदूक की गोली
ना हो दहशत की छाया
हो शालीमार में हर दम
बढ़ते चिनार की छाया।

राजस्थान की पावन धरती
जहाँ जलती जौहर की ज्वालायें
नम न करूं मैं उसको
यह राणा प्रताप की धरती॥
गंगा की लहरें हो,
हो बहता हुआ समंदर
मानवता दम तोड़ रही
बढ़ता आधुनिकता का आडंम्बर॥

जहां कौशल कारखानों में
नित नई वस्तुयें गढ़ता,
जहां मानव अपने श्रम से,
लोहा पिघलाता है॥

नहीं रही अब शिव की जटायें
नित काटे जाते हैं जंगल
गंगा की लहरों को
अब वश में करते कम्प्यूटर।

घटती जमीन अपनी है,
बढ़ती जाती आबादी,
पानी को प्यासी है, धरती
प्रदूषण की बढ़ती चिंगारी॥

Chapter 21

Mouth Parts, Food Habits in different Animals and Use of Certain Botanicals in Maintaining Oral Hygiene and Preventing Gum Diseases in Human Beings during COVID

*Chitra Arya, Bhumi Sarang and Arun Arya**

Department of Environmental Studies, Faculty of Science, The Maharaja Sayajirao University of Baroda, Vadodara – 390 002, Gujarat
**e-mail: sarojarun10arya@rediffmail.com*

ABSTRACT

Good oral hygiene is essential for the healthy teeth, gum and fresh breath. A number of methods are used in oral hygiene to prevent and cure oral diseases. It is of importance to look at the roles plants play in oral hygiene as a number of them have medicinal properties. Use of plants for oral hygiene is recommended since Egyptian civilization. In many Indian villages and African homes, teeth are cleaned in the morning by chewing the root or slim stem of certain plants until they acquire brush-like ends. Plants have been used for medicinal purpose long before prehistoric period. Dentine is sensitive, so its erosion can lead to your teeth being more sensitive to hot, cold or sweet foods and drink. Gum disease (periodontal disease) i.e. the infection or inflammation of the tissues that surround the teeth. These are most common cause of tooth loss in adults. And is main cause of bad breath. Depending on the severity, gum disease is generally divided into two types - gingivitis and periodontitis.

Animals including human beings have specialized mouth structures that help them capture, handle, and eat the food available to them in their environment. In nature one may find that that different shapes of beak help birds eat different kinds of food. They don't get tooth decay like we do.

They don't take carbohydrates such as starch and sugar and have very less chances of getting tooth decay when their main meal is nothing but raw meat and bones.

Smoking is a major risk factor for developing gum disease. Several studies have attempted to prevent or improve oral mucositis (OM) but have not resulted into treatment yet. During COVID times people are at home and they frequently consume eatables. However: Many people also use an antiseptic mouthwash three or four times each day to help prevent gum disease. In such cases, besides the use a toothbrush, twice the regular rinsing with an herbal antiseptic mouthwash helps to clean the teeth. A special plastic tongue scraper will be needed for tongue cleaning. The study evaluates the antimicrobial efficacy of Citrox made from dried, ground pulp of oranges and freshly prepared extracts of of caraway (Carum carvi L.) margosa (Azadirachta indica Juss.), tulsi (Ocimum sanctum L.), curry leaves or Murraya koenigii (L.) Spr., babul (Acacia nilotica L.), eucalyptus (Eucalyptus camaldulensis Dehnh.), sorrel (Hibiscus sabdariffa), mango (Mangifera indica L.), guava (Psidium guajava L.), rose (Rosa indica L.), Malabar nut (Justicia adhatoda L.) and Aloe (Aloe barbadensis Miller) and powder of cloves [Syzygium aromaticum(L.)Merr.] in maintaining oral hygiene.

Keywords: *Botanicals, Citrox, Cloves, Feeding habits, Oral hygiene, Gum diseases, COVID.*

21.1 Introduction

Good oral hygiene is necessary for the healthy teeth, gum and fresh breath. It helps to prevent and cure oral diseases. It is of importance to look at the roles plants play in oral hygiene as a number of them have medicinal properties. When compared to toothpaste, mouthwashes, denitrifies *etc.* plants used for oral hygiene stand out. In many African and Asian homes even today, teeth are cleaned in the morning by chewing a root or stem of certain plants until they acquire brush-like ends. The Babylonians recorded the use of chewing sticks in 7000 BC and its use ultimately spread throughout the Greek and Roman Empires, it was practiced by Egyptians, Jews and in the Islamic Empires. It is believed that the counterpart of the modern day toothbrush was unknown in Europe until about 300 years ago. Presently, chewing sticks is also common in South America, the Middle East and Asia. Ancient Egyptians are known to be responsible for the first artistic drawings that emphasize the importance of beauty and hygiene. An unclean body was thought to be impure. Pedanius Dioscorides, a Greek physician and surgeon (40–90 AD) whose writings served as a medical textbook, suggested for treatment of bad breath a mouthwash mixture of the decoction of the leaves of the olive tree, milk, the juice of pickled olives, gum myrrh with wine and oil, pomegranate peelings, nutgalls, and vinegar. The ancient Romans included teeth cleaning as part of their religious ceremonies. The patriarchy made their slaves to clean their teeth. The Romans included a secret ingredient in their mouthwash: human urine. They imported urine from Portuguese people because they thought it had more strength. Until the 18th century, urine continued to be an active ingredient in toothpaste and mouthwash because it contained ammonia which has cleansing abilities.

Oral health is essential to general well-being and is related to the quality of life that extends beyond the functioning of the craniofacial complex. The relationship between oral diseases and activities of the microbiota of the oral cavity is well established. Over 750 species of bacteria (50 per cent of which are still to be identified) colonize the oral cavity and a number of these are responsible in oral diseases (Jenkinson and Lamont, 2005) It has an impact on a person's speech, selection of food, quality of life, and well-being. In view of the prevalence of oral diseases, their impact on individuals and society, and the expense of their treatment, oral diseases may be considered a major public health problem and they are listed among the most common of the chronic diseases that affect mankind. Oral diseases are the fourth most expensive diseases to treat in certain countries (Sheiham, 2005).

Although oral diseases are largely preventable, they persist with high prevalence, reflecting widespread social and economic inequalities and inadequate funding for prevention and treatment, particularly in low-income and middle-income countries (LMICs). As with most non-communicable diseases (NCDs), oral conditions are chronic and strongly socially patterned. Children living in poverty, socially marginalized groups, and older people are the most affected by oral diseases, and have poor access to dental care. The statistics suggest that dental caries affect 60-90 per cent of school going children in developing countries (Petersen, 2003). The personal consequences of chronic untreated oral diseases are often severe and can include unremitting pain, sepsis, reduced quality of life, lost school days, disruption to family life, and decreased work productivity. By describing the extent and consequences of oral diseases, their social and commercial determinants, and their ongoing neglect in global health policy, we aim to highlight the urgent need to address oral diseases among other NCDs as a global health priority. Oral infection can spread by kissing. In ancient times people used to tie pungent smelling spices in form of 'Taviz' to keep the persons away from child face. For children use of sugar-free medicines is suggested. Some people chew **sugar-free** gum after each meal. Use of peppermint in betel leaves is popular among young and elderly people. Peppermint is supposed to be antibacterial (Kovács *et al.*, 2019) Chewing gum increases the flow of saliva. Saliva helps to flush the mouth to help clear any debris and acid remaining from the meal. According to the World Health Organization (WHO) report, dental caries, though exhibiting a declining trend in many parts of the industrialized world, are still an important public health concern in many developing countries.

21.1.1 Antimicrobial Mouth Rinses

Antimicrobial mouth rinses have also been suggested as adjuncts for mechanical plaque control methods. The most commonly used antiplaque agent is chlorhexidine gluconate. The use of chlorhexidine has some potential drawbacks like altered taste sensation, staining of teeth, and development of resistant bacteria that incapacitate its application on long-term basis. There exists a need to develop some innovative strategies that act against both dental caries and periodontal diseases simultaneously. One such strategy would be to explore the abundantly available medicinal plants in nature. The "naturally occurring" active ingredients in plant medicines restore health, with minimal harmful effects and maximum efficiency. The use of natural products is a comprehensive remedy that includes promotional and preventive strategies in the maintenance of health. Natural herbs used either exclusively or in combination are proven to be safe and effective in the management of various oral health problems such as halitosis, bleeding gums, mouth ulcers, and dental caries. Herbal products have the dual advantage of minimal side effects and being alcohol and/or sugar-free, which are the two other common ingredients found in other over-the-counter products.

21.2 Structure and Function of Oral Cavity and different Mouth Parts

The oral cavity or mouth is the first portion of the digestive tract and is bounded by the lips anteriorly, the faucets posteriorly, the cheeks laterally, the palate superiorly and a muscular floor inferiorly. In human anatomy, the mouth is the first portion of the alimentary canal that receives food and produces saliva. The tongue fills the available space in the floor of the mouth within the arch of the lower teeth. The oral cavity can be divided into two regions: (1) *The vestibule*, which is the space between lips or cheeks and the teeth; and (2) *the oral cavity proper*, the region medial to the teeth. The oral cavity is lined with moist stratified epithelium which shows characteristic variations of different regions. In addition to the primary role of opening to digestive system, it helps in respiration and in voice production. Throat, tongue, jaw and lips take part in this act.

The anterior portion of the digestive tract from the mouth to the pharynx or, in the absence of a pharynx, directly to the esophagus.In different animals the oral cavity develops wholly or partially from aninvagination of the ectoderm of the fore gut. In invertebrates special adaptations are often associated with the oral cavity, including the piercing and cutting formations in freeliving nematodes, the jaws in gastropod and cephalopod mollusks, and the mouth parts in arthropods.

In vertebrates the oral cavity is lined with a multilayered and sometimes horny epithelium. In fish and terrestrial vertebrates the oral cavity is bounded in front and on the sides by the jaws and teeth, and underneath by a muscular floor; the hyoid arch, which in most animals supports the tongue, is contained in the thick of the muscular floor. In fish and amphibians the roof of the oral cavity is reinforced by the bones of the base of the skull, which form the primary palate and the choanae; the latter open immediately into the oral cavity.

In amniotes the oral cavity is divided into a superior (respiratory) section and an inferior section (secondary oral cavity) by the development of a hard palate. In mammals the posterior part of the hard palate is the soft palate, which separates the oral cavity from the pharynx. In birds and more frequently in mammals, there are papillae, transverse ridges, or palatine plates on the palate, which are covered with a horny layer and which facilitate the mastication of food. In baleen whales the horny layer of these ridges is greatly enlarged and forms the baleen, or whalebone. In mammals the development of soft, mobile lips led to the formation of a cavity between the lips and the teeth called the vestibule of the mouth. In some mammals the vestibule forms lateral processes called cheek pouches. In vertebrates and man the oral cavity contains the tongue, teeth, and glands of the mouth.

The efferent ducts of the major salivary glands the parotid, sublingual, and sub-maxillary glands also open into the oral cavity. The blood supply, lymph drainage, and innervation of the walls of the oral cavity are closely associated with the vascular and nervous systems of the jaws. The taste organs and receptors are located in the mucous membrane. The oral cavity contains permanent microflora, which acts as a biological barrier to microbes, and incidental flora. Diseases of the oral cavity include gingivitis, stomatitis, leukoplakia, papilloma, and diseases of the teeth, jaws, tongue, lips, and salivary glands.

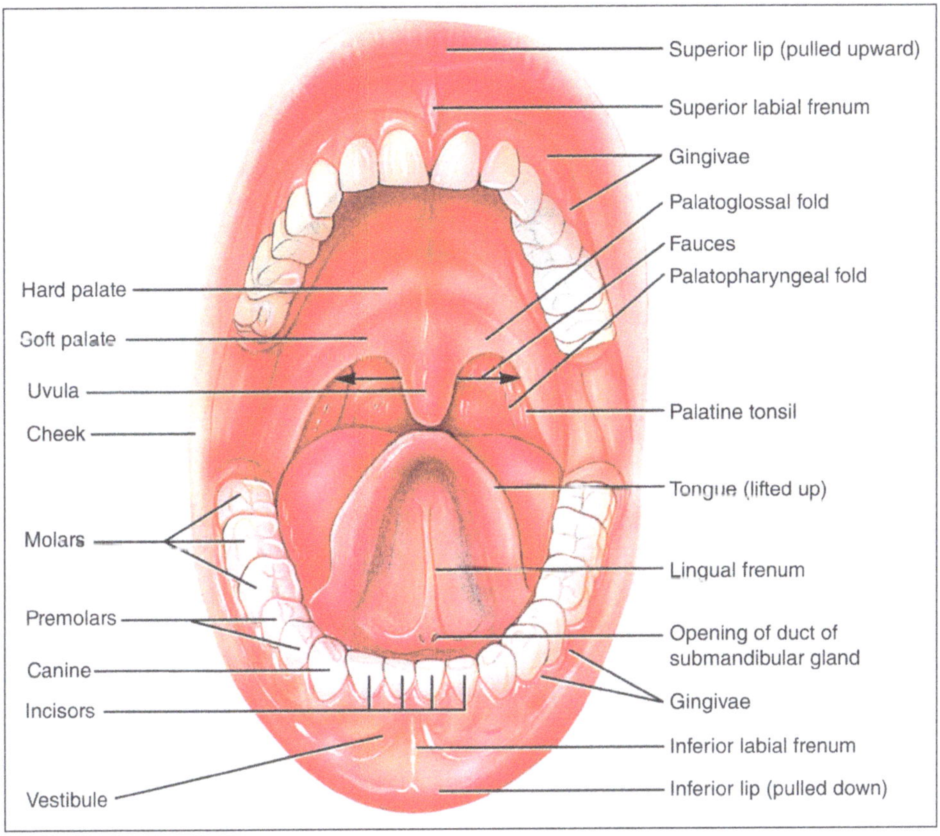

Figure 21.1: Oral Cavity Showing Mouth Parts in Human. (*Source*: Modified from Tortora, and Grabowski, 2000).

Infants are born with a sucking reflex, by which they instinctively know to suck for nourishment using their lips and jaw. The mouth also helps in chewing and biting food. For some disabled people, especially many disabled artists, who through illness, accident or congenital disability have lost dexterity, their mouths take the place of their hands, when typing, texting, writing, making drawings, paintings and other works of art by maneuvering brushes and other tools, in addition to the basic oral functions. Mouth painters hold the brush in their mouth or between their teeth and maneuver it with their tongue and cheek muscles, but mouth painting can be strenuous for neck and jaw muscles since the head has to perform the same back and forth movement as a hand does when painting.

The action of the tongue in mastication has been observed in subjects who have lost some of their teeth. The tongue takes part in all stages, defined as these of 'preparation' when the tongue becomes trough-like, 'throwing' the food between the teeth, 'guarding' the food from falling back from the teeth, 'sorting out' the particles and 'bolus formation', and in swallowing. The muscles of the right side of the tongue throw the food between the left grinding teeth and prevent it from escaping during chewing movements, and this explains why, in hemiplegia the patient is forced to chew on the paralyzed side.

In patients suffering from tongue injuries such as severe wounds or painful scars, or after operations on the tongue involving the removal of half its substance, a stomach tube is essential for feeding. In normal conditions it is equally easy to masticate on either side, but in cases of hemiplegia of the right side, the patient cannot, at the beginning of his illness, masticate on the left unparalysed side though the mandibular muscles on that side are still healthy and active. This is because he cannot throw the ingested food between the left grinding teeth, owing to the impaired 254 Shafik Abd-el-Malek function of the muscles of the right half of the tongue. The bite on the paralyzed side is weak as it has to be carried out by the muscles of mastication of the opposite side. In cases of epithelioma of the tongue, if the jaws are left intact, the patient can, after recovery from operation, only masticate on the affected side, for it is only to this side that his tongue can throw the food, though he can bite on the healthy side normally. Swallowing also may be difficult, the bolus escaping to the paralyzed side where it may be retained in the pyriform area.

Bacteria coating the human body, inside and out, produce proteins that affect our digestion, our health, and our susceptibility to diseases. They are so prevalent that the body is estimated to contain more cells in its micro biome – the bacteria, fungi, and other microbes – than it has human cells. The scientists have now compiled 200 000 genomes and 170 million protein sequences from more than 4 600 bacterial species in the human gut. Their new databases, the Unified Human Gastrointestinal Genome collection and the Unified Gastrointestinal Protein catalogue, reveal the tremendous diversity in our guts and pave the way for further micro biome research. The project revealed that more than 70 per cent of the detected bacterial species had never been cultured in the lab – their activity in the body remains unknown. The largest group of bacteria that falls into that category is the Comantemales, an order of gut bacteria first described in 2019 by the Bork Group at EMBL Heidelberg. "It was a real surprise to see how widespread the Comantemales are. This highlights how little we know about the bacteria in our gut," explains Almeida (2020).

A tooth consists of a crown, containing the pulp chamber and one or more roots, which contain the pulp canals. The anatomic crown is covered by enamel, the hardest biological substance known; the clinical crown is the portion of the crown exposed in the oral cavity. One or more cusps and ridges separated by grooves and sulci are present on the occlusal surface (upper or grinding surface) of premolars and molars. The incisal surface of incisors is flatter and thinner and that of canines is more pointed; these teeth function in biting and tearing. The surfaces of the crowns of premolars and molars that face the cheeks are the buccal or facial surfaces. The surfaces of the crowns of incisors and canines that face the lips are the labial or facial surfaces. The surfaces of the maxillary teeth that face the palate are the palatal surfaces; the surfaces of the mandibular teeth adjacent to the tongue are the lingual surfaces. The surfaces of the teeth in each arch that face the adjacent teeth are the proximal surfaces. The proximal surface of premolars and molars facing the posterior part of the oral cavity is the distal surface; that facing the anterior part of the oral cavity is the mesial (toward the midline) surface. This convention is maintained for the incisors and canines: distal, toward the adjacent tooth in the arch closer to the posterior teeth; mesial, toward the midline. Dentin supports the enamel, which covers the crown, forming the root. Dentin encloses the pulp, which through its blood and nerve supply and immunologic and regenerative functions maintains the vitality of the dentin and the cells that produce it, odontoblasts. Odontoblasts line the periphery of the pulp and secrete and mineralize the matrix components of dentin, predominantly collagen. Each odontoblast has a long apical or distal cytoplasmic, odontoblast process, that extends partway through the dentin in a dentinal tubule. The dentinal tubules are created as the dentin is deposited around the odontoblast processes and the odontoblasts gradually move deeper into the pulp. Some of the dentinal tubules also contain nerve endings, which in conjunction with the odontoblasts and other nerve endings associated with the odontoblasts are responsible for the sensation of pain in dentin.

21.3 Structure of Mouth Parts in Animals and their Food Habits

Scientists have recently estimated that there are about 8.7 million species on Earth. They believe that 1-2 million of those species are animals. It is estimates that 86 per cent of all land species and 91 per cent of all sea species have not yet been discovered. Scientists believe there are over seven billion humans in the world today. That's a large number, but there are more chickens (over 18 billion) than humans. At least we humans do outnumber cattle (1.4 billion) and sheep (1.1 billion). To get nutrition all these have a mouth, from where food is engulfed.

Clams a mollusk use their valves to get food. Oyster use both valve and their cilia in their bodies. The food is drawn into their shell by water current together with the beating of cilia and the opening of valves. Claws – cats, tigers and lions use their claws in getting food. Cats catch mice by using their claws. Tongue – cows and sheep place their lower jaw near the ground while grazing. They wrap their tongue around the grass and close their mouth. Frog and chameleon can throw their tongue to catch the prey. The whale's big mouth opens by the force of the animals, while in forward motion. Around 300 horny baleen plates hang down like curtains from the roof of the mouth. Cray fishes have mouth parts that crush and bite food into small pieces.

Owls, hawks, kingfishers and other birds use their beaks in getting food. The kingfishers with their powerful bills, perch on a tree branch near the water to catch fish. The pelicans use their large bills to catch fish. Other birds have small, short hard beaks for cracking nuts and seeds. Squid and octopus use their tentacles to obtain food. A squid has ten tentacles while the octopus has eight. They also have beak like teeth used to bite and tear their prey into pieces. Butterflies, bees and moths use the long fine sucking tubes of their mouths to sip nectar from flowers.

Grasshoppers, beetles and termites have biting mouth parts that tear the food into small pieces before swallowing them.

The static and dynamic characteristics of the jaws, tongue, and other structures surrounding the oral cavity of mammals strongly affect the types of environments in which a particular mammal can successfully function. A highly specialized mouth and tongue, such as that noted in anteaters by Sonntag (1920), is an extreme example of an ingestive apparatus that sets very narrow limitations on usable habitats. With reference to the primary focus of this chapter, liquid ingestion, less dramatic but equally significant constraints are found in adult animals unable to suck liquids, presumably because of their mouth characteristics. A specific liquid intake and intraoral manipulation mechanism may have great adaptive advantage for a given ecological niche. However, the possibility of evolutionary modification in the event of major, relatively rapid changes in habitat may be greatly reduced. Nonspecialized omnivores, having ingestive apparatus usable with a wide range of foods and liquid sources, avoid these problems. It may be possible that omnivores may not be able to exploit some habitats as efficiently as specialized feeders. Particular movements are associated with liquid ingestion by licking/lapping and by sucking, while suckling presents a special case. Specialization in form, spacing, and type of occlusion of the

According to a senior wildlife researcher, Adam Britton, crocodiles have their very own way of ensuring their teeth stay clean and white. They expose their teeth while in water. The water washes their teeth. And later the Sun sterilizes their teeth when they are basking in the glory of the Sun with their jaws open. But that's not all. Crocodiles are lucky enough to be endowed by nature with 40 sets of teeth throughout their life. And most crocodiles die before the limit of 40 sets is reached. Another unbelievable happening is the symbiotic or two-way relationship between the Nile crocodile and Egyptian plover.

While resting on land after a meal, the Nile crocodile leaves its jaws open, enough for the Egyptian plover to enter into its mouth. The Egyptian plover has no fear of entering the crocodile's mouth and performing the work of a dental hygenist, 'scaling' the crocodile's teeth. The bird eats whatever food remains that are stuck on the crocodile's teeth! That's an amazing wonder. What about wild cats, wolves and wild dogs? To start with, wild cats and wild dogs don't take mushy and soft food like pet cats and pet dogs. Their main diet consists of raw meat and bones. Wolves, wild dogs and wild cats clean their teeth when they gnaw on the bones of their kill.

The other thing about these wild animals is that they don't live that long, not long enough to suffer from excruciating tooth pain or very bad tooth decay. And of course, these smart creatures abstain from taking sweet and starchy food, the key culprits for tooth decay in humans and pet dogs or cats. In the case of wild animals, cavities happen because of injury. But in our case, it happens because of food trapped in between our teeth. So what is it that we can learn from wild animals and their 'dental care'? Practice good oral hygiene and abstain from taking too much sugary or starchy foods to prevent tooth decay.

21.3.1 Types and Evolution of Teeth in Mammals

During biting the lower jaw moves forwards so that the anterior and the lower incisors come together edge to edge. This observation casts doubt on Wood's theory that the lower incisor of the lagomorph is the lower homo logue of the posterior upper incisor.

It is suggested that the anterior upper incisors are worn and shortened by movement of the lower incisors upwards and backwards against their wear surface during biting. The transverse movement of the lower incisors during the grinding phase of chewing wears the posterior upper incisors and also maintains the chisel edge on the tips of the lower incisors. No movements have been seen whereby the remainder of the wear surface of the lower incisors is abraded: from examination of the teeth it is suggested that the level of this surface is maintained by being passed forwards across the cutting edge of the anterior upper incisors.

During chewing the lower cheek teeth move from without inwards on the side on which chewing is taking place during the grinding phase. It appears that the tongue is functionally divisible into three parts: the diastemal portion is concerned with the procuring and manipulation of food and its passage backwards to the cheek teeth; the intermediate portion, which has a dorsal shield and lies between the cheek teeth, is responsible for placing the food on the lower cheek teeth and keeping it to the side on which chewing is taking place and passing it progressively backwards; and the posterior portion situated behind the cheek teeth provides temporary storage for crushed food prior to swallowing, and with the soft palate and pharyngeal musculature in swallowing. The tongue movements and swallowing mechanism in the rabbit is compared with those of man.

Long ago, perhaps even before the celebrated mists of time, there was a great family of animals that lived in Africa. The story starts some 10 million years ago and then the family grew and spread out. Around three million years ago, a branch of it spilled into Europe and Asia. As the animals moved into new territories, they adapted to more northern climes. Eventually, some crossed the bridge of Beringia, migrating from north-east Asia into North America.

This is the story of the elephantines: of mammoths, *Loxodonta* and *Elephas*. The tale of the elephant's tooth is somehow different, a change in behaviour clearly precedes a change in anatomy. The most striking characteristics of living elephants – trunks and tusks – had appeared in their gomphothere ancestors by 20 million years ago. For a large animal with a short neck, the trunk was an extremely useful development, allowing these proboscideans to grasp leaves and bring them to the mouth, thus providing an evolutionary advantage.

The development of a trunk and the transformation of incisors into tusks were accompanied by a change in the shape of the skull. Inside the mouth, the teeth were also changing. A short jaw left little room for a full set of molars, while the teeth needed to be able to sustain a long lifetime's worth of heavy wear. According to evolution theory rather than having a whole set of premolars and molars crammed into the mouth at the same time – as in our mouth – there was just a single, large tooth occupying each side of the upper and lower jaw at any time. As this tooth wore down, another would be growing behind it, ready to slide into place when the worn-out tooth fell out, providing the animal with up to six sets of teeth in a lifetime.

The teeth of fossil gomphotheres and elephants preserve a signal of their diets. The ratio of different isotopes of carbon in the tooth enamel shows whether a particular individual was focusing more on browsing on leaves or eating grass. The grasslands of Africa first began to spread around 10 million years ago and isotope analysis reveals that late gomphotheres and early elephants switched to eating mainly grass around eight million years ago. In elephants, this switch is reflected in another change to their chewing teeth, which became three times as tall, with a proliferation of enamel ridges. But these adaptations to an abrasive diet appeared around five million years ago, three million years after that switch from soft leaves to tough grasses. With the degree of resolution we can achieve when looking far back into the past, it's often difficult to know what came first – a change in behavior or in anatomy. But in this case, it's very clear: the changes to teeth lagged millions of year after the change in diet.

In our evolutionary narratives, the organism itself often seems to play a passive role: a powerless victim, almost, of changes to its environment or mutations in its genes. But the tale of the elephant's tooth is somehow different, a change in behavior clearly precedes a change in anatomy (and the underlying genetic instructions for tooth development). Perhaps we shouldn't be surprised by this: developmental plasticity means that the final shape of an animal's body is determined not only by DNA but also by external factors. And animals are more flexible in the way they interact with their environments than we sometimes assume. As elephants show, the source of novelty in evolution can come from behavior rather than from genes. The Columbian mammoth is an extinct species of mammoth that inhabited North America as far north as the northern United States and as far south as Costa Rica during the Pleistocene epoch. It was one of the last in a line of mammoth species, beginning with *Mammuthus subplanifrons* in the early Pliocene. The Columbian mammoth evolved from the steppe mammoth, which entered North America from Asia about 1.5 million years ago. The pygmy mammoths of the Channel Islands of California evolved from Columbian mammoths. The closest extant relative of the Columbian and other mammoths is the Asian elephant

A fang is a long, pointed tooth. In mammals, a fang is a modified maxillary tooth, used for biting and tearing flesh. In snakes, it is a specialized tooth that is associated with a venom gland. Spiders also have external fangs, which are part of the chelicerae. Fangs are most common in carnivores or omnivores, but some herbivores, such as fruit bats, have them as well. They are generally used to hold or swiftly kill prey, such as in large cats. Omnivorous animals, such as bears, use their fangs when hunting fish or other prey, but they are not needed for consuming fruit. Some apes also have fangs, which they use for threats and fighting. However, the relatively short canines of humans are not considered to be fangs.

21.3.2. Oral Microbiome and Human Health

A newly born baby from a fairly sterile environment is bombarded with bacteria just after birth. Infants are born with a sucking reflex, by which they instinctively know to suck for nourishment using their lips and jaw. The mouth also helps in chewing and biting food. If your oral hygiene is not proper, your mouth may be loaded with bacteria and that can be a big problem for your baby. Transmitting harmful bacteria from your mouth to hers can happen as easily as kissing those perfect, tiny lips.

The bacterial flora in a healthy oral cavity vs a diseased one is distinctly different, suggesting there may be a profile for a core oral microbiome of health (Aas *et al.*, 2005). According to various studies, identical bacterial sequences have been discovered in the oral cavities of unrelated healthy individuals (Zaura *et al.*, 2009). Bik *et al.* (2010) performed a study based on the largest set of near full-length sequences per individual to date. The analysis identified 10 variables shared between 11 bacterial species. However, the same study also showed that significant interindividual differences exist, supporting the concept of both a core and variable microbiome within the oral cavity. The major genera with the largest representation in healthy oral cavities include the following: *Streptococcus, Veillonella, Granulicatella, Gamella, Actinomyces, Corynebacterium, Rothia, Fusobacterium, Porphyromonas, Prevotella, Capnocytophaga, Nisseria, Haemophilis, Treponema, Lactobacterium, Eikenella, Leptotrichia, Peptostreptococcus, Staphylococcus, Eubacteria,* and *Propionibacterium* (Aas *et al.*, 2005; Jenkinson and Lamont, 2005). Developing an in depth definition of health, and understanding molecular differences between health and disease, may give clinicians the ability to recognize and diagnose diseases at an earlier and reversible stage (Zaura *et al.*, 2009). Most patients infected with oral cancer practice poor oral hygiene (Meurman, 2010). In general, numerous studies conducted around the world have shown poor oral health and tooth loss to increase the risk of gastric, pancreatic, and other cancers. Inflammation is usually the first symptom of compromised oral health and it gets worse as health regresses. Approximately 15–20 per cent of human tumors contain pathogenic agents derived from inflammatory infections. Proper oral hygiene will maintain control of such inflammatory agents that may contribute to oral cancer (Meurman, 2010).

Figure 21.2: Indian Dental Association Celebrates National Oral Health Day on 1st August.

21.4 Problems in Human Teeth and other Oral Structures

21.4.1 Dental Plaque

As practitioners know, the buildup of a plaque on the surface of teeth can increase the risk of a patient developing caries or periodontitis over time (Manjunath *et al.*, 2011). Therefore, the removal of bacteria and their food sources is vital to the prevention and management of oral diseases. Twice daily brushing with a fluoride toothpaste and the daily use of an interdental brush are essential to maintaining good oral hygiene. The use of a mouth rinse as a supplement enables them to protect their teeth and rinse out any bacteria, when mechanical removal of these microorganisms may be particularly challenging. As a key ingredient in many current mouth rinse products, chlorhexidine is well-known for its efficacy at protecting teeth against a wide range of bacteria. The established practices to prevent dental caries and periodontal diseases are the use of fluorides in different forms and mechanical plaque control in combination with professional care, respectively. Basically major population will not have adequate dexterity and motivation that are necessary to maintain optimum oral hygiene. This is especially true in rural areas. Some bacteria utilize these ingredients and release byproducts that cause halitosis.

21.4 2 Teeth Loss

Worldwide, 60 per cent –90 per cent of school children and nearly 100 per cent of adults have dental cavities, often leading to pain and discomfort. Severe periodontal (gum) disease, which may result in tooth loss, is found in 15 per cent –20 per cent of middle-aged (35–44 years) adults. Oral care influences the general quality of life, and poor oral health can be linked to other systemic conditions. The impact that oral care has on a person has indicated the need for treatments. **Gingivitis** means inflammation of the gums. There are various types. However, most cases of gingivitis are caused by plaque. Natural products were suggested by Badria and Zidan (2004) to prevent dental caries.

21.4.3 Bad Breath

The terms halitosis, or bad breath are used to denote unpleasant breath odor. These terms are not synonymous with oral malodor, which has its origin only from the oral cavity. Halitosis should not be confused with odor associated with food intake, smoking, or morning breath on awakening (Quirynem and Steenberghe, 2006). Oil pulling therapy has been used extensively as a traditional Indian folk remedy for many years for strengthening teeth, gums, and jaws and to prevent decay, oral malodor, bleeding gums and dryness of throat, and cracked lips. Nearly 85 per cent of the cases of halitosis have the cause originating from the oral cavity. Gingivitis, periodontitis, and tongue coating are the predominant causative factors (Persson, 1990).Extra oral causes include ear-nose-throat pathology, systemic diseases like diabetes, metabolic, hormonal, renal, or hepatic disturbances, bronchial carcinoma, or gastroenterologic pathology. Oil pulling or oil swishing, in alternative medicine, is a procedure that involves swishing oil in the mouth for oral and systemic health benefits. It is not a new concept and it has been mentioned in the Ayurvedic text Charaka Samhita (Asokan *et al.*, 2020). *Eucalyptus globulus* was screened for active compounds and macrocarpals were isolated and identified as DPP-4 inhibitors. Among them, macrocarpal C, which is previously reported to have an antibacterial effect (Yamakoshi *et al.*, 1992) and to inhibit human immunodeficiency virus reverse transcriptase activity, showed a characteristic inhibition curve indicating a distinct inhibitory character from other macrocarpals (Kato *et al.*, 2017). It can be used to get relief from bad breath.

21.4.4 Scurvy/Disease of Jaw

The incidence of scurvy is extremely rare in industrialized countries due to technological developments including food processing and transportation. It is still present in certain populations with poor nutrition including elderly persons living alone, those living in poverty, institutionalized or chronically ill patients, alcoholics and psychiatric patients. Scurvy is less frequent in the pediatric population. Infants who are fed with evaporated or boiled milk and children with a poor diet or those with psychiatric or developmental disorders are at risk. In our patient scurvy resulted from lack of vitamin C rich diet (Weinstein 2001).

Scurvy is a clinical syndrome characterized by hemorrhage, hyperkeratosis, hypochondriasis and hematological abnormalities, but the presentation can vary by individual (Bary *et al.*, 2009). Scurvy has been known to exist for more than three millennia, with description of similar conditions recorded by the ancient Egyptians in the Ebers Papyrus (Major 1954). It was a major cause of morbidity and mortality in much of Europe during the great potato famine and United States civil war. Captain James Cook was one of the first to demonstrate that sailors who spent months at sea could avoid scurvy by maintaining a diet rich in vegetables. In 1753, scurvy and its prevention by citrus fruits was systematically described by surgeon James Lind.

Clinical manifestations can be seen within 8 to 12 weeks of irregular or inadequate intake or when the body pool falls below 350 mg. Early clinical manifestations include pallor, myalgia, bone pain, irritability, poor weight gain, easy bruising, petechiae, perifollicular haemarroges, cork screw hairs, gum disease, poor wound healing, mood changes and depression. In advanced stages of scurvy, the patient is miserable and tends to remain in a characteristic immobilized posture from subperiosteal pain, with flexion of hips and knees. Late manifestations include generalized edema, severe jaundice, fever, seizures, acute spontaneous bleeding and death. In infancy and childhood, the sternum may sink inward leaving a sharp elevation at the rib margins (scorbutic rosary) (Wilson 1994).

21.5 Use of Herbs as Tooth Brush or Paste

In ancient times vegetable oil and common salt was used to clean the teeth. Use of Neem and acacia or babool twigs to brush was common in villages. According to Dr. Chandrana the prop roots of vad (*Ficus benghalensis*) were used as natural tooth brush (Plate 22.1D). Now electronic brushes are becoming popular. Most of the paste contains Fluoride offers anticariogenic benefits; while silica or diatomaceous earth acts as an abrasive agent to prevent accumulation of plaque; alkaloids found in herbs exert analgesic action; oils exert analgesic, antiseptic, and carminative effects; and tannins act as astringent and form a protective coating on the tooth enamel that aids in caries prevention.

The herbal toothpaste has been created with 5 Ayurvedic ingredients *i.e.* neem, which helps in gum care, clove strengthen teeth, amla keeps germs away and basil helps in keeping good oral breath. The primary components of other toothpaste were almonds, miswak (*Salvadora*), clove papaya, menthol, pineapple, cinnamon, Triphala, black pepper, gum arabic and pomegranate. Eucalyptus extracts, tea tree oil, myrrh (gum resin with a bitter slightly pungent taste obtained from a tree *Commiphora abyssinica* of the family Burseraceae), Myrrha decreases inflammation, kills bacteria, and treats gum disease. sage and chamomile (*Matricaria chamomilla*) is also used (Singh *et al.*, 2011). Chung *et al.* (2006)suggested use of nutmeg and Park *et al.* (2003) used root bark of *Morus alba* to treat the bacterial infection. A preliminary list of studies on the antimicrobial efficacy of herbal extracts on dental caries (*S. mutans* and *L. acidophilus*) and plaque microorganisms (primary, secondary, and tertiary plaque colonizers) was prepared following the initial screening. The antimicrobial efficacy of the ten herbal extracts on dental caries and plaque microorganisms were examined (Rudraiah *et al.*, 2015).

21.6 Use of Herbal Extract as Mouth Wash

Mouth rinses are widely used as adjuncts to oral hygiene and in the delivery of active agents to the teeth and gums. The ability of these rinses to influence plaque formation and to alter the course of gingival inflammation has been extensively studied. Natural products have been used for folk medicine purposes throughout the world for thousands of years (Table 21.1). Many of these have pharmacological properties, such as antimicrobial, antiinflammatory and cytostatic effects. They have been recognized as useful for human medicine.

It was found that the aqueous extract of *Azadirachta indica* was effective against *Streptococcus mutans*, *S. sanguis*, *S. salivarius*, and *S. mitis*. The mean diameters of inhibition zone against these bacteria at 5 per cent, 10 per cent, and 50 per cent concentrations were 2.4 mm, 3.4 mm, and 3.8 mm (*S. mutans*); 2 mm, 3 mm, and 3.4 mm (*S. sanguis*); 1.5 mm, 2.5 mm, and 2.9 mm (*S. salivarius*); and 1.5 mm, 2 mm, and 2.7 mm (*S. mitis*). The mean inhibition zone against these bacteria increased as the concentration increased. The maximum efficacy of the extract was found against *S. mutans* at 50 per cent concentration (Prashant *et al.*, 2007). The chemicals present in the neem are mentioned in Table 21.1. Fluoride offers anticariogenic benefits; while silica acts as an abrasive agent to prevent accumulation of plaque; alkaloids exert analgesic action; oils exert analgesic, antiseptic, and carminative effects; and tannins act as astringent and form a protective coating on the tooth enamel that aids in caries prevention.

Agarwal *et al.* (2010) evaluated the antimicrobial efficacy of ethanolic extract of *Ocimum sanctum* or basil of family Lamiaceae against *S. mutans*, the principal bacteria in the causation of dental caries. The inhibition was found at 4 per cent concentration while using 75 μL (22 mm). The antimicrobial efficacy is attributed to the presence of eugenol. Sunitha *et al. (2012)* in their study found the ethanolic and aqueous extracts of *Murraya koenigii* L. to be effective against *S. mutans*. Murraya is

Table 21.1: Certain Plant and their Parts Used for different Oral Care

Sl.No.	*Plant/Part Used*	*Treatment*	*Active Compound*	*Reference*
1.	*Acorus calamus* Paste of the Rhizome	Applied to painful teeth and gums		Dash *et al.*, 2008
2.	*Allium sativum* The paste of the bulb	Used against gums and cavities of infected teeth	Sulphur containing Alicin	Bakri,and Douglas 2005
3.	*Ocimum sanctum* Powder of dry leaves along with salt	Antimicrobial efficacy	Presence of eugenol (1-hydroxy-2-methoxy-4-allylbenzene). ursolic acid and carvacrol	Agarwal *et al.* (2010)
4.	*Bombax ceiba* Gum	Used to treat tooth care,bleeding gums	Roct showed presence of n-triacontanol, β-sitosterol. New glycosides identified as 5,7 3,4 tetrahydroxy-6-methoxy flavon3-O-β-D glucopyranosyl-D-xylotyranoside sesquiterpenoids, and new sesquiterpene lactone,.	Rani *et al.*, 2007
5.	*Juglens regia* Oil and fruits	Making traditional tooth powder to cure toothache and Pyorrhoea	folates, niacin, pantothenic acid, pyridoxine, vit. A, C, E and vit K) and minerals (potassium, phosphorus, sodium, Fe, Cu, Mn, and zinc	Snafi 2018
6.	Cinamomum Stem bark juice	Applied to the teeth to treat tooth decay and toothache	Eugenol, Alpha- phellandrene	Sabnis and Daniel 1990
7.	Clove	Gum and toothpain	Eugenol	Gupta *et al.*, 2011
8.	*Citrus medica*	Due to high content of Vitamin C, used to treat bleeding gums in scurvy.Antifungal.	Vit. C aromatic oilcaretenoids	Rafiq *et al.*, 2018
9.	*Datura stramonium* Seeds mixed with butter are burnt	smoke is inhaled into the mouth	Alkaloidsscopolamine	Sabnis and Daniel 1990
10.	*Justicia adhatoda* The twigs of the plant	8 are used as tooth picks/brushes to treat Pyorrhoea	Anisotine and vasicoline, Pemirolast are very good inhibitors of Covid	Bag and Bag 2020
11.	Myrica esculenta The bark	chewed to relieve toothache	Glycerides of myristic	Sabnis and Daniel 1990

Sl.No.	*Plant/Part Used*	*Treatment*	*Active Compound*	*Reference*
12.	The extracts of Margosa (*Azadirachta indica*)	alkaloids exert analgesic action; oils exert analgesic, antiseptic, and carminative effects; and tannins act as astringent and form a protective coating on the tooth enamel that aids in caries prevention.	Alkaloids, saponins, flavanoids, sterols, resins, tannins,	Prashant *et al.*, 2007
13.	The stem bark of babul (*A. nilotica)*	effect on *E. coli* and antifungal activity against *Candida albicans*The tannins have an astringent effect on the mucous membrane and are claimed to form a protective layer on the enamel.	Of alkakoids, tannins, saponins, flavonoids, cardiac glycosides, and anthraquinones.	Dabur *et al.* (2007)
14.	aqueous extracts of *Aloe vera*	latex compound that has been found to be bacteriostatic in nature	Salicylic acids, amino acids, and saponins	Sunitha *et al.* (2012) George *et al. (2009)*
15.	Aqueous extracts of *Psidium guajava*	Extract used for selected early settlers of dental plaque	Polyphenols	Razak *et al.*, 2006

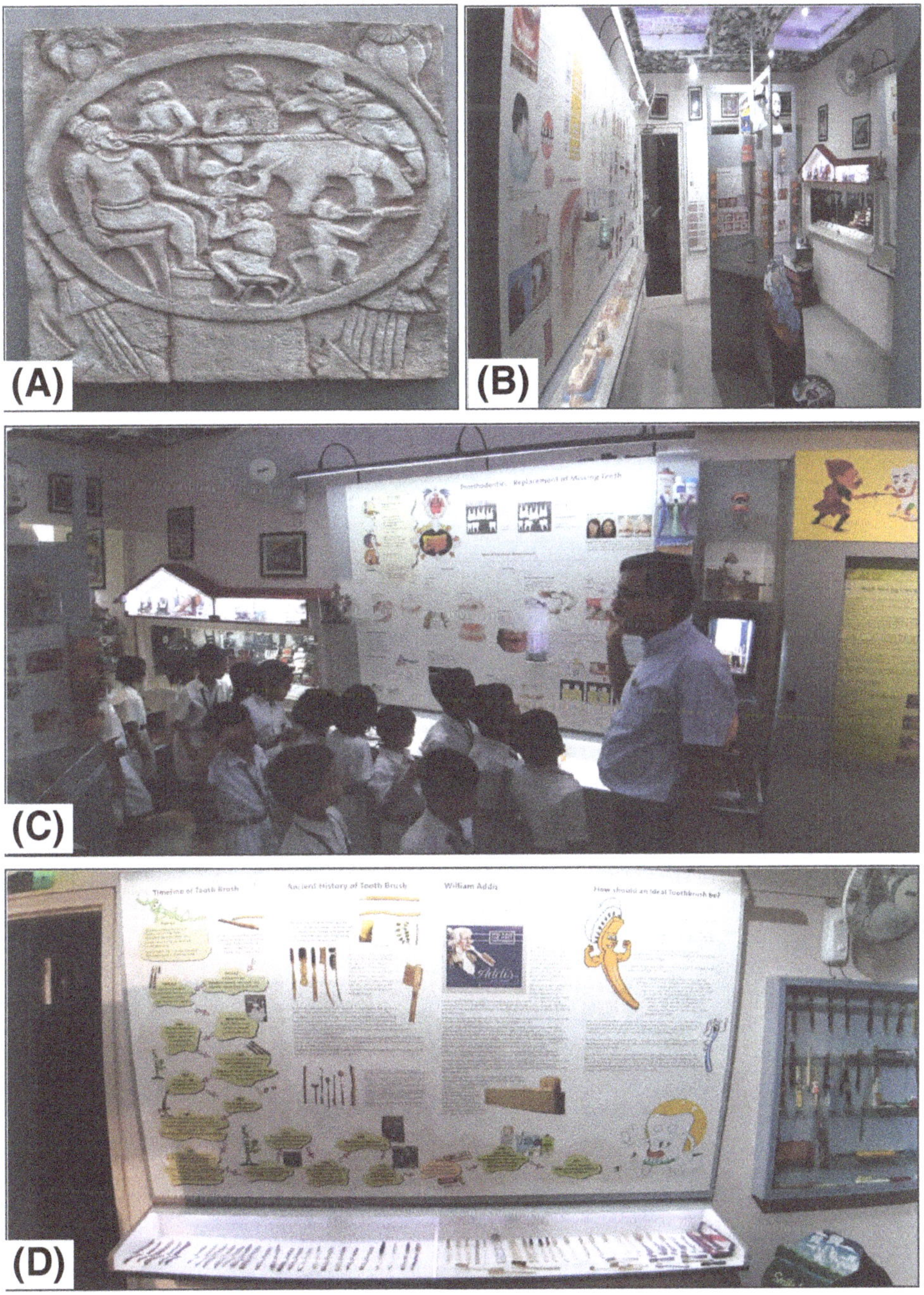

Plate 21.1

(A) A ceramic plate showing removal of teeth using the help of an elephant. Monkeys biting on tail part and elephant moves away from patient. (B) A gallery of dental museum. (C) Student visiting Dr. Chandrana's dental museum at vadodara. (D) A gallery of museum showing different tooth brushes.

Plate 21.2

(A) Dr. Yogesh Chandrana explaining the exhibits to visitor students; (B) A gallery of dental museum showing postage stamps; (C) Dr. Chandrana's showing a car with dental structure.

a commemorative name after Andrew Murray, Swedish student of Linnaeus. The mean diameter of the inhibition zone of the ethanolic and aqueous extracts was 16 mm and 13.05 mm, respectively. The ethanolic extract was effective in inhibiting *Lactobacillus brevis* (20.10 mm) and *Actinomyces viscosus* (17.05 mm). The aqueous extract inhibited *Lactobacillus casei* (16 mm). The antimicrobial efficacy of curry leaves is due to carotene, beta carotene, calcium, iron, phosphorus, zinc, folic acid, and riboflavin. These also produced other biological activities in oral cavity like induction of immunity, which indirectly reduces the risk of oral diseases.

Dabur *et al.* (2007) found the methanolic and aqueous extracts of *A. nilotica* to exert inhibitory effect on *E. coli*. The MIC of methanolic and aqueous extracts was 75 µg/mL and 18.75 µg/mL, respectively. Deshpande *et al.* (2013) have found the ethanol fraction of stem bark to be effective against *S. mutans*. Pai *et al.* (2010) found the aqueous extracts of *A. nilotica* to exert antifungal activity against *Candida albicans*.

Nagata *et al.* (2006) determined the effect of macrocarpals A, B, and, C, which are phloroglucinol derivatives of eucalyptus leaves. These were effective against *P. gingivalis*, which displayed the maximum sensitivity. The MIC of macrocarpals A and B was 1µg/mL while it was 0.5 µg/mL for macrocarpal C. Takarada *et al.* (2004) found eucalyptus oil to exert inhibitory effect on various oral bacteria including *S. mutans, Streptococcus sobrinus, Actinobacillus actinomycete mycomitans, Porphyromonas gingivalis*, and *Fusobacterium nucleatum*.

Nagarajappa *et al.* (2013) found the aqueous and ethanol extracts of *Hibiscus rosa-sinensis* derived using both the hot and cold extraction methods to exert inhibitory effect on *L. acidophilus* and *S. mutans*. The extract significantly inhibited the growth of *L. acidophilus* and *S. mutans* ($P \leq 0.05$) with MIC value of 25 µg/mL. The study demonstrated the therapeutic potential of the plant extract against two proven cariogenic bacteria.

Prashanth *et al.* (2007) found that the mean diameters of the inhibition zone produced by *Mangifera indica* against *S. mutans* at 10 per cent and 50 per cent concentrations were 1.5 mm and 2.9 mm, respectively. *M. indica* failed to inhibit the growth of the bacteria at 5 per cent concentration even after 48 h of incubation. The antimicrobial efficacy of *mango* was due to the presence of bitter gums, resins, and tannins. Although the published literature demonstrates the antimicrobial efficacy of *guava* (Ismail *et al.*, 2012; Gonçalves *et al.*, 2008), no studies were found on oral bacteria. *P. guajava* contained guajaverin, psidiolic acid, and other essential oil constituents such as 1.8-cineol, monoterpenes, ρ-cimen, and acetate of α-terpenil, which may offer antimicrobial benefits (Ismail *et al.*, 2012; Rudraiah *et al.*, 2015).

Antimicrobial potential of *Rosa damascene* and *Rosmarinus officinalis* was tested against *S. mutans, S. sanguinis*, and *S. sorbinus*. The MIC of *Rosa damascene* (flower extract) against these bacteria was > 8 mg/mL. The MIC of *R. officinalis* leaf extracts against *S. mutans, S. sanguinis*, and *S. sorbinus* was *4 mg/mL, 2 mg/Ml, and 4 mg/mL* (Tsai *et al., 2008).*

Sunitha *et al.* (2012) found the aqueous extracts of *Aloe vera* to inhibit growth of *S. mutans* (13.05mm), *S. mitis* (15.15mm), *L. brevis* (15.7 mm), *L. acidophilus* (17.95 mm), and *A. viscosus* (17 mm). The ethanolic extract was also found to be effective against *L. brevis* (13.25 mm) and *A. viscosus* (13.05 mm). George *et al. (2009)* found the gel to inhibit the growth of *S. mutans, L. acidophilus, Prevotella intermedia, Candida albicans, Enterococcus faecalis,* and *Peptostreptococcus anaerobius*. The antimicrobial activity of *Aloe vera* against these microorganisms was attributed to the constituents listed in Table 21.1. "Herbal shotgun" or "synergistic multitarget effects" are the terms used for the strategy of combining different extracts.

21.7 Use of Citrox as Mouth Wash

The long-term use of mouth rinses with chlorhexidine alone can result in adverse side effects such as tooth discolouration and impaired taste. Alcohol-based mouth rinse products can also dry out the mouth and lower saliva production, which enables oral bacteria to proliferate. An innovative, new mouth rinses comprised of safe and highly effective ingredients was tried. Citrox was manufactured using naturally-derived, non-toxic, hypo-allergenic, and non-corrosive ingredients, Citrox takes advantage of the bioflavonoids (plant-based compounds) found in plants. Scientists have discovered that bioflavonoids have the ability to modify the body's reaction to allergens, inflammation, viruses, and carcinogens. Not only can bioflavonoids protect the body against oxidative and free-radical damage, they can also increase the strength of the capillaries, thereby regulating their permeability to prevent ruptures and protect against infection. This helps to heal wounds and support a healthy immune system.

Citrox was created using dried, ground piths and pulp of bitter oranges. This is then mixed with other natural substances (organic acids) to produce a powerful antimicrobial, antioxidant, and anti-inflammatory agent. Citrox has now been developed as a key mouth rinse ingredient, as it has the ability to kill 99.9 per cent of bacteria, viruses, and fungi. This means that out of 10,000 potentially harmful microorganisms, only 0.1 of these could affect an individual following treatment of Citrox, making it a highly effective, natural, antimicrobial substance. Experts at Cardiff University believe that Citrox has numerous advantages as a key ingredient in oral healthcare products, including mouth rinses. Research has revealed that Citrox is effective against dangerous pathogens such as MRSA, C. Difficile, Hepatitis A, Hepatitis B, and *Streptococcus*. According to a study by Williams *et al.*, Citrox combined with other supplements is more effective than chlorhexidine alone at protecting the oral cavity from a wide range of bacteria known to cause oral diseases such as dental caries and periodontitis.

21.7.1 How does Citrox Benefit Us?

Citrox is effective against dangerous pathogens such as MRSA, C. difficile, Hepatitis A, Hepatitis B, and *Streptococcus.* According to a study by Williams *et al.* (www.curaprox.co.uk),

Citrox combined with other supplements is more effective than chlorhexidine alone at protecting the oral cavity from a wide range of bacteria known to cause oral diseases such as dental caries and periodontitis. This is due to the substance's ability to inactivate a variety of microorganisms by affecting multiple target sites, thereby surpassing traditional antibiotics that usually target one specific area.

Used as a vital mouth rinse ingredient, Citrox works by breaking down bacterial biofilm quickly and efficiently to ensure a fast-acting result with long-lasting, residual effects – a concentration of just 0.25 per cent Citrox solution can kill biofilm cells in less than ten minutes. The strength of Citrox is evident even in the presence of organic matter, and it is also less prone to bacterial resistance so that it can protect the human body against bacterial re-infiltration. Citrox is completely biodegradable, safe to handle, and eco-friendly. For additional peace of mind, Citrox is GMO (genetically modified organism) free, does not contain any petrochemicals or allergic substances, and is not tested on animals – thus providing a truly natural alternative to traditional mouth rinse ingredients.

21.8 Use of Herbs in Oral Health: Future Prospects

Loss of teeth because of periodontitis often causes discomfort, and compromises the esthetics and function. Moreover, recent studies suggest an association between chronic low-grade infections such as periodontitis and systemic health problems (preterm low birth weight, cardiovascular diseases, diabetes mellitus, and chronic obstructive pulmonary disease). The treatment of established dental diseases is very expensive and may not be considered a realistic strategy, especially for the economically deprived sections of the society. Treating dental caries alone, which is estimated at US$ 3513 per 1,000 children is more than the total health budget for children in many low-income countries. There is an immediate need for promoting preventive strategies that are socially acceptable, easily available, and at the same time be cost-effective. This calls for the evolution of innovative strategies that are robust, efficient, and feasible.

21.9 Conclusion

The use of herbal extracts for the control of oral/dental diseases is considered an interesting alternative to synthetic antimicrobials due to their lower negative impact and to overcome intrinsic (primary) resistance or secondary resistance to the drug during therapy. During ancient times use of betel leaves and peppermint was used after food intake. The use of volatile compounds from *Mentha piperata* were used to suppress growth of *Campylobacter jejuni* a common micro aerophyllic bacteria present in gastrointestinal tract (Kovacs *et al.*, 2009).The current evidence suggests all the plant extracts tried had antimicrobial efficacy against dental caries and periodontal pathogens. Most of these studies have been conducted using individual plant extracts on certain bacteria that are involved in either dental caries or periodontitis. The authors of all these studies have recommended further research before considering them for clinical use. Some authors have recommended the use of these plant extracts in combination (Abd El-Kalek and Mohamed, 2012). The research assessing the antimicrobial efficacy of a combination of these plant extracts against dental caries and periodontal pathogens is the need of the hour, and such research will aid the development of a novel, innovative method that can simultaneously inhibit the two most common dental diseases of mankind, besides lowering the development of drug resistance.

REFERENCES

Aas JA, Paster BJ, Stokes LN, Olsen I, Dewhirst FE (2005). Defining the normal bacterial flora of the oral cavity. J Clin Microbiol 43: 12. Avila M, Ojcius DM, Yilmaz O (2009).

Abd El-Kalek HH, and Mohamed EA. (2012) Synergistic effect of certain medicinal plants and amoxicillin against some clinical isolates of methecillin–resistant *Staphylococcus aureus* (MRSA) Int J Pharm Sci. 3: 387–98.

Agarwal P, Nagesh L, Muralikrishnan (2010) Evaluation of the antimicrobial activity of various concentrations of Tulsi (*Ocimum sanctum*) extract against *Streptococcus mutans*: An *in vitro* study. Ind. J Dent Res. 21: 357–9.

Almeida A. *et al.* (2020) A unified catalog of 204,938 reference genomes from the human gut microbiome. *Nat Biotech*. Published online 20 July 2020. DOI: 10.1038/s41587-020-0603-3.

Asokan, S. Kumar R.S., Emmadi P., Raghuraman R., SivakumarN (2011) Effect of oil pulling on halitosis and microorganisms causing halitosis: A randomized controlled pilot trial J. of Indian Soc. of Pedodontics and Preventive Dentistry 29(2) http: //www.jisppd.com on Friday, September 18, 2020, IP: 117.198.203.128]

Badria FA, Zidan OA. (2004) Natural products for dental caries prevention. J. of Medicinal Food 2004;7(3): 381–84.

Bakri IM, Douglas CWI. (2005) Inhibitory effect of garlic extract on oral bacteria, Archives of Oral Biology 2005;50(7): 645–51.

Bary A, Javaid BK, Rehman S, Naz S. (2009) Scurvy: presenting as musculoskeletal pain. J Coll Physicians Surg Pak. 19: 198–200.

Bik EM, Long CD, Armitage GC *et al.* (2010) Bacterial diversity in the oral cavity of ten healthy individuals. ISME J 4: 962–974. DOI: 10.1038/ismej.2010.30.

Chung JY, Choo JH, Lee MH, Hwang JK. (2006) Anticariogenic activity of macelignan isolated from *Myristica fragrans* (nutmeg) against *Streptococcus mutans*. Phytomedicine 2006; 13, (4): 261–66.

Dabur R, Gupta A, Mandal TK, Singh DD, Bajpai V, Gurav AM, *et al.* (2007) Antimicrobial activity of some Indian medicinal plants. Afr J Tradit Complement Altern Med. 2007;4: 313–8

Dash TR *et al.* (2008) Role of Medicinal Herbs in Oral Health Management. Phytotherapy Research 22(3): 308– 12.

Deshpande SN, Kadam DG. (2013) Phytochemical analysis and antibacterial activity of acacia nilotica against streptococcus mutans. Int J Pharm Pharm Sci. 2013;5: 236–8.

George D, Bhat SS, Antony B. (2009) Comparitive evaluation of the antimicrobial efficacy of aloe vera tooth gel and two popular commercial tooth pastes: An *in vitro* study. Gen Dent. 57: 238–41.

Gonçalves FA, Andrade Neto M, Bezerra JN, Macrae A, Sousa OV, Fonteles-Filho AA, (2008) Antibacterial activity of Guava, Psidium guajava linnaeus, leaf extracts on diarrhea-causing enteric bacteria isolated from Seabob Shrimp, *Xiphopenaeus kroyeri* (Heller) Rev Inst Med Trop Sao Paulo. 2008;50: 11–5.

Gupta C. Kumari A. Garg A.K. Catanzaro R. and Marotta F. (2011) Comparative study of cinnamon oil and clove oil on some oral microbiota ACTA BIOMED 2011; 82: 197-199

Ismail M, Minhas PS, Khanum F, Sahana VM, Sowmya C. Antibacterial activity of leaves extract of guava (*Psidium guajava*) Int J Res Pharm Biomed Sci. 2012;3: 1–3.

Jenkinson and HF, Lamont RJ. (2005) Oral microbial communities in sickness and in health. Trends in Microbiology 2005; 13(12): 589–95.

Kato,E. Kawakami K. and Kawabata J. (12017) Macrocarpal C isolated from *Eucalyptus globulus* inhibits dipeptidyl peptidase 4 in an aggregated form. Journal of Enzyme Inhibition and Medicinal Chemistry 2017, 33 (1): 106–109 https: //doi.org/10.1080/14756366.2017.1396458

Kovács,J. Felsy, KP Horváth,G Schmidt,J Dorn,A Ábrahám H. A. Cox,A Márk L. Emydy,L. Kovács T., and Schneider G. (2019) Stress Response and Virulence Potential Modulating Effect of Peppermint Essential Oil in *Campylobacter jejuni* Hindawi BioMed Research International Volume 2019, Article ID 2971741, 11 pp https: //doi.org/10.1155/2019/2971741

Major RH. A (1954) history of medicine. Springfield, In: Thomas CC; 1954. p. 5.

Manjunath BC, Praveen K, Chandrashekar BR, Rani RM, Bhalla A. (2011) Peridontal infections: A risk factor for various systemic diseases. Natl Med J India. 2011;24: 214–9.

Meurman JH (2010). Oral microbiota and cancer. J Oral Microbiol 2: 5195.

Nagarajappa R, Batra M, Sharda AJ, Asawa K, Sanadhya S, Daryani H, *et al.* (2013) Antimicrobial effect of *Jasminum grandiflorum* L. and *Hibiscus rosa-sinensis* L. Extracts against pathogenic oral microorganisms - An *in vitro* comparative study. Oral Health Prev Dent. 2013

Nagata H, Inagaki Y, Yamamoto Y, Maeda K, Kataoka K, Osawa K, (2006) Inhibitory effects of macrocarpals on the biological activity of *Porphyromonas gingivalis* and other periodontopathic bacteria. Oral Microbiol Immunol. 21: 159–63.

Palombo EA. (2011) Traditional Medicinal plant extracts and natural products with activity against oral bacteria: Potential application in the prevention and treatment of oral diseases. Evid Based Compliment Alternat Med 2011 : 680354.

Pai MB, Prashant GM, Murlikrishna KS, Shivakumar KM, Chandu GN. (2010) Antifungal efficacy of *Punica granatum, Acacia nilotica, Cuminum ciminum* and *Foeniculum vulgare* on *Candida* albicans: An *In vitro* study. Indian J Dent Res. 21: 334–6.

Park KM, You JS, Lee HY, Baek NI, Hwang JK, Kuwanon G (2003) An antibacterial agent from the root bark of *Morus alba* against oral pathogens. Journal of Ethnopharmacology 2003; 84(2-3): 181– 85.

Persson S, Edlund MB, Claesson R, Carlsson J. (1990) The formation of hydrogen sulfide and methyl mercaptan by oral bacteria. Oral Microbiol Immunol 5: 195-201.

Petersen PE. The World Oral Health Report (2003) Continuous improvement of oral health in the 21st century- -the approach of the WHO Global Oral Health Programme. Community Dent Oral Epidemiol. 31(Suppl 1): 3–23.

Prashant G M, Chandu GN, Murulikrishna KS, Shafiulla MD. (2007) The effect of mango and neem extract on four organisms causing dental caries: *Streptococcus mutans, Streptococcus salivavius*, streptococcus mitis, and *Streptococcus sanguis*: An *in vitro* study. Indian J Dent Res. 2007;18: 148–51.

Quirynem M, van Steenberghe D. (2006) Oral malodor. In clinical periodontology. In: Newman, Takei, Klokkevold, Caranza, editors. 10th ed. Philadelphia: Saunders -An imprint of Elsevier: 2006. p. 330-42.

Rafiq S. Kaul R. Sofi SA Bashir N. Nazir F. Ahmad and Gulzar A. N. (2018) Citrus peel as a source of functional ingredient: A review J. of the Saudi Society of Agricultural Sciences 17, (4): 351-358.

Razak FA, Othman RY, Rahim ZH. (2006) The effect of *Piper betle* and *Psidium guajava* extracts on the cell-surface hydrophobicity of selected early settlers of dental plaque. Journal of oral Science 48(2): 71–5.

Rudraiah B.C. Nagarajappa R, Suma S. and Thakur R. (2015) Herbal extracts in oral health care - A review of the current scenario and its future needs. *Pharmacogn Rev*. 9(18): 87–92. doi: 10.4103/0973-7847.162101: 10.4103/0973-7847.162101

Sabnis S.D. and Daniel M. (1990) A phytochemical Approach to Economic Botany. Kalyani Pub. New Delhi pp. 487

Sahni PS, Gillespie MJ, Botto RW, Otsuka AS. (2002) In vitro testing of xylitol as an anticariogenic agent. General dentistry 50(4): 340–43.

Sheiham A. (2005) Oral health, genral health and quality of life. Bull World Health Organ. 2005;83: 644–5

Singh N, Jain A, Sinha N, Chauhan A, Rehman R. Application of Four-Handed Dentistry in Clinical Practice: A Review. Int J Dent Med Res 2014;1(1): 8-13.

Singh O, Khanam Z, Misra N, Srivastava MK. (2011) Chamomile (*Matricaria chamomilla* L.): An overview. *Pharmacogn Rev*. January 5(9): 82-95. doi: 10.4103/0973-7847.79103.

Snafi A.E. (2018) Chemical constituents, nutritional, pharmcological and therapeutic importance of Juglans regia- A review IOSR Journal Of Pharmacy www.iosrphr.org 8, (11): 01-21

Sonntag, C. F. (1920) The comparative anatomy of the tongues of the mammalia: I. General description of the tongue, *Proc. Zool. Soc. London*, 115–129.

Sunitha JD, Patel S, Madhusudan AS, Ravindra SV. (2012) An *in vitro* antimicrobial activity of few plant extracts on dental caries microorganisms. Int J A PS BMS. 2012;3: 294–303.

Takarada K, Kimizuka R, Takahashi N, Honma K, Okuda K, Kato T. A comparison of the antibacterial efficacies of essential oils against oral pathogens. Oral Microbiol Immunol. 2004;19: 61–4.

Tortora, G.J. and Grabowski, S.R. (2000) *Principles of Anatomy and Physiology*, 9th edition, Wiley, New York. Reproduced by permission of John Wiley and Sons.)

Tsai TH, Tsai TH, Chien YC, Lee CW, Tsai PJ. (2008) *In vitro* antimicrobial activities against cariogenic streptococci and their antioxidant capacities: A comparative study of green tea versus different herbs. Food Chem. 2008;110: 859–64.

Weinstein M, Babyn P, Zlotkin S. (2001) An orange a day keeps the doctor away: scurvy in the year 2000. Pediatrics. 108: E55

Wilson JD. (1994) Vitamin deficiency and excess. In: Isselbacher KJ, Braunwald E, Wilson JD, Fauci AS, Kasper DL, editors. Harrison's principles of internal medicine. 13th ed. United States of America: McGraw-Hill Inc; pp. 472–80.

www.curaprox.co.uk

Website: www.jisppd.com DOI: 10.4103/0970-4388.84678

Yamakoshi Y, Murata M, Shimizu A, and Homma S. (1992) Isolation and characterization of macrocarpals B-G antibacterial compounds from Eucalyptus macrocarpa. Biosci Biotechnol Biochem 1992;56: 1570–6. Nishizawa M, Emura M, Kan Y, *et al.* (1992) Macrocarpals: HIV-RTase inhibitors of Eucalyptus globulus. Tetrahedron Lett 1992;33: 2983–6.

Zaura E, Keijser BJ, Huse SM, and Crielaard W (2009). Defining the healthy "core microbiome" of oral microbial communities. BMC Microbiol 259: 12.

Chapter 22

Preventive Methods of Human Infection Due to COVID in Industries and Work Place

Amita Arya[1] and Sunil Kumar Singh[2]

[1]Microbiologist, [2]Anesthesiologist,
House No-124, Sector-15, Indira Nagar Lucknow – 226 016, Uttar Pradesh
e-mail: amitaarya0210@gmail.com

ABSTRACT

The 1918 influenza pandemic was the most severe pandemic in recent history. It was caused by an H1N1 virus with genes of avian origin. Although there is no universal consensus regarding where the virus originated, it spread worldwide during 1918-1919. In the United States, it was first identified in military personnel in spring 1918. Since 1918, the world has experienced three additional pandemics, 1957, 1968, and 2009 – each less severe than the 1918 pandemic but raising the question of whether a high severity pandemic on the scale of 1918 could occur in modern times. Many experts have thought so and have recommended global preparation for possible pandemics as current preparedness is seen to warrant improvement. Today, considerable advancement has been made in health technology, disease surveillance, medicine, drugs and vaccines, medical care, and pandemic planning. Flu vaccines are now produced and updated yearly. And yearly vaccination is recommended for the population at large. Diagnostic tests are available for identifying flu and we now have "rapid molecular assays" to provide flu test results more efficiently and more accurately than in the past.

The most common symptoms of COVID-19 are fever, dry cough, and tiredness. Other symptoms that are less common and may affect some patients include loss of taste or smell, aches and pains, headache, sore throat, nasal congestion, red eyes, diarrhea, or a skin rash. Stay home and self-isolate even if you have minor symptoms such as cough, headache, mild fever, until you recover. The paper describes various preventive methods like cleaning and disinfecting the surfaces, recording body temperature, precautions needed to be used in industries and work places.

Keywords: *Preventive methods, Human infection, COVID-19, Industries, Work place, Sanitization.*

22.1 Introduction

Influenza is a highly infectious viral illness. The name "influenza" originated in 15th century Italy, from an epidemic attributed to "influence of the stars." The first pandemic, or worldwide epidemic, that clearly fits the description of influenza was in 1580. At least four pandemics of influenza occurred in the 19th century, and three occurred in the 20th century. The pandemic of "Spanish" influenza in 1918–1919 caused an estimated 21 million deaths worldwide. The first pandemic of the 21st century occurred in 2009–2010. Smith, Andrewes, and Laidlaw isolated influenza A virus in Ferrets in 1933, and Francis isolated influenza B virus in 1936. In 1936, Burnet discovered that influenza virus could be grown in embryonated hens' eggs. This led to the study of the characteristics of the virus and the development of inactivated vaccines. The protective efficacy of these inactivated vaccines was determined in the 1950s. The first live attenuated influenza vaccine was licensed in 2003.

22.1.1 Influenza Virus: Structure

Influenza is a single-stranded, helically shaped, RNA virus (Figure 22.1) of the orthomyxovirus family. Basic antigen types A, B, and C are determined by the nuclear material. Type A influenza has subtypes that are determined by the surface antigens hemagglutinin (H) and neuraminidase (N). Three types of hemagglutinin in humans (H1, H2, and H3) have a role in virus attachment to cells. Two types of neuraminidase (N1 and N2) have a role in virus penetration into cells. The nomenclature to describe the type of influenza virus is expressed in this order: 1) virus type, 2) geographic origin where it was first isolated, 3) strain number, 4) year of isolation, and 5) virus subtype.

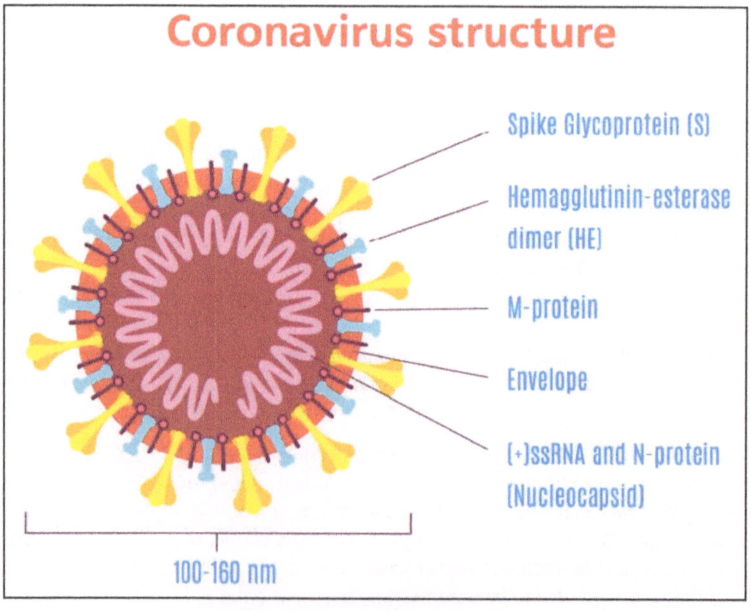

Figure 22.1: Corona Virus
(*Source*: Tok and Tatar, 2017).

Influenza A causes moderate to severe illness and affects all age groups. The virus infects humans and other animals. Influenza A viruses are perpetuated in nature by wild birds, predominantly waterfowl. Most of these viruses are not pathogenic to their natural hosts and do not change or evolve. Influenza B generally causes milder disease than type A and primarily affects children. Influenza B is more stable than influenza A, with less antigenic drift and consequent immunologic stability. It affects only humans. Influenza C is rarely reported as a cause of human illness, probably because most cases are subclinical. It has not been associated with epidemic disease.

22.1.2 Influenza Virus: Pandemic

The 1918 influenza pandemic was the most severe pandemic in recent history. It was caused by an H1N1 virus with genes of avian origin. Although there is not universal consensus regarding where the virus originated, it spread worldwide during 1918-1919. In the United States, it was first identified in military personnel in spring 1918. It is estimated that about 500 million people or one-third of the world's population became infected with this virus. The number of deaths was estimated to be at least 50 million worldwide with about 675,000 occurring in the United States. Mortality was high in people younger than 5 years old, 20-40 years old, and 65 years and older. The high mortality in healthy people, including those in the 20-40 year age group, was a unique feature of this pandemic. At that time there was no vaccine to protect against influenza infection and no tested antibiotics were known to treat secondary bacterial infections that can be associated with influenza infections.

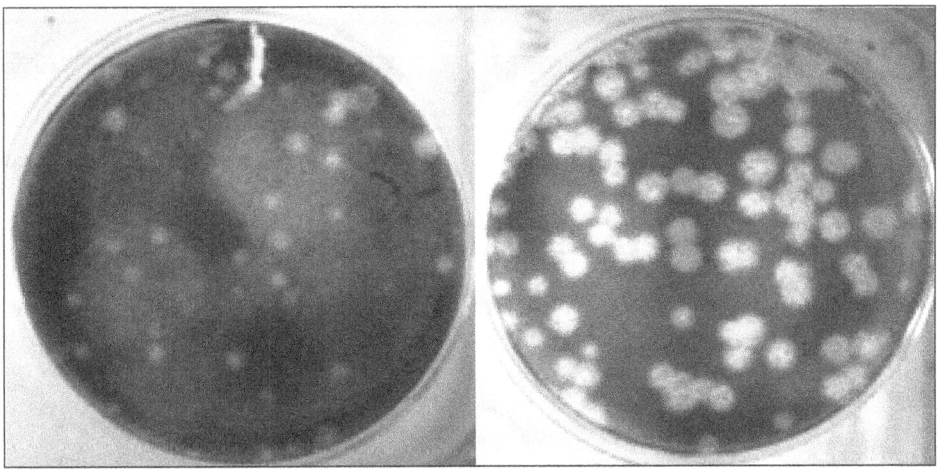

Figure 22.2: The Picture Shows Replication of a Human Seasonal Flu Virus called Tx/91 in Cell Culture. The picture on the right shows how when the polymerase (PB1) gene of this same virus is exchanged with that of the 1918 virus, the resulting virus' ability to replicate (*i.e.*, make copies of itself) is greatly enhanced. Photo credit: Terrence Tumpey, CDC.

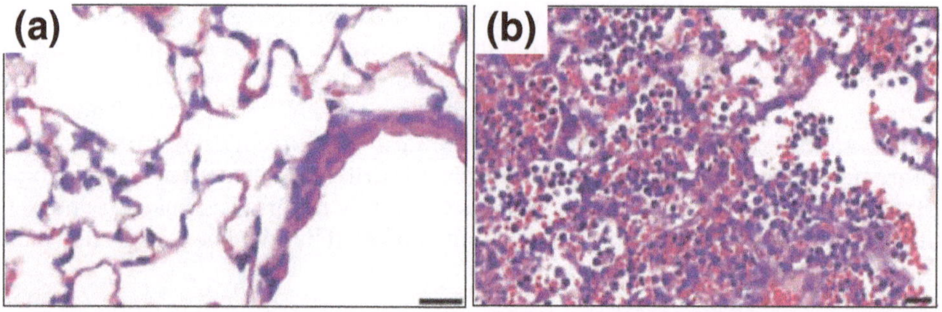

Figure 22.3: The 1918 Virus was Extremely Virulent.

Image (a) shows mouse lung tissue infected with a human seasonal H1N1 flu virus. Image (b) shows the impact of the 1918 virus in mouse lung tissue. The 1918 virus replicates quickly and causes severe disease in the lung tissues of mice. In 1918, the virus caused severe disease in the lungs of people infected, as well. Photo credit: CDC, *Science*.

Table 22.1: Occurrence of Major Influenza Outbreaks

Year	*Sub Type*	*Extent of Outbreak*
1889-1890	H2N8	Severe Pandemic
1900- 1903	H3N8	Moderate epidemic
1918-1919	H1N1 (HswN1) Spanish flu	Severe Pandemic
1933-1935	H1N1 (H0N1)	Mild epidemic
1946-1947	H1N1	Mild epidemic
1957-1958	H2N2 (Asian Flu)	Severe Pandemic
1068-1969	N3N2 (Hong Kong flu)	Mild epidemic
1977-1978	H1N1 (Russian flu)	Mild epidemic
2009-2010	H1N1	Pandemic

Source: Sastryet *et al.*, 2016.

Dr. Tumpey's work to reconstruct the complete 1918 virus began in the summer of 2005. To reduce risk to colleagues and the public, he was required to work on the virus alone and only after hours when fellow colleagues had exited the laboratories for the day and gone home. A biometric fingerprint scan was required for access into the BSL-3E laboratory, and the virus storage freezers were only accessible via an iris scan of his eyes. Dr. Tumpey was required to take a prescribed prophylactic (preventative) daily dose of the influenza antiviral drug, oseltamivir, as an additional safety precaution to prevent him from becoming infected. Using reverse genetics, Dr. Tumpey took the plasmids created by Dr. Palese for each of the 1918 virus' eight gene segments and inserted them into human kidney cells. The plasmids then instructed the cells to reconstruct the RNA of the complete 1918 virus. For multiple weeks in July 2005, colleagues and collaborators asked Dr. Tumpey if he had the 1918 virus and if it had appeared in cell-culture yet.

On the day the 1918 virus appeared in his cell-culture, Dr. Tumpey knew history had been made, and in fact, a historic virus had been brought back from extinction. He sent a playful, Neil Armstrong-inspired email later that day to colleagues and collaborators, which simply said "That's one small step for man, one giant leap for mankind." Everyone then knew what had been accomplished. He had become the first man to reconstruct the complete 1918 virus. The next step was to study it and unlock its deadly secrets. Laboratory studies on the reconstructed 1918 virus began in August 2005. A report of this work, "Characterization of the Reconstructed 1918 Spanish Influenza Pandemic Virus", was published in the October 7, 2005, issue of Science. To evaluate the 1918 virus' pathogenicity (*i.e.*, the ability of the virus to cause disease and harm a host), animal studies involving mice were conducted. The mice were infected with the 1918 virus, and measures of morbidity (*i.e.*, weight loss, virus replication, and 50 per cent lethal dose titers) were collected and documented. The fully reconstructed 1918 virus was striking in terms of its ability to quickly replicate, *i.e.*, make copies of itself and spread infection in the lungs of infected mice. For example, four days after infection, the amount of 1918 virus found in the lung tissue of infected mice was 39,000 times higher than that produced by one of the comparison recombinant flu viruses.

Furthermore, the virus was highly lethal in the mice. Some mice died within three days of infection with the 1918 virus, and the mice lost up to 13 per cent of their body weight within two days of infection. The 1918 virus was at least 100 times more lethal than one of the other recombinant viruses tested (Tumpey *et al.*, 2005). Experiments indicated that 1918 virus' HA gene played a large role in its severity. When the HA gene of the 1918 virus was swapped with that of a contemporary human seasonal influenza A (H1N1) flu virus known as "A/Texas/36/91" or Tx/91 for short, and combined with the remaining seven genes of the 1918 virus, the resulting recombinant virus notably did not kill infected mice and did not result in significant weight loss (Tumpey *et al.*, 2005). Within four days post infection, mice infected with the 1918 virus experienced similar lung complications, suggesting that this was a unique aspect of the 1918 virus' severity.

The impact of the 1918 virus on lung tissue was also studied using a human lung cell line known as Calu-3 cells. The amount of 1918 flu virus was measured in the cells at 12, 16 and 24 hours post infection and these results were compared to those produced by recombinant viruses with a combination of genes from the 1918 virus mixed with genes from contemporary human seasonal flu viruses. Real time reverse transcriptase PCR can detect and quantify the specific HA and NA genes.

For reference, there are four bio-safety levels that correspond to the degree of risk posed by research, with 1 posing the least risk and 4 posing the greatest risk. Work on human viruses is done in safety 3 labs. Each bio-security level also corresponds with specific laboratory practices and techniques, personnel training requirements, laboratory equipment, and laboratory facilities that are appropriate for the operations being performed. Each bio-security level includes considerations for what is known as "primary" and "secondary" barriers. Examples of primary barriers include safety cabinets, isolation chambers, gloves and gowns, whereas secondary barriers include considerations such as the construction of the facility and

HEPA filtration of air in the laboratory. The specific criteria for each biosafety level are detailed in the CDC/NIH publication. A BSL3 laboratory with enhancements includes a number of primary and secondary barriers and other considerations. For example, all personnel must wear a powered air purifying respirator (PAPR), double gloves, scrubs, shoe covers and a surgical gown. They also must shower before exiting the laboratory. In addition, all work with the virus or animals must be conducted within a certified Class II biosafety cabinet (BSC), and airflow within the laboratory is directionally controlled and filtered so that it cannot accidentally exit the laboratory.

22.1.3 Vaccinomics: The Development of Preventive Vaccines against Viruses

Vaccinomics is based on the use of cutting edge, high-dimensional (so called "omics") assays and novel bioinformatics approaches to the development of next-generation vaccines and the expansion of our capabilities in individualized medicine. Vaccinomics will allow us to move beyond the empiric "isolate, inactivate, and inject" approach characterizing past vaccine development efforts, and toward a more detailed molecular and systemic understanding of the carefully choreographed series of biological processes involved in developing viral vaccine-induced "immunity." This enhanced understanding will then be applied to overcome the obstacles to the creation of effective vaccines to protect against pathogens, particularly hypervariable viruses, with the greatest current impact on public health. The development of vaccines, which prime the immune system to respond to future infections, has led to global declines in morbidity and mortality from dreadful infectious communicable diseases. However, many pathogens of public health importance are highly complex and/or rapidly evolving, posing unique challenges to vaccine development. Several of these challenges include an incomplete understanding of how immunity develops, host and pathogen genetic variability, and an increased societal skepticism regarding vaccine safety.

As the world waits for vaccine for COVID-19 Maharaja Sayajirao invited Russian bacteriologist Waldemar Haffkine to conduct his vaccine inoculations for bubonic plague in 1897. After testing on rats, rabbits and finally on himself, he got good results. Freedom fighter and chief justice of Baroda state Shri Tyabji took his daughter (Figure 22.4) Sharifa to inspire the local community.

22.2 Human Health in Ancient Indian Texts

It is believed that the ancient *rishis* or seers of India received the gift of Ayurveda from their Hindu gods about 5,000 years ago. Essential information for how to achieve a balanced and healthy life was recorded in their sacred texts, the *Veda*s, specifically the *Atharva Veda*.

It is said that the Hindu god Brahma, one of the chief triumvirate gods of Hinduism, created Ayurveda. He then transmitted this knowledge to his son, Daksha Prajapati. Daksha passed it down to the twin Vedic gods Ashwini Kumaras. The twin gods became the physicians of the gods, and the Devas of Ayurveda. The twin gods presented Ayurveda to Indra, the king of gods. Indra had three physicians as

Figure 22.4: Haffkine with his Arm around Young Sharifa, and Abbas Taybji seen on his Right. (Picture courtsey: Maharani Radhikaraje Gaekwad, Vadodara).

his disciples, namely Acharya Bharadwaj, Acharya Kashyapa and Aacharya Divodas Dhanvantari. From Bharadwaj's teaching, his student Agnivesha developed the fundamental Ayurvedic text of internal medicine. Agnivesha's disciple, Acharya Charak then revised this body of work. This started the tradition of passing down the knowledge of Ayurveda from gods to sages. The *Mahabharata*, India's epic narrative, also tells of the incarnation of Vishnu in the being of Dhanvantari. During the great cosmic churning of the ocean for the celestial nectar of immortality, Dhanvantari emerged, and Vishnu commissioned him to help humanity cure diseases. One of the most prominent *Rishis* was Bharadwaja who lived around the time of 700 BC.

The Great Three Classics of Ayurveda consist of

Charaka Samhita,

Sushruta Samhita, and

Ashtanga Hridayam Sangraha.

The *Charaka Samhita* is believed to have thrived between the second century BCE and the second century CE. The original texts of this book were thought to be written by Agnivesha. He was one of the disciples of Punarvasu Atreya, an Ayurvedic scholar. Agnivesha and his co-disciples created the *Samhitas,* drawing from the knowledge they received from Atreya and adding their comprehension on the subject. Agnivesha's *Samhita* was of particular interest because of its unique and detailed content. Charaka later annotated Agnivesha's work, and focused more on the diagnosis of a disease and channeled Ayurveda as a means of preventing and curing illnesses. Charaka also detailed the medicinal value and qualities of over 10,000 herbal plants.

Sushruta Samhita explains the concept and practice of surgery in Ayurveda. Modern scholars and researchers suggest that the *Sushruta Samhita* was created approximately in the middle of the first millennium BCE. It is believed to be authored by Sushruta, one of the students of Divodasa. *Sushruta Samhita* is composed of 184 chapters and presents 1,120 health conditions, 300 types of operations that require 42 different surgical procedures, 121 various kinds of instruments, and 650 kinds of medicine derived from animals, plants, and minerals.

Vagbhata *wrote the Ashtanga Hridayam Sangraha* after some years of the *Charaka* and *Sushruta Samhita*. It mainly focuses on *Kayachikitsa*, the branch of Ayurveda that specializes in internal medicine. It was in this body of knowledge that the *dosha* and their subparts were presented in detail.

The Lesser Three Classics of Ayurveda consist of the *Sharngadhara Samhita*, *Bhava Prakasha*, and *Madhava Nidanam*. *Sharngadhara Samhita* was written by Sharngadhara and is valued for how it specified and explained pharmacological formulations utilized in *Panchakarma*. It is also in this book that the diagnosis of a person's health status is made through their pulse. *Bhava Prakasha* was created around the 16th century, which is one of the later Ayurvedic texts. It also deals with *Kayachikitsa*, and explains the qualities of various food, plants, and animals in respect of their medicinal and health benefits. *Madhava Nidanam* emerged around 700 CE and is valued for discussing diseases that involve women and children, toxicology, and conditions of the throat, nose, and ears.

22.3 Spanish Flu Pandemic

The 1918 H1N1 flu pandemic, sometimes referred to as the "Spanish flu," killed an estimated 50 million people worldwide, including an estimated 675,000 people in the United States (Johnson and Mueller 2002; Jester *et al.*, 2018, 2019,). An unusual characteristic of this virus was the high death rate it caused among healthy adults 15 to 34 years of age. A comparable death rate has not been observed during any of the known flu seasons or pandemics that have occurred either prior to or following the 1918 pandemic (Anonymous, CDC 2000). Over crowd was also responsible for spread of disease (Figure 22.6).

Why Pigs are the most Common Mixing Vessels ?

Both A 2-3 and A2-6 sialic acid receptors are found on the same respiratory epithelial cells of pigs and swine flu strains have specificity for both the receptor types.

Hence pigs can be infected simultaneously by human, swine and avian strains, thus serving as a mixing vessel.

Reassortment between the segments of various strains can take place inside the same swine cells.

Avian Flu

Birds are the primary reservoir of influenza viruses

All influenza subtypes (16H types and 9N type) are found in birds and some of the subtypes can be transmitted to mammals (*i.e.* H1 H2 N3) or in humans.

Figure 22.5: This 1951 Photo Shows Johan Hultin (on left) and Fellow University Colleagues during his Initial Attempt to obtain the 1918 Virus from Bodies of Victims Buried in Permafrost at the Brevig Mission Burial Site (Photo credit: Johan Hultin).

Figure 22.6: Crowded Conditions and the Movement of Troops during World War I likely Contributed to the Spread of the 1918 Virus around the World. (*Source*: www.museumsyndicate.com/item.php?item=56784#).

Usually the avian flu strains are highly infectious with PBIF2 proteins. These help in RNA Transcription and replication. Mitochondrion induces apoptosis.

22.4 Influenza Pathogenesis: Clinical Features

- ✰ Respiratory transmission of virus
- ✰ Replication in respiratory epithelium with subsequent destruction of cells

- Viremia rarely documented
- Virus shed in respiratory secretions for 5-10 days

Influenza **Incubation period 2 days (range 1-4 days)**

- 50 per cent of infected persons develop classic symptoms
- Abrupt onset of fever, myalgia, sore throat, nonproductive cough, headache

Following respiratory transmission, the virus attaches to and penetrates respiratory epithelial cells in the trachea and bronchi. Viral replication occurs, which results in the destruction of the host cell. Viremia has rarely been documented. Virus is shed in respiratory secretions for 5–10 days. The incubation period for influenza is usually 2 days, but can vary from 1 to 4 days. Influenza illness can vary from asymptomatic infection to severe. In general, only about 50 per cent of infected persons will develop the classic clinical symptoms of influenza.

"Classic" influenza disease is characterized by the abrupt onset of fever, myalgia, sore throat, nonproductive cough, and headache. The fever is usually 101°–102°F, and accompanied by prostration (bedridden). The onset of fever is often so abrupt that the exact hour is recalled by the patient. Myalgias mainly affect the back muscles. Cough is believed to be a result of tracheal epithelial destruction. Additional symptoms may include rhinorrhea (running nose), headache, substernal chest burning and ocular symptoms (*e.g.*, eye pain and sensitivity to light).

Systemic symptoms and fever usually last from 2 to 3 days, rarely more than 5 days. They may be decreased by such medications as aspirin or acetaminophen. Aspirin should not be used for infants, children, or teenagers because they may be at risk for contracting Reye syndrome following an influenza infection. Recovery is usually rapid, but some patients may have lingering asthenia (lack of strength or energy) for several weeks.

22.5 Influenza Complications

- Pneumonia
  - Secondary bacterial
  - Primary influenza viral
- Reye syndrome
- Myocarditis (inflammation of the heart)
- Death is reported than less than 1 per 1,000 cases

The most frequent complication of influenza is pneumonia, most commonly secondary bacterial pneumonia (*e.g.*, *Streptococcus pneumoniae*, *Haemophilus influenzae*, or *Staphylococcus aureus*). Primary influenza viral pneumonia is an uncommon complication with a high fatality rate. Reye syndrome is a complication that occurs almost exclusively in children taking aspirin, primarily in association with influenza B (or varicella zoster), and presents with severe vomiting and confusion, which may progress to coma due to swelling of the brain. Other complications include myocarditis and worsening of chronic bronchitis and other chronic pulmonary diseases. The majority of deaths typically occur among persons 65 years of age and older.

22.5.1 Impact of Influenza – United States, 1976-2007

- The number of influenza-associated deaths varies substantially by year, influenza virus type and subtype, and age group
- Annual influenza-associated deaths ranged from 3,349 (1985-86 season) to 48,614 (2003-04 season), with an average of 23,607 annual deaths
- Persons 65 years of age and older account for approximately 90 per cent of deaths
- 2.7 times more deaths occurred during seasons when A (H3N2) viruses were prominent

An increase in mortality typically accompanies an influenza epidemic. Increased mortality results not only from influenza and pneumonia but also from cardiopulmonary and other chronic diseases that can be exacerbated by influenza.

Among children 0–4 years of age, hospitalization rates have varied from 100 per 100,000 healthy children to as high as 500 per 100,000 for children with underlying medical conditions. Hospitalization rates for children 24 months of age and younger are comparable to rates for persons 65 and older. Children 24-59 months of age are at less risk of hospitalization from influenza than are younger children, but are at increased risk for influenza-associated clinic and emergency department visits. Healthy children 5 through 18 years of age are not at increased risk of complications of influenza. However, children typically have the highest attack rates during community outbreaks of influenza.

22.5.2 Laboratory Diagnosis

- Clinical and epidemiological characteristics
- Isolation of influenza virus from clinical specimen (*e.g.*, throat, nasopharynx, sputum)
- Significant rise in influenza IgG by serologic assay

The diagnosis of influenza is usually suspected on the basis of characteristic clinical findings, particularly if influenza has been reported in the community.

Virus can be isolated from throat and nasopharyngeal swabs obtained within 3 days of onset of illness. Culture is performed by inoculation of the amniotic or allantoic sac of chick embryos or certain cell cultures that support viral replication. A minimum of 48 hours is required to demonstrate virus, and 1 to 2 additional days to identify the virus type. As a result, culture is helpful in defining the etiology of local epidemics, but not in individual case management.

Serologic confirmation of influenza requires demonstration of a significant rise in influenza IgG. The acute-phase specimen should be taken less than 5 days from onset, and a convalescent specimen taken 10–21 days (preferably 21 days) following onset. Complement fixation (CF) and hemagglutination inhibition (HI) are the serologic tests most commonly used. The key test is HI, which depends on the ability of the virus to agglutinate erythrocytes and inhibition of this process by specific antibody. Diagnosis requires at least a fourfold rise in antibody titer. Rapid diagnostic testing for influenza antigen is available, but because these tests fail to detect many patients with influenza, CDC recommends antiviral treatment with oseltamivir or zanamivir as early as possible for patients with confirmed or suspected influenza who have severe, complicated, or progressive illness; who require hospitalization; or who are at greater risk for serious influenza-related complications.

22.5.3 Influenza Epidemiology

- ☆ Reservoir
  - ❖ Human, animals (type A only)
- ☆ Transmission
  - ❖ Respiratory
  - ❖ Probably airborne
- ☆ Temporal pattern
  - ❖ Peak December – March in temperate climate
  - ❖ May occur earlier or later
- ☆ Communicability
  - ❖ 1 day before to 5 days after onset (adults)

Influenza occurs throughout the world. Humans are the only known reservoir of influenza types B and C. Influenza A viruses may infect both humans and animals. There is no chronic carrier state.

Transmission

Influenza is primarily transmitted from person to person via large virus-laden droplets (particles more than 10 microns in diameter) that are generated when infected persons cough or sneeze. These large droplets can then settle on the mucosal surfaces of the upper respiratory tracts of susceptible persons who are near to infected persons. Transmission may also occur through direct contact or indirect contact with respiratory secretions such as when touching surfaces contaminated with influenza virus and then touching the eyes, nose or mouth.

Temporal Pattern

Influenza activity peaks from December to March in temperate climates, but may earlier or later. During 1982–2012, peak influenza activity in the United States occurred most frequently in January (17 per cent of seasons), and February (47 per cent of seasons). However, peak influenza activity occurred in March, April, or May in 17 per cent of seasons. Influenza occurs throughout the year in tropical areas.

Communicability

Adults can transmit influenza from the day before symptom onset to approximately 5 days after symptoms begin. Children can transmit influenza to others for 10 or more days.

22.6 Novel Corona Virus COVID-19

The current COVID-19 pandemic is unprecedented, but the global response draws on the lessons learned from other disease outbreaks over the past several decades. As part of WHO's response, the R and D Blueprint was activated to accelerate diagnostics, vaccines and therapeutics for this novel corona virus. The blueprint aims to improve coordination between scientists and global health professionals, accelerate the research and development process, and develop new norms and standards to learn from and improve upon the global response.

On 30 January 2020, following the recommendations of the Emergency Committee, the WHO Director-General declared that the outbreak constitutes a Public Health Emergency of International Concern (PHEIC). Building on the response to recent outbreaks of Ebola virus disease, SARS-CoV and MERS-CoV, the R and D Blueprint has facilitated a coordinated and accelerated response to COVID-19, including an unprecedented program to develop a vaccine, research into potential pharmaceutical treatments and strengthened channels for information sharing between countries. Social distancing was promoted.

Total confirmed cases on 15th October, 2020 was 38,394 169 with 1089047 deaths in 235 countries. The downward trend in India's corona virus cases continued on Monday, as the country recorded 55,342 fresh COVID-19 cases, taking its tally past the 71.7-lakh mark; 706 deaths in 24 hours, 1,09,856 total deaths. According to WHO for COVID-19 quarantine might sound like a frightening situation, but it is an effective public health measure.

22.6.1 The Top 5 Updates on Corona Virus Cases (as on 15/10/2020)

1. Almost 62.2 lakhs people have been cured from COVID-19 till now, government data shows, pushing up India's recovery rate to 86.8 per cent.
2. The Covid death rate in India, which has been reporting between 800 and 900 deaths a day since early August, is 1.5 per cent.
3. India's daily positivity rate stood at 5.2 per cent as the government tested 10.7 lakh people on Monday. A total of 8.8 crore tests have been done till now. With an average of about 9 lakh tests being conducted per week for the past month, India's average weekly positivity rate has fallen from 8.4 per cent to 6.3 per cent.
4. Though Maharashtra, Karnataka, Kerala and Tamil Nadu continued to dominate the list of states with maximum daily cases, their daily caseload has reduced significantly with the exception of West Bengal (3,583). The eastern state has been reporting it's highest ever 24-hour percentage rise in cases for the past fortnight. Together these states account for 52.5 per cent of all cases in the country.

5. Maximum deaths were reported from Maharashtra, Karnataka, Tamil Nadu, West Bengal and Uttar Pradesh, which together accounted for almost 57 per cent of all fatalities.

22.7. Preventive Methods for Covid-19

If COVID-19 is spreading in your community, industry or work place, stay safe by taking some simple precautions, such as physical distancing, wearing a mask, keeping rooms well ventilated, avoiding crowds, cleaning your hands, and coughing into a bent elbow or tissue. **Maintain at least a 1-metre distance between yourself and others** to reduce your risk of infection when they cough, sneeze or speak. Maintain an even greater distance between yourself and others when indoors. The farther away, the better.

- ☆ Make wearing a mask a normal part of being around other people. Isolation for 14 days was suggested as incubation and spread of virus may be up to 14 days.
- ☆ **Prof. Tom Jefferson** is a physician, researcher and campaigner for access to randomized controlled trial data at the Cochrane Institute in Rome. The review lead by him deduced that respiratory virus spread can be reduced by hand washing. The study found no evidence that the more expensive, irritating and uncomfortable N95 respirators were superior to simple surgical masks. It also reported that there is insufficient evidence to support screening at entry ports and social distancing (spatial separation of at least one meter between those infected and those non infected) as a method to reduce spread during epidemics. The Cochrane Collaboration is a global independent network of researchers, health professionals, patients, careers, and other people interested in health. There is evidence from a trial carried out in India in 2005 that washing hands and personal hygiene works for all respiratory illnesses. This trial was carried out in poor areas and it was found that it protected pregnant women and children. This is the most important and basic point. Clean water, sanitation and personal hygiene. If there is no water, use sand or anything that causes attrition and cleanses the hand.

22.7.1 Laxity in Use of Sanitizers and Mask may be Fatal

According to WHO guidelines the use of physical distancing, wearing mask on road and work place and proper sanitizers is recommended. It is surprising that most of the sanitizers do not have 70 per cent ethyl alcohol. A survey undertaken in Vadodara revealed that in different commercial disinfectants 62 to 95 per cent ethanol was being used. Not only Ethyl alcohol but isopropyle alcohol, *Aloe vera*, neem extract, glycerin and many such products are added to market their product by saying Ayurvedic, ecofriendly and organic.

The best way is to wash the hands is with any soap for 20 sec. The COVID virus has wall made up of chemical called lipid, and just like any soap helps to remove the oil from cloth the use of soap makes hole in virus outer cover and virus is unable to cause a disease. The wearing of gloves all the time is not preferred as within a short time we will not only touch different objects but our body parts as well. An experiment about human behavior was recorded in an office besides having the habit of shake hand, 10 out of 27 persons touched their eyes in one hour and 12 out of 27 touched their nose. So it is very essential that we wash our hands frequently. If we are out use of sanitizers or hand rub is suggested.

Table 22.2: Few Sanitizers Commonly in Use in Vadodara and Lucknow in July 2020

Sl.No.	*Sanitizer*	*Composition*	*Result*
1.	Lifebuoy Immunity boosting hand sanitizer	95 per cent Ethyl alcohol, Isopropyle alcohol, Niacinamide, perfumed gel base	Effective
2.	Sterigard By Ezygard, Vadodara	Isopropyle alcohol, N- propanol, Mecetronium ethyl sulphate	""
3.	Color drops Arnish Chamical Industries, Ahmedabad	Isopropyle alcohol, Glycerine Chlorhexidine, perfume, gelbase	"
4.	Ethyster Sterlife Ahmedabad	70 per cent Ethyl alcohol I.p. (Denat)Purified water	Very Good
5.	XX (local brand) in use at Env. Studies dept. MSU	Ethyl alcohol, Neem oil, *Aloe vera* gel base	Not good
6.	XY(local brand) in use at a shop in Indira nagar, Lucknow	75 per cent Ethanol 25 per cent water	Not good

22.7.2 Chemicals as Disinfectants

Only Ethanol with 70 per cent conc. is advocated as addition of 30 per cent water will provide more reactive time for microorganisms to be killed, hence it is better than 95 or 60 per cent alcohol. Less than 70 is also not good. If things are to be disinfected like phone, mobile, gate handle, medical implements *etc.* Isopropyle alcohol is more effective than ethyl alcohol. But use of isopropyle alcohol makes the skin more dry hence not preferred.

Ethanol

Fermentation yields a solution that is only about 12–15 percent alcohol because higher concentrations are toxic to the yeast cells. This solution can be distilled, however, to raise the ethanol content to as high as 95 percent. Fermentation is a relatively expensive method of making ethanol. Industrial ethanol is more commonly synthesized by the high-temperature catalytic addition of water to ethylene (C_2H_4).

Isopropyl Alcohol

2-propanol is made by indirect hydration of propylene (CH_2CHCH_3). Isopropyl alcohol is commonly used as an industrial solvent and as a rubbing alcohol applied to the skin. Although isopropyl alcohol is more toxic than ethanol, it has less of a drying effect on the skin. Similarly fumes of methanol are harmful and not recommended for use.

Alcohol reacts with microbes and denatures them by dehydration. *Alo vera* has an opposite effect and hence will prolong drying. Once alcohol is capable enough use of neem and other things not needed. By adding small vitamin B3 (Niacinamide) it is an immunity enhancer is misleading.

22.7.3 Why a Mask is Needed ?

It is also important that humidity helps to survive Viruses for longer time. Since the conc. of viruses may be more on door handle, mobile *etc.* regular cleaning is suggested. Sneezing by an infected person can produce 10 μm droplets which travel easily to 3 meter at a velocity of 20 meter per second. One ml of the droplets may have 100 lakhs of viruses.

Chances of infection are prevented by wearing a mask. The N95 mask is certainly better than cotton or surgical one but it should be used once only. Hence clean/washed/cotton mask is suggested for everyone. For many years, scientists weren't sure whether wearing a mask was effective at preventing the spread of viruses. However, recent studies suggest they can help. One looked at how masks could help people with the seasonal flu limit spreading it when they exhale small droplets containing the virus. Overall, researchers found masks led to a more than threefold reduction in how much virus people sprayed into the air. Analyzing data from thousands of Japanese schoolchildren, found that "vaccination and wearing a mask reduced the likelihood of developing seasonal influenza."

If you're considering wearing a mask to protect yourself against infections, there are three types you should know about.

1. Surgical face masks are fairly loose-fitting, disposable masks approved by the Food and Drug Administration for use as medical devices. Doctors, dentists, and nurses often wear them while treating patients.

   These masks prevent large droplets of bodily fluids that may contain viruses or other germs from escaping via the nose and mouth. They also protect against splashes and sprays from other people, such as those from sneezes and coughs. But the average masks you can buy from a local drugstore aren't enough to filter out viruses.

2. Respirators, also called N95 respirator masks, are designed to protect the wearer from small particles in the air, like viruses. They're certified by the Centre for Disease Control and prevention, USA and National Institute for Occupational Safety and Health.

   The name comes from the fact that they can filter 95 per cent of airborne particles, according to the CDC. N95 masks are also often used when painting or handling potentially toxic materials.

   Respirators are selected to fit your face. They must form a perfect seal so no gaps allow airborne viruses in. Healthcare workers use them to protect against airborne infectious diseases, such as tuberculosis and anthrax.

Unlike regular face masks, respirators protect against both large and small particles. Overall, respirators are considered much more effective at preventing the flu virus than regular face masks.

3. **Cloth face coverings or masks** can be used in public settings, such as grocery stores, where you may be in close contact with others and it's difficult to maintain your distance. They don't offer the same level of protection as surgical face masks or respirators. However, when worn by the public at large, they can still help reduce the community spread of viruses. This is because they help prevent people without symptoms from transmitting viruses via their respiratory droplets.

While talking we are not supposed to remove mask. When removing the cloth face mask, try to avoid touching your nose, mouth, and eyes. Cloth face masks shouldn't be used by children under 2 years old, people who have trouble breathing, and people who are unable to remove their own masks.

- ✰ You could spread COVID-19 to others even if you do not feel sick. The mask is meant to protect other people in case you are infected.
- ✰ Continue to keep about two meters distance between yourself and others. The mask is not a substitute for social distancing.

22.7.4 How to Wear a Mask ?

- ✰ Clean your hands before you put your mask on, as well as before and after you take it off.
- ✰ Make sure it covers both your nose, mouth and chin.

Here are some specifics on what type of mask to wear and when, depending on how much virus is circulating where you live, where you go and who you are?

- ✰ Wear a fabric mask unless you're in a particular risk group. This is especially important when you can't stay physically distanced, particularly in crowded and poorly ventilated indoor settings.
- ✰ Wear a surgical mask if you: Are over 60, have underlying medical conditions,
- ✰ Are feeling unwell, and/or looking after an ill family member.
- ✰ For health workers, medical masks are essential personal protective equipment when engaging with patients with suspected, probable or confirmed COVID-19. Respirator masks (such as FFP2, FFP3, N95, N99) should be used in settings where procedures generating aerosols are performed and must be fitted to ensure the right size is worn.
- ✰ Surgical mask has three layers. The innermost layer is an absorbent layer- which absorbs moisture from the wearers breath. Middle layer is a melt-blown material that acts as a filter. The outer layer is made up of material that repels liquid. Pleats increase surface area so that nose and chin can be covered adequately.

22.8 How to Make the Environment Safer?

- **Avoid the 3Cs: spaces that are closed, crowded or involve close contact.**
  - Outbreaks have been reported in restaurants, choir practices, fitness classes, nightclubs, offices and places of worship where people have gathered, often in crowded indoor settings where they talk loudly, shout, breathe heavily or sing.
  - The risks of getting COVID-19 are higher in crowded and inadequately ventilated spaces where infected people spend long periods of time together in close proximity. These environments are where the virus appears to spreads by respiratory droplets or aerosols more efficiently, so taking precautions is even more important.
- **Meet people outside.** Outdoor gatherings are safer than indoor ones, particularly if indoor spaces are small and without outdoor air coming in. **Avoid crowded or indoor settings.**
- **Open a window.** *Increase the amount of* 'natural ventilation' when indoors.

22.8.1 Follow the Basics of Good Hygiene

- **Regularly and thoroughly clean your hands with an alcohol-based(70 per cent) hand rub or wash them with soap and water for a min. 20 sec.** This eliminates germs including viruses that may be on your hands.
- **Avoid touching your eyes, nose and mouth.** Hands touch many surfaces and can pick up viruses. Once contaminated, hands can transfer the virus to your eyes, nose or mouth. From there, the virus can enter your body and infect you.
- **Cover your mouth and nose with your bent elbow or tissue when you cough or sneeze**. Then dispose of the used tissue immediately into a closed bin and wash your hands. By following good 'respiratory hygiene', you protect the people around you from viruses, which cause colds, flu and COVID-19.
- **Clean and disinfect surfaces frequently especially those which are regularly touched,** such as door handles, faucets and phone screens.

22.8.2 Know How it Spreads ?

- There is currently no vaccine to prevent corona virus disease (COVID-19).
- The best way to prevent illness is to avoid being exposed to this virus.
- The virus is thought to spread mainly from person-to-person.
  - Between people who are in close contact with one another (within about 6 feet).
  - Through respiratory droplets produced when an infected person coughs, sneezes or talks.
  - These droplets can land in the mouths or noses of people who are nearby or possibly be inhaled into the lungs.

- ❖ Some recent studies have suggested that COVID-19 may be spread by people who are not showing symptoms (Asymptomatic).

22.8.3 What to Do if you Feel Unwell?

- ☆ **Know the full range of symptoms of COVID-19.** The most common symptoms of COVID-19 are fever, dry cough, and tiredness. Other symptoms that are less common and may affect some patients include loss of taste or smell, aches and pains, headache, sore throat, nasal congestion, red eyes, diarrhea, or a skin rash.
- ☆ **Stay home/self-isolate even if you have minor symptoms such as cough, headache, mild fever**, until you recover.
- ☆ **If you have a fever, cough and difficulty breathing, seek medical attention immediately. Call by telephone first, if you can** and follow the directions of your local health authority.
- ☆ **Keep up to date on the latest information from trusted sources, such as WHO or your local and national health authorities.** Local and national authorities and public health units are best placed to advise on what people in your area should be doing to protect themselves.

22.8.4 Wash your Hands quite Often

- ☆ Wash your hands often with soap and water for at least 20 seconds especially after you have been in a public place or after blowing your nose, coughing, or sneezing.
- ☆ It's especially important to wash:
  - ❖ Before eating or preparing food
  - ❖ Before touching your face
  - ❖ After using the restroom/After leaving a public place
  - ❖ After handling your mask
- ☆ If soap and water are not readily available, **use a hand sanitizer that contains at least 70 per cent alcohol**. Cover all surfaces of your hands and rub them together until they feel dry.
- ☆ Avoid touching your eyes, nose, and mouth with unwashed hands.

22.8.5 Avoid Close Contact and Crowding

- ☆ **Inside your home:** Avoid close contact with people who are sick.
  - ❖ If possible, maintain two meter between the person who is sick and other household members.
- ☆ **In office or industry:** Maintain physical distance between yourself and people sitting in the room. Remember that some people without symptoms may be able to spread virus.
  - ❖ Stay at least 2 meter (about 2 arms' length) from other people.

- Keeping distance from others is especially important for people who are at higher risk of getting very sick.
- Place should be well ventilated. A.C. should not be preferred.

☆ The regular health check up and breathing exercisizes should be included in daily working schedule.

22.8.6 Cleaning and Disinfecting the Working Area

☆ Alcohol, chloroform, ether, hydrogen peroxide, sodium hypochlorite, iodine, Formaldehyde, BPL, detergent, ionizing radiations all are viricidal.

☆ Rooms/offices/work place should be clean, ventilated and avoid use of curtains and carpets. Floor should be cleaned with phenyl at least twice daily.

☆ Floor and walls up to 2.5 m should be disinfected. 1 per cent Sod. Hypochlorite is effective for 10h only. Bleaching powder 2 per cent can be used for surface disinfection.

☆ Clean and disinfect frequently touched surfaces **daily**. This includes tables, doorknobs, light switches, countertops, handles, desks, phones, keyboards, toilets, faucets, and sinks.

☆ If surfaces are dirty, clean them. Use detergent or soap and water prior to disinfection.

☆ Try to use registered disinfectants only. If disease is detected in workers the cleaning should be done properly by lock downing the whole area for at least 3-5 days.

☆ Waste disposal should be done properly in closed containers.

22.9 Conclusion

Vedas are the oldest literature, it is said that the Hindu god Brahma, created Ayurveda. The 1918 H1N1 flu pandemic, or "Spanish flu," caused a havoc worldwide. Characterization of pathogenic influenza virus was done by Tumpey *et al.* (2005). Getting a flu vaccine can save healthcare resources for the care of patients with COVID-19. **One should be alert for the symptoms.** Watch for fever, cough, shortness of breath *etc.* The paper brings first hand information about masks, frequent sanitization of hand, disinfecting working area, and healthy practices to keep the environment at work places and industries safe.

REFERENCES

Anonymous CDC (2000)Summary of Notifiable Diseases — United States, 2000. <http: //www2.cdc.gov/mmwr/summary.html

Ann Reid *et al.* (1999) Origin and evolution of the 1918 "Spanish" influenza virus hemagglutinin gene PNAS. February 1999 (96): 1651-1656.

Ann Reid *et al.*, 2004 "Novel Origin of the 1918 Pandemic Influenza Virus Nucleoprotein Gene" Journal of Virology. 78(22): 12462-12470

Christopher B. *et al.* (2001) "Sequence of the 1918 pandemic influenza virus nonstructural gene (NS) segment and characterization of recombinant viruses bearing the 1918 NS genes" PNAS. 98. (5): 1746-2751.

Dwivedi S. (2020) New Coronavirus Cases In India, Lowest In Nearly 2 Months Edited by Nandini Gupta Updated: October 13, 2020 11: 36 am IST

Jeffery K. T. Reid A.H., Lourens, R.M. Wang R., Jin G., FanningT.G. (2005) "Characterization of the 1918 influenza virus polymerase genes" Nature. 437 (6): 889-893

Jester B. *et al.* (2018) Readiness for Responding to a Severe Pandemic 100 Years After 1918. Am J Epidemiol. 2018. 187(12): 2596-2602. DOI: 10.1093/aje/kwy165.

Jester B. *et al.* (2019) Historical and clinical aspects of the 1918 H1N1 pandemic in the United States. Virology 2019. 527: 32-37.

Johnson P. and Mueller. J. (2002) Updating the accounts: global mortality of the 1918–1920 "Spanish" influenza pandemic. Bull. Hist. Med. 2002. 76(1): 105-115. external icon

Sastry AS Bhat KS and Janagond A (2016) Essentials of Medical Microbiology. Pub by Jaypee Brothers Medical Publishers(P)New Delhi pp 638

Taubenberger, K. (1997) Ann Reid, Amy Krafft *et al.*, Initial Genetic Characterization of the 1998 "Spanish' Influenza Virus. *Science*. 275, (5307): 1793-1796

Tumpey TM, Baster CF, Aguilar PV,(2005) Characterization of the Reconstructed 1918 Spanish Influenza Pandemic Virus" *Science*. 310(5745): 77-80.

Chapter 23

COVID-19: Pre and Post Lockdown Effects on Environment, Challenges and Management in Indian Context

Bansari Parikh, Garvita Parikh, and Hemen Dave*

Institute of Research and Development, National Forensic Sciences University, Gandhinagar – 3820 07

****e-mail: hemen.dave@gfsu.edu.in***

ABSTRACT

A new holocaust due to novel corona virus has threatened the whole mankind with its worldwide spread. It originated in December 2019 in Wuhan, China from bats. To restrain the spread of this virus many countries as well as India imposed a complete lockdown. In India it is imposed from 25th of March 2020 along with this travel restrictions, work from home policy, etc. As it is known, over the last two decades India has observed swift industrial growth which has assuredly upgraded the living standard of people living in India but dark side of that is pollution and deterioration of environmental quality. This lockdown was proved to be beneficial for environment; in this study observed and reported environmental repercussion of lockdown are summarized for Indian context. In Indian metro cities pollution levels have dwindled and all the organisms like animals and birds roam freely due to reduced pollution levels. Blue skies are again visible in cities like Delhi where there was so much problem of smog.

The present chapter provides quantitative assessment of environmental consequences due to lockdown along with the new-fangled challenges emerged due to Covid-19 pandemic. Pandemic showed some negative impacts such as increase in plastic waste as well as bio-medical waste generation throughout world. The onset of novel corona virus lead to increase in utilization of protective shields, food sterilization, toilet papers, face masks, hand gloves, cleaning products and hand sanitizers etc. resulting to disposal problem at end of lifecycle generating tonnes of plastic and other waste. This increment can cause pollution in oceans, poor quality of soil and harm to living being if not managed properly. According to WHO in the world about 16 million hand gloves and 89 million masks are used per day. Similarly, biomedical waste generation increased throughout the world. India is generating 775.7 tonnes of biomedical waste per day currently, where as if we compare that with origin and epicentre in China; the Wuhan city produce 240 tonnes of medical waste per day. The chapter further provides emphasis on these challenges and various approaches are suggested to minimize the negative consequences of Covid-19 in Indian context.

Keywords: *COVID-19, Lockdown, Environment, Challenges, Management, Indian context.*

23.1 Introduction

In the year 2020, world faced a new challenge due to pandemic of novel corona virus SARS-CoV-2, which is also named as COVID-19. World Health Organization (WHO) declared it as a pandemic on 11th March 2020, and subsequent to that most countries imposed strict lockdown. In India complete lockdown imposed for prevention and progression of COVID-19 progression from 25th March, 2020 for 21days which was extended till 31st May, 2020 (Showkathali *et al.*, 2020); despite that India became worst affected Asian country with continuous increase in COVID-19 cases. India reached to second position in the world with respect to recorded COVID-19 cases on 6th September 2020 and considering data for September first half India ranked number one in the world respect to COVID-19 cases and resultant deaths (Times of India, 2020), ("India has the world's second highest number of Covid cases | Deccan Herald," n.d.) ("September surge," n.d.). Today, India as well as the whole world is fighting a multi-front war against COVID-19 and consequences of lockdown, learning and try to re-emerge. The lockdown has slowed down the economy and brought several changes in environment.

23.2 COVID-19 Timeline from Global and Indian Perspectives

A breakout of an obscure illness pneumonia of undiscovered origin arose in Wuhan, China in late 2019. The virus responsible for the spread of the disease was called as SARS-CoV-2 and the communicable disease was named as COVID-19 by WHO (He *et al.*, 2020a). This infectious disease is caused by a new strain of the CoV for which treatment is not yet available (Paital *et al.*, 2020).

In Wuhan, China 27 cases of pneumonia of unknown cause were recognized which were all associated to the seafood market on 31st December 2019 (Sohrabi *et al.*, 2020). Gradually the cases increased and spread worldwide as it was a communicable disease. As the cases were increasing day by day the W H O declared a global emergency. Wuhan was the epicentre for the spread of the novel corona virus. On 13 January 2020 Thailand reported its first patient who travelled from Wuhan. In the US, the very first case was noted on 20th January 2020 (Holshue *et al.*, 2020). In mid-January, then the virus was recorded in Bangkok (Wu *et al.*, 2020). In India the very case was noted on 30th January 2020 who was a student from Wuhan (Sohrabi *et al.*, 2020). The spread of COVID-19 across the world and India and consequent actions taken are represented in Tables 23.1 and 23.2.

Table 23.1: Global Timeline for COVID-19 Pandemic and Consequent Actions

Sl.No.	Dates	Particular
1	31 December, 2019	A cluster of pneumonia cases of unknown etiology in Wuhan, China
2	7 January, 2020	A novel corona virus was identified and named temporarily as 2019-nCoV
3	20 January, 2020	First case in Pacific Northwest of Washington, USA
4	28 January, 2020	First suspected case was found in Brazil
5	31 January, 2020	First case confirmed of UK in York
6	31 January, 2020	First case in Spain
7	31 January, 2020	First case confirmed in Italy, Government suspended all flights to and from China and declared emergency
8	31 January, 2020	USA declared health emergency
9	2 February, 2020	USA put travel restrictions to and from China
10	13 February, 2020	First death was noted in Spain
11	25 February, 2020	First confirmed cased found in Brazil
12	03 March, 2020	UK unveiled their 'Corona virus Action Plan'
13	10 March. 2020	PM increased quarantine lockdown
14	11 March, 2020	WHO declared COVID-19 as pandemic
15	13 March, 2020	'State of Alarm' was declared nationwide for 15 days in Spain
16	13 March, 2020	Elections were postponed for one year in UK
17	14 March, 2020	No new cases were detected in Italy
18	15 March, 2020	Nationwide lockdown was declared in Spain
19	17 March, 2020	100,000 cases of COVID-19 were reported
20	19 March, 2020	First death of health worker in Spain
21	24 March, 2020	Almost 7000 people in Bergamo had COVID-19
22	28 March, 2020	3904 confirmed cases and 114 deaths in Brazil
23	28 March, 2020	Non-essential activities were banned
24	11 April, 2020	Highest death toll in world surpassing Italy was observed in USA
25	23 April, 2020	'State of Alarm' extended until 9 May in Spain
26	24 April, 2020	More than 50,000 cases reported in Brazil
27	28 April, 2020	Confirmed cases crossed 1 million in USA
28	03 May, 2020	More than 100,000 noted, the number of cases doubled in less than 10 days
29	04 May, 2020	Freedom to movement of citizens was re-established in Italy
30	05 May, 2020	UK death toll became highest in Europe and second in world
31	09 May, 2020	More than 10,000 Deaths reported.
32	11 May, 2020	Phase-1 of De-escalation (opening of small shops allowed)
33	18 May, 2020	Phase-2 of De-escalation
34	27 May, 2020	100,000 Americans died because of COVID-19
35	01 June, 2020	Primary schools reopened in UK
36	01 June, 2020	More than 30,000 cases reported in Brazil

Sl.No.	Dates	Particular
37	05 June, 2020	Brazilian Government shuts down official website with COVID-19 daily reports
38	15 June, 2020	Phase-3 of De-escalation
39	07 July, 2020	Brazilian PM was tested positive for COVID-19
40	08 July, 2020	The number of cases passed 3 million in the USA
41	17 July, 2020	Spain banned gathering of more than 10 people in public and private places
42	23 July, 2020	The number of cases passed 4 million in the USA
43	27 July, 2020	A pet cat tested positive for corona virus in UK
44	08 August, 2020	3 million infections and 1 lakh deaths in Brazil
45	03 September, 2020	4 million infections and 12,500 lakhs deaths in Brazil
46	04 September, 2020	10,476 cases in a day in Spain
47	22 September, 2020	USA recorded 200,000 deaths due to COVID
48	06 October, 2020	5 million Infections in Brazil
49	15 October, 2020	The governor of Campania, Italy closed school and universities until 30 October, 2020
50	16 October, 2020	The number of cases passed 8 million in the USA
51	21 October, 2020	Spain crossed 1 million cases (First country in western Europe)
52	23 October, 2020	Highest number of cases (85,085) recorded in a day in USA

Table 23.2: India Timeline for COVID-19 Pandemic and Consequent Actions

Sl.No.	Dates	Particular
1	30 January, 2020	1st case: A women medical student studying at Wuhan University - was reported in Kerala's Thrissur district
2	3 February, 2020	Two cases were reported from Alappuzha and Kasaragod dist. of Kerala and had a travel history from Wuhan China
3	4 February, 2020	India cancelled existing visas for Chinese and foreigners who had visited China in the last two weeks.
4	8 February, 2020	Screening of people returning from china started
5	27 February, 2020	759 Indians and 43 foreign brought from Wuhan.
6	2 March, 2020	Two more cases were reported in it a 45-year-old man in Delhi who had travelled back from Italy and a 24-year-old engineer in Hyderabad who had a travel history from Dubai.
7	6 March, 2020	Screening of all the international passengers started at airports.
8	11 March, 2020	50 COVID-19 cases were reported in India which include 13 states and UTs in India. Of these 16 patients have travel history from Italy and four had travel history from US. And other 22 cases were from local transmission and community transmission.
9	12 March, 2020	India reported its first death after a 76-year-old man died from Kalburgi, Karnataka.
10	13 March, 2020	Odisha declared COVID-19 a 'disaster' and ordered the immediate closure of cinema halls, educational institutions, public swimming pools and gyms.

Sl.No.	*Dates*	*Particular*
11	15 March, 2020	Maharashtra showed total confirmed cases to 100.
12	16 March, 2020	Delhi government ordered the closure of all gymnasiums, spas, night clubs and theatres until March 31.
13	22 March, 2020	A 14-hour voluntary lockdown called **'Janata Curfew'** was observed in India; India reported a total of 360 cases.
14	25 March, 2020	A nationwide lockdown was imposed till April 14 as India reported 606 case and 10 deaths.
15	28 March, 2020	India reported 1000 confirmed cases and declares that it has been entered the third stage of pandemic.
16	5 April, 2020	100 confirmed death cases
17	14 April, 2020	10,000 confirmed cases were recorded, as the nationwide lockdown was further extended till May 3.
18	22 April, 2020	20,000 confirmed case.
19	1 May, 2020	The nationwide lockdown was further extended till May 17.
20	7 May, 2020	50,000 confirmed cases were reported.
21	16 May, 2020	India saw 4,987 cases, the highest single-day spike till date; India with total 85,940 cases
22	17 May, 2020	Nationwide lockdown further extended till May 31
23	25 May, 2020	Domestic flight service resume with only 30 per cent of regular schedules.
24	29 May, 2020	Active cases 89,422 and death cases 5164
25	8 June, 2020	More then 2,50,000 COVID-19 cases and 7200 deaths; Unlock 1.0; Service resumes
27	1-31st July, 2020	Unlock 2.0 with more ease in restriction
28	9 July, 2020	Active cases 269789 and deaths 21129
29	1-31st, August	Unlock 3.0 with no night curfew and opening of yoga institutes, gyms *etc.*
30	9 August, 2020	Total active cases 628,747 and death cases 43379
31	26 August, 2020	The serum institute of India started its trial of Covishield
32	September 1-30, 2020	Unlock 4.0 Metro services to resume from September 7, 2020 in a graded manner. Social/academic/sports/entertainment/cultural/religious/political functions and other congregations will be permitted with a ceiling of 100 people. Open air theatres were allowed to reopen from September 21, 2020
33	12 September, 2020	Total active cases 958,316 and total death toll 77472
34	26 September, 2020	Total active cases 960,969 and death cases 93379
35	1-31st October, 2020	Unlock 5.0 Reopening of schools and the opening of cinema halls from October 15th onwards.
36	29 October, 2020	Total active cases 603,687 and death toll 120527
37	30 October, 2020	Total active cases 594386 and total death cases 121090

23.3 Was Lockdown a Necessity for India?? Mathematical Models

Lockdown has been continuously criticized as it took troll of economy however for COVID-19 there is no effective treatment, no vaccine and even no understanding about its epidemiological cycle and hence lockdown and social distancing was the only available option with government across the globe. India also has implemented lockdown a nationwide lockdown to reduce COVID-19 transmission; its spread across the country has been studied and analysed by variety of mathematical modelling and forecast created with respect to peak of COVID-19 cases and time period for complete vanish. Most mathematical models justify necessity of lockdown in absence of insight for treatment of COVID-19. Outcome of a mathematical modelling which predicts dynamic of COVID-19 across India indicates that COVID-19 spread reduced due to lockdown and appearance of peak has been delayed due to lockdown. Further study reported necessity of restrictive social distancing and contact tracing during and post lockdown for elimination of COVID-19 (Sarkar *et al.*, 2020). In another modelling study indicate India will have upward trajectory with respect to COVID-19 cases and flattening is expected in October 2020 with infection of more than 1.2 million people across the India with more cases in highly populated states (Kumar *et al.*, 2020). In mathematical modelling based on modified susceptible-exposed-infectious-recovered framework modified for the effect of social distancing it was found that curve for the epidemic COVID-19 will reach to peak in third week of June; for transmissibility with reproduction no. of 1.8 with social distancing 78 per cent to 97 per cent reduction in infection can be achieved compared to baseline scenario of no social distancing (Patrikar *et al.*, 2020).

In genetic programming based mathematical modelling, it was found that confirmed cases of COVID-19 rise from 80,000 to 1, 42,000 and death cases rise from2,500 to 4200 in time period of ten days, 13th May to 23rd May,2020. This model also put focus on necessity of lockdown and strict social distance to arrest the COVID-19 spread (Salgotra *et al.*, 2020). Peak of 43,000 infected active cases in mid of May, 2020 was predicted by Susceptible-Exposed-Infectious-Recovered (SEIR) model, and further the model justify role of lockdown in controlling of COVID-19 spread (Pai *et al.*, 2020). A mathematical that incorporates effect of lockdown and studies the effect of social distancing indicates effectiveness of lockdown at locations where higher percentage of symptomatic infection cases exists, necessity of mass testing, and possibility of sharp increase in COVID-19 cases if lockdown measures removed (Sardar *et al.*, 2020). In a compartmental mathematical model stimulation based on epidemic data till 30th April, 2020 predicted peak to appear after 60 days for COVID-19 cases and after that there will be a plateau in curve and it will persist for long period in India (Samui *et al.*, 2020). According to deep learning model used to forecast COVID-19 cases in India and USA will increase next one month (Shastri *et al.*, 2020). In a predicative model for initial spread of COVID-19 in India noticed that for COVID-19 pandemic male are affected more than female, COVID-19 active cases are decreasing in all age group, with youngest ones are the largest number of infected adults (Ghosh, 2020).

23.4 COVID-19 Pandemic in India: Challenges and Management

Alongside of other administrative and economic challenges arise due to COVID-19 and Lockdown, India Faced following challenges in respect of COVID-19 pandemic.

- Inadequate healthcare facility 735917 government hospital beds with 0.55 per 1000 capita hospital beds as per national health profile 2019.
- COVID-19 Cases growth rate (Median value) at national level is found to be 13.74 during first phase, 6.88 during second phase, 5.58 during third phase and 4.88 during forth phase of lockdown.
- The recovery rate was very low being 13.16 per cent in first phase;
- Ratio of COVID-19 cases to hospital beds which indicates healthcare infrastructure availability in states with highest spread of COVID-19 was found to be 1.2 in Maharashtra, 0.77 in Delhi and Gujarat.
- The national ratio of COVID-19 cases to hospital beds is found to be 0.24 as of 31st May, 2020.
- India has very low-test ratio being 3.19 per thousand people as on 31st May 2020. ICMR approved COVID-19 testing laboratories in India has capacity of 1, 25,000 test per day and till 31st May 37, 37,027 samples had been tested. With respect to testing Jammu and Kashmir was at highest position whereas Bihar was at lowest position.
- To prevent spread of COVID-19 in slum population was a major challenge, as per the last census data 42 per cent of total population of India lives in slum area, with 50 per cent slum population accommodated in a single room, without option for social distancing in highly unhygienic ambience.
- COVID-19 pandemic disclosed negative side of unplanned urbanization, major outburst cluster of COVID-19 was associated with highly populated cities situated in northern and western India (Ghosh *et al.*, 2020).
- In scientific study, for India average recovery time of COVID-19 patients found to be 25days, 50 per cent get recovery in 25 days, 40 per cent get recovery in 20 days whereas only 4 per cent of the total patients get recovery in 10days which is major challenge for India with limited healthcare facilities (Barman *et al.*, 2020).
- In a scientific study it is found that majority of COVID-19 patients are asymptomatic in range of 26 to 76 per cent (Singh and Misra, 2020).
- Age wise break up of COVID-19 patients in India indicates 87.26 per cent of total patients fall between 20-69 years and age bracket of 0-20 years contribute by 32.36 per cent on total COVID-19 patients which can become a super carriers for COVID-19 (Ragavi and Geetha, 2020).

Despite above mentioned challenges, poor infrastructure and healthcare facilities mortality were less in India due to COVID-19 in comparison to developed western world. Also recovery rate of India is higher than those of many other countries having similar situation of COVID-19 Pandemic (Jain *et al.*, 2020a). Also

India has on record COVID-19 infection rate of 1.7 (one patient infecting 1.7 people) comparatively lower than in other countries in hot zones (Sarkar *et al.*, 2020). Many aspects are accountable for this low infection/low mortality rate and high recovery including suspicion of diagnosis, timely detection and treatment, warmer climate, 21 days lockdown of India being one of the most stringent compared to other countries, BCG Vaccination, Inherent immunity of Indian (Singh and Misra, 2020), Race and ethnicity, Use of Hydroxychloroquine (Jain *et al.*, 2020a). First patient of COVID-19 was reported in Kerala on 30th January 2020, a student returned from Wuhan (Samui *et al.*, 2020). The scanning of each and every person coming from China and started by Indian Government from 8th January,2020 which is much earlier than WHO declared COVID-19 a public health emergency (Jain *et al.*, 2020a; Salgotra *et al.*, 2020). Before lockdown confirmed cases of COVID-19 were 499, whereas globally 332930 confirmed cases were reported (Saha *et al.*, 2020). Thus COVID-19 struck India later than other countries whereas timely lockdown and governments efforts delayed infections and mortality in India (Jain *et al.*, 2020a).

Further to address above mentioned challenges, Indian government took various initiatives such as launch of mobile application "Aarogya Setu" by Health Ministry, Conversion of Railway Coaches to healthcare facility by Indian railways, and importantly 2.1 billion US$ budget for health sector was allotted to fight against COVID-19 (Pai *et al.*, 2020). Further India's national medical council launched a pool testing method for RT-PCR diagnosis for quick identification and instant isolation of infected person, ICMR launched countrywide COVID-19 laboratories and few private laboratories with proper protocols for health and safety. Other department of Indian government also provided support *i.e.* map of geo-spatial distribution of COVID-19 cases across country developed by Indian Space Research Organization (ISRO), required technology transferred by Defence Research and Development Organisation (DRDO) to private companies along with further focus on developing critical care medical equipments (Pai *et al.*, 2020). Council of Scientific and Industrial Research (CSIR) with five branched approach geared up research to mitigate COVID-19 situation in its 38 laboratories while Department of Science and Technology (DST) started schemes to promote research, inventions and start-ups to address COVID-19 situation (Pai *et al.*, 2020).

23.5 Impacts of COVID-19 and Lockdown on Environment

With gradual increase in the cases of COVID-19 most of countries implemented lockdown due to which there was decrease in all the human activities including vehicular movement, industrial operation and waste discharge, cross-boundary movement, recreational activity, celebrations *etc.* to a greater extent along with social-distancing being practised in most countries (Zambrano-Monserrate *et al.*, 2020). This led to decrease in pollution level all around the world with positive impacts on environment. Decrease in air pollutant emissions and thus lower emissions of GHGs, NO_2, SO_2, $PM_{2.5}$ and PM_{10} concentrations were observed, besides this there was rise in the concentration of stratospheric O_3 at global level (Zambrano-Monserrate *et al.*, 2020). Further there were some other impacts like clean beaches and other water bodies, lower noise pollution, wildlife roam freely into their natural habitat *etc.* The pandemic exhibits a varying outcome human development, in the regard that, it has created a havoc globally and also impacted the environment positively (Chakraborty and Maity, 2020).

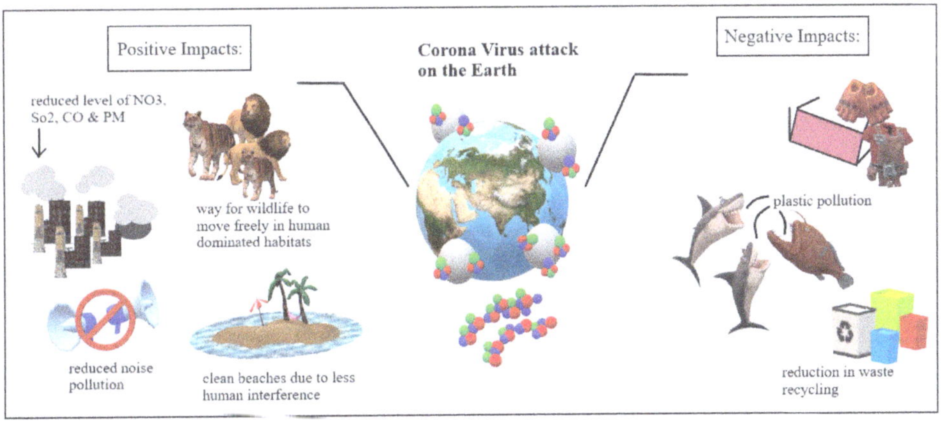

Figure 23.1: Positive and Negative Impacts of COVID-19 Pandemic on Environment.

As presented in Figure 23.1, Covid-19 crises not only have positive impacts, but it also have some perilous effects on environment such as volume of waste has increased due to lack of recycling (YVONNE, 2020), fishery export has been stopped resulting accumulation of more organic waste, natural ecosystem has been left unmonitored which causes illegal poaching, deforestation and hunting (Hamwey, 2020) and agriculture is also kept halted due to lockdown which has resulted in extensive use of chemicals and fertilizers to increase crop life and its production (UNDP, n.d.) which further runoff to water bodies and polluting and cause eutrophication. Another is the improper handling of bio medical waste in health care facility with increase in quantity of medical waste not securely disposed in environment and has capacity to infect patients, health workers *etc.* On the other side plastic usage has been increased in the form of protective gears as well as more packaging material was used due to online shopping and home delivery (Hamwey, 2020). All these accounts are causing negative impact on our environment and put forwards challenge for managing the waste during time of this pandemic. This pandemic has also led to serious and critical losses in economies all around the globe. It is difficult to estimate the effects and time span of economic catastrophe on individual households arising from the pandemic because many dubieties surround the crisis period (Martin *et al.*, 2020). Besides, the COVID-19 outbreak also seriously impacted international as well as domestic trade (Bashir *et al.*, 2020).

23.6 Positive Impacts of COVID-19 Pandemic on Air Quality

The pandemic has not only affected economy it has also severely impacted transportation, industrial operation, and cross-boundary movement. Local and international transportation has reduced significantly which has reduced the consumption of fossil fuels and utilization of energy. As a result of which air quality improved in most countries, environment has rehabilitated and flourishing again (Shehzad *et al.*, 2020). These impacts on air environment will be impermanent unless we apply long-term actions to slash emissions (Fedunik-Hofman, 2020). The major impact is seen with respect to air quality and the concentration of Ozone in the stratosphere. In Wuhan, China the Air Quality Index (AQI) decreased around 20-25 per cent, and P.M.$_{2.5}$ was reduced by 24 per cent after lockdown (He *et al.*, 2020a). These pollutants are mainly emitted from burning fossil fuels in vehicles, industries, and electricity generation. Also, there was an increment observed in the ozone concentration which was 116.6 per cent in Wuhan (Lian *et al.*, 2020). In Rome, the reduction in NO_2 concentration was observed to be around 26-35 per cent while in Madrid it was found to be 51 per cent. In UK, the reduction in NO_2 concentration was observed up to 60 per cent. The major city in Spain like Barcelona showed a reduction of 20-30 per cent in NO_2 levels. While in France, Italy and the USA the emission reduction was observed to be around 30 per cent (Muhammad *et al.*, 2020). Because of a partial lockdown implemented in Brazil there was reduction in the concentration of NO_2 and CO, 16.8-53.8 per cent and 30.3-48.5 per cent respectively. While the concentration of O_3 increased up to 67 per cent (Siciliano *et al.*, 2020). The air quality of Turkey got better by 2.5 times than previous because of the reduced vehicular activities (Aydýn *et al.*, 2020).

23.6.1 COVID-19 Pandemic and Air Quality in Indian Context

There is a major problem of air pollution observed and experience in India especially in Delhi. It is estimated that approximately 55 per cent deaths occur due to air pollution every year in India. A little increase in the concentration of P.M.$_{2.5}$ that is 1μg/m^3 corresponds to an increase of 15 per cent in death rate caused by the novel corona virus (Mehmood *et al.*, 2020). Ambient air quality and meteorological conditions can influence transmission of viral diseases; spread of droplet mediated virus diseases can be favoured by dry and cold climate whereas warm and hot climate observed to increase transmission like Influenza and SARS (Kumar, 2020).

With respect to India relation between meteorological parameters and spread of COVID-19 has been analysed in various studies. At initial phase of the pandemic, blast of new cases was observed in countries having ambient temperature > 18°C and absolute humidity in range of 3 to 9 g/m^3; it was believed that for India rise in temperature and high absolute humidity due to approaching summer may slow down spread of COVID-19, which not proved true (Kumar, and Kumar 2020). In a study carried out with respect to air quality and meteorological parameters and COVID-19 cases, positive association between daily COVID-19 cases and temperature was found while mixed association with respect to relative and absolute humidity (Kumar, 2020). The study further reports reduction of PM_{10}, $PM_{2.5}$ and NO_2 in ambient air of six mega cities of India by 60 to 45 percentages during lockdown which may have reduced the risk of COVID-19 spread, because of less availability of aerosols (Kumar, 2020).

A study carried out for Mumbai, Maharashtra, India indicates active cases due to COVID-19 have noteworthy correlation with meteorological parameters such as temperature (minimum temperature T_{min}), dew point (DP_{max}), relative humidity (RH_{max}, RHavg, RH_{min}) as well as surface pressure (Pmax, Pavg, Pmin); relative humidity and pressure parameter being most noteworthy (Kumar and Kumar, 2020).

A generalized additive models with regression utilized to check how meteorological factors temperature and humidity affects spread of COVID-19 across India which indicate significant role of meteorological factors in spread of COVID-19 particularly daily temperature and relative humidity (Goswami *et al.*, 2020).

The metro cities of India, Delhi and Mumbai showed a significant reduction in NO_2 concentration that is 60 per cent and 34 per cent respectively (Fedunik-Hofman, 2020). The concentration of O_3 increased remarkably by 37.35 per cent in Delhi while in Mumbai an increment of 20.65 per cent was observed (Kumari and Toshniwal, 2020). Even the water of river Yamuna improved as there was very low amount of industrial pollutants in the river during lockdown (Biocon, 2020). In Delhi, India when the concentration of particulate matter was compared to the average of three years, a decrease of about -57 per cent and -33 per cent was observed in the concentration of PM_{10} and $PM_{2.5}$ respectively (Mahato *et al.*, 2020).

In Chandigarh, during the lockdown phase the primary source for pollution were the residential areas, whereas the pollutants from straw and coal burning were the secondary sources. Because of this a decrease in particulate matter level was observed. The concentration of $PM_{2.5}$ was noted to be reduce by 23.4 per cent and concentration of PM_{10} was reduce by 22.8 per cent (Mor *et al.*, 2020). A notable increase was observed in the concentration levels of O_3. The increased concentrations through the first phase, second and third phase of the lockdown were noted as 38.7 per cent, 91.4 per cent and 128.9 per cent throughout the entire lockdown, only the first phase showed a decreased concentration because of stern restrictions on all human activities (Mor *et al.*, 2020). It was reported that the AQI in northern, southern, eastern, central and western India was decreased by 44 per cent, 33 per cent, 29 per cent, 15 per cent and 32 per cent respectively. Later on when the lockdown began, the pollution created by particulate matter decreased by an average of 35 per cent and the concentration of NO_2 decreased to 60 per cent. During the same time span, it was observed that the surface ozone concentration increased by 1.5-2 per cent (Arora *et al.*, 2020).

It was noticed that there was some decrease in the concentration of GHGs due to pandemic, this decrease may have a little impact on the whole concentration that have gathered in the atmosphere for years (Mahipal, 2020). It was also observed that the concentration of ozone increased during this period of lockdown. The concentration of O_3 increased by 45.56 per cent in Kolkata (Bera *et al.*, 2020). The reason behind the increase in the concentration O_3 is reduced levels of NO (Kumari and Toshniwal, 2020). In megacities and north India, degradation of air quality due to increase in particulate matter concentration frequently creates emergency and respiratory system related health issues. Increase in air pollution when lockdown relaxed may worst the situation by more spread and increase in death due to COVID-19 (Kumar, 2020).

23.6.2 Other Positive Impacts of COVID-19 on Environment

Due to increased industrialisation and urbanisation, wildlife movement is restricted in human dominated areas. But because of this novel corona virus lockdown was implemented all around the world, which has restricted humans to their homes. This has given wildlife a way to roam openly in their own habitat (Rupani *et al.*, 2020). It is known that beaches are the national assets for many countries as it is the main attraction for the tourists. Nevertheless, it is also evident that beaches are very much exploited and plastics dumping and other waste disposal which are not cleared and get gathered in the benthic zone. As social-distancing was a compulsion because of corona virus it retrieved the beaches as a result of which the shores are clean and water crystal clear (Biocon, 2020). Because of lockdown there was no traffic, no pollution from industries, no honking from vehicles everything was calm. In Delhi, most of the residential areas now have 30-40 dB (A) of noise level instead of 55 dB (A) (Gandhiok and Ibrar, 2020). some scientists observed that the high-frequency sound by big machineries in industries, vehicles has reduced by 50 per cent in the countries affected by the pandemic. A major drop was on observed in New York and Singapore (TOI.COM, 2020).

23.6.3 Negative Impacts of COVID-19 on Environment

COVID-19 crises have many negative impacts as well. There is rapid increase in demand and usage of plastic/polymer products such as masks, gloves, sanitization bottles and PPE kits *etc.* having short product life or meant for one-time use. This widespread use of protective equipment all over the world leads to plastic waste management and disposal problems (Klemeš *et al.*, 2020) and also affecting the ecosystem if not properly disposed (Hamwey, 2020). On the other side the biomedical waste generated from the hospitals and quarantine centres cause danger as face masks, infected shoes, expired medicines, ointments, gloves *etc.* are made of materials which takes too much time to get decompose in environment. Due to this it causes several health related issues to living organisms as well as it accumulate in environment and can harm the environmental components also. According to a report given by Ministry of Ecology and Environment, Wuhan has produced 240 tons of waste per day during the outbreak and data from Thailand shows the increase in plastic waste to 6,300 tons on daily basis. The illegal waste disposal has been increased by 300 per cent in United Kingdom (Duer, 2020). For India, study indicated that Mumbai city generated 24,889 kg waste per day (Reddy, 2020).

23.7 Plastic Pollution and its Negative Impacts

Plastic is one of the major products and having a wide range of application with incredibly low cost and large life duration makes it especially important material of our life. But the COVID-19 pandemic has boosted the single-use plastic pattern and the fear in public leads to more production and consumption of personal protective gears such as face masks, hand gloves, general equipment like syringes and sanitizer bottles *etc.*, hence increasing plastic pollution (Sarkodie and Owusu, 2020). Also used plastic product in COVID-19 pandemic is pathogen-contaminated and is difficult to handle and dispose (Klemeš *et al.*, 2020). In normal circumstances, according to Oceans Asia around 300 million tons of plastic waste is produced worldwide every year and of this 8 million enters in the aquatic ecosystem (Carol, 2020). During the lockdown period the sale of protective gears and groceries has raised by 20 per cent in supermarkets.

At Global scale the disposable of face masks alone has set a high record from $800 million in 2019 to $160 billion in 2020 (UNCTAD 2020) and as per data provided by WHO, 16 million gloves are used per month (Iyer, 2020). In more than 15 countries in Africa alone, the total face masks and hand gloves used per day is seven hundred million as based on an average of 2 face masks daily per capita (Nzediegwu and Chang, 2020). In Barcelona, face masks and gloves generation increases by 350 per cent (Sarkodie and Owusu, 2020). In India 1.5 crore masks are manufactured on daily basis

During lockdown period there has been rise in home delivery and online purchase that increases the use of plastic for packaging of groceries and food items (Klemeš *et al.*, 2020) which leads to increase in the household plastic waste (Shakil *et al.*, 2020). This not only causes plastic pollution as it will also cause decrease in renewable resources as their will be more transportation and supply of goods to individual residences which require more fossil fuels to get utilize (Yvonne, 2020). In Singapore 5.7 million residents generated household waste of 1,470 tons only from packaging and food delivery. On one side there is increase in plastic usage, on the other side COVID -19 pandemic has ensued stoppage of waste recycling. As many countries have reduced the recycling process to stop the spread of COVID-19 such as the USA had stopped the recycling of waste in some of their cities (Colin, 2020).

23.8 Biomedical Waste and Associated Impacts

Due to COVID-19 the focus has turned towards the medical waste which includes waste from hospitals, quarantine centres and private clinics ("COVID-19 AND THE ENVIRONMENT: IMPACT AND RESPONSE," 2020). This waste is hazardous and dangerous to public health as well as environmental health as it possess disposal items such as PPE kits that include facial medicated masks, hand gloves, and shoes and also other medical waste as blood soaked cotton plugs, drain bags, empty ampules, urine bags, *etc.* This waste when mixed with the household waste can create a harmful impact on the health of living beings and also can cause negative impact on environment (Varmani, 2020).

In Wuhan, China the medical waste generated during the pandemic was 246 tons per day (Eroðlu, 2020). In Barcelona the biomedical waste such as face masks and gloves increased by 350 per cent and this generate around 1,200 tons of medical waste (Sarkodie and Owusu, 2020). Same as in Bangladesh the medical centres produce 18,70,000 kg of hazardous medical waste (National Institute of Preventive and Social Medicine (NIPSOM), 2020).

According to a EPCA the biomedical waste produced in Delhi, is around 350 tons per day (10,500 tonnes per month) (Sharagi and Pillai, 2020). Whereas, Delhi has capacity to treat and dispose 2,220 tons of medical waste per month similarly other states of India has different capacities such as Haryana has 288 tons per month and Uttar Pradesh has 1,656 tons per month *etc.* Central Pollution Control Board has developed a new guidelines for the disposal of biomedical waste generated from COVID-19 healthcare facility; however capacity criteria must be kept in mind otherwise it may lead to hazardous situation in various parts of country (Thomas, 2020).

23.9 Other Impacts of COVID-19 Pandemic

The pandemic not only brings negative impacts on human beings as it also harms the wildlife around us. The improper disposal of protective gears such as face masks, hand gloves and other lifesaving equipment's eventually end up in the sea. In case of improper disposal, when rain is there this all gears flow with the runoff into the sea and pollutes the water because masks are made up of polypropylene and polyethylene that will degrade slowly into smaller microplastic pieces (Patrício *et al.*, 2020) and also it will cause threat to marine life and wildlife (Nikolia, 2020). Another similar situation was found in Columbia where degradation of nonwoven synthetic material causes microplastic pollution in the water and soil (Silva *et al.*, 2020). According to UN Environment Programme about 75 per cent of corona virus plastic waste is clogging the landfills and are found floating on the sea.

In- situ and *Ex-situ* conservation are phasing the negative impact; risk associated with the protected species during the pandemic. As in many countries due to lockdown environmental protection workers are staying at home leaving the protected areas unobserved due to their absence people are doing illegal hunting, illegal deforestation, fishing and poaching in national parks, wildlife sanctuaries, and in marine zones. The ecotourism activity has also stopped which has left the natural ecosystem at a risk of encroachment and illegal harvesting. Also in several destination ecotourism was only the source of economic mainstay which raises the unemployment among the people living there, it causes many families to harvest resources from ecosystem as to feed their family with food (Hamwey, 2020).

Various consequences due to lockdown and COVID-19 pandemic have been studied scientifically, addressed, and reported in various scientific literatures which are summarized as below.

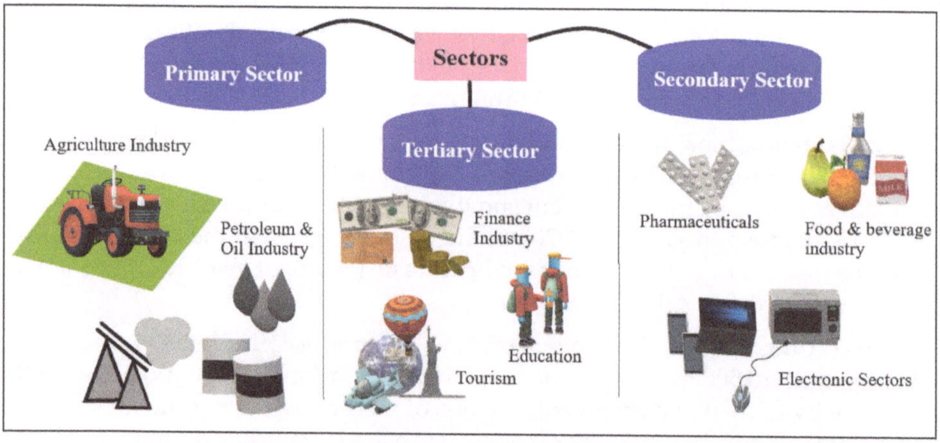

Figure 23.2: Sectors of Economy Affected by COVID-19 Pandemic.

23.10 Socioeconomic Impacts of COVID-19 and Lockdown

This pandemic is not only a health calamity but it has seriously impacted global economies and societies in depths (United Nations, 2020). The reason for the loss of many jobs and the decreased manpower is nothing but an outcome of physical-distancing, self-isolation, and travel restrictions across all the sectors. All the institutions have closed down and manufacturing of products and other commodities has also reduced (Koon, 2020). This outbreak of COVID-19 has impacted each and every fraction of the population and it is constantly affecting population taking into consideration poor people, old people, disabled, youth and locals (United Nations, 2020).

There are three major sectors which are impacted by COVID-19 and that are- Primary sector, Secondary sector, and Tertiary sector. The Primary economic sector includes agriculture and petroleum and oil. The on-going pandemic has interrupted the agricultural activities all around the world. The prices of agricultural products have dropped by 20 per cent as there is a global crash in demand of food from restaurants and hotels. Also, panic buying is generating difficulty in shortage (Nicola *et al.*, 2020). Similarly, the pandemic has lowered the demand and price crude oil. The oil prices have reduced which a direct effect that is positive supply shock as manufacturing costs decrease; and an indirect effect because of reduced revenue in net oil-exporting countries and the other reason being demand slipped because of lockdown (Taghizadeh-Hesary *et al.*, 2016).

The Secondary economic sector includes manufacturing industry, further the manufacturing industry includes electronics sector, pharmaceutical industry, food and beverage manufacturing industry and chemical industry. The major part of the economy in the whole world is the manufacturing industry which accounts to about 16 per cent of global GDP in the year 2018 ("Impact of COVID-19 on Global Manufacturing Industry," 2020). Because of the pandemic they are facing trouble regarding the availability of raw materials. China accounts for around 85 per cent of total elements used in smart phones and around 75 per cent in the case of television so as a result this sector is noticeably affected by the pandemic. Many countries including India depend on China for the active pharmaceutical ingredients (API). India imports nearly 80 per cent of API from China while USA gets 13 per cent of their products manufactured in China ("Impact of COVID-19 on the Process Manufacturing Industry; 2020," 2020).

Food and Beverages manufacturing industries are facing reduced consumption along with supply chain issues. Also, the supply of availability of raw material to the manufacturing facility has stopped the production and has forced the manufacturers to shut down the facility. It is estimated that the food and beverage manufacturing industry has suffered a total loss of 22 per cent in turnover globally ("Impact of COVID-19 on the Process Manufacturing Industry; 2020," 2020). Also due to shortage in workforce and supply of raw materials the production of chemical industry have been decreased by 40-60 per cent. In February, the global production of chemicals dropped by 2.4 per cent. Nevertheless, the paints and coatings industry felt the highest decline of 7.3 per cent and 9.4 per cent respectively (GEP, 2020).

The Tertiary economic sector includes health care, education system, finance industry, tourism, and consumer goods. Most of the world's population is under lockdown due to this pandemic. Because of which people are suffering from depression because of social rejection, financial crisis and enforced isolation. It is observed that quarantine can cause negative psychological effects on humans like depression, anger, confusion and post-stress traumatic symptoms (Nicola *et al.*, 2020). The governments of every country have temporarily closed all the educational institutes in order to contain the spread of COVID-19. Many students varying from school-going to postgraduates are impacted due to this. According to the available data, these have impacted over 91 per cent of global students (Kasrekar and Wadhavane, 2020). Hence, e-learning strategies are promoted all around the world.

The decrease in global stock market has released an unstable environment with critical cash-flow levels. Central Banks have negotiated to ensure the cash-flow is maintained and reduce the economic shock, with several leaders to overcome these effects (Nicola *et al.*, 2020). Also according to the World Travel and Tourism Council, 50 million jobs are at risks in this sector. In India, tourism contributes 5.06 per cent of total GDP in the year 2016-17 but it is ignored (Goswami and Soundararajan, 2020). If this industry wants to grow in such situation, it should establish a common safety and sanitation standards. The prevailing fear of the virus spread amid the buyers have also impacted domestic utilization causing lower supplier margins. India's Seafood Industry comprises of 30 per cent export. China has a share of 22 per cent in India's Export industry which has moved toward lower demand leading to excessive stock ("Impact of COVID - 19 On Finance Industry," 2020).

23.11 COVID-19 Challenges in Indian Context

COVID-19 pandemic and consequent lockdown has severely impacted India economy. In a study it is estimated that there will be severe impact on economy due to lockdown; Indian economy may lessen by 10-20 per cent (Singh and Misra, 2020). Also, lockdown has drastically dropped community mobility across India; drop of 46.3 per cent reported for visits to parks, 66 per cent for transit stations, 56.7 per cent for workplaces and 51.2 per cent for grocery and pharmacy units, 73.3 per cent for retails and recreational mobility, while increase of 23.8 per cent reported for mobility towards residential places (Saha *et al.*, 2020). Further in the study spatiotemporal variation in community mobility across India is reported, which indicate that just after one day of lock down *i.e.* on 25th march retail and recreation mobility dropped by 70.51 per cent, grocery and pharmacy mobility dropped by 64.26 per cent, parks mobility dropped by 46.17 per cent, transit stations mobility dropped by 65.6 per cent, workplaces mobility dropped by 60.03 per cent. Chandigarh had highest 85 per cent decrease in overall mobility followed by 82 per cent for Delhi, 71 per cent for Andaman and Nicobar, 71 per cent for Maharashtra whereas lowest drop observed in north east states, 34 per cent for Mizoram, 39 per cent for Arunachal Pradesh, 41 per cent for Bihar and Manipur (Saha *et al.*, 2020).

Healthcare systems are foremost affected across the world as well in India. Being second highest populated country, in India healthcare infrastructure need to serve 1.3 billons people whereas as per report of world economic forum India is at 150th rank in global healthcare. With respect to average household income, being at 112th position out of 164 countries, large population of India depends upon government healthcare facilities (Salgotra *et al.*, 2020). Government healthcare infrastructure was already inadequate India, before lockdown with total 735917 government hospital beds with 0.55 per 1000 capita hospital beds; whereas 12 states- Manipur, Assam, Maharashtra, Odisha, Haryana, Madhya Pradesh, Chhattisgarh, Uttar Pradesh, Gujarat, Jharkhand, Andhra Pradesh and Bihar where 70 per cent of total population is residing had per capita hospital beds availability less than the national level of 0.55 (Ghosh *et al.*, 2020). COVID-19 pandemic created an additional requirement for isolated beds and likewise for healthcare equipment specially the 'ventilator', which resulted huge disparity across the country; availability of isolated beds per 1000 persons is 0.43 for Andhra Pradesh, 0.36 for Madhya Pradesh, 0.41 for Tamil Nadu, 0.31 for Telangana, whereas isolated beds are in shortage availability being 0.02 per 1000 person in state of Uttar Pradesh, Bihar, Karnataka (Ghosh *et al.*, 2020). As per the Health Ministry of India, elderly people are most vulnerable to COVID-19 Pandemic having 63 per cent casualty in age group of above 60 years whereas 30 per cent causality in age group of 40–60 years; but fortunately in India availability of bed is 5.18 per 1000 old population (Ghosh *et al.*, 2020).

Covid-10 pandemic drown major attention of healthcare systems and thus less focus have been provided to other non-emergent aspects. It is reported that during pandemic COVID-19 hospital admission for acute myocardial infarction and other cardiovascular care decreased all over the world which may be due to fear, unavailability of transportation, decreased in air pollution during lockdown or this may be true decrease (Ramakrishnan *et al.*, 2020) (Showkathali *et al.*, 2020) and at a time out of hospital cardiac arrest increased. For India such study involving 27 states and 257 enrolling centres is under progress to analyse impacts of COVID-19 pandemic and for necessary recommendations needed for acute myocardial infarction and other cardiovascular care (Ramakrishnan *et al.*, 2020). In another study it is reported that due to lockdown hospital admission for acute coronary syndrome decreased by 43 per cent and there was delay in presentation for admitted patients (Showkathali *et al.*, 2020). COVID-19 affected solid organ transplant program worldwide and, in this respect, living and deceased donor liver transplant (DDLT) in India nearly stopped, being only one DDLT performed throughout the country in month of April 2020. With gradual relaxation in curfew measures in second week of May, 2020 only two cases of DDLT performed throughout India (Pahari *et al.*, 2020).

India is also facing challenges due to COVID-19 pandemic in combating against Tuberculosis (TB). India being country with one fourth of TB cases and one third global drug resistant TB cases accounts 2.6 million cases at every 10 million cases worldwide. It is further reported in the study that in India there was drop of 78 per cent in diagnosis of new TB cases in April 2020 compared to last year due to COVID-19 and direct observed therapy centres for TB and National Tuberculosis Elimination Program (NTEP) impair significantly (Iyengar and Jain, 2020; Jain *et al.*, 2020b). Both TB and COVID-19 spread by close contact and having similar clinical symptoms though transmission differ slightly; high number to TB cases in India further imposed challenges for management of TB and COVID-19 in India (Iyengar and Jain, 2020). Further it is pointed out that TB patients may develop resistant to multiple drugs and are super susceptible for COVID-19 Infection (Jain *et al.*, 2020b). Lockdown resulted closing of liquor shops and supply chain of licit/illicit substances which resulted to a unique challenge faced by India's healthcare infrastructure dedicated to treatment of opioid use disorder (Pandit *et al.*, 2020).

Lifestyle factors and other diseases play important role in COVID-19 mortality. Prevalence of diabetes found highest in Indian COVID-19 patients, in a scientific study of 176 cases from total 206 deaths reported till 10th April 2020 it was found that in 50.5 per cent death cases any pre-existing co-morbidities was there. Highest being diabetes (27.8 per cent) followed by hypertension (22.1 per cent), respiratory disease (COPD and asthma) (13.6 per cent), and cardiovascular disease (6.2 per cent). 13 per cent of decease cases had both diabetes and 2.8 per cent had diabetes, hypertension and CVD. Similar results reported by Integrated Disease Surveillance Program of Ministry of Health and Family Welfare, India by analysing death of 15,962 patients of COVID-19; in case of 57 per cent death more comorbidities found while 43 per cent deaths are due to COVID-19 only (Singh and Misra, 2020). Thus, to reduce mortality due to COVID-19 in India would be challenging due to prevalence of diabetes because as per estimates every one out of six adults suffering from diabetes in the world are from India. Further it is reported that weight gain during lockdown and de-stabilization of glucose control due to lack of physical activity can increase diabetes complications and risk; an estimate shows 7 per cent increase in risk of diabetes due to weight gain during lockdown (Singh and Misra, 2020).

23.12 Conclusion

The worldwide concern due to upsurge of COVID-19 pandemic seriously affecting global public health forced countries all over the globe to implement some sort of lockdown. It is estimated that around 210 countries and territories are affected due to this outbreak. The lockdown has extremely affected global economy with many visible/invisible impacts on human civilization and environment. Nevertheless, lockdown has shown some sort of positive impact on the environment and at a time there are direct indirect negative impacts on environment. The chapter provides specific emphasis of positive impacts on environment due to stopped large-scale industrial production, energy-use and GHGs emissions, and low traffic has led to reduction in pollution of environment. However, it cannot be assured that this reduction will continue in the future if lockdown is over; in fact, poor air quality in post lockdown may boost spread of COVID-19 spread. Along with other positive impacts on environment COVID-19 pandemic can seriously impinge on environment if challenges due to increase plastic waste and biomedical waste generation not addressed properly. Further the chapter provides insight on socioeconomic aspect of pandemic with special emphasis on impacts of COVID-19 on India' healthcare infrastructure.

REFERENCES

Arora, S., Bhaukhandi, K.D., Mishra, P.K. (2020) Coronavirus lockdown helped the environment to bounce back. Sci. Total Environ. 742, 140573. https: //doi. org/10.1016/j.scitotenv.2020.140573

Aydýn, S., Nakiyingi, B.A., Esmen, C., Güneysu, S., Ejjada, M. (2020) Environmental impact of coronavirus (COVID-19) from Turkish perceptive. Environ. Dev. Sustain. https: //doi.org/10.1007/s10668-020-00933-5

Barman, M.P., Rahman, T., Bora, K., Borgohain, C. (2020) COVID-19 pandemic and its recovery time of patients in India: A pilot study. Diabetes Metab. Syndr. Clin. Res. Rev. 14, 1205–1211. https: //doi.org/10.1016/j.dsx.2020.07.004

Bashir, M.F., Ma, B., Shahzad, L. (2020) A brief review of socio-economic and environmental impact of Covid-19. Air Qual. Atmosphere Health. https: //doi.org/10.1007/s11869-020-00894-8

Bera, B., Bhattacharjee, S., Shit, P.K., Sengupta, N., Saha, S. (2020) Significant impacts of COVID-19 lockdown on urban air pollution in Kolkata (India) and amelioration of environmental health. Environ. Dev. Sustain. https: //doi.org/10.1007/s10668-020-00898-5

Biocon, (2020) Impact of COVID-19 on Environment. URL, Biocon 4th June 2020 https: //bioconblog.com/2020/06/04/impact-of-covid-19-on-environment/

Carol K, (2020) Another Side Effect of COVID-19: The Surge in Plastic Pollution. http;//earth.org 6th July 2020.

Chakraborty, I., Maity, P. (2020) COVID-19 outbreak: Migration, effects on society, global environment and prevention. Sci. Total Environ. 728, 138882. https: //doi.org/10.1016/j.scitotenv.2020.138882

Colin, S. (2020) Coronavirus pandemic disrupts recycling sector. Resour. Recycl. News. URL https: //resource-recycling.com/recycling/2020/03/17/coronavirus-pandemic-disrupts-recycling-sector/(accessed 11.1.20).

COVID-19 AND THE ENVIRONMENT: IMPACT AND RESPONSE, 2020.

Duer, J. (2020) The plastic pandemic is only getting worse during COVID-19.World economic Forum 1st July 2020.

Emily A, (2020)UBC researchers develop biodegradable medical mask.Waste and Recycling magazine 22nd may, 2020.

Eroðlu, H. (2020) Effects of Covid-19 outbreak on environment and renewable energy sector. Environ. Dev. Sustain. https: //doi.org/10.1007/s10668-020-00837-4

Food Waste: An Opportunity for Change, (2020)

Fedunik-Hofman, L. 2020. What impact will COVID-19 have on the environment? Aust. Acad. Sci.4/5/2020

Gandhiok, J. Ibrar, M,(2020) Covid-19: Noise pollution falls as lockdown rings in sound of silence. Times of India April 23rd 2020.

GEP, (2020) COVID-19'S IMPACT ON THE CHEMICAL INDUSTRY. URL https: //www.gep.com/blog/mind/covid-19s-impact-on-the-chemical-industry

Ghosh, K., Sengupta, N., Manna, D., De, S.K. (2020) Inter-state transmission potential and vulnerability of COVID-19 in India. Prog. Disaster Sci. 7, 100114. https: //doi.org/10.1016/j.pdisas.2020.100114

Ghosh, S., 2020. Predictive model with analysis of the initial spread of COVID-19 in India. Int. J. Med. Inf. 143, 104262. https: //doi.org/10.1016/j.ijmedinf.2020.104262

Goswami, A., Soundararajan, N. (2020) Impact Of COVID-19 On Tourism In India.

Goswami, K., Bharali, S., Hazarika, J. (2020)Projections for COVID-19 pandemic in India and effect of temperature and humidity. Diabetes Metab. Syndr. Clin. Res. Rev. 14, 801–805. https: //doi.org/10.1016/j.dsx.2020.05.045

Hamwey, R. (2020) Environmental impacts of coronavirus crisis, challenges ahead | UNCTAD [WWW Document]. URL https: //unctad.org/news/environmental-impacts-coronavirus-crisis-challenges-ahead (accessed 10.31.20).

Hamwey, R. (2020) Environmental impacts of coronavirus crisis, challenges ahead.

He, F., Deng, Y., Li, W. (2020a.) Coronavirus disease 2019: What we know? J. Med. Virol. 92, 719–725. https: //doi.org/10.1002/jmv.25766

He, G., Pan, Y., Tanaka, T. (2020b) The short-term impacts of COVID-19 lockdown on urban air pollution in China. Nat. Sustain. https: //doi.org/10.1038/s41893-020-0581-y

Holshue, M.L., DeBolt, C., Lindquist, S., Lofy, K.H., Wiesman, J., Bruce, H., Spitters, C., Ericson, K., Wilkerson, S., Tural, A., Diaz, G., Cohn, A., Fox, L., Patel, A., Gerber, S.I., Kim, L., Tong, S., Lu, X., Lindstrom, S., Pallansch, M.A., Weldon, W.C., Biggs, H.M., Uyeki, T.M., Pillai, S.K. (2020) First Case of 2019 Novel Corona virus in the United States. N. Engl. J. Med. 382, 929–936. https: //doi.org/10.1056/NEJMoa2001191

Impact Of COVID - 19 On Finance Industry, (2020) URL https: //www.iifl.com/blogs/impact-covid-19-finance-industry

Impact of COVID-19 on Global Manufacturing Industry, (2020)

Impact of COVID-19 on the Process Manufacturing Industry; (2020)

India now has the world's second highest number of Covid-19 cases; tops Brazil's corona virus tally | Deccan Herald [WWW Document], n.d. URL https: //www.deccanherald.com/national/india-now-has-the-worlds-second-highest-number-of-covid-19-cases-tops-brazils-coronavirus-tally-883184.html (accessed 10.31.20).

India's Covid cases world's 2nd highest(2020) | India News - Times of India [WWW Document], n.d. Times India. URL https: //timesofindia.indiatimes.com/india/indias-covid-cases-worlds-2nd-highest/articleshow/77956328.cms (accessed 10.31.20).

India to generate 775.5 tonnes of medical waste daily by (2020) Study [WWW Document], n.d. Econ. Times. URL: https: //economictimes.indiatimes.com/industry/healthcare/biotech/healthcare/india-to-generate-775-5-tonnes-of-medical-waste-daily-by-2020-study/articleshow/63426284.cms (accessed 9.6.20).

Iyer, P. (2020) During the pandemic, strides made over the last year in dealing with single-use plastic have suffered a setback.

Iyengar, K.P., Jain, V.K. (2020) Tuberculosis and COVID-19 in India- double trouble! Indian J. Tuberc. https: //doi.org/10.1016/j.ijtb.2020.07.014

Jain, V.K., Iyengar, K., Vaish, A., Vaishya, R. (2020a) Differential mortality in COVID-19 patients from India and western countries. Diabetes Metab. Syndr. Clin. Res. Rev. 14, 1037–1041. https: //doi.org/10.1016/j.dsx.2020.06.067

Jain, V.K., Iyengar, K.P., Samy, D.A., Vaishya, R. (2020b) Tuberculosis in the era of COVID-19 in India. Diabetes Metab. Syndr. Clin. Res. Rev. 14, 1439–1443. https: //doi.org/10.1016/j.dsx.2020.07.034

Kasrekar, D., and Wadhavane, G. (2020) Impact of COVID-19 on Education System in India. www. Latest laws. Com. 16th may 2020.

Klemeš, J.J., Fan, Y.V., Tan, R.R.and Jiang, P. (2020) Minimising the present and future plastic waste, energy and environmental footprints related to COVID-19. Renew. Sustain. Energy Rev. 127, 109883. https: //doi.org/10.1016/j.rser.2020.109883

Koon, E. O. (2020) The Impact of Socio-cultural Influences on the COVID-19 Measures–Reflections From Singapore. J. Pain Symptom Manage. 60, e90–e92. https: //doi.org/10.1016/j.jpainsymman.2020.04.022

Kumar, A., Rani, P., Kumar, R., Sharma, V., and Purohit, S.R. (2020) Data-driven modelling and prediction of COVID-19 infection in India and correlation analysis of the virus transmission with socio-economic factors. Diabetes Metab. Syndr. Clin. Res. Rev. 14, 1231–1240. https: //doi.org/10.1016/j.dsx.2020.07.008

Kumar, G., and Kumar, R.R. (2020) A correlation study between meteorological parameters and COVID-19 pandemic in Mumbai, India. Diabetes Metab. Syndr. Clin. Res. Rev. 14, 1735–1742. https: //doi.org/10.1016/j.dsx.2020.09.002

Kumar, S. (2020) Effect of meteorological parameters on spread of COVID-19 in India and air quality during lockdown. Sci. Total Environ. 745, 141021. https: //doi.org/10.1016/j.scitotenv.2020.141021

Kumari, P., and Toshniwal, D. (2020) Impact of lockdown measures during COVID-19 on air quality– A case study of India. Int. J. Environ. Health Res. 1–8. https: //doi.org/10.1080/09603123.2020.1778646

Lian, X., Huang, J., Huang, R., Liu, C., Wang, L., Zhang, T. (2020) Impact of city lockdown on the air quality of COVID-19-hit of Wuhan city. Sci. Total Environ. 742, 140556. https: //doi.org/10.1016/j.scitotenv.2020.140556

Lokhandwala, S., Gautam, P,(2020) Indirect impact of COVID-19 on environment: A brief study in Indian context. Environ. Res. 188, 109807. https: //doi.org/10.1016/j.envres.2020.109807

Mahato, S., Pal, S., Ghosh, K.G. (2020) Effect of lockdown amid COVID-19 pandemic on air quality of the megacity Delhi, India. Sci. Total Environ. 730, 139086. https: //doi.org/10.1016/j.scitotenv.2020.139086

Manufacturing 15 mn masks a day, no dearth of gloves, sanitizers: Govt, (2020). BS Web Team Press Trust India.

Martin, A., Markhvida, M., Hallegatte, S., Walsh, B. (2020) Socio-Economic Impacts of COVID-19 on Household Consumption and Poverty. Econ. Disasters Clim. Change. https: //doi.org/10.1007/s41885-020-00070-3

Mandy Z, (2020) Coronavirus leaves China with mountains of medical waste [WWW Document]. South China Morning Post. URL https: / /www.scmp.com/news/ china/society/article/3074722/coronavirus-leaves-china-mountains-medical-waste (accessed 9.6.20).

Mehmood, K., Saifullah, Abrar, M.M., Iqbal, M., Haider, E., Shoukat, H.M.H. (2020) Can PM2.5 pollution worsen the death rate due to COVID-19 in India and Pakistan? Sci. Total Environ. 742, 140557. https: //doi.org/10.1016/j. scitotenv.2020.140557

Mor, S., Kumar, S., Singh, T., Dogra, S., Pandey, V., Ravindra, K. (2020) Impact of COVID-19 lockdown on air quality in Chandigarh, India: Understanding the emission sources during controlled anthropogenic activities. Chemosphere 263, 127978. https: //doi.org/10.1016/j.chemosphere.2020.127978

Mahipal, N. (2020) The POSITIVE and NEGATIVE impact of COVID-19! URL https: //www.inventiva.co.in/trends/natania/the-positive-negative-impact-of-covid-19/

National Institute of Preventive and Social Medicine (NIPSOM), Dhaka, Bangladesh, Hossain, D.I. (2020) Pandemic COVID-19 and Biomedical Waste Handling: A Review Study. J. Med. Sci. Clin. Res. 08. https: //doi.org/10.18535/jmscr/ v8i5.88

Nicola, M., Alsafi, Z., Sohrabi, C., Kerwan, A., Al-Jabir, A., Iosifidis, C., Agha, M., Agha, R. (2020) The socio-economic implications of the coronavirus pandemic (COVID-19): A review. Int. J. Surg. 78, 185–193. https: //doi.org/10.1016/j. ijsu.2020.04.018

Nikolia A, (2020) Coronavirus plastic waste polluting the environment. D.W.com

Nzediegwu, C., and Chang, S.X. (2020) Improper solid waste management increases potential for COVID-19 spread in developing countries. Resour. Conserv. Recycl. 161, 104947. https: //doi.org/10.1016/j.resconrec.2020.104947

Pahari, H., Shellagi, N., Nath, B. (2020) Deceased Donor Liver Transplantation in India in the COVID-19 Era: Current Scenario and Future Perspectives. Transplant. Proc. https: //doi.org/10.1016/j.transproceed.2020.06.004

. Pai, C., Bhaskar, A., Rawoot, V. (2020) Investigating the dynamics of COVID-19 pandemic in India under lockdown. Chaos Solitons Fractals 138, 109988. https: //doi.org/10.1016/j.chaos.2020.109988

Paital, B., Das, K., Parida, S.K. (2020) Inter nation social lockdown versus medical care against COVID-19, a mild environmental insight with special reference to India. Sci. Total Environ. 728, 138914. https: //doi.org/10.1016/j.scitotenv.2020.138914

Pandit, P.M., Bhatia, G., Sarkar, S. (2020) Treatment Experience for Opioid Use Disorder During COVID-19 in India: Learning from Patients. J. Subst. Abuse Treat. 108128. https: //doi.org/10.1016/j.jsat.2020.108128

Patrício S., A.L., Prata, J.C., Walker, T.R., Campos, D., Duarte, A.C., Soares, A.M.V.M., Barcelò, D., Rocha-Santos, T. (2020) Rethinking and optimising plastic waste management under COVID-19 pandemic: Policy solutions based on redesign and reduction of single-use plastics and personal protective equipment. Sci. Total Environ. 742, 140565. https: //doi.org/10.1016/j.scitotenv.2020.140565

Patrikar, S., Poojary, D., Basannar, D.R., Faujdar, D.S., Kunte, R. (2020) Projections for novel coronavirus (COVID-19) and evaluation of epidemic response strategies for India. Med. J. Armed Forces India 76, 268–275. https: //doi.org/10.1016/j.mjafi.2020.05.001

Ragavi, V., Geetha, N.K. (2020) A drill down analysis of the pandemic COVID-19 cases in India using PDE. Mater. Today Proc. https: //doi.org/10.1016/j.matpr.2020.05.595

Ramakrishnan, S., Jabir, A., Jayagopal, P.B., Mohanan, P.P., Nair, V.K., Das, M.K., Mandal, M., Roy, D., *et al.* (2020) Pattern of acute MI admissions in India during COVID-19 era: A Cardiological Society of India study - Rationale and design. Indian Heart J. https: //doi.org/10.1016/j.ihj.2020.09.004

Reddy, A. 2020. Delhi to Vijayawada, India has started dumping Covid-19 infected waste in public places.The Print 5th Sept. 2020.

Research, G., n.d. Asia Pacific Medical Waste Management Market Share (2020-2026)/ : Report [WWW Document]. Graph. Res. URL https: //www.graphicalresearch.com/industry-insights/1355/asia-pacific-medical-waste-management-market (accessed 9.6.20).

Rupani, P.F., Nilashi, M., Abumalloh, R.A., Asadi, S., Samad, S., Wang, S., 2020. Coronavirus pandemic (COVID-19) and its natural environmental impacts. Int. J. Environ. Sci. Technol. https: //doi.org/10.1007/s13762-020-02910-x

Saha, J., Barman, B., Chouhan, P., 2020. Lockdown for COVID-19 and its impact on community mobility in India: An analysis of the COVID-19 Community Mobility Reports, 2020. Child. Youth Serv. Rev. 116, 105160. https: //doi.org/10.1016/j.childyouth.2020.105160

Salgotra, R., Gandomi, M., Gandomi, A.H., 2020. Time Series Analysis and Forecast of the COVID-19 Pandemic in India using Genetic Programming. Chaos Solitons Fractals 138, 109945. https: //doi.org/10.1016/j.chaos.2020.109945

Samui, P., Mondal, J., Khajanchi, S., 2020. A mathematical model for COVID-19 transmission dynamics with a case study of India. Chaos Solitons Fractals 140, 110173. https: //doi.org/10.1016/j.chaos.2020.110173

Sardar, T., Nadim, S.S., Rana, S., Chattopadhyay, J., 2020. Assessment of lockdown effect in some states and overall India: A predictive mathematical study on COVID-19 outbreak. Chaos Solitons Fractals 139, 110078. https: //doi.org/10.1016/j.chaos.2020.110078

Sarkar, K., Khajanchi, S., Nieto, J.J., 2020. Modeling and forecasting the COVID-19 pandemic in India. Chaos Solitons Fractals 139, 110049. https: //doi.org/10.1016/j.chaos.2020.110049

Sarkodie, S.A., Owusu, P.A., 2020. Impact of COVID-19 pandemic on waste management. Environ. Dev. Sustain. https: //doi.org/10.1007/s10668-020-00956-y

September surge: India No.1 in Covid cases and deaths in first 15 days | India News - Times of India [WWW Document], n.d. Times India. URL https: / /timesofindia. indiatimes.com/india/september-surge-india-no-1-in-covid-cases-and-deaths-in-first-15-days/articleshow/78149166.cms (accessed 10.31.20).

Shakil, M.H., Munim, Z.H., Tasnia, M., Sarowar, S., 2020. COVID-19 and the environment: A critical review and research agenda. Sci. Total Environ. 745, 141022. https: //doi.org/10.1016/j.scitotenv.2020.141022

Shastri, S., Singh, K., Kumar, S., Kour, P., Mansotra, V., 2020. Time series forecasting of Covid-19 using deep learning models: India-USA comparative case study. Chaos Solitons Fractals 140, 110227. https: //doi.org/10.1016/j. chaos.2020.110227

Shehzad, K., Sarfraz, M., Shah, S.G.M., 2020. The impact of COVID-19 as a necessary evil on air pollution in India during the lockdown. Environ. Pollut. 266, 115080. https: //doi.org/10.1016/j.envpol.2020.115080

Showkathali, R., Yalamanchi, R., Sankeerthana, M.P., Nandha Kumaran, S., Shree, S., Nayak, R., Oomman, A., Mahilmaran, A., 2020. Acute Coronary Syndrome Admissions and Outcome during COVID-19 Pandemic–Report from Large Tertiary Centre in India. Indian Heart J. https: //doi.org/10.1016/j. ihj.2020.09.005

Siciliano, B., Dantas, G., da Silva, C.M., Arbilla, G., 2020. Increased ozone levels during the COVID-19 lockdown: Analysis for the city of Rio de Janeiro, Brazil. Sci. Total Environ. 737, 139765. https: //doi.org/10.1016/j.scitotenv.2020.139765

Singh, A.K., Misra, A., 2020. Impact of COVID-19 and comorbidities on health and economics: Focus on developing countries and India. Diabetes Metab. Syndr. Clin. Res. Rev. 14, 1625–1630. https: //doi.org/10.1016/j.dsx.2020.08.032

Singhal T., 2020. A Review of Coronavirus Disease-2019 (COVID-19). https: //doi. org/10.1007/s12098-020-03263-6

Sohrabi, C., Alsafi, Z., O'Neill, N., Khan, M., Kerwan, A., Al-Jabir, A., Iosifidis, C., Agha, R., 2020. World Health Organization declares global emergency: A review of the 2019 novel coronavirus (COVID-19). Int. J. Surg. 76, 71–76. https: //doi.org/10.1016/j.ijsu.2020.02.034

Shrangi V. and Pillai S., 2020. Better segregation of biomedical waste needed: EPCA. The Hindustan times 31st July 2020.

Taghizadeh-Hesary, F., Yoshino, N., Mohammadi Hossein Abadi, M., Farboudmanesh, R., 2016. Response of macro variables of emerging and developed oil importers to oil price movements. J. Asia Pac. Econ. 21, 91–102. https: //doi.org/10.1080/13547860.2015.1057955

TIMESOFINDIA.COM, 2020. Good news: Noise pollution has dropped by 50 per cent during the coronavirus pandemic.

Thomas, A. 2020. Upload Covid-19 waste disposal data on app, says top court.

UNCTAD (2020) Growing plastic pollution in wake of COVID-19: how trade policy can help, 2020.

UNDP, n.d. COVID-19 and the Environment: Impact and Response | UNDP in the Arab States [WWW Document]. UNDP. URL https: //www.arabstates.undp.org/content/rbas/en/home/presscenter/articles/2020/covid-19-and-the-environment—impact-and-response.html (accessed 10.31.20).

United Nations, 2020. Socio-Economic impact.

Varmani, M. 2020. Bio-medical waste management during COVID-19.Team India Blogs, June 22, 2020.

WHO, 2020. Novel Coronavirus (2019-nCoV).

Wu, J., Cai, W. Watkins, D. Glanz, J. 2020. How the Virus got out. New Yorks Times. di Editorial Team Digital Health Global, 23/03/2020

YVONNE, 2020. The Unexpected Environmental Consequences of COVID-19 [WWW Document]. Voices Youth. URL https: //www.voicesofyouth.org/blog/unexpected-environmental-consequences-covid-19 (accessed 10.31.20).

Zambrano-Monserrate, M.A., Ruano, M.A., Sanchez-Alcalde, L., 2020. Indirect effects of COVID-19 on the environment. Sci. Total Environ. 728, 138813. https: //doi.org/10.1016/j.scitotenv.2020.138813

Chapter 24

Management of the Pest *Plutella xylostella* Linnaeus, 1758 (Lepidoptera: Plutellidae) by Artificial Rearing Method

Dolly Kumar and Harshita Sharma

Division of Entomology, Department of Zoology, The Maharaja Sayajirao University of Baroda, Vadodara – 390 002, Gujarat

e-mail: dollymsu@gmail.com

ABSTRACT

Agriculture fields of Vadodara are surrounded by many insect pests such as Helicoverpa armigera (Hubner, 1808), Spodoptera litura (Fabricius, 1775), Earias insulana (Boisduval, 1833), Aphis gossypii (Glover, 1877), Bemisia tabaci (Gennadius, 1889). One such major insect pest of agricultural fields is Plutella xylostella Linnaeus, 1758 (Lepidoptera: Plutellidae) commonly called Diamondback moth or DBM. It is a pest of Cruciferous crops such as cabbage, cauliflower, broccoli, etc. Most of these are economically important edible crops. The caterpillar stage is the damaging stage feeding on all the phases of growing plants. Fourth and fifth instars being voracious feeders. The population of this insect becomes abundant in September and October, March and April; but heavy rains can destroy the moth population. Farmers start controlling the insect by the spraying of insecticides as soon as they see the infestation. A wide range of pesticides is used by farmers to control this pest in the agricultural fields of Vadodara. The majority of the farmers are using chemical insecticides, some are using botanicals and very few use Bacillus thuringiensis (Bt). The use of chemical insecticides is expensive and results in the increased cost of the crop and financial load on the farmers. The DBM has become very difficult to manage as it has developed high levels of resistance to several groups of organophosphates, carbamates, and synthetic pyrethroids including Bt. (Uthamasamy et al., 2011). An investment from 4 billion dollars to 5 billion dollars was reported from different parts of Australia just to manage and control this pest (Zalucki et al., 2012). The present work was carried out from September 2017 to October 2019.

Various larval stages were collected from agricultural fields in Vadodara and multiple generations were reared in the lab and used for laboratory experiments. The temperature was kept constant with variation in diet and humidity, devising a standard laboratory rearing method. Such a study helps in providing an uninterrupted supply of insect pests for laboratory experiments.

Efficacy of Bio-insecticides, Botanicals, Microbials, etc. can be tried in the laboratory as an alternative control method. Insect resistance can also be checked for developing new molecules of chemical pesticides which can help to delay resistance and manage the pest. This will aid in the reduction of insecticide usage in the Agri-fields which in turn can lower environmental pollution resulting from the overuse of insecticides.

Keywords: *Agriculture, Artificial diet, DBM, Insecticides, Rearing.*

24.1 Introduction

A survey of agricultural fields of Vadodara revealed many insect pests infesting economically important edible crops namely cotton, castor, cabbage, cauliflower, brinjal *etc.* These economically important crops were damaged by insects mainly belonging to the orders Orthoptera (*Grylotalpa fossor, Schistocerca gregaria*), Hemiptera (*Aphis gossypii, Pyrilla perpusilla*), Thysanoptera (Thrips), Coleoptera (*Holotrichia insularis,*), Lepidoptera (*Helicoverpa armigera, Plutella xylostella*) (Table 24.1). One such major insect pest in the agricultural fields of Vadodara is *Plutella xylostella* Linnaeus, 1758 (Lepidoptera: Plutellidae), also called as diamondback moth or DBM. *Plutella xylostella* L. has a host range of 120 plant species including plants from the families Malvaceae, Brassicaceae, Fabaceae, *etc.* (Talekar and Shelton, 1993). Even in the agricultural fields of Vadodara *Plutella xylostella* was found destroying crops mainly belonging to the family Brassicaceae damaging cabbage, mustard and cauliflower.

Table 24.1: Examples of Insect Pests Damaging Economically Important Crops

Order	Example of insects pests
Hemiptera	*Aphis gossypii, Pyrilla perpusilla, Dysdercus cingulatus*
Orthoptera	*Gryllotalpa orientalis, Schistocerca gregaria, Poekilocerus pictus*
Thysanoptera	Thrips
Coleoptera	*Holotrichia insularis, Oryctes rhinoceros,*
Lepidoptera	*Helicoverpa armigera, Spodoptera litura, Earias insulana, Plutella xylostella, Pectinophora gossypiella, Chilo suppressalis, Pieris brassicae*

Although in Gujarat, farmers are to a certain extent also implementing Integrated Pest Management (IPM). But most of the time there is total dependence on insecticides. There are various reasons like the unavailability of an integrated approach for the control of all the insect pests and lack of awareness amongst farmers on the alternatives to insecticides. Certain control methods of IPM other than chemical control are also used by farmers in Vadodara (Table 24.2). Also Biological control is used (Table 24.3) and the commonly used biological control methods by farmers in Vadodara include Trichocards, Bt seeds, NPV. Naturally occurring biocontrol agents are dragonflies, damselflies, beetles and spiders. Pheromonal control (Table 24.4), Botanical control (Table 24.5) are used in different agricultural fields of Vadodara.

Table 24.2: IPM Practices in the Agricultural Fields of Vadodara

Control Method	*Agricultural Fields Around Vadodara*	*Status in Vadodara*
Mechanical control	Dabhoi (30 km. South East of Vadodara)	Using mechanical methods like hand-picking of insects, mainly caterpillars
Physical control	Waghodia (8 km East from Vadodara)	Using a barrier like a row made up of plastic or polyester. Physical barriers such as ant fences applied to parallel fields to keep ants away, also to avoid the mealybug population
Traps	Padra (15 km. South West of Vadodara)	Light traps for attracting insect pests towards the light. Used in paddy fields against insects such as grasshoppers. Here, flypaper or sticky boards are used to capture the insect pests
Mulching	Channi (10 km towards North from Vadodara)	Used against aphids, whiteflies, thrips. The process of mulching also helps to retain soil moisture
Cultural control	Savli (30 km North from Vadodara)	Control by this method is observed. It includes crop rotation, trap cropping, intercropping

Table 24.3: Naturally Occurring Biocontrol Agents in the Agricultural Fields of Vadodara

Category	*Examples*
Predators	*Pantala flavescens*- Dragonfly, *Agriocnemis pygmaea*- Damselfly (Odonata), *Coccinella septempunctata*, *Cicindela sexguttata* (Coleoptera), *Chrysoperla carnea* (Neuroptera), *Mantis indica* (Dictyoptera)
Parasite/Parasitoids	*Trichogramma chilonis* (Hymenoptera)
Pathogens	*Bacillus thuringiensis*, *Pseudomonas fluorescence* (Bacteria), NPV- Nuclear Polyhedrosis virus (Virus), *Beauveria bassiana*, *Trichoderma* sp. (Fungi)

Table 24.4: Pheromonal Lures Used in the Agricultural Fields of Vadodara

Type	*Used against*
Heli lure	*Helicoverpa armigera* (Cotton bollworm, American bollworm)
Earias lure	*Earias vitella*, *Earias insulana* (Spotted bollworm, Spiny bollworm)
Scripo lure	*Sciropaphaga excertpalis* (White top borer)
Spodo lure	*Spodoptera litura* (Tobacco cutworm, Cotton leafworm)
Lucin lure	*Leucinodes orbanalis* (Eggplant fruit borer)

Table 24.5: Botanicals Used in the Agricultural Fields of Vadodara

Compound	*Source*
Nicotene	*Nicotiana tabacum* leaves
Rotenone	*Derris involute* roots
Pyrethrum	*Chrysanthemum* flowers
Azadirechtin	*Azadirachta indica* leaves, seeds, Neem Seed Kernal

Survey also throws light on the careless and indiscriminate use of pesticides in the agricultural fields of Vadodara; to control the harmful insects damaging the economically important crops (Table 24.6).

Table 24.6: Insecticides Used in the Agricultural Fields of Vadodara

Insecticides	*Trade Names*
Organophosphates	Malathion, Chloropyriphos, Phosphates, Phosphorothiolate
Carbamate	Carbaryl, Aldicarb, Carbofuran, Bendiocarb
Synthetic pyrethroids	Tetramethrin, Cypermethrin, Cyhalothrin, Fenvalerate
Neonicotinoids	Acetamiprid, Clothianidin
Diamides	Chlorantraniliprole, Cyantraniliprole

Knowledge of different life stages of any pest of interest is a pre-requisite before starting the actual study. To rear the organism, we must make ourselves familiar with its life cycle. A temperature range between 20°C to 30°C and humidity between 50 per cent to 65 per cent is known to be best for the growth and multiplication of insect pests (Chaudhari *et al.,* 1999). Hence the laboratory life cycle of *Plutella xylostella* was observed on the natural diet *i.e.* on cabbage leaves and kept at a constant temperature and humidity of 26± 4°C and 70±10 per cent respectively. Total development time from the egg to adult stage averages 24 days. Diamondback moth eggs are oval and yellow. Females deposit up to 300 eggs in the initial 2-3 days. Eggs are deposited in small groups in depressions on the surface of foliage, or sometimes on other plant parts. The first instar larva emerges in 2 to 4 days. A total of four instar stages is present. A larva or caterpillar is having a green to yellow body with distinct body segments and short hairs. The larva, when disturbed, reacts violently and it may also drop off the plant, where it hangs on a silken thread. The feeding habit of early instar (1st and 2nd) larvae is mining inside the leaf. Late instar (3rd and 4th) eat by creating holes in the leaves by biting and later molt and turn into a pupa. Every instar stage lasts for 2 to 3 days. Total larval duration is of 12-14 days. The next stage is the dormant pupal stage. Last 4th instar takes around 3 days to become pupa. The pupation in DBM is distinct from other Lepidoptera pupae where a hard and definite cocoon is formed, but in DBM a very soft and loose kind of silk cocoon is usually formed. The color of the pupa is green to brown. Pupation lasts from 4 to 7 days after which the adult male and female moth emerge. Male survives for around 12 days while female survives for around 16 days. The adult is a small size grey-brown moth with clavate antennae. The band at the back in both males and females, covering the thorax and abdomen forms the shape of a diamond that is why the common name of this insect is diamondback moth or DBM.

Generally, large quantities of insecticides have been used to control *Plutella xylostella*. Therefore the insect has become resistant to many classes of insecticides and as a result causes up to 80 per cent loss in marketable yield. DBM in India is found to infest important crucifers like cabbage, cauliflower, radish, mustard. DBM population is influenced by abiotic factors like, temperature, rainfall and humidity. It has also been observed that microbial insecticides are better in delaying resistance development. (Uthamasamy *et al.*, 2011). The overuse chemical insecticides has resulted in the development of resistance in the pest to almost all insecticides. Along with it, the use of commercial, nuclear transgenic crops expressing *Bacillus thuringiensis* (Bt) toxins has escalated in recent years because of their advantages over the chemical insecticides. But DBM has been proved to have become resistant to different strains of Bt as well (Liu and Tabashnik, 1997).

There is a strong need to work on alternative methods to control the insects in the fields. To do so, a large population of the pure strain of insects is required to be bred in the laboratory. Six to seven generations of insects can be reared and bred in the laboratory so that novel modes of controlling insect pests can be tried and devised. The insect selected in this study was DBM. For efficient rearing of DBM following objectives were undertaken:

- ☆ Diet optimization to find out an effective lab diet for DBM. Standardize artificial diet for breeding the larvae of *Plutella xylostella* (Linnaeus, 1758) for an uninterrupted good food available to the insect pests in the lab
- ☆ To study the effect of different humidity conditions on the larval stages at a constant temperature
- ☆ To find out efficient humidity conditions while rearing the insects.

Thus an Artificial Rearing Method to aid in the management of the pest *Plutella xylostella* Linnaeus, 1758 (Lepidoptera: Plutellidae) was developed in the laboratory.

24.2 Experimental Section

24.2.1 Collection and Preservation

To conduct different experiments initial requirement was to have insects in the laboratory. Hence all stages of the larvae of *P. xylostella* were collected from the agricultural fields in and around Vadodara (Table 24.2). Collection was made during morning times from 8 am to 10 am. The most abundant population was during the months of October-November (early winter) and February-March (late winter). Larvae of the DBM was found damaging mainly cruciferous vegetables like cabbage, cauliflower and broccoli. Various cruciferous vegetation was observed in the fields. Larvae of the DBM were found damaging almost all parts of the plants. Larvae once collected from the fields were transferred in the lab.

24.2.2 Rearing in the Laboratory Conditions

The larvae collected from the field were stage-wise separated (1st, 2nd, 3rd, 4th instar) and transferred from small plastic boxes (used for collection) to larger plastic boxes (used for rearing) with the help of camel brush size 0 and 1. The round plastic boxes were having cabbage leaves. The cabbage grown in organic farms was utilized to feed larvae in the laboratory conditions. This was done to ensure that leaves eaten by larvae remain insecticide free. The cabbage leaves were everyday changed and larvae transferred on fresh cabbage. The pupae formed were transferred to adult cages so that the adult emerging from it can be separated from larvae. The adults were allowed to lay eggs on small folded butter paper under whose crevices eggs were laid by the insect. Eggs were collected later and transferred to different boxes. The neonates emerging from the eggs were transferred to fresh cabbage leaves to facilitate its nutrition and growth. The insect *Plutella xylostella* was reared for its complete life-cycle maintained on natural diet *i.e.* cabbage leaves for several generations (Figure 24.1). The insect pests were kept in BOD maintained at 26 ± 4°C temperature and 70±10 per cent relative humidity where it completed its various life stages (Figure 24.2). Careful rearing was done in the lab with sterile environment to prevent contamination on natural and artificial diet.

Figure 24.1: Natural Diet (Young cabbage leaves) for Rearing *Plutella xylostella* L.

First instar larvae or neonates of *Plutella xylostella* were used from the laboratory population to see its survival on different artificial diets as well as the impact of humidity and temperature on the larval growth and development. Diet and humidity were varied to know the best suited diet and humidity for increasing the DBM population. A total of three diets and a total of four humidity conditions were tested for DBM.

Rearing was done in culture trays. In these culture trays, there were separate cells for each larva. A total of three artificial diets were made. Temperature and humidity were maintained in the BOD incubator. Temperature and Humidity were kept constant at 26°C and 70 per cent, respectively. Percent survival was recorded. Observations were recorded till adult formation.

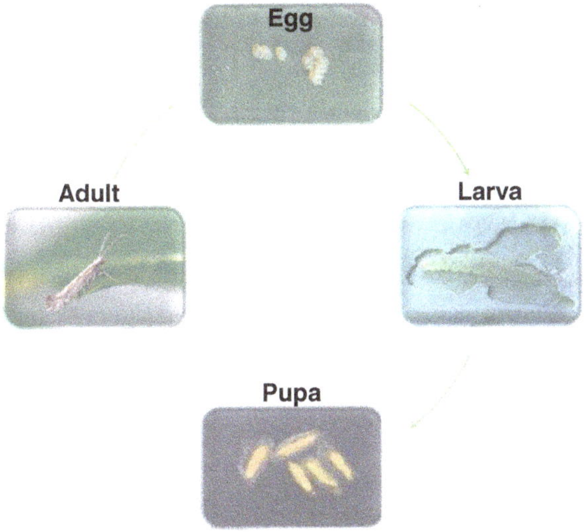

Figure 24.2: Stages in the Life Cycle of *Plutella xylostella* L.

Artificial diets were prepared for 500 ml diet (Table 24.7). The diet ingredients were divided into two parts and then mixed. In first part water is boiled with agar and yeast is added to it, all other ingredients are added to second part. Lastly both the parts are mixed and fresh diet is poured into insect trays.

Table 24.7: Diet Composition of *Plutella xylostella* L.

Sl.No.	*Ingredients*	*Diet 1*	*Diet 2*	*Diet 3*
1	Wheat germ	15g	15g	15g
2	Casein	17.5g	17.5g	17.5g
3	Cabbage leaves	15g	15g	15g
4	Sucrose	17.5g	17.5g	17.5g
5	Becosule	5g	5g	5g
6	Ascorbic acid	2ml	2ml	2ml
7	I-Inositol	0.09ml	0.09ml	0.09ml
8	Wesson's salts	5g	5g	5g
9	Methyl-P	0.75ml	0.75ml	0.75ml
10	Formaldehyde (37 per cent)	0.25ml	0.25ml	0.25ml
11	Cholesterol	1.25g	1.25g	1.25g
12	Oil (Linseed)	3.5ml	3.5ml	3.5ml
13	KOH	2.5ml	2.5ml	2.5ml
14	Chickpea flour	-	15g	15g
15	Streptomycin	-	0.5g	0.5g
16	Bavistin	-	-	0.3g
17	Agar	11.25g	11.25g	11.25g

- ☆ Diet 1- *The* 1st diet for *P. Xylostella* culture consisted of ingredients inspired by the diet of Hsiao and Hou, 1978 (Diet1).
- ☆ Diet 2 and Diet 3- Further modification in diets were done after looking into the problems and challenges faced in the previous diet. Streptomycin was added to Diet 2 to stop the microbial infection observed. In Diet 3, along with Streptomycin, Bavistin was added to enhance the anti-microbial property of the diet. This was done as microbial infection in the form of fungus was observed in Diet 2.

Humidity was varied at 35 per cent, 70 per cent, 90 per cent. A condition was also tested for a room where humidity was not maintained and 33-58 per cent came out to be the range of the varying humidity conditions there (variable humidity). Maintaining the temperature at a constant of 26°C; Larval growth was observed at regular intervals. Fresh cabbage leaves were given to larvae every 24 hrs.

Complete life cycle was traced at each humidity condition. Thus, we obtained the following results after rearing under different conditions of diet and humidity.

24.3 Results

Out of the three diets, diet 3 was found to be the best diet at temperature of 26°C and 70 per cent humidity. In diet 3, 60.94 per cent survival was observed (Table 24.8). Diet 3 was found to be the best diet for survival followed by 18.75 per cent in diet 2 and 6.25 per cent diet 1, Diet3>Diet2>Diet1 (Figure 24.3). Larval growth index for *Plutella xylostella* was again found to be best in diet 3(5.54) followed by diet 2 (1.56) and diet 1 (0.42) (Table 24.9). Larval Growth Index is used as a parameter to check the efficiency of a diet or condition on the growth and development of larvae. A higher value indicates better conditions for the larvae to survive and grow.

Table 24.8: Experimental Overview Showing Per cent Survival of *Plutella xylostella*

Diet	*1st Instar Larvae/ Cell*	*No. of Cells/Tray*	*Larvae/ Tray*	*No. of Trays*	*Total Larvae Released*	*No. of Adults Emerged*	*Per cent Survival*
Diet 1	2	32	64	2	128	8	6.25
Diet 2	2	32	64	2	128	24	18.75
Diet 3	2	32	64	2	128	78	60.94

Table 24.9: Larval Growth Index of *Plutella xylostella*
Larval Growth Index (LGI) = Percent pupation/Larval period (days)

Diet	*Per cent Survival*	*Larval Period (Days)*	*Larval Growth Index (LGI)*
1	6.25	15	0.42
2	18.75	12	1.56
3	60.94	11	5.54

Figure 24.3: Per cent Survival of Larvae on different Diets.

To reduce the usage of the natural host, the experiment was set to test three different artificial diets. As we know that streptomycin possess 'antibacterial' property and Bavistin possess 'antifungal' property proved to be beneficial in limited amounts for making a good diet for *P. xylostella*. A comparative analysis between the diet appearance before and after reveals severe infection in diet 1, little infection in diet 2 while no infection in diet 3 (Figures 24.4 and 24.5).

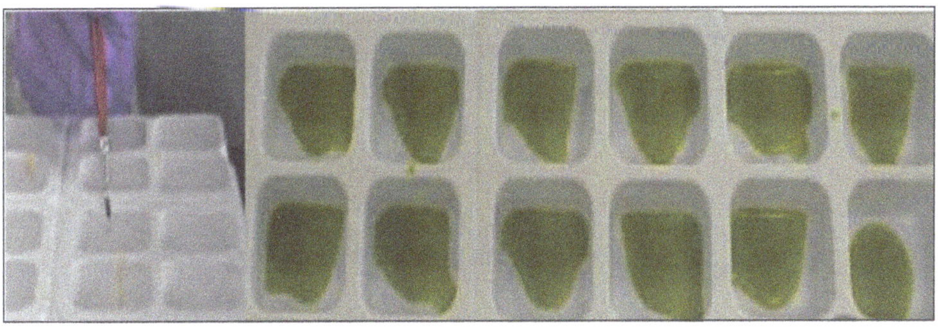

Figure 24.4: Transferring of Neonates on Fresh Diet.

Infection in Diet-1 **Infection in Diet-2** **No infection in Diet-3**

Figure 24.5: Post One Day Diet Appearance and Infection.

Out of the various humidity conditions, 35 per cent humidity was found to be the best at a constant temperature of 26°C when fed on young and fresh cabbage leaves. A complete survival was observed where best humidity was 35 per cent followed by 70 per cent, 90 per cent and 33-58 per cent (varying humidity) (Table 24.10) Humidity condition at 35 per cent and 70 per cent proved to be the best for lab rearing of DBM (Figure 24.6).

Table 24.10: Percentage Survival of *Plutella xylostella* on different Humidity

RH (per cent)	*Neonates*	*Total Adults*	*Per cent Survival*
35	20	20	100
70	20	19	95
90	20	15	75
33-58	20	16	80

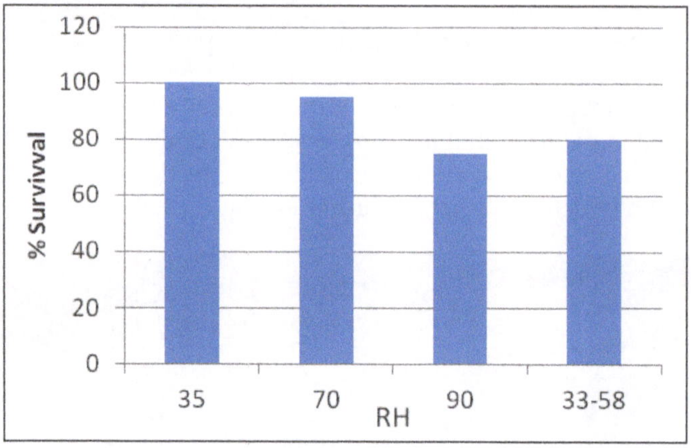

Figure 24.6: Per cent Survival of *Plutella xylostella* in different Humid Conditions.

24.4 Conclusion

An effective artificial diet for diamondback moth is necessary to rear inside the laboratory. Artificial diet reduces the labor dependence of growing and waiting for the natural plants. At the same time, artificial diet can fulfill all nutritional requirements of the insects. Artificial diet for DBM is a challenge and most of the places conducting research work on DBM are using natural diet. In this study, an already proposed diet has been modified for DBM. 60.94 per cent survival could be achieved in the 3rd Diet. This can be further improvised for getting a higher percent survival. Humidity condition between 35 per cent -70 per cent can be considered best for rearing DBM.

Rearing on artificial diet gives good results and mass rearing of the DBM can be done in the lab. Large population of insects can be utilized for the trial of newer environment friendly methods to control the insect pest

References

Chaudhari, G. B., Bharpora, P. M., Patel, J. J., Patel, K. I., and Patel, J. R. (1999). Effect of weather on activity of cotton bollworm in middle Gujarat. *Journal of Agrometoerology*, 1(2): 137–142.

Hsiao, J. H. and Hou R. P. (1978). Artificial rearing of the diamondback moth, *Plutella xylostella* (L.), on a semi-synthetic diet. *Institute of Zoology, Academia Sinica*, 17(2): 97–102.

Liu, Y. B., Tabashnik, B. E., and Pusztai-Carey, M. (1997). Inheritance of Resistance to the *Bacillus thuringiensis* Toxin Cry1C in the Diamondback Moth. *Applied and Environmental Microbiology*, 63(6): 2218-2223.

Talekar, N. S., and Shelton, A. M. (1993). Biology, ecology, and management of the diamondback moth. *Annual review of Entomology*, 38 (1): 275-301.

Uthamasamy, S., Kannan, M., Senguttuvan, K., and Jayaprakash, S. A. (2011). Status, damage potential and management of diamondback moth, *Plutella xylostella* (L.) in Tamil Nadu, India. *In: Proc.* The sixth international workshop on management of the diamondback moth and other crucifers insect pests. (eds. R. Srinivasan, Anthony M. Shelton, Hilda L. Collins) AVRDC- The World Vegetable Center. 21-23 March. Taiwan. pp 270–279.

Zalucki, M. P., Shabbir, A., Silva, R., Adamson, D., Shu-Sheng, L., and Furlong, M. J. (2012). Estimating the economic cost of one of the world's major insect pests, *Plutella xylostella* (Lepidoptera: Plutellidae): just how long is a piece of string? *Journal of Economic Entomology*, 105(4): 1115-1129.

Chapter 25

Insecticide Resistance in a Model Insect Pest *Spodoptera litura* F., 1775 (Lepidoptera: Noctuidae)

Rupa Sule and Dolly Kumar*

Division of Entomology, Department of Zoology, Faculty of Science, The Maharaja Sayajirao University of Baroda, Vadodara – 390 002, Gujarat
**e-mail: dollymsu@gmail.com*

ABSTRACT

Over use of insecticides to control insect pests in the agricultural fields in and around Vadodara has led to the accumulation of chemicals in soil, air and water; subsequently polluting them. Excessive and careless use of insecticides in the fields results in insects developing resistance towards the toxic chemicals. Survey showed that agricultural fields were mainly sprayed by insecticides namely Organophosphates e.g. Malathion, Chlorpyriphos; Carbamates like Methomyl, Carbofuran; Neonicotinoids like Imidacloprid, Dinotefuran; Synthetic Pyrethroids like Cypermethrin, Deltamethrin; Spinosyns like Spinosad, Spinetoram; Anthranilic diamides like Coragen, Flubendiamide for controlling the insect pests. Insect pests were mainly from the Orders Thysanoptera (Thrips tabaci, Scirtothrips dorsalis), Hemiptera (Aphis gossypii, Bemisia tabaci) Lepidoptera (Helicoverpa armigera, Spodoptera litura, S. exigua) and Coleoptera (Megasoma sp., Rhinotia hemistictus). These insects were damaging the economically important crops like cotton, castor, wheat, paddy maize and vegetable crops like cabbage, cauliflower, tomato and brinjal.

Looking at the scenario, one such polyphagous pest Spodoptera litura (Fabricius, 1775) was reared and bred in the laboratory on artificial diet to obtain mother cultures. Laboratory experiments were performed on mother culture. Third instar Larvae of Spodoptera litura were subjected to insecticides belonging to four different groups: Cypermethrin 25EC a Synthetic Pyrethroid, Chlorpyriphos 20EC an Organophosphate, Spinosad 45 SC a Spinosyn and Coragen 20SC a Diamide.

Five generations of S. litura were exposed to each insecticide. When exposed to insecticides, first generation of larvae indicated mortality towards all the four insecticides. After three generations, efficacy of insecticides indicated reduced mortalities at different concentrations. This signposted towards a phenomenon called "Resistance". Out of the four insecticides tested; Spodoptera litura showed maximum resistance towards Coragen followed by Cypermethrin, Chlorpyriphos and Spinosad. This is an alarming situation and should be dealt with utmost priority as the price of insecticides is affecting the value of produce and also the health of the environment.

Keywords: *Insecticides, Insect pests, Resistance, Spodoptera litura.*

25.1 Introduction

The major impacts of nationwide urbanization are clearly seen in rural areas, livelihood and indicated by environmental pollution and generation of waste. The accelerated growth has created challenges for the Mother Nature in terms of hazardous waste generated by various factors like industrial fumes, over use of pesticides in the form of insecticides. One of the major concerns of the century has been the prodigious consumption of pesticides for controlling the high amount of pest pressure. The various natural resources which are affected by usage of pesticides are air, water and soil. The agricultural activities like use of nitrogenous fertilizers in India have served as potential contaminant of groundwater (Khanna and Gupta, 2018). Hence urbanization and industrialization had some adverse effects on environment in the form of air, water and soil pollution by over usage of pesticides. As the agriculture revolutions took place, there were introduction of insecticides in all the nooks and corners of the world. The introduction of synthetic insecticides – organophosphate (OP) insecticides in the 1960s, carbamates in the 1970s and pyrethroids in the 1980s and the introduction of herbicides and fungicides in the 1970s–1980s contributed greatly to pest control and agricultural output (Aktar, 2009). The pesticide usage is very high in states like Maharashtra, Punjab, Uttar Pradesh, West Bengal, Gujarat, and Karnataka (Abhilash and Singh, 2009). The main use of pesticides is for economically important crops like cotton, paddy, wheat, castor, maize and tomato. Indiscriminate pesticide use is detrimental to the environment and human health and increases insects' resistance to pesticides (Srinivasan, 2012). With the positive impacts of all types of agriculture revolution, there were negative impacts too. They included the slow build-up of pest pressure, control of crops by all types of insect pests, the greed of farming leading to the introduction of pesticides. The use of pesticides is preferred by most farmers because of the myth of providing a quick solution to the detrimental problem of pests. When there is heavy pest pressure, the pesticides are applied for getting a quick control. In this whole process, it is forgotten that the effects of pesticides are much more as compared to the quick solution provided. This includes resistance development, toxic effects on plants, long term effects on farmer's health, contamination of the environment. Even the non-target animals ranging from micro-organisms to different kinds of birds, animals, plants, and weeds are affected due to the indiscriminate use of pesticides (Cedergreen and Streibig, 2005). As the farmers think "if little is good, a lot more would be better". The rampant use of pesticides caused tremendous loss to the environment and health of human beings. The pest controlling chemicals caused a subsequent imbalance in natural resources, as pesticides persisted and soaked in the environment for a longer time. The insecticide resistance was a serious problem which had gained impetus because of extensive use of insecticides. Previous exposure with insecticides can confer resistance to newly introduced insecticides through cross-resistance reducing the effectiveness of new insecticides (Rehan, Saleem, and Freed, 2011). The problem of the development of resistance to insecticides is more acute in *Spodoptera litura* (Fabricius) (Lepidoptera: Noctuidae) because of its polyphagous nature and rapid multiplication (Ramakrishnan *et al.*, 1984).

S. litura is resistant to a wide range of insecticides, which has led to sporadic outbreaks of the pest and failure of crops (Ahmad *et al.*, 2007).The cutworm *Spodoptera litura* Fab. (Lepidoptera, Noctuidae) is well known as a serious cosmopolitan pest with extensive host range of economically important crops such as cotton, groundnut, soybean, tomato, sweet potato. Base line data on the susceptibility of the target pest to these toxicant was the most important factor for insecticide use especially for monitoring the development of resistance. Resistance to insecticides has become a major problem associated with the chemical control of insect pests. Previous exposure and selection with insecticides was conferring cross resistance to newly introduced insecticides. Due to all such reasons it had become very important to study the insecticide resistance in the polyphagous pest *Spodoptera litura*, in laboratory conditions. The study was planned in Vadodara, Gujarat, where there was enormous use of insecticides in the fields of castor, cotton, cabbage, tomato and brinjal.

The current study is carried out on laboratory culture of *Spodoptera litura*. This pest was brought from the nearby fields of Vadodara, Gujarat. The insecticides are classified into various groups according to the toxicity levels *i.e.* extremely toxic, highly toxic, moderately toxic and slightly toxic. During the extensive surveys in the agricultural fields near Vadodara, it was found that common insecticides used to control *S.litura* are Deltamethrin, Cypermethrin, Chlorpyriphos, Coragen, Bifenthrin, Lambda Cyhalothrin, Endosulfan, Spinosad. The present study was planned using four insecticides namely Spinosad (Tracer®), Cypermethrin 25EC (synthetic pyrethroid belonging to group 3A) Chlorpyriphos 20EC, (belonging to Organophosphate group 1B), Spinosad 45SC (belonging to Spinosyn, group 5)

Thus the indiscriminate use of various types of insecticides and toxicity levels conferred by insecticides lead to taking up this research study. Following the reports of various insecticides which are sprayed by farmers in fields, this work was planned as a laboratory study in controlled environment. This study is expected to be fruitful in implementing integrated pest management strategies. It will prove to be a helpful data for the agriculture professionals as well as scientists working towards a common cause of resistance development in pests like *Spodoptera litura, Helicoverpa armigera, Plutella xylostella, Scirpophaga incertulas, Platydera gossypiella, Earias insulana, Percallia ricini, etc.* The purpose of this study is to take into account the development of resistance of *Spodoptera litura* a major pest of crops. Other pests such as those belonging to family Hemiptera, *Aphis gossypii, Aphis craccivora* and *Myzus persicae* (Family: Aphididae); *Bemisia tabaci* (Family: Aleyrodidae); *Phenacoccus solenopsis* (Family: Pseudococcidae) are considered as a serious threat to crops which were causing the major infestation to cotton, brinjal, cauliflower, cabbage, chickpea in the agricultural fields of Vadodara (Kataria and Kumar, 2012). Most studies were performed on populations collected from nearby locations, but this study takes into account the level of resistance developed in laboratory conditions. *Spodoptera litura* was reared for at least three generations on artificial diet to ensure the robustness and infection free condition of the culture. It was continuously monitored for six months and observations were recorded on daily basis for natural occurring mortality. These results can be extrapolated in field conditions, as the whole life cycle of the pest completes in natural surroundings and there are full chances of development of resistance, if rotation of insecticides is not followed (Rotation of Organophosphates and Carbamates with bio pesticides).

25.2 Experimental Section

25.2.1 Collection and Preservation

A site survey was done in some parts of Vadodara, Gujarat and populations of *Spodoptera litura* were collected from fields of nearby regions of Vadodara in Gujarat. The information on sprays occurring in these fields was recorded beforehand, taking help of the local farmers at the time of collecting the populations of the pest. A mixed culture containing mostly smaller instars like second and third instars were collected in separate bowls along with healthy leaves for survival. Pupae and adults were collected in plastic jars with holes. Tissue papers were kept in these containers and moist conditions were maintained so that the collected culture did not desiccate due to dry conditions. The collection of pest was done in the early morning between 6: 00 am to 8: 00 am. The pest was then reared in laboratory conditions at 27±2°C and 65-70 per cent relative humidity (RH). Incubators were used, to maintain constant conditions for the survival of test insect before conducting bioassays. The culture was reared for at least three generations so as to ensure the health and infection free nature. After successful rearing, next generation was selected for testing of insecticide through monitoring bioassay.

25.2.2 Rearing in Laboratory Conditions

Larvae of *Spodoptera litura* were collected from castor and cotton fields (200-300 larvae). They were reared in controlled laboratory conditions *i.e.* 25±2°C, 65-70 per cent RH and a photoperiod of L: D, 14: 10. It was reared on artificial diet. The diet was poured in a plastic container which had partitions in it. The larvae were carefully transferred on diet by using brush. The diet was changed at regular intervals. All the lab paraphernalia used for the whole process was pre- sterilized to avoid fungal and bacterial infections. The transferring of larvae becomes very crucial; hence utmost care was taken. Until pupation, the larvae were kept on artificial diet. Rearing in container was feasible as there was no cannibalism observed in *Spodoptera litura.* After complete formation of pupae, they were transferred to bowl. The completion of pupal stage leads to the beginning of adult emergence. As soon as adult emergence started, healthy male and female adults were released in oviposition pots in the ratio of 2: 2(male: female). Adult diet was also provided by using honey solution. The moths were provided with water and honey solution. Another method was used for rearing *i.e.* rearing on natural diet. The freshly laid egg masses from fields along with the leaves. The eggs were kept in Petri dishes (11 cm dia.) and were covered with fine muslin cloth and secured with rubber bands. The larvae were kept in rearing jars (15 cm × 13 cm) covered with muslin cloth and secured with rubber bands. They were daily supplied with fresh cabbage leaves for feeding. The adults were also kept in rearing jars (15 cm × 13 cm), supplied with a piece of folded paper for oviposition and a cotton swab dipped in 50 per cent honey solution was hanged from the top in order to provide feeding material for adults. The honey solution was renewed after every 48 hours.

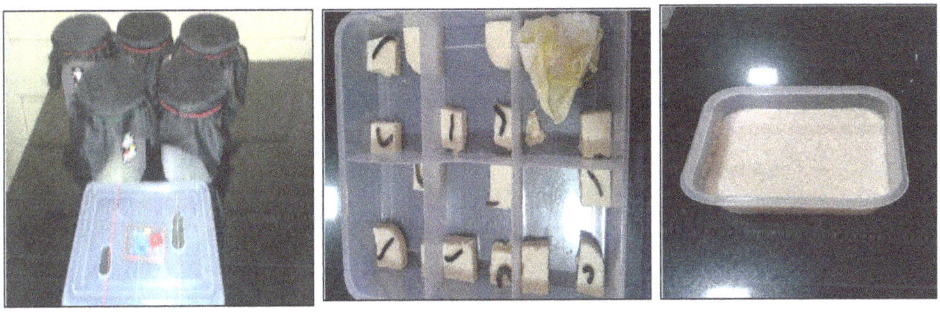

Figure 25.1. Rearing of *Spodoptera litura* in Lab Conditions (27± 2, °C 65 per cent RH).

The Petri dishes having *Spodoptera litura* eggs and rearing jars containing larvae and adults were kept in B.O.D. incubator maintained at 27 ± 2°C temperature and 65 ± 2 per cent RH. The culture was maintained and the second-generation adults (male and female) were kept in jars. Both types of rearing *i.e* natural diet and artificial diet were done, so as to ensure the survival of larvae for the testing against insecticide. The larvae which were reared on natural diet had many challenges like plant health, virus in plant material, fungus development, changing of diet on regular intervals. In contrast, those reared on artificial diet had different challenges like, diet developing infections, larvae preference for artificial diet *etc.* The larvae reared on artificial diet were then selected for second generation. The larval health was observed on regular intervals by different parameters like larvae weight, formation of pupae, type of wings in adults, number of egg masses and longevity of adult. Table 25.1 indicates the health parameters indicating healthy and infestation free culture.

Table 25.1: Health Parameters for Rearing of *Spodoptera litura*

Larvae (Total 10 nos)	*Days for Completion of Life Cycle*	*Larval Weight (g) Early Instar*	*Larval Weight (g) Late Instar*	*Weight of Food Consumed (g)*	*Adult Longevity (Days)*
Set-1	27	0.12	0.42	1.88	4
Set-2	25	0.13	0.58	1.96	5
Set-3	26	0.14	0.65	1.85	5
Set-4	27	0.11	0.55	2.10	4
Set-5	28	0.15	0.49	1.95	5
Set-6	27	0.12	0.50	2.00	4

25.2.3 Insecticide Selection

The insecticides were selected based on the market survey as well the field survey done. From the field survey done in the surrounding fields of Vadodara, the farmers were interviewed for the type of sprays, irrigation provided to the fields, maintenance, chemicals used if any and also the pests attacking the crops. In this survey, the data of insecticides majorly used in various crops were listed down. As the farmer was having less knowledge about the chemistries of these insecticides, they were only asked for the names of chemicals used for the sprays. The farm owners were informed beforehand, about the research work background for getting the whole information. There were fields of Cotton, Castor, Maize, Pigeon pea, Tobacco which were surveyed. The common insecticides used for these fields were Deltamethrin, Cypermethrin, Chlorpyriphos, Coragen, Bifenthrin, Lambda Cyhalothrin, Endosulfan, Spinosad. Thus, the insecticides belonging to IRAC (Insecticide Resistance Action Committee) groups were Organophosphates, Pyrethroids, Spinosyns, Diamides and Cyclodienes. Out of these the insecticides extensively used by the farmers are Pyrethroids, Organophosphates, Diamides and Spinosyns. Another survey was done to know the locations from where these pesticides were available in the market *i.e.* pesticides shops from where the farmer used to buy these chemicals. This was done to know the exact information about the insecticides and to know the approximate amount the farmer is spending in lure of getting a healthy crop. Fortunately, or unfortunately, the results of this survey were eye openers. The farmers were spending a whole lot of money just to maintain the crop. In response to this what they are getting in long run was only "Resistance development" to the same insecticides sprayed in the field. The survey done for the local pesticide market shops created a great interest and was a very useful piece of information, for selection of insectides for the bioassays. The commercial formulations of all the insectides were very easily available in the market.

25.2.4 Selection of Larvae for Bioassay

As described in the rearing procedure, the larvae of fourth generation were selected for resistance monitoring bioassay. The larvae selected for the experiment were pre-checked for any type of infections. After ensuring the healthy nature of larvae, third instar was selected for the bioassay. These were separated from mother culture and kept for starvation before initiating the experiment.

25.2.5 Leaf Dip Bioassay

The traditional leaf dip bioassay was conducted in laboratory conditions. Primary stock solutions of insecticides were calculated and bracketing was done to arrive the different concentration on third instar larvae of *S. litura*. Test solution was prepared by using the commercial formulation of the four insecticides. Different ppm concentrations were made, using serial dilution process. Healthy and infestation free cotton leaves were collected from field and they were washed in laboratory using distilled water. Leaf discs of five centimetres were cut. These leaf discs were dipped in the test solutions for ten seconds with gentle agitation and were placed on tissue papers for drying with adaxial surface. Natural drying was performed by giving enough time. The leaf discs were placed in petri plates having moist filter paper to avoid desiccation of leaves in ten replicates. The larvae was kept for starvation for one hour before exposing it to testing. On each leaf disc, three 3rd instar larvae (F1 generation) were released, using fine camel hair brush. All the test units were kept in controlled environmental conditions, humidity chamber (25±2°C, 65-70 per cent). The humidity chamber was properly checked to ensure the correct working according to the parameters set inside. Untreated check was also kept in which the leaf discs were treated with distilled water. After 72 hours, the test units were taken out of the chamber and brought to laboratory conditions. Anything unusual was captured in data sheet.

25.3 Results

The whole study of insecticide resistance monitoring was done in laboratory conditions, maintaining range of temperature 27°C to 29°C. Humidity conditions were maintained at 65 RH to 70 per cent RH. The collection of pests of *Spodoptera litura* was done from the fields of Vadodara during the year 2016 to 2017 mainly during Kharif seasons. During my field survey, it was observed that the major crops cultivated in the district were rice, wheat, sorghum (Juwar), yellow peas, grams, oil seeds, groundnut, tobacco, cotton and sugarcane.

25.3.1 Insecticide Resistance in *Spodoptera litura*: The Generation Study

The bioassays were performed in the year 2017 (March to August) and 2018 (June to September). The insecticides solutions were made considering the field recommended doses. Commercially available Cypermethrin 25EC was diluted to make different concentrations in ppm and was exposed to 3rd instar larvae of *Spodoptera litura*. In the first generation, the larvae showed expected results (Mortalities in concentrations starting from 0.125, 0.25 and 0.5 ppm.) The second, third and fourth generation also showed the mortalities in these concentrations. But in fifth generations, there was a decrease in mortalities, which indicated a low level of resistance. After completion of one insecticide, the culture was not exposed to any other insecticide to ensure its good health. A close watch was kept on the mother culture after the completion of insecticide resistance experiments. Moving forward in experimental procedures, the pest was exposed to Chlorpyriphos 20EC at different concentrations. In the fifth generation, there was a very low level of resistance observed at 0.5 ppm. This indicated the onset of resistance. But the resistance developed against Chlorpyriphos 20EC was low as compared to that developed by Cypermethrin 25 EC. The third insecticide was Spinosad (Tracer®) 45SC. Insecticide resistance was developed at 1 ppm showing mortality of 50 percent. But Spinosad being a biopesticide was considered to be an insecticide with low toxicity values. The last insecticide tested was Coragen 20SC. This insecticide showed alarming results. The data indicated the onset of resistance from the third generation itself.

25.3.2 Observation of Mortalities

Separate brushes were used for untreated check unit and treated units so as to avoid contamination. The test units were first checked for any kind of fungal/

bacterial infections. Larvae were considered to be dead, if there was no movement after contact with brush. Larvae was considered to be moribund, if it showed less and uncoordinated movement as compared to untreated check. Larvae were considered live if it showed normal movement when compared to untreated check.

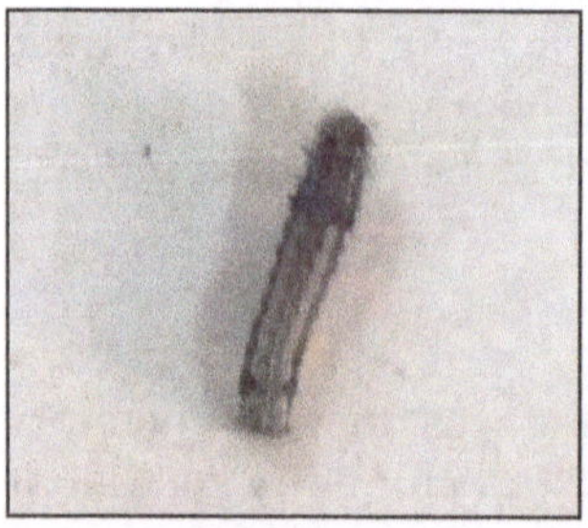

Live Larvae

Moribund Larvae

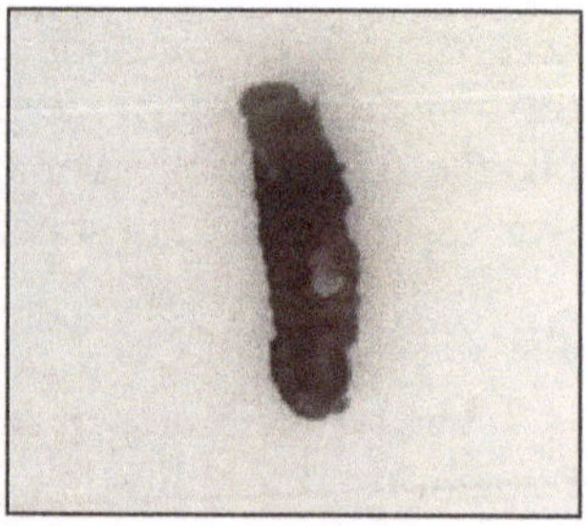

Dead Larvae

25.3.3 Data Analysis

Larval mortalities were recorded at 96 hours. The larvae were considered dead if they failed to make a coordinated movement when prodded with probe. Data was analysed for control mortalities using Abbott's (1925) formula. The data was further analysed by the probit analysis method through POLO-PC Program of LeOra, 2003. The resistance developed in the pest was concluded by decreasing mortalities and further from lethal concentration estimated like LC50, LC90 and other important statistical parameters. The larvae which survived to the different dosages (ppm) of testing insecticide were then mixed and continued to fourth generation for testing on the same dosages (ppm) The whole process was repeated for five generations. Hence the observations were taken insecticide wise and data was analysed.

25.3.3.1 Insecticide Cypermethrin 25EC

Table 25.2: Mortality Values of *S. litura* over Five Generations on Exposure to Cypermethrin 25EC

Sl.No.	*Concentration (ppm)*	*Percent Mortality (per cent)*				
		G-1	*G-2*	*G-3*	*G-4*	*G-5*
1	2	100.00	100.00	100.00	100.00	100.00
2	1	100.00	96.67	93.33	93.33	93.33
3	0.5	53.33	53.33	53.33	50.00	46.67
4	0.25	13.33	13.33	16.67	13.33	13.33
5	0.125	3.33	0.00	0.00	0.00	0.00
6	0.0625	0.00	0.00	0.00	0.00	0.00
7	Untreated check	0.00	0.00	0.00	0.00	0.00

* G: Generation.

Table 25.3: LC Estimates of Cypermethrin 25EC over Generations

LC Estimates	**G-1*	*G-2*	*G-3*	*G-4*	*G-5*
LC50	0.43	0.45	0.45	0.48	0.49
LC90	0.82	0.83	0.90	0.92	0.94
Slope ± Std Error	4.51±0.64	4.86±0.72	4.31±0.64	4.50±0.64	4.48±0.60
Chi square	4.99	0.82	0.83	0.68	1.08
Significance	0.28	0.93	0.93	0.95	0.89

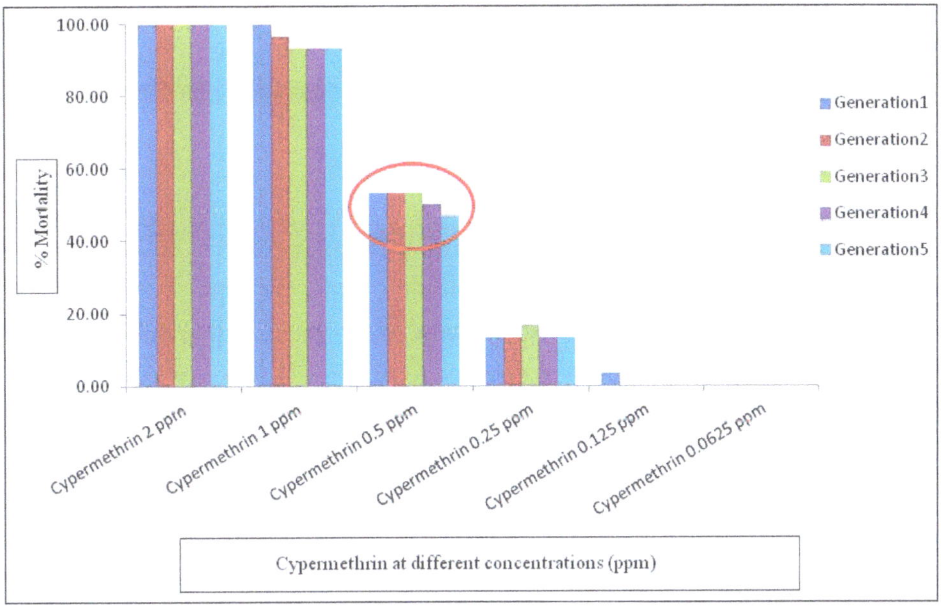

Figure 25.2: Graph Indicating Resistance in *S. litura* against Cypermethrin 25EC.

25.3.3.2 Insecticide Chlorpyriphos 20EC

Table 25.3: Mortality Values of *S. litura* over Five Generations on Exposure to Chlorpyriphos 20EC

Sl.No.	*Concentration (ppm)*	*G*-1*	*G-2*	*G-3*	*G-4*	*G-5*
1	6.25	100.00	100.00	100.00	100.00	100.00
2	1.25	100.00	100.00	96.67	96.67	93.33
3	0.25	96.67	93.33	93.33	93.33	96.67
4	0.05	53.33	56.67	53.33	50.00	46.67
5	0.01	16.67	20.00	16.67	20.00	20.00
6	0.002	0.00	0.00	0.00	0.00	0.00

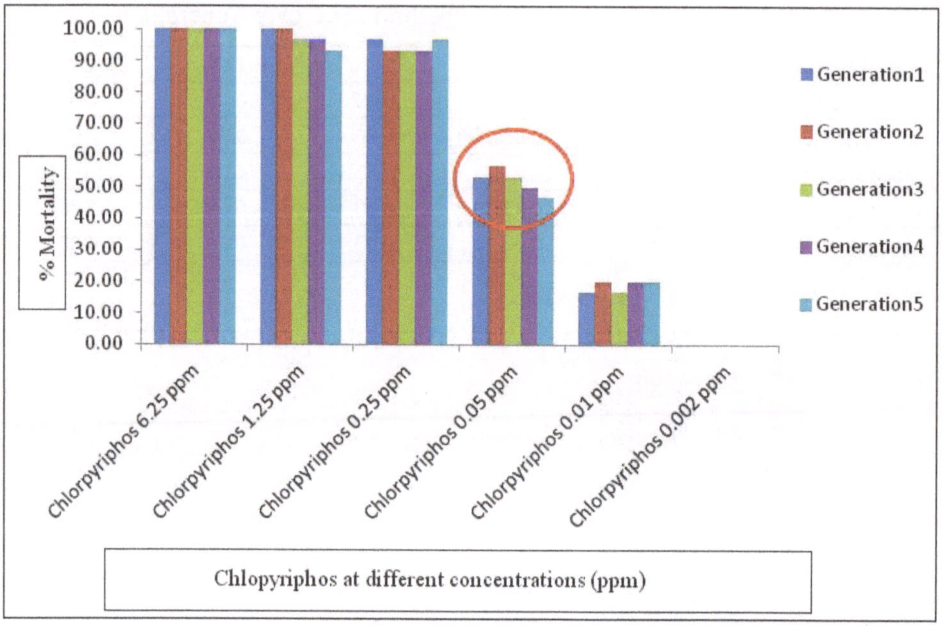

Figure 25.3: Graph Indicating Resistance in *S. litura* against Chlorpyriphos 20EC.

25.3.3.3 Insecticide Spinosad 45SC

Table 25.4: Mortality Values of *S. litura* over Five Generations on Exposure to Spinosad 45 SC

Sl.No.	Concentration (ppm)	G*-1	G-2	G-3	G-4	G-5
1	30	100.00	100.00	93.33	96.67	100.00
2	10	83.33	90.00	76.67	86.67	90.00
3	3	63.33	66.67	60.00	66.67	70.00
4	1	46.67	56.67	50.00	46.67	43.33
5	0.3	6.67	10.00	3.33	6.67	6.67
6	0.1	3.33	0.00	3.33	6.67	3.33
7	0.03	0.00	0.00	0.00	3.33	0.00
8	0.01	0.00	0.00	0.00	0.00	0.00
9	Untreated check	0.00	0.00	0.00	0.00	0.00

* G: Generation.

For Cypermethrin 25EC, the LC values observed in all the generations show a gradual development of resistance LC50 and LC90 in first generation for Cypermentrhin 25EC @ 0.5 ppm were 0.43 and 0.82 respectively. In the second generation, these values show a slight increase i.e 0.45 and 0.83 respectively. While in third and fourth generation,these attain a value of 0.45, 0.90 and 0.48, 0.92

Table 25.5: LC Estimates of Spinosad 45SC against *S. litura* over Generations

Treatment (Spinosad)	*Heterogeneity (Chi square)*	*LC 50*	*LC 90*	*Fiducial Limit (Upper)*	*Fiducial Limit (Lower)*	*Slope ± Error*	*Intercept ± Error*
Generation 1	6.00	1.68	10.31	21.72	6.30	2.79 ± 0.35	-0.63 ± 0.22
Generation 2	6.71	1.33	6.87	13.70	4.36	3.08 ± 0.40	-0.38 ± 0.23
Generation 3	5.64	2.24	17.78	41.48	10.19	2.44 ± 0.31	-0.85 ± 0.22
Generation 4	3.44	1.51	11.82	26.57	6.88	2.45 ± 0.30	-0.44 ± 0.20
Generation 5	3.65	1.38	7.06	14.07	4.49	3.10 ± 0.40	-0.43 ± 0.23

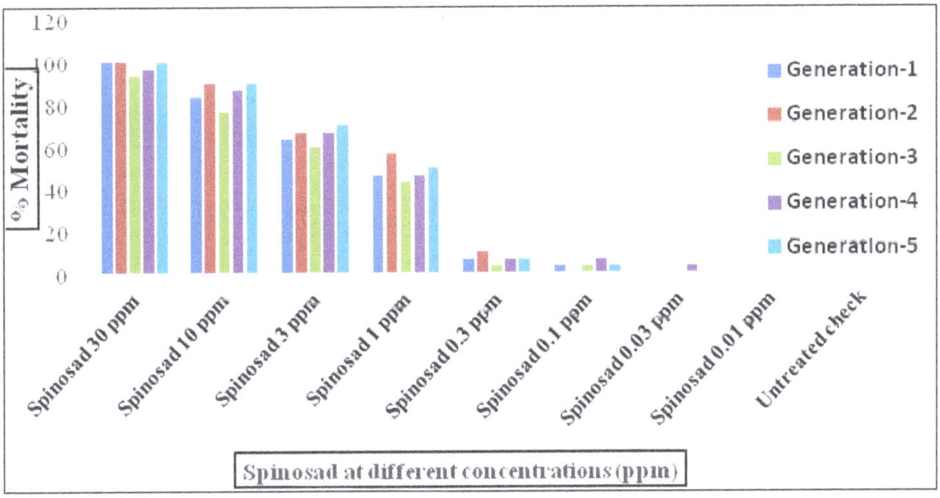

Figure 25.4: Graph Indicating Resistance in *S. litura* against Spinosad 45 SC.

respectively. When the bioassay was repeated in fifth generation, the LC50 and LC90 values indicate 0.49 and 0.94 respectively. Chlorpyriphos 20EC at different concentrations *i.e.* 6.25, 1.25, 0.25, 0.05, 0.01 and 0.002 ppm was exposed for five generations. The mortalities obtained in fifth generation were 100, 93.33, 96.67, 46.67 and 20.00 per cent respectively. The LC50 and LC90 values indicated less amount of resistance being developed in the pest. In the fifth generation, LC50 value and LC90 values were 0.05 and 0.34. There was an onset of resistance observed in Chlorpyriphos 20EC @ 0.05. All these values indicate onset of resistance in both the insecticides in laboratory conditions. For Spinosad 45SC, in the first generation, the LC50 and LC90 values were 1.68 and 10.31 respectively. The second generation indicated values 1.33 and 6.87 respectively. In the third and fourth generation 2.24, 3.34 and 1.51, 11.82 indicated low amount of resistance developing in the culture. In the fifth generation, 1.38 and 7.06 values also indicated on developemnt of resistance. The upper and lower Fiducial limits were found to be 1.17 and 2.24 for first generation.

25.4 Discussion

The current studies showed onset of resistance which may be considered as an alarming situation. At the same time, due to its less toxicity and biological properties. Spinosad being a bio pesticide may prove to be useful by keeping it one of the insecticides in rotation. These were the results of resistance monitoring against three insecticides *i.e.* Cypermethrin 25 EC, Chlorpyriphos 20 EC and Spinosad 45 SC. Repeated use of the same class of pesticides to control a pest can cause undesirable changes in the gene pool of a pest leading to a form of selection pesticide resistance. Studies on development of resistance in *Spodoptera litura*, showed onset of resistance in the fifth generation. This was observed for the insecticide Coragen® 20SC, indicated by the mortality value of 36.67 per cent at 1ppm in contrast to 40.00 per cent in the fourth generation at same concentration. There was an onset of resistance developed in *S. litura*, when exposed to Chlorpyriphos @ 0.05 ppm. In the fifth generation, there was a decrement in the values of mortality for the same concentration *i.e.* 0.05 ppm. Thus, there was moderate amount of resistance developed against Chlorpyriphos in third instar larvae in fifth generation. When the pest was exposed to Cypermethrin, it indicated the development of resistance from the fourth generation itself. Hence the sustainable agriculture not just includes agricultural practices, but also the deep studies of development of resistance development in agriculturally important pests. Hence such type of studies performed in laboratory conditions can serve as a powerful base for making the farmer understand the true importance of the correct dosage to be applied to their fields. Nowadays, most of the farmers rely on chemical methods to control pest attacks since fully organic farming is not possible if their fields are surrounded by other fields in which insecticidal sprays are performed. In such cases, the farmer can be just educated about the right dosage of insecticides as well as showing importance of chemical methods combined with biological methods to control insect pest. The industrialization is the backbone of any country's growth on the economy facet. But what are we giving back to the nature in response to the money earned?

Residues of pesticides? Is this really what we call progress, or just a hallucination????

25.5 Significant Findings and Conclusions

- ✰ The resistance developed in *Spodoptera litura* for four insecticides was in the order: **Cypermethrin 25EC > Chlorpyriphos 20EC > Spinosad 45SC**
- ✰ These results can be extrapolated to field conditions, as the whole life cycle of the pest completes in natural surroundings and there are full chances of development of resistance, if rotation of insecticides is not followed (Rotation of Organophosphates and Carbamates with biopesticides).
- ✰ Rotation of less toxic insecticides with bio-pesticides as well as usage of bio-products must be adopted to get maximum control which will lead

to high yield. This will cause less harm to the farmers as well as people consuming food ultimately.

- Insecticide sprays in the fields are used non-judiciously by the farmers. They are not having proper knowledge about the label of the particular insecticide and symbols placed on the labels, which indicate toxicity level of the same.
- Though this is difficult to make the farmer understand, it proves to be helpful to the scientists and agriculture professionals worldwide. They can slowly educate the farmer by showing the long-term effects of such insecticides used in extensive and haphazard manner. The best alternative to this is switching to organic farming or using bio-pesticides which are having less adverse effect on environment and more positive and healthy effects on plants.
- Information regarding the correct application of pesticides and the use of advanced technologies for target delivery of pesticide, as well as intensive training on selective application of the correct pesticides at the correct time for the correct pests, should be disseminated to the user group.
- The current studies showed onset of resistance in Chlorpyriphos, Cypermethrin and Spinosad in different generation with varying mortality values. This can be considered as an alarming situation. Spinosad being a bioinsecticide can be used in the rotation programs.
- Hence from these studies, it is concluded that the resistance and generation turn over holds a strong relation. As the generations of pest pass by, the insecticides create a resistant effect on the pest and it then becomes robust and slowly ineffective against the same pesticides
- Farmer awareness training programmes, flash mobs, interactive sessions with famers are the ways and means to educate the farmer to address their issues related to pest attack, development of resistance, application of pesticides, understanding the labels for dosage of insecticides.
- Thus with reference to the studies done so far in the area of resistance development against various insecticides, safety practices, pesticide use regulations, the correct application technologies and integrated pest management form the basis of developmental strategy for reducing the exposure to harmful pesticides

REFERENCES

Abhilash, P. C., and Singh, N. (2009). Pesticide use and application: An Indian scenario. *Journal of Hazardous Materials, 165*(1–3): 1–12.

Ahmad, M., Iqbal Arif, M., and Ahmad, M. (2007). Occurrence of insecticide resistance in field populations of Spodoptera litura (Lepidoptera: Noctuidae) in Pakistan. *Crop Protection, 26*(6): 809–817.

Cedergreen, N., and Streibig, J. C. (2005). The toxicity of herbicides to non-target aquatic plants and algae: Assessment of predictive factors and hazard. *Pest Management Science, 61*(12): 1152–1160.

Kataria, R., and Kumar, D. (2012). Occurrence and Infestation Level of Sucking pests/ : Aphids on various host plants in Agricultural Fields of. *International Journal of Scientific and Research Publications*, 2(7): 1–6.

Khanna, R., and Gupta, S. (2018). Agrochemicals as a potential cause of groundwater pollution/ : A review. *International Journal of Chemical Studies*, 6(3): 985–990.

Ramakrishnan, N., Saxena, V. S., and Dhingra, S. (1984). Insecticide-resistance in the population of Spodoptera litura (F.) in Andhra Pradesh. *Pesticides, 18*(9): 23–27.

Rehan, A., Saleem, M. A., and Freed, S. (2011). Baseline susceptibility and stability of insecticide resistance of *Spodoptera litura* (F.) (Lepidoptera: Noctuidae) in the absence of selection pressure. *Pakistan Journal of Zoology, 43*(5): 973–978.

Srinivasan, R. (2012). Integrating biopesticides in pest management strategies for tropical vegetable production. *Journal of Biopesticides, 5*(1): 36–45.

Chapter 26

Role of Fomites in Fungal and Viral Disease Transmission

Arun Arya

Department of Environmental Studies, Faculty of Science, The Maharaja Sayajirao University of Baroda, Vadodara – 390 002, Gujarat
e-mail: sarojarun10arya@rediffmail.com

ABSTRACT

It is evident that not all viruses behave alike toward a solid under identical conditions. Under most natural conditions viruses with a low isoelectric point appear to be more poorly adsorbed to most solid surfaces. This must be taken into consideration when evaluating concentration or treatment systems which involve adsorption. This phenomenon is also important in determining the transport of viruses in the environment. Certain enteric viruses appear to adsorb less readily to soils and aquatic sediments than others. Thus their potential transport to groundwater may be greater and they may be less likely to settle in surface waters. To take into consideration these differences in virus behavior, it is perhaps best to use viruses with widely varying isoelectric points or marked differences in hydrophobicity to evaluate the extremes in virus interaction with a given surface which could occur. The viruses most commonly associated with food- and waterborne outbreaks of gastroenteritis are the noroviruses. The lack of a culture method for noroviruses warrants the use of cultivable model viruses to gain more insight on their transmission routes and inactivation methods. Inactivation of suspended viruses was temperature and time dependent in the range from 0 to 100°C. UV-B radiation from 0 to 150 mJ/cm^2 caused dose-dependent inactivation, with a 3 D (D = 1 log_{10}) reduction in infectivity at 34 mJ/cm^2 for both viruses. Inactivation of viruses and fungi is possible by the use of 70 per cent ethanol. Sodium hypochlorite solutions were only effective at >300 ppm. For all treatments, detection of viral RNA underestimated the reduction in viral infectivity. Norovirus was never more sensitive than the animal caliciviruses and profoundly more resistant to low and high pH. Overall, both animal viruses showed similar inactivation profiles when exposed to heat or UV-B radiation or when incubated in ethanol or hypochlorite.The incomplete inactivation is achieved by ethanol and the high hypochlorite concentration.

Money in form of currency notes and coins is handled by persons of varying health and hygiene standards and also under varied environmental conditions. Potential spread of pathogenic microflora is possible through these fomites. The coins were found to have association of fungi like Aspergillus niger, Mucor heimalis, and Rhizopus stolonifer and bacteria like E. coli and Salmonella enteritidis. These organisms can survive on such paper and metallic surfaces for few days.

Keywords: *Fomites, Fungi, Virus, Transmission, Viability.*

26.1 Introduction

The role of fomites in the transmission of disease remains a controversial subject. Some epidemiological studies have suggested that contaminated surfaces may play a role in the spread of respiratory viruses (Hendley *et al.*, 1973) and laboratory studies have supported this hypothesis. Other studies have implicated environmental surfaces in the transmission of bacteria. However, the role of environmental surfaces in the transmission of disease remains an issue of scientific debate and fundamental information concerning the microbial transfer rates from environmental surfaces to the hands and from the hands to the mouth remains scarce. Separate studies have shown that microorganisms are more efficiently transferred from nonporous than from porous surfaces. Transfer of *Escherichia coli* from a laminate surface to fingers was 40 per cent up to two hours after the contamination event (Scott and Bloomfield 1990a) while the transfer efficiency of *E. coli* from a damp cloth to the human hand was only 0.47 per cent (Mackintosh and Hoffman 1984). It was showed that viral transfer from porous surfaces was poorer than from stainless steel although percent transmission was not described. The efficacy of transfer of rhinovirus from a donor's fingertips to a recipient's hands via door knobs was as high as 22 per cent (Pancic *et al.*, 1980). Studies do not provide direct comparisons of the transmission rates from surfaces to hands of bacteria and viruses. This information may be important to determine which types of pathogens will be most affected by environmental sanitation practices. Navlani *et al.* (2020) reported potential health risk from fungi present on Indian and foreign coins.

Worldwide annually there are 1.7 million deaths from diarrheal diseases and 1.5 million deaths from respiratory infections. Viruses cause an estimated 60 per cent of human infections, and most common illnesses are produced by respiratory and enteric viruses. Unlike bacterial disease, viral illness cannot be resolved with the use of antibiotics. Prevention and management of viral disease heavily relies upon vaccines and antiviral prevention methods. Both vaccines and antiviral medications are only 60 per cent effective. Additionally, to date there are no vaccines or antiviral drugs for most common enteric and respiratory viruses with the exception of influenza virus and hepatitis A virus (HAV). Consequently, viral disease spread is most effectively deterred by preclusion of viral infection. Increases in population growth and mobility have enhanced pathogen transmission and intensified the difficulty of interrupting disease spread. Control of viral disease spread requires a clear understanding of how viruses are transmitted in the environment. For centuries it was assumed that infectious diseases were spread primarily by the airborne route or through direct patient contact, and the surrounding environment played little or no role in disease transmission. Up until 1987 the Centers for Disease Control and the American Hospital Association focused on patient diagnosis due to the belief that nosocomial infections were not related to microbial contamination of surfaces. Over the years studies have changed the perspective on viral transmission to include a more complex multi-factorial model of disease spread. There is now growing evidence that contaminated fomites or surfaces play a key role in the spread of viral infections.

Viral transmission is dependent on interaction with the host as well as interaction with the environment. Viruses are probably the most common cause of infectious disease acquired indoors. The rapid spread of viral disease in crowded indoor establishments, including schools, day care facilities, nursing homes, business offices, and hospitals, consistently facilitates disease morbidity and mortality.

Fomites consist of both porous and nonporous surfaces or objects that can become contaminated with pathogenic microorganisms and serve as vehicles in transmission. During and after illness, viruses are shed in large numbers in body secretions, including blood, feces, urine, saliva, and nasal fluid. Fomites become contaminated with virus by direct contact with body secretions or fluids, contact with soiled hands, contact with aerosolized virus (large droplet spread) generated via talking, sneezing, coughing, or vomiting, or contact with airborne virus that settles after disturbance of a contaminated fomite (*i.e.*, shaking a contaminated blanket). Once a fomite is contaminated, the transfer of infectious virus may readily occur between inanimate and animate objects, or vice versa, and between two separate fomites (if brought together). Rusin *et al.* (2002) demonstrated that 65 per cent of virus could be transferred to uncontaminated hands and 34 per cent to the mouth. The nature and frequency of contact with contaminated surfaces vary for each person depending on age, personal habits, type of activities, personal mobility, and the level of cleanliness in the surroundings. Viral transfer and disease transmission is further complicated by variations in virus survival on surfaces and the release of viruses from fomites upon casual contact. Virus survival on fomites is influenced by intrinsic factors which include fomite properties or virus characteristics and extrinsic factors, including environmental temperature, humidity, *etc.*

If viruses remain viable on surfaces long enough to come in contact with a host, the virus may only need to be present in small numbers to infect the host. After contact with the host is achieved, viruses can gain entry into the host systems through portals of entry or contact with the mouth, nasopharynx, and eyes. Host susceptibility to viruses is influenced by previous contact with the virus and the condition of the host immune system at the time of infection.

There are many complex variables that influence virus survival on fomites, viral transfer from fomites, and viral infection of the host. As a result, direct experimental evidence of viral transmission via fomite has been very difficult to generate due to a variety of uncontrollable variables and the unpredictability of human infection. An example of the difficulty in producing illness in the host after exposure was indicated in the Gwaltney study using rhinovirus. Over a 10- year period, Gwaltney intranasally challenged 343 adults without rhinovirus antibodies and infected 95 per cent of the participants. However, only 30 per cent of the individuals who became infected displayed disease symptoms. Generally, the majority of laboratory and clinical evidence is considered indirect; however, fomite transmission data are supported by both epidemiological studies and intervention studies. Epidemiological data indicating transmission via fomite are also difficult to evaluate. This difficulty stems from problems in distinguishing between different routes of transmission, such as person-to-person transmission or autoinoculation. Currently, laboratory studies, epidemiological evidence, and disinfection intervention studies have generated strong indirect and circumstantial evidence that supports the involvement of fomites as a vehicle in respiratory and enteric virus transmission. Studies from a variety of disciplines investigating viruses clearly support the following: (i) most respiratory and enteric viruses can survive on fomites and hands for varying lengths of time; (ii) fomites and hands can become contaminated with viruses from both natural and laboratory sources; (iii) viral transfer from fomites to hands is possible; (iv) hands come in contact with portals of entry for viral infection; and (v) disinfection of fomites and hands interrupts viral transmission.

26.2 Viability of Viruses

The potential for a virus to be spread via contaminated fomite depends first on the ability of the virus to maintain infectivity while on the fomite surface. Viruses are obligate parasites; therefore, the level of viral infectivity on a fomite can only decrease over time. Several studies have demonstrated that viruses can remain infective on surfaces for different time periods. The length of time a virus remains viable depends on a number of complex variables. In general, UV exposure and pH have minimal effects on viral survival in indoor environments. Viral survival may increase or decrease with the number of microbes present on a surface. Increasing amounts of microbes can protect viruses from desiccation and disinfection, but deleterious effects may also result from microbial proteases and fungal enzymes. Typically, viral presence on fomites may decrease with surface cleanliness and increase with surface usage. However, some cleaning products or disinfectants are ineffective against viruses and can result in viral spread or cross-contamination of surfaces. Easily measured and predictable factors that influence viral survival on surfaces include fomite properties, initial viral titer, virus strain, temperature, humidity, and suspending medium. Intrinsic factors, like fomite properties, virus strain, and viral inoculation titer, consistently impact the total virus survival end point in days/hours. The majority of viruses remain viable longer on nonporous surfaces; however, there are exceptions. Astrovirus survives for 90 days on porous paper but only 60 days on nonporous aluminum. Initial inoculation titer can prolong viral survival on environmental surfaces. Brady *et al.*, found that the viral survival decay rate increased with inoculum titer: a 104 virus Skin, hands, door knob, towels, medical cart items, carpets, curtains, lockers, bed covers, bed rails, drinking cup, water jug handle, lampshade, drinking glasses.

Virus survival on fomites can also vary significantly within viral type and strain. Typically, nonenveloped enteric viruses remain viable longer on surfaces than enveloped respiratory viruses. The enteric viruses HAV, astrovirus, and rotavirus can all remain infective on surfaces for 2 months or longer. In contrast, respiratory viruses usually remain viable for several hours to several days.

26.3 Viability of Fungal Spores

Germination is commonly used to monitor spore viability, but many fungal spores do not germinate readily. Rapid viability tests are needed. Methods were developed to monitor viability, based on enzyme activity and ATP content of fungal spores and hyphae. ATP content, measured by the luciferin-luciferase system, was much higher in living spores than in dead spores. Hydration was an additional aid in differentiating the ATP content of viable spores compared with dead spores. Approximately 40 µg of spores were required to measure the ATP level. Preparation from 25 µg of living spores or 100-150 µg fresh hyphae contained sufficient phosphatase or glucosidase to give a positive fluorescence test for hydrolysis of methylumbelliferyl-phosphate or methylumbelliferyl-glucoside at 37° C in 2 h. Dead spores and hyphae did not fluoresce. These methods were used to test the viability of : *Tilletia controversa, T. caries, Ustilago scitaminea, Puccinia striiformis, Pisolithus tinctorius, Rhizopogon colossus, and Elaphomyces granulatus spores,* and *hyphae of T. caries, Rhizoctonia solani, Hymenochaete tabacina, Phellinus weirii, Haematostereum sanguinolentum and Gaeumannomyces graminis var. tritici.* These techniques should be useful in determining quickly the effects of fumigants and fungicides on fungal spores and hyphae (Yu *et al.*, 1984).

Five methods of inducing oospore germination in *P. manshurica,* the causal agent of soyabean downy mildew, were attempted, but none was successful. Five different staining methods were therefore tested for their suitability for determining oospore viability. Tetrazolium chloride and nitroblue tetrazolium chloride primarily stained oospores which were cytologically abnormal and appeared to be degenerating: the dyes were presumed to be excluded from the cytoplasm in cytologically normal oospores by the oospore wall or plasmalemma. Strong autofluorescence of dead/ degenerating oospores in the fluorescein diacetate test made this an unreliable technique for assessing viability. Oospore nuclei were not stained by a modified Feulgen technique, probably because the stain did not penetrate the oospore wall. Phloxine B consistently stained degenerating oospores and a small proportion of oospores appearing cytologically normal under the light microscope. Parallel experiments with live and dead yeast cells (*Saccharomyces cerevisiae*) showed that only dead cells became stained, presumably because the semi-permeability of live cells excludes the stain. Phloxine B is thus considered a promising stain for determination of oospore viability in *P. manshurica.*

In 1945 the late G. R. Bisby reported that sporophores of the agaric *Schizophyllum commune* Fr. collected by A. H. R. Buller at Winnipeg in 1909 and 1911 and freeze-dried in 1910 and 1912 had retained their viability up to 35½ years. Pure cultures were obtained and in June 1945, two months after submitting his communication for publication, they developed sporophores.

Regional variation in spore deposition and viability was studied for two fungi, *Fomitopsis rosea* (Alb. and Schwein.: Fr.) P. Karst. and *Phlebia centrifuga* P. Karst., both confined to old growth spruce forests in the boreal zone. Seven regions in

Sweden were studied along a north south transect in which the historical impact from forestry increases and the amount old forests decreases towards the south. The two southernmost regions were located outside the distribution border of the species. Spore deposition was measured species specifically as heterokaryotisation of homokaryotic mycelia growing on wood discs. There was a significant decline in spore deposition towards the south for both species. *F. rosea* deposited an average amount of 111 spores m^{-2} 24 h^{-1} in the northernmost region compared to less than 1 spore in the four southernmost regions. The corresponding values for *P. centrifuga* were 27 spores m^{-2} 24 h^{-1} in the north compared to less than 2 spores in the 4 southernmost regions. No deposition was found south of the distribution borders. The viability of spores from local populations within each region was measured as germination success on nutrient media. Individual fruiting bodies from large populations in the north generally produced spores with higher germinability than fruiting bodies from geographically isolated populations in the central and southern regions. However, there was a high variation among the southern populations. Our data suggest that some populations in mid and south Sweden may suffer from negative genetic effects, possibly associated with fragmentation and loss of habitat. Thus, the combination of low spore deposition and low germinability of spores may be a threat to the long term persistence of *F. rosea* and *P. centrifuga* in southern Sweden (Edman *et al.*, 2004).

When considering fungal contamination in indoor habitats, the ability to colonize is not only dependent on spore survival, but also on the ability for spores to adhere to a surface, germinate, and grow. Germination and hyphal growth are established through polarized growth (Kwon *et al.*, 2011). In *A. niger*, the stabilization of polarity axes during germination is dependent on the Rho GTPase RacA. A *racA* deletion displays a hyperbranching phenotype which results in compact colonies (Kwon *et al.*, 2011). Spore adhesion to surfaces is facilitated by proteins in the cell wall that help fungi to grow on a wide-range of substrates (*e.g.*, from quartz used on windows to silicone and polycarbonate used in medical/scientific instruments). Furthermore, fungal growth is surface-associated, which can induce biocorrosion. In fact, fungal-induced biocorrosion has led to major problems in spacecraft safety such as those in the Mir space station (Novikova *et al.*, 2006). Understanding whether polar growth impacts spore revival and subsequent surface-associated growth is important to better control fungal contamination in the spaceflight context.

Pigments, such as melanin are involved in different cellular processes from adhesion to virulence (Eisenman and Casadevall, 2012), and are known to help cells against radiation-induced stress and ROS (Cockell and Knowland, 1999). Studies concerning the role of pigmentation in radiation resistance of fungal spores have been performed with monochromatic UV-C (254 nm) and pulsed light up to 1770 J/m^2 on inoculated agar with concentrations up to 10^7–10^8 spores/ml (Esbelin *et al.*, 2013). This study shows that spores of *A. niger* are extremely resistant to space radiation. Spores were able to withstand high doses of X-ray (LD_{90} = 360 Gy), cosmic radiation (helium-ion LD_{90} = 500 Gy; and iron-ion LD_{90} = 100 Gy), and UV-C radiation (LD_{90} = 1038 J/m^2). *A. niger* spore resistance to UV-C is particularly interesting, given that it is even higher than that of other radiation-resistant microorganisms (*e.g.*, *D. radiodurans*). Air-drying the spores made them more susceptible to X-ray radiation. Moreover, wild-type spores treated with 1000 J/m^2 UV-C were shown to have decreased biofilm formation ability (81 per cent reduction). It is important to note that the ionizing radiation doses used in this study (up to 1000 Gy) are multiple times higher than doses expected from traveling in interplanetary space. For example, a 360-day round-trip to Mars would yield a dose of 0.66 ± 0.12 Gy (Zeitlin *et al.*, 2013). It is therefore unlikely that *A. niger* spores become easily inactivated due to space radiation alone. We thus recommend that current planetary protection guidelines are revisited to address the high resistance of fungal spores in space travel scenarios. In addition, further studies are needed in order to address fungal spore resistance to other space environmental factors such as vacuum, changes in pressure, and extreme temperature fluctuations.

26.4 Effect of Humidity on Survival of Fungi

Humidity is essential requirement for the growth of fungi. Bilgrami and Verma (1978) reported 95-100 per cent RH supports best growth of most fungi and humidity below 80 to 85 per cent is inhibitory for them. However, Pathak and Dhawan (1989) stated that the most preferred RH for the fungi is 75to 80 per cent. Some fungi called xerophiles associated with stored products can grow at RH 75 to 86 per cent. The conidia of powdery mildew fungi have moisture present in them and no external moisture is needed for germination of these members. The growth of mushrooms occurs during rainy season, as the basidiospores germinate and forms basidiocarp in presence of moisture in atmosphere. A large number of aphyllophores produce perennial fruiting bodies and these emit huge quantity of basidiospores, which are blown by wind and germinate when conditions are moist.

26.5 Effect of Humidity on Survival of Viruses

The higher inactivation coefficient found among respiratory viruses indicates a faster decay rate or decreased survival on surfaces. Variations in virus survival may also occur within a viral family or strain, as seen between the 12-h survival of coronavirus 229E and the 3-h survival of coronavirus OC43. Consequently, variations in fomite composition, initial viral inoculation, and virus type can dramatically influence the amount of time the virus survives on a surface. Extrinsic environmental factors, such as temperature, humidity, and surrounding viral medium, have a varying effect on viral decay rate, depending on the viral strain. In the Abad *et al.*, study, media changes had no noticeable effect on enteric virus survival (HAV and rotavirus); Humidity influences the viral desiccation rate. Humidity in the United States can range from 14 to 94 per cent in outdoor environments. The study found that rhinovirus exhibits optimum survival at 50 per cent humidity (Sattar *et al.*, 1987).

26.6 Effect of pH and Presence of Salts

In general, pH has a minimal effect on virus survival in an indoor environment. Hurst *et al.* (1980) measured the effects of several environmental conditions on virus persistence in soil. They found that the temperature and virus adsorption to soil is more important for virus survival than the effect of pH. Enteric viruses are able to survive severe conditions in the gastrointestinal tract, such as low pH in stomach or high bile concentration. They are generally most stable near pH 7, but prefer low pH's (3–5) rather than alkaline pH's (9–12). It was shown that noroviruses are able to persist at pH 2.7 and room temperature more than 3 h (Duizer *et al.*, 2004). As the surface charge of viral particles varies depending on the pH, disruption of electrostatic interactions between viral particles and environmental surfaces commonly consists of changing the pH, and thus affect viral persistence in the environment (Gerba, 1984).

The viral adsorption to the surfaces can be influenced by ionic strengths; therefore, salts are commonly used to favour the attachment of viruses to different types of surfaces. Theoretically, adsorption of viral particles is better in high ionic strength. Enteric viruses are destabilized and inactivated by water lacking salt ions, *e.g.* Mg^{2+}. On the contrary, increased concentrations of salts (*e.g.* NaCl) are antiviral for many viruses.

26.7 Negative Influence of Other Microorganisms on Virus Survival

One proposed mechanism of viral loss in the environment is inactivation by the direct or indirect action of microorganisms. Environmental isolates of bacteria with antiviral ability have been found on several occasions. These microbes are able to produce metabolites, which adversely affect viral particles, or can use the viral capsid as a nutrient source (Deng and Cliver 1995; Ward 1982). The inactivation of viruses by bacterial cultures is temperature dependent. The lower the temperature of the mixed waste, the longer the virus is able to persist. Moreover temperature could strongly influence microbial activity, and thus influence viral persistence (Deng and Cliver 1992).

Ward (1982) investigated the influence of mixed-liquor suspended solids (MLSS) of inactivated sludge on a poliovirus 1 survival. The first experiment was conducted in order to determine the effect of MLSS on the recovery of poliovirus after a different period of time. The second experiment was designed to find out the role of different MLSS components in virus loss. Results of these studies indicate that MLSS contain a component(s) which can inactivate poliovirus 1. These components were pelleted during centrifugation, destroyed by autoclaving, and removed by filtration. Consecutively, residual activity of an MLSS supernatant fraction was also studied to confirm the absence of antiviral activity in non-living heat-sensitive material in the MLSS. Subsequent increasing activity of this supernatant, coupled with the previous results, strongly indicates the antiviral activity of some microbial species. Deng and Cliver (1992) demonstrated the antiviral effect of several bacterial cultures from swine manure slurry and mixed septic tank effluent. These cultures were identified as *Micrococcus luteus*, *Staphylococcus epidermidis*, *Bacillus* sp. and *Streptococcus sanguis* group. A comparison of poliovirus 1 inactivation in raw mixed waste, autoclaved mixed waste, and bacterium-free filtrate of raw mixed waste demonstrated that virus inactivation is related, at least in part, to microbial activity in similar environmental conditions. Inhibition of poliovirus 1 inactivation by protease inhibitors suggested that antiviral activity of mixed waste was partially due to proteolytic enzymes produced by bacteria in wastes.

Bacteria may also produce substances that inactivate viruses by processes other than enzymatic ones, *e.g. Pseudomonas aeruginosa* can produce substances with molecular weights below 500 Da which appear to inactivate viral particles. Substances with such low molecular weights cannot act enzymatically and they are referred to as virolytic substances (Cliver and Hermann 1972; Deng and Cliver 1992). Deng and Cliver (1995) studied the role of the microbial activity of animal wastes in inactivation of HAV. Ten out of 31 bacterial isolates were able to efficiently inactivate HAV. The inactivation capacity of four of the nine culture filtrates was significantly reduced by incubation with selected protease inhibitors before the virus was added. These inhibitors did not affect the activities of the other five culture filtrates. Fractions prepared by ultrafiltration (nominal molecular weights <1,000 Da) from two of these cultures inactivated HAV suggested that their mode of action was not enzymatic.

26.8 Transmission of Respiratory Viruses

Several different viruses cause respiratory infections, including respiratory human parainfluenza virus (1 thru 4) (HPIV), influenza virus (A and B), human coronavirus (SARS, OC43). It is generally accepted that respiratory viruses are spread person to person via aerosol transmission. Nevertheless, current scientific evidence also suggests that fomites are an important vehicle in the spread of respiratory viruses. By using an aerosolized source, HPIV1 was found to infect only 2 of 40 children Therefore, HPIV transmission by aerosol was considered improbable; however, transmission may have taken place by surface contamination or close contact. Respiratory viruses cause sneezing and coughing, which expel an estimated 10^7 infectious virions per ml of nasal fluid. Nasal secretions can travel at a velocity of over 20 m per second and a distance greater than 3 m to contaminate surrounding fomites. A hospital in Taiwan used reverse transcriptase PCR to detect coronavirus on hospital phones, doorknobs, computer mouses, and toilet handles during an outbreak of severe acute respiratory syndrome (SARS).

Studies have proven that RSV, HPIV, influenza virus, coronavirus, and rhinovirus can remain viable on fomites for several hours to several days. Avian influenza virus was detected on several surfaces for over 6 days. A study by Hendley *et al.* (1973) found that 1 in 2.7 hospital grand round attendees rubbed their eyes and 33 per cent picked their nose within a 1-hour observation period. Indirect evidence from clinical and laboratory studies clearly supports the involvement of fomites in respiratory virus infection. A study by Gwaltney *et al.* (1982) observed that 50 per cent of subjects developed infections after hand shake.

26.9 Conclusion

Fomites are solid surfaces which consist of both porous and nonporous surfaces or objects that can become contaminated with pathogenic microorganisms and serve as vehicles in transmission. During and after illness, viruses are shed in large numbers in body secretions, including blood, feces, urine, saliva, and nasal fluid. Fomites become contaminated with microbes including viruses by direct contact with body. Once a fomite is having pathogenic agent it can serve as resting place and can help in spread of disease for a varied period of its survival. Many fungal spores can survive with thick sporopollenin wall and melanin help in survival and pathogenicity. Studies have proven that RSV, HPIV, influenza virus, corona virus, and rhinovirus can remain viable on fomite for several hours to several days. Avian influenza virus was detected on several surfaces for over 6 days. Indirect evidence from clinical and laboratory studies clearly supports the involvement of fomites in respiratory virus infection.

REFERENCES

Ainsworth G.C. (1962) Longevity of *Schizophyllum commune*.II *Nature* volume 195, pages1120–1121

Cliver, D. 0. and Herrmann, J. E. (1972) Proteolytic and microbial inactivation of enteroviruses. Water Res. 6, 797-805.

Cockell, C. S., and Knowland, J. (1999). Ultraviolet radiation screening compounds. *Biol. Rev. Camb. Philos. Soc.* 74, 311–345. doi: 10.1111/j.1469-185x.1999.tb00189.x

Cortesao M. Hass A.D. Unterbusch R. FujimoriA schutze T, Meyer V. and Moeller R. (2020)*Aspergillus niger* Spores Are Highly Resistant to Space Radiation. Front. Microbiol., 03 *April 2020 | https: //doi.org/10.3389/fmicb.2020.00560*

Deng M.Y. and Cliver D.O. (1992) Inactivation of poliovirus type 1 in mixed human and swine wastes and by bacteria from swine manure. Appl Environ Microbiol. 58(6): 2016–2021.

Deng, M. Y. and Cliver, D. O. (1995) Persistence of inoculated Hepatitis A virus in mixed human and animal wastes. Appl. Environ. Microbiol. 61, 87-91.

Duizer *et al.*, Erwin Duizer, Paul Bijkerk, Barry Rockx, Astrid de Groot, Fleur Twisk, Marion Koopmans 2004Inactivation of Caliciviruses. Applied and Environmental Microbiology. 70(8) p. 4538–4543 **DOI:** 10.1128/AEM.70.8.4538-4543.2004

Edman M., Gustafsson M., Stenlid J. and Ericson L. (2004) Abundance and viability of fungal spores along a forestry gradient – responses to habitat loss and isolation?Oikos 104(1): 35-42https: //doi.org/10.1111/j.0030-1299.2004.12454.x

Eisenman, H. C., and Casadevall, A. (2012). Synthesis and assembly of fungal melanin. *Appl. Microbiol. Biotechnol.* 93, 931–940. doi: 10.1007/s00253-011-3777-2

Esbelin, J., Mallea, S., Ram, A. F., and Carlin, F. (2013). Role of pigmentation in protecting *Aspergillus niger* conidiospores against pulsed light radiation. *Photochem. Photobiol.* 89, 758–761. doi: 10.1111/php.12037

Gwaltney, J. M., and Hendley, J. O. (1982). Transmission of experimental rhinovirus infection by contaminated surfaces. *American Journal of Epidemiology, 116,* 828–833.

Gerba C.P. (1984). Applied and Theoretical Aspects of Virus Adsorption to Surfaces. Advances in Applied Microbiology, 30: 133-168

Hendley, J. O., Wenzel, R. P., and Gwaltney, J. M. (1973). Transmission of rhinovirus colds by self-inoculation. *New England Journal of Medicine, 288,* 1361–1364.

Hurst C. J. Gerba C.P. and Cech I. (1980) Effects of Environmental Variables and Soil Characteristics on Virus Survival in Soil. APPLIED AND ENVIRONMENTAL MICROBIOLOGY, Dec. 40: 1067-1079

Kwon, M. J., Arentshorst, M., Roos, E. D., Van Den Hondel, C. A., Meyer, V., and Ram, A. F. (2011). Functional characterization of Rho GTPases in *Aspergillus niger* uncovers conserved and diverged roles of Rho proteins within filamentous fungi. *Mol. Microbiol.* 79, 1151–1167. doi: 10.1111/j.1365-2958.2010.07524.x

Mackintosh, C.A. and Hoffman, P.N. (1984) An extended model of transfer of microorganisms via the hands: differences between organisms and the effect of alcohol disinfection. *Journal of Hygiene, Cambridge* 92, 345– 355.

Navlani D. Maheshwari A. Thakkar V. Taunk R. Chaudhary N. and Arya A. (2020) Fungal flora associate with Indian and Foreign coins and their potential health risk. Acta Scientific microbiology. 3(3): 1-4

Novikova, N., De Boever, P., Poddubko, S., Deshevaya, E., Polikarpov, N., Rakova, N., *et al.* (2006). Survey of environmental biocontamination on board the International Space Station. *Res. Microbiol.* 157, 5–12. doi: 10.1016/j.resmic.2005.07.010

Pancic, F., Carpentier, D.C. and Came, P.E. (1980) Role of infectious secretions in the transmission of rhinovirus. *Journal of Clinical Microbiology* 12, 567– 571.

Roongruangsree, U. T., Kjerulf-Jensen, C., Olson, L. W., Lange, L. (1988) Viability tests for thick walled fungal spores (ex: oospores of *Peronospora manshurica*). Journal of Phytopathology1988 123 3 244-252 DOI 10.1111/j.1439-0434.1988.tb04474.x

Rusin P. Maxwell S, Gerba C. (2002) Comparative surface to hand and fingertip to mouth transfer efficiency of gram positive bacteria, gram negative bacteria, and phage Journal of Applied Microbiology, 2002 – 93(4): 585-592

Sattar, S. A., Karim, Y. G., Springthorpe, V. S., and Johnson-Lussenburg, C. M. (1987). Survival of human rhinovirus type-14 dried onto nonporous inanimate surfaces–effect of relative-humidity and suspending medium. *Canadian Journal of Microbiology*, 33, 802–806.

Scott, E. and Bloomfield, S.F. (1990a) The survival and transfer of microbial contamination via cloth, hands and utensils. *Journal of Applied Bacteriology* 68, 271– 278.

Yu, S.Q.Trione E.J. and ChingT.M. (1984) Biochemical Determination of the Viability of Fungal Spores and Hyphae. Mycologia Vol. 76, No. 4 (Jul. - Aug., 1984), pp. 608-613

Zeitlin, C., Hassler, D. M., Cucinotta, F. A., Ehresmann, B., Wimmer-Schweingruber, R. F., Brinza, D. E., *et al.* (2013). Measurements of energetic particle radiation in transit to mars on the mars science laboratory. *Science* 340, 1080–1084. doi: 10.1126/science.1235989

Chapter 27

Changing Life and Changing Lifestyle Post COVID

K. Shruti, Nirmal Chaudhary, Soumyadeep Choudhury, Minu Kumarswami, Vrij Shah and Neha Singh

Department of Environmental Studies, Faculty of Science, The Maharaja Sayajirao University of Baroda, Vadodara – 390 002, Gujarat
e-mail: neha8277@gmail.com

ABSTRACT

Corona virus pandemic caused a seismic tear in the lives of our world. With every continuous update about new cases being pumped out from every corner of the world it has pushed us one more step towards fighting and preparing to escape it.The suspenseful state of anticipation of the dreaded future for our loved ones and us had been a challenge lately. The pandemic has pushed everyone to take a step back and made us realize what **HOPE** *is and that it will pull us out of difficult times to come out stronger than ever. Everyone has problems and takes the challenge of life to live its fullest. The story of RNA containing a pathogenic virus unfolded from Wuhan city of China in December last year and engulfed about 200 countries affecting health of millions of countrymen all across the developed Europe and America to Brazil and India. The deadly disease not only affected the economy of nations but loss of life due to severe pathogenic potential of virus.*

With almost no one outside and only sounds that can be heard are birds chirping, calming rain drops, slow whispering of trees and all the bliss that we were missing out nature had found a way to make us realize and relieve again. The world is on a Pause and has changed many of our lives. This change in urban life has resulted into new normal. No theatre no movies and public functions teaching online through mobile and internet is what we are observing now. People are working from home lot of vehicular pollution is reduced and a change in behavior of wildlife animals is seen. Just like birds in cage now man has put himself in isolation and trying to save from deadly virus.

Keywords: *People, Virus, Health life style, Wild life, Anthropause.*

27.1 Introduction

"If life were predictable it would cease to be life, and be without flavour"

— *Eleanor Roosevelt*

Over the past few months, the life has come to a halt. The world went into lockdown to control the spread of COVID-19. Brought about by the most tragic circumstances, this period of unusually reduced human mobility and we are facing isolation in different phases. There are two major influences: staying at home (which includes digital-education, smart working, limitation of outdoors and in-gym physical activity) and stockpiling food, due to the restriction in grocery shopping. In addition, the interruption of the work routine caused by the quarantine could result in boredom, which in turn is associated with a greater energy intake (Moynihan, 2015). In addition to boredom, hearing or reading continuously about the COVID-19 from media can be stressful. Stress leads subjects toward overeating, especially 'comfort foods' rich in sugar, defined as "food craving" (Yýlmaz, 2020; Rodríguez, 2015). Those foods, mainly rich in simple carbohydrates, can reduce stress as they encourage serotonin production with a positive effect on mood (Ma *et al.*, 2017). However, this food craving effect of carbohydrates is proportional to the glycemic index of foods that is associated with the increased risk of developing obesity and cardiovascular diseases, beyond a chronic state of inflammation, that has been demonstrated to increase the risk for more severe complications of COVID-19 (Wu *et al.*, 2020 Muscogiuri, 2020). People are doing exercise and yoga to keep them fit.

27.1.1 Changing Life in Cities and Villages

People were used to get up early in the morning in different cities and were busy in daily schedule going office or in industry. Children were passing their time in schools and colleges. The main priority is changed now people are worried about their health. Meditation, yoga, exercise and a healthy balanced diet are the ultimate goals to escape the difficult times. Along with the modern medicine the homeopathic, ayurvedic, and even the home remedies have spring into action to battle the viral infection. People are not leaving any stone left unturned.

The social distancing has made us stay away from each other but as long as our hearts are connected we are together. Lockdown has made us realized who we truly are, what are our goals and priorities and to whom do we value the most ? We are able to realize the sacrifices that our parents have made to fulfill our demands and in return they only want quality time with us. The shower of love and care and all the positivity just helps to boost our strength each day.The tough phase also helped us to strengthen our talents and weaknesses to kick start our life once again in full throttle to a better future. For some it's a new course, for some it's an experiment while some become adventurous and start to understand nature better while someone just want to have a calm and peaceful mind for their organized and optimal work.Self realization due to lockdown phase is a resting phase to bounce back strong into reality.

Countless stories have surfaced the web about all of us breaking our barriers and doing things out of our comfort zone. People are coming out and helping others in need and for others it may be gender equality of shared responsibility of home and office. For some it may be breaking boundaries to find out a cure for the virus infection while for some it may be keeping everyone safe from virus and fighting it at all costs. Wearing of masks is new normal people are caring to sterilize their hands with soap and water or any alcoholic disinfectant. In cities community kitchens are working to provide help and food to needy ones both homeless and elders unable to cook themselves.

Not much has changed in the life of a villager as the government had allowed agricultural work as usual. Lot of migration of labourers has occurred from different cities as industries were shut and no transport was allowed. Poor migrant labourers have to walk on foot many kilometers to reach their respective villages. The lack of regular workers has also affected the harvesting of crops.

27.1.2 Looking at small things

The little joy and happiness surrounding us that we earlier used to miss are now the best things of the day making us smile and feel nice. Be it peacefully listening to birds chirping, enjoying the sunrise and sunset and sometimes the pit pat patter of rain drops falling on the window panes. The small moments of our lives or well the un-noticed things in the busy and hectic schedule of our lives have returned back. In this quarantine period many of us have discovered the joys of small things and have found happiness in anything we do or in anything that gives us positive vibes. For some positivity is in enjoying the morning and evening tea with a splendid sunrise and sunset, while for some it maybe painting, dancing, listening to music, cooking, baking *etc.* and for some it can be as simple as reading a nice book without any disturbance. Positivity keeps the body, mind and soul healthy. The calm and peaceful mind is must for a healthier oneself.

27.2 Changing Behavior of Birds and Wildlife during Lockdown

Anecdotal observations indicate that many animal species are enjoying the newly afforded peace and quiet, while others, surprisingly, seem to have come under increased pressure. Here, we highlight how the international research community can use these extraordinary circumstances to gain unprecedented mechanistic insight into how human activity affects wildlife. Scientists are calling it as Anthropause phase. We outline urgent steps different stakeholder groups need to take to ensure this opportunity is not missed, and introduce global collaborative research initiatives that are currently forming to facilitate coordination. Scientific knowledge gained during this devastating crisis will allow us to develop innovative strategies for sharing space on this increasingly crowded planet, with benefits for both wildlife and humans. Possible effects on wildlife Social media abound with posts sharing surprising wildlife encounters during lockdown. As we gaze out of our windows, or relish a brief walk in the park, nature appears to have changed, especially in urban environments. There not only seem to be more animals than usual, but there are also some unexpected visitors. People have reported sightings of pumas in downtown Santiago, Chile, of dolphins in untypically calm waters in the harbour of Trieste, Italy, and of jackals in broad daylight in urban parks in Tel Aviv, Israel. Hidden from view, animals may also start roaming more freely across the world's oceans, following reductions in vessel traffic and noise-pollution levels. Citizens saw honorable prime minister feeding the peacock by hands in his residence. We are observing a number of birds outside from our balcony or windows.

Two major things have happened currently during lockdown, 1. No movement by air, road or rail and 2. People movement is reduced on roads and public parks, gardens, forests *etc.* For some species, the pandemic has created new challenges. For example, various urban-dwelling animals, like rats, stray dogs, cats, gulls or monkeys, have become so reliant on food discarded or provided by humans that they may struggle to make ends meet under current conditions. Temples are closed which were common places for food to such animals. Interestingly, at certain places people are helping them to provide some food.

At the same time, reduced human presence in more remote areas may potentially expose endangered species, such as rhinos or raptors, to increased risk of poaching or persecution. Unprecedented circumstances as expanding human populations are transforming environments at unprecedented rates, understanding the linkages between human and animal behaviour is of critical importance. It is key to preserving global biodiversity, to maintaining the integrity of ecosystems, and to predicting global zoonoses and environmental change. This knowledge is not only worth billions of dollars, but it is also vital for shaping a sustainable future. So far, however, researchers have had to rely predominantly on purely observational approaches. Scientists have long sought to quantify how humans impact various aspects of animal biology, such as population levels, reproductive and mortality rates, movement and activity patterns, foraging behaviour, and stress responses.

Lockdown effects have been drastic, sudden, and widespread. Countries have also responded in broadly similar ways across large parts of the world, presenting invaluable replicates of this perturbation. So, how exactly can we make the most of these exceptional circumstances?

27.3 Changing Lifestyle

27.3.1 Realigning Daily Routines

People who were working full-time missed on the family time, leisure and fun activities but with the lockdown in effect they have started changing their routines - Enjoying all three meals with the family, catching a short nap occasionally, playing with pets and engaging in evening games with the family. First they were in panic mode but slowly started managing the workload along with enjoying the family time. Everyone is free to twist, turn and bend their schedule to their own comfort level. But sometimes work takes up more time than generally in the office. The management becomes easier and the things are solved faster when you are face to face with the situation in an office or an industry. With the opening of lockdown the industries and offices are working with maintaining social distancing till further orders. Many offices has allowed work from home and many big multinational are in a mood to adopt it as new normal and this will relieve them of heavy office expenditures.

People are giving enough attention to some physical exercise or yoga to get body relaxed and fit. Some people started regular evening prayers at home. Because of closedown of parks and gyms a large number of persons related to sports were unable to do their daily practice. Even the Olympics scheduled to be held in Tokyo, Japan were postponed.

This new condition may compromise maintaining a healthy and varied diet, as well as a regular physical activity. For example, limited access to daily grocery shopping may lead to reduce the consumption of fresh foods, especially fruit, vegetables and fish, in favour of highly processed ones, such as convenience foods, junk foods, snacks, and ready-to-eat cereals, which tend to be high in fats, sugars, and salt. Moreover, psychological and emotional responses to the COVID-19 outbreak (Wang *et al.*, 2020; Montemurro 2020), may increase the risk of developing dysfunctional eating behaviors. It is well known how the experience of negative emotions can lead to overeating, the so-called "emotional eating" (van Strien 20180). In order to contrast and respond to the negative experience of self-isolation, people could be more prone to look for reward and gratification physiologically associated with food consumption, even overriding other signals of satiety and hunger(Singh 2014).

On the other hand, negative experiences may lead to eating restriction, due to the physiological stress reactions that mimic the internal sensations associated with feeding induced satiety. Finally, lifestyle may be substantially changed due to the containment measures, with the consequent risk of sedentary behaviours, modification in smoking and sleeping habits. Of interest, different studies reported an association between sleep disturbances and obesity due to increase the secretion of pro-inflammatory cytokines by the increased visceral adipose that could contribute to alter the sleep–wake rhythm (Pugliese 2020).

In addition, also diet seems to influence the quality of sleep, in fact very recently in a cross-sectional study included 172 middle aged adults it has been reported that good sleepers had higher adherence to the Mediterranean diet (MD) and lower body mass index (BMI) compared to poor sleepers (Muscogiuri, 2020). Considering the smoking, there are a significant association exists between SARS-CoV-2 infection and air pollution, and in this context in smokers, more severe COVID-19 symptoms occur. Low physical activity levels have been suggested to interact both with body fat and appetite dysregulation (Panahi and Tremblay, 2018). Eating habits and lifestyle modifcation may threaten our health. Maintaining a correct nutrition status is crucial, especially in a period when the immune system might need to fight back. In fact, subjects with severe obesity (BMI≥40 kg/m2) are one of the groups with the higher risk for COVID-19 complications (Anonymous 2020). Obesity is an expansion of the adipose tissue, which produces cytokines and contributes to a pro-inflammatory milieu (Hauner, 2018). Moreover, in regards to pulmonary physiology, subjects with obesity have decreased expiratory reserve volume, functional capacity and respiratory system compliance. In patients with high abdominal fat, pulmonary function is further compromised in the supine position by decreased diaphragmatic excursion, making ventilation more difficult (Dietz and Santos-Burgoa 2020).

27.3.2 Starting Something New

"Anyone who stops learning is old, whether at twenty or eighty "- This quote of Henry Ford tells us that learning is a continuous thing. For many people with work only in the interludes and the stressful situation and a great deal of time on our hands many of us have started something or the other be it baking, cycling, dancing, singing or even learning needle work or playing instruments to writing inspiration or sharing funny pieces of write-ups on face book and WhatsApp. The lockdown period is just not for safety but also for all of us to rejoice, relieve and learn something new.

" I'm still learning " – Michelangelo

27.3.3 Change in School and College Life

In a pandemic young adults are more likely to be infected, increasing the potential for Universities to be explosive disease outbreak centers. Outbreak management is essential to reduce the impact in both the institution and the surrounding community. Simplest healthy behavior was adopted along with the online teaching and various online learning resources to continue education in changing situations. Online resources such as Webinars, lecture recordings, forum tutorials allow for off-campus education, online recordings, virtual learning environment, blogs, web conferencing and discussion forums all are utilized to assist in the delivery of lessons. Having a contingency and communication plan for teaching key sections may provide the needed continuity for students and faculty. For a long time permission was not allowed to schools to prevent the explosion of Covid cases due to their assembly at one place. Now permission is granted with good standard operating procedures. The school premices were used to arrange for community kitchen, distribution of food grains or places to be used as isolation wards along with hospitals.

27.3.4 Rediscovering Connections with Family and Friends

The family and friends connections have only grown deeper and deeper day by day the lockdown has started. All the zoom call, Google meets, video calls and conference calls with hours and hours of gossip was part of daily routine. Chatting and playing with your cousins over online games and enjoying with your relatives over cooking competitions were considered as the new way of stepping into a new path of togetherness.

The family time increases increasing the bonds of affection, love and care. All the daily activities are now the fun tasks of the day creating memorable memories and tasty dishes to last a lifetime. The sibling prank wars to miss your conversations, the twinning to the arguments over who wins the race to the remote, all the precious moments are relieved. Doing all the things that was once not possible due to go-go-go time - the busy schedules was practiced on daily basis.

"To reminisce with my old friends, a chance to share some memories, and play our songs again" – Ricky Nelson

Reminiscing old memories are valuable experiences for the future generations. It preserves family history, bring families closer together, helps improve their quality of life, reduces symptoms of depression and increases self-esteem, promotes physical and mental health and also help to lower blood pressure and heart rates, works through unresolved conflicts to live a healthy present, eliminates boredom providing laughter and entertainment, improves communication and as well as reduces stress. Flipping through the old albums and pictures, watching favourite movies, relieving holidays, traditions, enjoying your choicest foods, playing favourite songs *etc.* all these helped to relieve the special moments and made the moments memorable for ever.

27.4 Staying Engaged

" In times of great stress or adversity, it's always best to keep busy, to plow your energy into something positive " – Lee Iacocca

Keeping oneself busy in doing positive activities are a stress buster. Some have taken up the fitness challenge while some have taken cooking challenges. For some it can be making YouTube and Tik-Tok for fun and entertaining challenges while some are enjoying making memes but the important thing is staying engaged. Don't let the inhibitions inhibit you in anyway.

Loneliness and social isolation are associated with poor mental and physical health and may increase the likelihood of common mental disorders (depressive and anxiety disorders), substance use and cognitive decline (Courtin and Knapp 2017; Leigh-Hunt *et al.*, 2017). At this moment, people around the globe have been urged to self-isolate and refrain from social interaction due to the COVID-19 pandemic. From public health and preventative care perspectives, there is a pressing need to provide individuals, communities and health agencies with information and interventions to maintain the healthiest possible lifestyle while in isolation. Healthy lifestyle (HL) behaviours have been consistently associated with reduced all-cause mortality, and increased lifespan and wellbeing (Larsson *et al.*, 2017). Unhealthy behaviours (poor-quality diet, lack of physical exercise, tobacco and alcohol use) are major contributors to the global burden of disease (Martýnez 2020) and have also been associated with worse outcomes across psychiatric disorders (Firth *et al.*, 2019). Moreover, it is increasingly acknowledged that unhealthy lifestyles may be a driving force in the epidemic of common mental disorders. Evidence suggests that the current pandemic-related, mandatory self isolation may trigger depression and post-traumatic stress disorder (PTSD) (Brooks *et al.*, 2020) and that being a healthcare worker or having COVID-19 is risk factors for stress-related psychiatric disorders. Given the lack of effective treatments for COVID-19, non-pharmacological interventions (NPIs) are mandatory to decrease disease transmission. NPIs include personal restrictions and physical-distancing policies, such as mass confinement and compulsory home isolation.

27.5 Self Care or Keeping Fit

During the last few decades, researchers have shown an increased interest in the search for complementary therapies to maintain health through changes in diet and lifestyle. Yoga is one such process that affects not only physical health but also mental state. Yoga includes various asanas, or control of posture, and pranayamas, or breathing exercises. Yoga is estimated to be a 5000-year-old discipline originating in India (Shankar 2010). One of the most widely used breathing programs derived from yoga is Sudarshan Kriya (SK) (Janakiramaiah *et al.*, 2005).It is a set of breathing techniques taught by the Art of Living Foundation. Sudarshan is a Sanskrit word; Su stands for "proper" and darshan means "vision." Kriya is a yogic practice that is meant to purify the body; thus, SK is a simple breathing technique that harmonizes the body and mind. In addition to breathing techniques, it emphasizes the importance of prayers, asanas (yoga postures), and meditation. SK has 4 components. The practitioner's eyes are kept closed through all the components. The first and second components are done in vajrasana (ie, sitting with the legs folded and the spine erect). The first component is a 3-stage pranayama, with Ujjayi breath and specific arm positions. Ujjayi is a slow and deep breathing technique, done at the rate of 2 to 4 breaths per minute with a specific ratio of inhalation, exhalation, and breath holds. The second breathing component is Bhastrika, done with vigorous and faster breathing at the rate of approximately 20 to 30 respiratory cycles per minute. Up-and-down arm movements are used to increase the force of inhalations and exhalations. After Bhastrika, the sitting posture is changed from vajrasana to sukhasana (*i.e,* sitting with the legs crossed). The third component is Om chanting in which "Om" is chanted 3 times with prolonged expiration.

The fourth and central component is the healing breath technique. It involves 3 rounds of an exercise that uses cyclical breathing with slow, medium, and fast rates in succession. In each round, slow breaths are approximately 20 respiratory cycles per minute; medium breaths are approximately 40 to 50 respiratory cycles per minute; and fast breathing is approximately 60 to 80 cycles per minute. This breathing pattern is rotated during SK in the above-described specific order(Telles *et al.*, 1995,Rama *et al.*, 1998). SK has many physical and emotional benefits. These breathing techniques have the potential to relieve anxiety, depression, posttraumatic stress disorder (PTSD), chronic pain, and many stress-related medical illnesses. In addition, it may be useful in treatment of behavioral disorders because they generally are linked to stress(Brown *et al.*, 2005).

"Self care is so important. When you take time to replenish your spirit, it allows you to serve others from the overflow. You cannot serve from an empty vessel " – Eleanor Brown

Practicing self-care is essential when it comes to taking care of our emotional health and well-being. This is a time of incredible anxiety and stress. Focusing on what makes us feel nourished, on what gives us meaning, is part of easing those feelings and giving us a more solid foundation. Self-care includes myriad practices that you find both enjoyable and that in some way promote your physical, emotional, spiritual, or mental health. Prioritizing sleep, working out, intuitive eating, high protein snacks for an energy boost, keeping stress busting foods on hand, whipping up a healthy mock tail, practice positive self talk, practicing breathing focused meditation, chilling with music, crawl up on the sleeves of creativity, enjoying a home spa *etc.* all this helps to revitalize oneself.

27.8 New Normal: Lessons to be Learnt from COVID

We are just in a 'snow-day loop' that isn't ending any time soon, with the honeymoon period well and truly over. But end it will. Maybe overindulging in food and drink which is not helping your mood or self-esteem at this point but this can be tried and moderated. Simple mapping out our daily schedule starting with getting up, making bed, breakfast, lunch, and dinner. Daily 45 mins exercise, creative time, gardening, video calls, self care *etc.* Do family group activities balanced with time on your own. Be criticl in your precision about how you consume news: only allow one or two short periods per day, only from reliable sources, and absolutely not at bedtime. Take a whole day off the news every few days! Be vigilant about screen-time; time yourself and stick to short periods. If you do feel anxious, sad or irritated, allow yourself to feel this and go with it, being compassionate with yourself. Then move onto something else; get out in your area and be with nature, listen to rainforest meditations. Calm yourself as you would a child.

Absorbing activity is the cornerstone of Positive Psychology – doing stuff that will put you 'in flow' – *i.e.*, you are so engaged in the task you forget to worry about what else is happening. Some days you will do better than others. Allow off days, if you have to, as a time-out from routine, keeping a loose one going so it's easier to hop back onto the structure next day. Less ticks, but ticks for all that. Really, if you've managed to get up, make your bed, wash and eat you've achieved everything! Then, congratulate yourself! Above all, be your own best supporter, and keep reminding yourself, as a mantra, 'This Too Shall Pass'. It may not be easy, but try to not make it any more difficult. It'll be better for everyone when, we will enter the next phase after Covid-19. In this dark phase of life I, nor Me, We and Us together only can fight for all of us. The entire world has a chance to start on a healthier note by bringing the positive lifestyle changes for a healthy body and mind. With the wave of fear and death, there is always a ray of light to help us keep moving forward.

" Yesterday is not our's to recover, but tomorrow is our's to win or lose."

– Lynden B Jonson

27.9 Conclusions

While we may be in lockdown at present, or feeling relaxed tomorrow we should prepare ourselves mentally for the changes going to take place post-COVID-19.

These include following new normals:

Embrace Contactless Greetings

Planning to shake hands? Think again. Immediately after COVID-19 struck, there were visuals of world leaders ditching the handshake and using the 'Namaste' greeting. The Namaste is a form of greeting used in India that does not involve contact with the other person, unlike a handshake.

Changing the Way we Learn

With schools and colleges being shut, there is a lot of interest around online classes. Teachers across the world are using the power of the internet to deliver educational content to a student's house. Real-time online classes are being held on apps like Zoom that have ensured that the absence of a physical classroom does not stop the learning of students.

Practice Frequent Washing of Hands

Are you washing your hands frequently? One thing the COVID-19 pandemic has taught us is to maintain very high standards of personal hygiene. People are now understanding how simple acts like washing hands with soap and water may save their lives.

Avoid Public Spitting, Use Mask

In India, it is common to see spit stains on government buildings, railway stations and roads. While the government tried to spread awareness by putting up posters and even imposing fines, no one cared. But now that everyone has to wear masks in public places and maintain personal hygiene, spitting could be a thing of the past.

Better Air Quality

Three to four weeks into the lockdown, environmentalists across the world noticed something strange – the ozone layer was healing itself. Not just that, air quality in cities like Delhi and Mumbai has improved drastically during the lockdown. But we are feeling fresh air all around us. Let us keep environment clean.

Be more Grateful

The COVID-19 pandemic has brought us closer than ever before. Everyone is united in efforts to contain the virus. We have realized that if we don't prepare ourselves well, then we can lose what we have now. This thought that we may lose it if we are not careful has made us more appreciative of the present and thankful for what we have. Let us live in harmony with nature.

REFERENCES

Anonymous 2020 People Who Are at Higher Risk for Severe Illness | Coronavirus | COVID-19 | CDC. https: //www.cdc.gov/coronavirus/2019-ncov/need-extra-preca utions/groups-at-higher-risk.html. Accessed 7 May 2020.

Anonymous 2020 Nature Ecology and Evolution | VOL 4 | September 2020 | 1156–1159 | www.nature.com/natecolevol

Brown RP, Gerbarg PL. 2015 Sudarshan Kriya yogic breathing in the treatment of stress, anxiety, and depression, I: neurophysiologic model. J Altern Complement Med.11(1): 189-201.

Brooks SK, Webster RK, Smith LE *et al.*, The psychological impact of quarantine and how to reduce it: rapid review of the evidence. Lancet 2020;395: 912–920.

Courtin E, Knapp M. Social isolation, loneliness and health in old age: a scoping review. Health Soc Care Community 2017;25(3): 799–812.

Dietz W, Santos-Burgoa C.2020 Obesity and its implications for COVID-19 mortality. Obesity. https: //doi.org/10.1002/oby.22818.

Engin AB, Engin ED, Engin A. 2020 Two important controversial risk factors in SARS-CoV-2 infection: obesity and smoking. Environ Toxicol Pharmacol. https: //doi.org/10.1016/j.etap.2020.103411.

Firth J, Siddiqi N, Koyanagi A *et al.*, The Lancet Psychiatry Commission: a blueprint for protecting physical health in people with mental illness. Lancet Psychiatry 2019;6: 675–712.

Hauner H. 2005 Secretory factors from human adipose tissue and their functional role. Proc Nutr Soc. 2005;64: 163–9.

Janakiramaiah N, Gangadhar BN, Naga Venkatesha Murthy PJ, Harish MG, Subbakrishna DK, Vedamurthachar A. 2000 Antidepressant efficacy of Sudarshan Kriya Yoga (SKY) in melancholia: a randomized comparison with electroconvulsive therapy (ECT) and imipramine.J Affect Disord. 2000;57(1-3): 255-259.

Larsson SC, Kaluza J, Wolk A. Combined impact of healthy lifestyle factors on lifespan: two prospective cohorts. J Intern Med 2017;282: 209–219.

Leigh-Hunt N, Bagguley D, Bash K *et al.*, An overview of systematic reviews on the public health consequences of social isolation and loneliness. Public Health 2017;152: 157–171.

Ma Y, Ratnasabapathy R, Gardiner J. 2017 Carbohydrate craving: not everything is sweet. Curr Opin Clin Nutr Metab Care. 2017;20: 261–5.

Martýnez V. Balanza, B. Atienza–Carbonell3, F. Kapczinski4, R. B. De Boni 2020 Lifestyle behaviours during the COVID-19 – time to connect ACTA PSYCHIATRICA SCANDINAVICA 2020: 141: 399–400

Montemurro N. 2020 The emotional impact of COVID-19: from medical staf to common people. Brain Behav Immun. https: //doi.org/10.1016/j.bbi.2020.03.032.

Moynihan AB, van Tilburg WAP, Igou ER, Wisman A, Donnelly AE, Mulcaire JB. 2015 Eaten up by boredom: consuming food to escape awareness of the bored self. Front Psychol.;6: 369.

Muscogiuri G, Pugliese G, Barrea L, Savastano S, Colao A.2020 Obesity: the "Achilles heel" for COVID-19? Metabolism. 108: 154251.

Muscogiuri G, Barrea L, Aprano S, Framondi L, Di Matteo R, Laudisio D, 2020 Sleep quality in obesity: does adherence to the mediterranean diet matter? Nutrients. 12: 1364.

Panahi S, Tremblay A.2018 Sedentariness and health: is sedentary behavior more than just physical inactivity? Front Public Heal. 6: 258.

Promila Sharma, Ashish Thapliyal, Tribhuwan Chandra, Saumya Singh, Himani Baduni, Syed Mohsin Waheed,2015 Rhythmic Breathing : Immunological,

Biochemical, and Physiological Effects on Health Health advancesjournal.com 18 ADVANCES, WINTER, 29, (1)18-25

Pugliese G, Barrea L, Laudisio D, Salzano C, Aprano S, Colao A,2020 Sleep apnea, obesity, and disturbed glucose homeostasis: epidemiologic evidence, biologic insights, and therapeutic strategies. Curr Obes Rep. 2020;9: 30–8.

Rama S, Ballentine R, Hymes A. Science of Breath: A Practical Guide. Honesdale, PA: The Himalayan Institute Press; 1998.

Rodríguez-Martín BC, Meule A. Food craving: new contributions on its assessment, moderators, and consequences. Front Psychol. 2015;6: 21.

Shankar SS. Patanjali Yoga Sutras. Bangalore, India: Art of Living Press; 2010.

Singh M. Mood, food and obesity. Front Psychol. 2014;5: 1–35.

Wu C, Chen X, Cai Y, Xia J, Zhou X, Xu S, *et al.*, Risk factors associated with acute respiratory distress syndrome and death in patients with coronavirus disease 2019 Pneumonia in Wuhan, China. JAMA Intern Med. 2020. https: //doi. org/10.1001/jamainternmed.2020.0994.

Telles S, Nagarathna R, Nagendra HR. Autonomic changes during "OM" meditation. Indian J Physiol Pharmacol. 1995;39(4): 418-420.

van Strien T. Causes of emotional eating and matched treatment of obesity. Rep: Curr Diab; 2018.

Wang C, Pan R, Wan X, Tan Y, Xu L, Ho CS, *et al.*, Immediate psychological responses and associated factors during the initial stage of the 2019 coronavirus disease (COVID-19) epidemic among the general population in China. Int J Environ Res Public Health. 2020;17: 1729.

Yýlmaz C, Gökmen V. Neuroactive compounds in foods: occurrence, mechanism and potential health effects. Food Res. 2020;128: 108744.

Chapter 28

Pollution, Changing Lifestyle after COVID-19 and Sustainable Development

V.S. Patel

Consulting Engineer and Visiting Professor, Department of Environmental Studies, Faculty of Science, The Maharuju Sayajirao University of Baroda, Vadodara – 390 002, Gujarat

e-mail: vpatel_usa@hotmail.com

ABSTRACT

The COVID-19 pandemic, is an on going worldwide disease caused by the corona virus. The deadly virus has claimed more than 1,162,098 (on 26th October, 2020). The recent virus pandemic has changed life of millions of people worldwide. Pollution levels are affected and some good news has been achieved in lesser air and water pollution levels during lockdown stage. Use of solar power has gained momentum in different countries. The E.U. requires solar producers to recycle products, and similar laws are in the works in some other parts of the world, including Japan and India. In the U.S., so far, only the State of Washington requires solar panel recycling; the majority of old solar panels in the country end up in landfills now, wasting valuable materials such as silicon and risking the spread of toxic components such as lead. An experiment in giving buildings smart energy analytics saved $95 million in energy costs in 6,500 buildings. If adopted nationwide, the savings could be enormous.

In 2015 the global community adopted the 17 Global Goals for Sustainable Development to improve people's lives by 2030. Goal 2 – Zero Hunger – pledges to end hunger, achieve food security, improve nutrition and promote sustainable agriculture, and is the priority of the World Food Programme. Every day, WFP and its partners work to bring us closer to a zero hunger world. The world has made great progress in reducing hunger: there are 300 million fewer hungry people than in 1990-92, despite a 1.9 billion increase in the world's population. But there is still a long way to go, and no one organization can achieve Zero Hunger if it works alone.

Whether bacterial, viral or parasitic, virtually every disease pathogen that has affected people over the last several thousand years is still with us, because it is nearly impossible to fully eradicate them. The only disease that has been eradicated through vaccination is smallpox. Mass vaccination campaigns led by the World Health Organization in the 1960s and 1970s were successful, and in 1980, smallpox was declared to be fully eradicated. So success stories like smallpox are exceptional. It is rather the rule that diseases come to stay. Hopefully COVID-19 will not persist for millennia. But until there's a successful vaccine available and likely even after that, no one is safe. The paper describes pollution, alternative sources of energy and sustainable development strategies in changing COVID times.

Keywords: *Pollution, Changes life style, COVID-19, Virus, Sustainable development.*

28.1 Introduction

"There is nothing permanent in life except change," said philosopher Heraclitus. Others have called change or the same old thing all the time wears out a person's soul as 'the flavor of life'. Along these lines, changes (mix or reshuffle) in the legislature every once in a while should not shock anybody; however changes in the political field are frequently seen with doubt.

Way of life is a word that shows how we bear our life, how we act in this life, how we connect with others and it likewise implies the manner in which we consider each part in our lives. We do not utilize bicycle to go to different places, now we use bike and vehicles to arrive at any spot. Our way of life is transforming from the crude one quite a while back to a cutting edge way of life in this globalization period. We have realized that with industrialization our way of life has changed. We simply utilize our mobiles, tablets, PC or mini-computer to do accounting and communication with each other. Really with increase in population and massive industrialization there is increase in pollution. The anthropogenic pollution from vehicles, industries and agricultural activities has resulted into loss of many human lives.

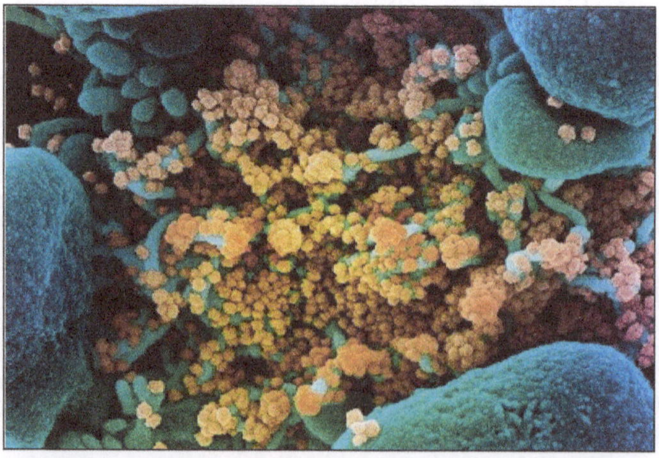

Figure 28.1: Corona Virus Particles in Lung Tissues.

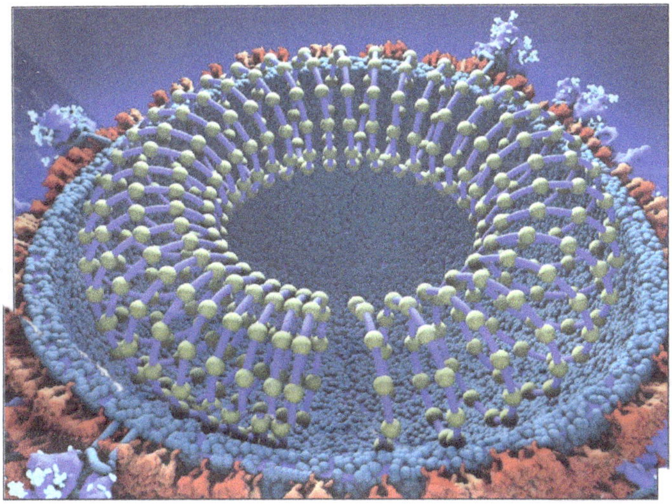

Figure 28.2: Architecture of CORONA Virus Particle Showing Spikes.

Figure 28.3: People Protecting Infection by Covering their Face with Mask.

28.1.1 Pollution from Electroplating Industries in Mumbai

Air pollution has become a popular issue in the terms of social aspects (Muthukaruppan *et al.*, 2010). Plantation is carried out to mitigate the air pollution issue in the region (Bora and Joshi, 2014; Thakar and Mishra, 2010). Electroplating is the process of deposition of a fine layer of one metal on another through electrolytic process to impart various properties and attributes, such as abrasion and wear resistance, corrosion protection, enhanced surface hardness, lustre, colour, aesthetics and value addition (Akolkar *et al.*, 2015). Electroplating uses electric current and electrochemical reaction for making metal coatings. It is considered to be a major polluting industry because it discharges toxic. The effluent from electroplating industries have high concentration of heavy metals like Iron, Chromium, Copper and Nickel (Singh, 2014). There are many registered electroplating units, located in Mumbai Metropolitan Region (MMR). These plants are divided into region I, II, III, IV. Many of these industries are operated at a small scale and hence environmental compliance and waste management is not given due importance. Due to the high treatment costs, the wastewater, solid waste and gaseous emissions from these industries are released without any prior treatment leading to deterioration of nearby land, air and water.

Small scale entrepreneurs owning these electroplating industries and the employees are either ignorant of health impacts or environmental effects of effluent discharge or are worried that it would be a financial burden to them (Singh, 2014). Electroplating industry has been generating a huge amount of waste in the forms of wastewater, spent solvent, spent process solutions and sludge (Freeman, 1988). Solid waste includes residues such as cleaning powder, buffing compounds generated during the pre-treatment process and spent anodes during the plating process. Liquid wastes includes spent chemicals and solutions such as acids, alkali, cleaning agents, bath chemicals comprising plating chemicals as well as additives such as brighteners, levellers *etc.* and rinse waters. Gaseous wastes include solvents and vapours from hot pre-treatment and process baths. Acid alkali mist and Volatile Organic Compounds (VOCs) are the gaseous wastes from electroplating industries. In some cases, mists and VOCs may contain metals in addition to process chemicals. Air emissions from electroplating industries may contain toxic organic materials such as trichloroethylene and trichloroethane (Singh, *et al.*, 2016). Around 30 per cent of the used solvent and degreasing chemicals release VOC (Cushine, 1985). The minimization of the air emissions to be adopted is related to the optimum bath temperature control and the preventive maintenance of the suction and filtration unit or scrubber (Ramesh *et al.*, 2009).

28.1.2 Atmospheric Pollution in Delhi and NCR

Pollution refers to the contamination of the earth's environment with materials that interfere with human health, quality of life or the natural functioning of the ecosystems. The major forms of pollution include water pollution, air pollution, noise pollution and soil contamination. Other less-recognized forms include thermal pollution and radioactive hazards. It is difficult to hold any one particular form responsible for maximum risk to health; however, air and water pollution appear to be responsible for a large proportion of pollution related health problems.

Of late, the air pollution status in Delhi has undergone many changes in terms of the levels of pollutants and the control measures taken to reduce them. Delhi (or the National Capital Territory of Delhi), is jointly administered by the central and state governments. It accommodates nearly 167.5 lakhs people (2011 Census). Metros across the world bear the major brunt of environmental pollution; likewise, Delhi is at the receiving end in India. A study funded by the World Bank Development Research Group was carried out in 1991-1994 to study the effects of air pollution (Cropper *et al.*, 1997). During the study period, the average total suspended particulate (TSP) level in Delhi was approximately five-times the World

Health Organization's annual average standard. Furthermore, the total suspended particulate levels in Delhi during this time period exceeded the World Health Organization's 24-h standard on 97 per cent of all days on which readings were taken. The study concluded that the impact of particulate matter on total non-trauma deaths in Delhi was smaller than the effects found in the United States of America, but found that a death associated with air pollution in Delhi caused more life-years to be lost because these deaths were occurring at a younger age. A report by the Ministry of Environment and Forests, India, in 1997 reviewed the environmental situation in Delhi over concerns of deteriorating conditions (Anonymous, 1997). Air pollution was one of the areas of concern identified in this study. It was estimated that about 3000 metric tons of air pollutants were emitted every day in Delhi, with a major contribution from vehicular pollution (67 per cent), followed by coal-based thermal power plants (12 per cent). There was a rising trend from 1989 to 1997 as monitored by the Central Pollution Control Board (CPCB). The concentrations of carbon monoxide from vehicular emissions in 1996 showed an increase of 92 per cent over the values observed in 1989, consequent upon the increase in vehicular population. The particulate lead concentrations appeared to be in control; this was attributable to the de-leading of petrol and restrictions on lead-handling industrial units. Delhi has the highest cluster of small-scale industries in India that contribute to 12 per cent of air pollutants along with other industrial units.

The PM_{10} standard is generally used to measure air quality. The PM_{10} standard includes particles with a diameter of 10 μm or less (0.0004 inches or one-seventh the width of a human hair). These small particles are likely to be responsible for adverse health effects because of their ability to reach the lower regions of the respiratory tract. According to the Air Quality Guideline by the WHO, the annual mean concentration recommended for PM_{10} was 20 μg/m^3, beyond which the risk for cardiopulmonary health effects are seen to increase (WHO, 2005). Major concerns for human health from exposure to PM_{10} include effects on breathing and respiratory systems, damage to lung tissue, cancer and premature death. Elderly persons, children and people with chronic lung disease, influenza or asthma are especially sensitive to the effects of particulate matter. The urban air database released by the World Health Organization in September 2011 reported that Delhi has exceeded the maximum PM_{10} limit by almost 10-times at 198 μg/m^3, trailing in the third position after Ludhiana and Kanpur. Vehicular emissions and industrial activities were found to be associated with indoor as well as outdoor air pollution in Delhi (Goyal and Khare, 2011).

Assessment of ambient air quality trend analysis at Hapur, a town near the capital city Delhi, was done by Tyagi *et al.* (2020) with respect to the concentration of different pollutants. This study mainly focuses on the extent of air pollution and air quality index for a period before lockdown between 1st to 22nd March 2020, including one day of Janata curfew (22nd March 2020); first lockdown (25 March -13 April, 2020); 2nd (15 April – 3 May 2020); 3rd (4 – 17 May, 2020) and 4th (18– 31 May, 2020). There was little improvement in air quality during lockdown first (wrt AQI, PM_{10} and $PM_{2.5)}$ and fourth (wrt CO during 16-24 h) as compared to 21 days of the pre-lockdown period. On March 22, 2020; AQI was moderate in Hapur, which remained moderate and little improved in lockdown first but again increased in consecutive lockdowns. The range of $PM_{2.5}$ concentration in pre-lockdown and lockdown 1, 2, 3, 4 was reported as 14–98μg/m3, 19–60 μg/m³, 18-74 μg/m³, 22–65 μg/m³, and 13–73 μg/m³ respectively with exceedance of standard limit between 4.7 per cent to 28.5 per cent of days. The range of PM_{10} concentration in pre-lockdown, and lockdown 1, and 4 was reported as 22- 233 μg/m³. The range of SO_2 concentration in pre-lockdown, and lockdown 1, was reported as 1-16 μg/m3 and 6-11 μg/m³, respectively. The range of NO_2 concentration in lockdown 1, 4 was reported as 35-72 μg/m³, and 48-69 μg/m³ respectively. The ratio of $PM_{2.5}$ and PM_{10} is also calculated to analyze the impact of fine particles in the composition of particulate matter due to lockdowns, which is observed ranging between 0.3 and 0.6. The results displayed reduced levels during lockdown-1, in comparison to others during the study period, which can be seen from the percentage of days they exceeding the standard limits for different pollutants. This case study disclosed unique findings unlike big cities as no much impact of lockdown 2,3 and 4 probably due to pollution generated from transport, vehicular movement, industries, construction and demolition activities, biomass and excessive burning, road dust which was curbed down due to the lockdown 1 restrictions and as result air quality was improved. Under the nationwide lockdown, the good air quality index was observed globally. The rampant increase in vehicles and construction work in urban areas contributes considerably to pollution.

28.2 Pollution and Climate Change

Climate change has long been seen as an inherently global issue, but as international conferences and accords keep striking a wall of national interests and old fashion geopolitics, scholars from different disciplines have started to look beyond the international scale for an answer (Hochstetler and Viola 2011). One geographical entity that has received considerable attention is cities. It is argued that in an increasingly urbanized world, the cities is where concrete action will have to take place, in order to curve escalating greenhouse gas emissions (Corfee-Morlot *et al.*, 2009).

Changes can be categorized under two main types. Changes that take place in nature we have little or no control over. We cannot, for instance, switch the time of tides, which anyway, wait for no one. The other kind of change is the one we witness either in political, social or other fields including the area of personal life. These are changes over which one can exercise some degree of control, changes which can be guided by oneself or others. India is the world's poorest people who are hardest hit by climate change effect such as devastating droughts or high flood devastate families who are dependent on agriculture for their food as well as their income As climate change makes weather patterns less predictable, it is the poorest who are suffering the worst. and other extreme weather events. Impacts are already being seen in unprecedented heat waves, cyclones, floods, Stalinization of the coastline and effects on agriculture, fisheries and health.

Evidence is gathering that human activities are changing the climate. This **'climate change'** could have a huge impact on our lives. The following factors make the change in climate important for us.

- ☆ Increase in Global Carbon Dioxide
- ☆ Increase in Methane levels
- ☆ Production of other Green house gases
- ☆ A Rise in Global Sea Level
- ☆ A change in ecosystem
- ☆ Human Health effects
- ☆ An increase in disease levels

Ministry for Environment and Forests has published on New Delhi, - November 16th 2010 (The 4x4 Assessment examines the implications of climate change for India in 2030s deduced from a Regional Climate Model Had RM3 (Hadley Centre Regional Model Version 3) run for A1B scenario). That :

- ☆ **Climate change scenarios for 2030**s indicate an overall warming for all the regions in focus. The net increase in annual temperatures in 2030s with respect to1970s ranges between 1.7°C – 2.2°C, with extreme temperatures increasing by 1 - 4°C, with maximum increase in coastal regions. The extreme maximum and minimum temperatures are also projected to increase in 2030s with respect to 2070s.
- ☆ **All the regions are projected to experience** an increase in precipitation in 2030s with respect to 1970s and the increase is maximum in the Himalayan region and minimum increase in the North Eastern region. The extreme precipitation events are likely to increase by 5-10 days in all the regions.
- ☆ **Rise in Sea level along the Indian coast** has been rising at the rate of 1.3mm/year and is likely to rise in consonance with the global sea level rise in the future. Further projections indicate that the frequency of cyclones is likely to decrease in 2030s, with increase in cyclonic intensity.
- ☆ **Change in Agriculture pattern - Irrigated rice in all the regions** are likely to gain in yields marginally due to warming as compared to the rainfed crop as the irrigated rice tends to benefit from CO_2 fertilization effect. Maize and sorghum are projected to have reduced yields in all the regions. The Coconut productivity is projected to rise in the western coast and reduce in the eastern coastal region. Observations indicate a reduction in apple production in the Himalayan region, which is likely to continue in the future. Colder country on and above arctic circle (66.5° North) have longer and warmer summer days may produce more grain.
- ☆ In case of marine fisheries some species will gain in yields, as the warming favours their productivity such as Sardines. Some species like Indian mackerel are likely to move upwards to the northern latitudes thus maintaining their yields. Species like Threadfin breams, may shift their spawning seasons adjusting to the season which optimally favours spawning temperatures,
- ☆ **With overall warming,** the thermal humidity Index is projected to increases in all the regions, especially in the months of May and June, leading to stress to the livestock and hence reduction in its milk productivity.

- ☆ Water yield (which is a function of precipitation, total surface run off, evapo -transpiration and soil properties), is projected to increase in the Himalayan region in 2030s by 5-20 per cent, owever, water yields are likely to be variable across the North Eastern region, Western Ghats, and Coastal region. In some places in these regions, it is projected to increase and in some places it is projected to decrease.
- ☆ Moderate to extreme drought severity is projected in 2030s for the Himalayan region, as compared to the other regions. All the regions are likely to experience flooding which are exceeding existing magnitudes by 10 per cent to 30 per cent.
- ☆ **Forest Ecosystem**- Change is projected for 8 per cent 18 per cent, 56 per cent, and 30 per cent of the vegetation grids and increase- in Net Primary Productivity by 23 per cent, 20 per cent, 57 per cent, and 31 per cent is projected in Western Ghats, North eastern region, Himalayan region, and the Coastal region.
- ☆ **Human Health Issues**– Malaria is projected to spread in new areas in Jammu and Kashmir in the Himalayan region. In the North eastern region opportunities for transmission is likely to increase for a longer period. In the Western Ghats, no change is observed between in 2030s and the trends observed in 1970s. However, in the Coastal region, especially in the eastern coast a marked decrease in number of months is projected in which that the malaria transmission window would be open. Due to global warming new virus will developed. This academic inspect on human life.

28.3 Anthropogenic Pollution and COVID-19

To Day the world environment is very happy due to lockdown. Entire world feel pollution free. The air becomes very clean. There is no pollution in air. Sky is very clean. The visibility is increased. One can see the sky is very far and visible full of stars. The trees are dancing in clean air. The fruits are germinating very well. The mangos and other fruits and vegetables are grown very well. The animals are moving freely. There is no noise and pollution of automobiles. The noise level DB is decreased. Birds are singing early morning before sun rise and enjoying noise free environment. The water is in river and Lakes are now very clean. Even Ganga water becomes pollution free in last two months where we are trying since long.

The environment has taken the pollution by itself. Selfish mankind has destroyed the environment considerable last 50 - 100 years. Science has developed in all fields. Men have gone to space. Man can travel at lightning speed. Science has altered DNA. Everything is now possible with help of science. But environment has taken his own revenge due to pollution on earth. The Corona virus, Covid -19 outbreak appears on earth as a pandemic and reiterated the call for countries to take immediate actions and scale up response to treat, detect and reduce transmission to save people's lives.

People are dying due to corona infection. The death of world people now reached more than one lakh (on 6th April 2020). 7 April 2020 is (the World health day) to celebrate the work of nurses and midwives and remind world leaders of the critical role they play in keeping the world healthy. Nurses and other health workers are at the forefront of COVID-19 response - providing high quality, respectful treatment and care, leading community dialogue to address fears and questions and, in some instances, collecting data for clinical studies. Quite simply, without nurses, there would be no response. In this International Year of the Nurse and the Midwife, World Health Day will highlight the current status of nursing and around the world.

WHO and its partners will make a series of recommendations to strengthen of the nursing and midwifery workforce. Currently, there is no vaccine available to protect against 2019- novel Corona Virus (Figure 28.1 and 28.2). It is not certain how long the virus that causes COVID-19 survives on surfaces, but it seems to behave like other corona viruses. Studies suggest that corona viruses (including preliminary information on the COVID-19 virus) may persist on surfaces for few to several days or months. **Science is looking for an answer.** As on 26th October 2020at 18.20 GMT, the total infected in 215 countries with virus are 43,583,782 and 1,162,098 deaths, in one day total cases reported were 35,078.

Government of India is taking all necessary steps to ensure that we are prepared well to face the challenge and threat posed by the growing pandemic of COVID-19 the Corona Virus. With active support of the people of India, we have been able to contain the spread of the Virus in our country. The most important factor in preventing the spread of the virus locally is to empower the citizens with the right information and taking precautions as per the advisories being issued by Ministry of Health and Family Welfare.

28.4 Effect of Lockdown on Environmental Pollution

Some of the immediate repercussions of the COVID-19 pandemic included mass cancellations of flights, no industrial activities, and fewer vehicles on the roads. These activities are amongst the highest greenhouse gas emissions emitters,and their cancellation led to a direct reduction in greenhouse gas emissions in the short run. Nitrogen dioxide (NO_2) can be classified as potent greenhouse gas as it is 300 times more effective at trapping heat in the atmosphere than carbon dioxide. It is emitted from the engines of cars, buses, trucks, and factories.The WHO states that if the gas is present in a higher concentration than 200 $\mu g/m^3$, it can cause inflammation of the breathing pipe in humans, leading to respiratory diseases like asthma. As a result of the pandemic, the concentration of NO_2 has fallen drastically all over the world. For example, in China, during the month of January, the levels of NOx were about 250 to 500 micromoles per sq. m. In February, however, the levels had abated to 0 to about 187.5. Other examples include New York, where scientists at Columbia University reported a 5 to 10 per cent drop in CO_2 emissions. The same thing was witnessed in other countries such as Italy, UK, Spain. In India, this was observed in highly polluted cities such as Mumbai, Delhi, Chennai, Bangalore and Kolkata *etc.* The significant reduction in pollution has allowed people to get the sight of the Himalayas.

The lockdown has also resulted in significant depletions of $PM_{2.5}$, one of the most harmful forms of air pollution, is included in the category of group one carcinogen and is so small that it can travel from your lungs into the bloodstream, which can cause respiratory diseases as well as heart attacks and premature deaths. The World Health organization has estimated that about 4 m deaths are caused every year, world wide, due to the deadly $PM_{2.5}$. Considering the lockdown, the $PM_{2.5}$ has depleted significantly over the world. This is the reason why we can see the beautiful blue sky. To add to this, a scientist from Stanford University researched regarding the $PM_{2.5}$ levels , he collected data from Chinese cities, observed the impacts of falling $PM_{2.5}$ on mortality rates. The research concluded that the falling levels in just one month, saved more than 77,000 lives due to the decrease in air pollution. Not only can we see improvements in the air but also in water bodies. The stoppage of industrial waste has definitely caused a positive change on the water quality of rivers. For example, the rivers Ganges and Yamuna are cleaned up in two months of lockdown. As industrial and commercial activities halted, the rivercleaned itself, allowing countless Indian and migratory birds to flock to its banks.

Lockdown in Italy hasresulted in the canals of Venice to be crystal clear, pollutionfrom motorboats has been vastly reduced (Fig. 7).

28.5 Effect of Lockdown on Change in Behaviour of Wildlife

The wildlife has been affected due to lockdown as well, a lot of videos went viral over the internet , displaying rarely seen animals coming out to the deserted streets. As the lockdown means mainly two things- less traffic on roads and reduced pollution, which can give wildlife space to thrive. Animals that were spotted out in the manmade surroundings include, penguins in Cape Town, kangaroos in Adelaide, jellyfish in the canals of Venice, deer and 'big spotted cat' in London, goats in Wales as well as elephants in south India. There is evidence that wild bees will benefit from the depleted air pollution, as it can disrupt their ability to smell flowers from a distance. However, people who work in the informal economy are unemployed and as a result unable to make money for a living are being forced to poach and hunt animals for meals. There has been a spike in poaching, logging as people who worked to make ends meet each day to be fed. While countries grope their way to procure protective equipment, a disturbingly large amount of medical waste has to be safely disposed off. Even with changes as to how PPE should be managed, used, treating a large number of patients with serious conditions would cause a tremendous amount of biohazard medical waste. The treatment and disposal processes, which include landfill and incineration, can also raise negative climate and other environmental impacts. Although the lockdown has been beneficial to the environment in the short run, it has been observed that the economy and environment thrive in inversely proportional manners, that is, when the economy depresses the environment prospers and vice versa. Whilst we overcome the environmental problems, we need to transition to a healthier economy that benefits environment in the long term, as well as promoting sustainable growth that ensures the growth of economy and supports environment.

28.6 Bats Harbour a Pool of Corona Viruses Related to Pandemic Culprit

Viruses closely related to SARS-CoV-2, the virus causing the COVID-19 pandemic, have been circulating in horseshoe bats, ready to jump to humans, for decades – and maybe even longer. David Robertson at the University of Glasgow, UK, and his colleagues analyzed the RNA of 68 corona viruses, including SARS-CoV-2 and the virus that causes severe acute respiratory syndrome, or SARS (Boni *et al.*, 2020). This analysis shows that horseshoe bats (*Rhinolophus* spp.) host an expanding lineage of viruses that, like SARS-CoV-2, can infect humans. The team estimates that the ancestor of SARS-CoV-2 split 40 to 70 years ago from the closely related bat virus RaTG13. Though the two viruses are highly similar genetically, RaTG13 doesn't infect humans. The analysis also suggests that viruses in the lineage are ready to jump to humans directly from bats. But SARS-CoV-2 might have first hopped to another species that humans are more exposed to, rather than spreading straight from bat to human.

28.7 Sustainable Development

Ideally, humankind should get to a point where at least the minimum Human development Index (HDI) is achieved and live below the maximum ecological footprint per capita. Living above the minimum HDI would guarantee that human needs such as education or health are satisfied. An ecological footprint represents the maximum limit of consumption per person according to Earth ecological capacity. Living below it wouldn't compromise the future generations, as the planet would be able to regenerate itself. But the fact is that every year the Earth overshoot day comes earlier. This day represents the date when humankind gets in debt with the planet. This is because our demand for ecological resources in a given year exceeds, what the planet can regenerate in that same year. We are keeping this deficit because we're using more ecological resources than the planet can handle to lose, At the same time, we're also not taking proper care of our waste. Today's consumption habits are a big threat to sustainable development.

Figure 28.4: Shows a Village of Uttarakhand Washed Away in Floods.

Figure 28.5: Another Hill Station Dehradun.

The degradation of ecosystem has resulted into floods (Figure 28.4) and earthquake like disasters the irreversible biodiversity loss has resulted. The pollution and tree felling has polluted the environment as drawn by a student (Figure 28.6). Better conservation strategies need to be adapted to protect the threatened flora and fauna. Today's framework on sustainable development is quite strong although there is still a huge way to go. The latest IPCC report demonstrated that big changes will need to happen quickly regarding the reduction of CO_2 emissions to keep the Earth's temperature below 2°C and prevent its devastating impacts.

Figure 28.6: Sustainable Development in a Drawing.

Figure 28.7: Beautiful City Venice with Clear Water.

There are many actors working with different audiences in different areas of sustainability. They share the same goal – to raise awareness on this topic and to create conditions for it to grow and develop. One of the main players is the United Nations, where different teams actively work on multiple campaigns such as #beat plastic pollution or #solve different, apart from organizing the meetings between the world leaders. On the business side, the World Business Council for Sustainable Development (WBCSD) helps its member companies to accelerate their business transition to create a sustainable world. There are also some certifications that reward the business with the best practices for the planet, such as the B-crop improvement, the rainforest alliance, the Fair trade Foundation or the conscious Capitalism Movement, at the same time, entities like the Elen Mac Arthur Foundation are opening the way when it comes to the circular economy and how societies and business can align how they use natural resources with the way nature does it. Aligning businesses operation across their supply chains is also allowing different and ecological business models to develop- such as growing mushrooms from coffee leftovers.

The word sustainability is derived from the Latin *sustinere* (*tenere*, to hold; *sus*, up). Dictionaries provide more than ten meanings for *sustain*, the main ones being to "maintain", "support", or "endure". However, since the 1980s *sustainability* has been used more in the sense of human sustainability on planet Earth and this has resulted in the most widely quoted definition of sustainability and sustainable development, that of the Brundtland Commission of the United Nations on March 20, 1987: "sustainable development is development that meets the needs of the present without compromising the ability of future generations to meet their own needs."

At the 2005 World Summit it was noted that this requires the reconciliation of environmental, social and economic demands - the "three pillars" of sustainability. This view has been expressed as an illustration using three overlapping ellipses indicating that the three pillars of sustainability are not mutually exclusive and can be mutually reinforcing.

The UN definition is not universally accepted and has undergone various interpretations. What sustainability is, what its goals should be, and how these goals are to be achieved is all open to interpretation. For many environmentalists the idea of sustainable development is an oxymoron as development seems to entail environmental degradation. Ecological economist Herman Daly has asked, "what use is a sawmill without a forest?" From this perspective, the economy is a subsystem of human society, which is itself a subsystem of the biosphere, and a gain in one sector is a loss from another. This can be illustrated as three concentric circles.

A universally accepted definition of sustainability is elusive because it is expected to achieve many things. On the one hand it needs to be factual and scientific, a clear statement of a specific "destination". The simple definition "sustainability is improving the quality of human life while living within the carrying capacity of supporting eco-systems", though vague, conveys the idea of sustainability having quantifiable limits. But sustainability is also a call to action, a task in progress or "journey" and therefore a political process, so some definitions set out common goals and values. The Earth Charter speaks of "a sustainable global society founded on respect for nature, universal human rights, economic justice, and a culture of peace." The Brundtland report also called as 'our common future' gave the most recognized and widely accepted definition of the term sustainable Development in 1987.

28.7.1 Artificial Carbon Sequestration

Artificial carbon sequestration refers to removal of carbon dioxide from the atmosphere to slow or reverse atmospheric Carbon pollution. The idea is to keep carbon emissions produced by human activities from reaching the atmosphere by capturing and diverting them to a secure storage or to remove carbon from the atmosphere by various means. To understand the need for artificial sequestration, we have to consider the fact that the current average CO_2 is 400 ppm which has skyrocketed after industrial revolution from 280 ppm and that has led to 0.7^{o} increase in the average global temperature. Even if we suddenly halt all the emissions, it could take as long as 1,000 years for environmental measures like sea level and ocean surface temperature to return to pre-industrial levels. This particular poster will discuss the most plausible options for carbon sequestration. 1.) This method is a geological sequestration where the carbon dioxide is pumped into underground chambers such as old oil reservoirs, aquifers and coal seams that are unable to be mined. 2.) Iron fertilization is an intentional introduction of iron to iron-poor areas of the ocean surface to stimulate phytoplankton production. This is intended to enhance biological productivity and accelerate carbon dioxide (CO_2) sequestration from the atmosphere. 3.) Last of all plausible options is Carbon Capture Utilization(CCU) where the carbon is captured and utilized for conversion into other substances or products with higher economic value (*e.g.* plastics, concrete, biofuel) while maintaining the carbon neutrality of the whole processes. Artificial Carbon Sequestration is a promising technology for mitigating increasing concentrations of carbon dioxide (CO_2) in the atmosphere. In 2018 the total CO_2 capture capacity was 40 million metric tons. It surely doesn't neglect the fact that we would have to cut back on our carbon emissions but while it's almost impossible that we would hit 2°C, and even less so 1.5°C, without some sort of negative emissions technology. Also if implemented properly it could do wonders for country like ours to phase out from the carbon based economy

28.7.2 Hybrid Vehicles and Reduction in Pollution Level

The hybrids have a second electric motor, they burn less fuel than conventional cars and emit lower levels of greenhouse gases during operation. That's great, but what about the hybrid manufacturing process? In 2007, a report commissioned by an auto industry trade group insisted that when you factor in the waste generated during production, the notoriously gas-guzzling Hummer is actually greener than the Prius. Hybrid cars do, in fact, require more energy to produce than conventional cars, emitting more greenhouse gases and burning more fossil fuels during the manufacturing process. The production of hybrid batteries, in particular, requires much more energy than producing a standard car battery and results in higher emission levels of gases like sulfur oxide (Burnham *et al.*, 2006) Building a hybrid car is almost exactly the same as building a conventional car, requiring high-tech and highly automated assembly lines. This type of manufacturing process requires tremendous inputs of energy, particularly the forging of materials like steel, aluminum, glass and plastic. Interestingly, lightweight vehicles can sometimes be more energy-intensive (Moon *et al.*, 2006). Experts estimate that 10 to 20 percent of a vehicle's total lifetime greenhouse gas emissions are released during the manufacturing stage alone.

Regenerative braking lets hybrids generate and store their own energy to power the vehicle at low speeds and while idling. Unfortunately, both nickel-hydride batteries and the newer lithium-ion batteries rely on the mining of nickel, copper and so-called rare earth metals. The production of lithium-ion batteries account for 2 to 5 percent of total lifetime hybrid emissions and nickel-hydride batteries are responsible for higher sulfur oxide emissions, roughly 10 kg per hybrid compared with about 1 kg for a conventional vehicle (Samaras 2008; and Burnham *et al.*, 2006).

There are additional environmental concerns related to those rare earth metals, like those used in the magnets of hybrid batteries. In recent years, rare earth metals like lithium have been imported almost exclusively from China, which was able to lower its prices enough to monopolize the industry (Strickland, 2009). One of the reasons China could sell lithium so cheaply was because it widely ignored environmental safeguards during the mining process. In the Bayan Obo region of China, for example, miners removed topsoil and extracted the gold-flecked metals using acids that entered the groundwater, destroying nearby agricultural land. Even the normally tight-lipped Chinese government admitted that rare earth mining has been abused in some places. A regulator at the Ministry of Industry and Information Technology in China went so far as to tell The New York Times, "This has caused great harm to the ecology and environment" (Bradsher, 2009). Although hybrid vehicle production is more energy-intensive and results in higher production emissions, hybrid vehicles are still the greener choice overall. Hybrid cars are much more fuel efficient than conventional vehicles 16.5 km/liter) compared with 11.4 km/liter) for a gas-only vehicle — they require far less gas to cover the same distance. The 23.1 kg of CO_2 is emitted every 161 km, while a conventional car will emit 34 kg. The trouble with plug-in hybrids is that electricity isn't always cleaner than gasoline. More than 45 percent of electricity in the U.S. is generated by coal-powered plants.

28.7.3 Sustainable Development and Use of Renewable Energy Sources

The main sources for the country's power production include solar, hydropower, petroleum and natural gas. It is projected that the country's energy requirements will substantially increase. The country's energy policies must therefore ensure a robust and efficient energy system that is secure and sufficient. There are multiple energy sources, and the technical processes for harnessing, usage and impact on the environment vary from one energy type to another.

Complete energy decarbonization would result in an unprecedented change to the world's "social metabolism", altering not only the amount of fossil fuel based energy the world consumes, but also entire social and economic systems involved in resource extraction, processing, delivery and use (Shelley, 2018). These systems include the human societies that have been built around providing the labour and capital necessary for fossil fuel exploitation, as well as the natural environments that have supported, and been degraded by fossil fuel extraction and combustion (Gonzalez and Toledo, 2017). The energy transformation necessary to mitigate climate change will profoundly alter societies and the natural world (Clark *et al.*, 2013). Energy decarbonization has the potential to bring much more justice, opportunity, and sustainability to communities around the world. However, this will not happen without an intentional focus on the full social metabolism of the energy system. The electricity generation, industrial processes, and transportation consume the most energy as well as contribute the most to greenhouse gas (GHG) emissions, in particular carbon dioxide (CO_2) (Raimi *et al.*, 2018). Interestingly, energy development is essential and a primary input for almost all economic activities and is therefore vital for improvement in quality of life. Virtually, all human economic sectors utilized energy for its growth; such sectors includes industrial, commercial, transportation, telecommunications, wide range of agricultural and household services which has compelled us to focus our attention to ensure its continuous supply to meet our ever increasing demand (Al-Baijali and Shamayleh, 2018).

The post 1992 UN Conference on Environment and Development (UNCED) held in Rio de Janeiro, focused on the restrictions on the release of greenhouse gases, that can only be achieved through renewable sources of energy. Renewable Energy refers to energy (like solar, wind, geothermal, ocean tide, biomass) obtained from energy sources whose utilization do not result in the depletion of the earth's resources. However, renewable energy is replaceable on a human time scale, having the inherent ability to reappear, or replenish themselves by recycling, reproduction or replacement, hence remained augmentable in flow. Other energy sources are non-augmentable having some impact on the environment. Fossil-fuel, coal, oil and natural gas do substantially more harm than renewable energy source by most measures including air and water pollution, damage to public health, wildlife and habitat loss, water use, land use and global warming emissions (Raimi *et al.*, 2018). In contrast, most renewable energy sources produce little to no global emission even when including "Life Cycle" emission of clean energy (*i.e* the emission from each stage of technology's life-manufacturing, installation, operation and decommissioning). The IPCC (2011) opened that, the global warming emission associated with renewable energy are minimal. For instance, burning natural gas for electricity releases between 0.6-2 pounds of carbon dioxide, equivalent to per Kilowatt-hour (CO_2E/KWh), Coal emits between 1.4-3.6 pounds of CO_2E/KWh; while Wind is responsible for only 0.02-0.04 pounds of CO_2E/KWh on a life-cycle basis, Solar 0.07-0.2 CO_2E/KWh. Increasing the supply of renewable energy would allow us to replace carbon intensive energy sources and significantly.

A) Wind Energy

Wind energy is a manifestation of the solar energy and globally, wind power is one of the most expanded renewable and sustainable energy sources. Wind has the potential to supply a significant portion of world energy needs and provides a small but growing segment of electrical energy, reaching 2 per cent to 3 per cent globally (with higher proportions in some countries, such as nearly 26 per cent in Denmark, 16 per cent in Spain and Portugal, and 12 per cent in Ireland). Harnessing power from the wind is one of the cleanest and most sustainable ways to generate electricity as it produces no toxic pollution or global warming emissions. The Green Piece International Estimates that if the trend continuous, wind power could supply 10 per cent of the world's electricity and create 1.7million new jobs by 2020 (Christopherson, 2013). This effort would reduce global emissions of CO_2 by 10 billion metric tons. On a more perspective, every 10,000 Megawatts of wind generated capacity reduces carbon dioxide emissions by 33 million metric tons if it replaces coal or 21 million metric tons if it replaces mixed fuels.

B) Solar Energy

It has the maximum potential of all sources of depleting fossil fuels as it has the maximum potential of all the sources of renewable energy (Paul, 2017). Three technologies are used to generate electricity from solar radiation: photovoltaic (PV) cells, which generate electricity directly; concentrating solar power thermal systems, which use a liquid to transfer absorbed heat to a steam generator that drives a turbine; and solar towers, which are effectively chimneys in which rising hot air powers turbine generators. Solar energy technologies have been deployed in both small-scale (mainly rooftop) applications and in large-scale electrical production. Like wind, the sun provides a tremendous resource for generating clean and sustainable electricity.

C) Biomass for Electricity

This is energy sources from green plants. Green plants capture solar energy through the process of photosynthesis and convert it into organic matter. This organic matter is known as biomass and is basically a form of solar energy, which is converted into chemical energy by the green plants *i.e.* solar energy in the form of chemical energy. Bio energy is generated when this biomass is burned in the form of wood, charcoal and agricultural waste or animal. Biomass power plants share some similarities with fossil fuel power plants; since both involved combustion of a feedstock to generate electricity. Biomass plants raise similar but not identical concerned about air emissions and water use as fossil fuel plants. On the contrast, the feedstock of biomass plants can be sustainably produced, while fossil fuels are non-renewable. Its production sources include energy crops (Switchgrass), agricultural waste, manure, forest products and waste, and urban waste. It has a very high energy content and its burning is non-polluting according to Paul (2017).

D) Hydrogen

It can be produced from fossil fuels, but it is usually obtained from water by means of electrical energy which in turn may be generated from geothermal energy, wind energy, solar energy *etc.* The total hydroelectric power potential in Nigeria is about 8,824MW and only 24 per cent large scale and 4 per cent small hydroelectric power potential have been exploited (Oyedepo, 2012; Nwagbo, 2017), hence more of the country's hydro energy capacity remained unexploited. Hydrogen energy can be used for generating electricity for operating domestic appliances and mini industry. Some of the full advantage of this increasing momentum include. Hydrogen is versatile. Technologies already available today enable hydrogen to produce, store, move and use energy in different ways. A wide variety of fuels are able to produce hydrogen, including renewables, nuclear, natural gas, coal and oil. It can be transported as a gas by pipelines or in liquid form by ships, much like liquefied natural gas (LNG). It can be transformed into electricity and methane to power homes and feed industry, and into fuels for cars, trucks, ships and planes. Hydrogen can enable renewable to provide an even greater contribution. It has the potential to help with variable output from renewable, always well matched with demand. Hydrogen is one of the leading options for storing energy from renewable and looks promising to be a lowest-cost option for storing electricity over days, weeks or even months. Hydrogen and hydrogen-based fuels can transport energy from renewable over long distances from regions with abundant solar and wind resources. Hydrogen can be used much more widely. Today, hydrogen is used mostly in oil refining and for the production of fertilizers. For it to make a significant contribution to clean energy transitions, it also needs to be adopted in sectors where it is almost completely absent at the moment, such as transport, buildings and power generation.

28.8 Towards More Inclusive Water Management

As much as it is a domestic, agricultural and industrial necessity, water is also a basic human right and should be managed as a public good. However, policy and practice related to water management have failed to create inclusive solutions due to blinkered disciplinary thinking about a resource that plays multiple sociocultural, environmental, economic and ecological roles. It is crucial for decision-makers to engage with interdisciplinary approaches to create truly democratic water systems. As a fundamentally interdisciplinary subject and realm, water needs to be analysed through multiple lenses and managed using a capacious toolbox. In order to understand it, comprehension of the hydrological cycle is a prerequisite. This makes it evident that an interdisciplinary systems thinking is needed to govern water resources. However, the sector is currently managed in silos by institutions that use reductionist approaches. Ideas of some critical sociological thinkers can contribute to the more inclusive management of our water resources and truly ensure water for all.

Figure 28.8: River Vishwamitri Passing through Vadodara.

Modern water management is dominated by a bias towards engineering and economic approaches. This domination of certain knowledge systems over others in a subject that is essentially interdisciplinary is reminiscent of Gramsci's concept of hegemony (Calhoun *et al.*, 2012). Water resource departments in most Indian states recruit only civil and mechanical engineers. Their decision-making tools are based on merely economic criteria such as cost–benefit analysis, and the few participatory processes like stakeholder consultations are reduced to box-ticking efforts. This also leads to the hegemony of certain powerful classes over others. For example, the stakeholders of a hydropower project typically include panchayats, non-governmental organizations, community welfare groups (that include farmers, users of irrigation facilities, women welfare groups, self-help groups, and village youth welfare groups), hydropower developers and the state energy department.

Due to the domination of ideas of profit maximization over considerations of equity or ecology, there is neither benefit-sharing with locals nor their inclusion in decision-making around sustainability concerns. For instance, locals often find out about a life-altering hydropower project only after the arrival of machinery and workers in the area (Buechler *et al.*, 2016. Rivers are not human artifacts; they are natural phenomena, integral components of ecological systems. They are not pipelines to be cut, turned around, welded and rejoined (Umamaheshwari 2015).

Cullet *et al.* (2012) also critique the popular reductionist perception of water as "a resource to be exploited for various water services, and not as a resource that has certain basic economic, social, environmental and cultural value." Taking water out of this holistic context leads to its current perception as a resource to be tapped, used, and consumed in ways that make only scientific and economic sense. This hegemonic "scientism" also serves political interests; dominant values can be "politicized" by hiding them behind a smokescreen of "science," preventing questions to protect the interests of the powerful. Figure 28.9 show the flooded path of a town and its alternative solution may be bamboo path. This will provide a clean passage and people will be prevented from water related diseases.

Figure 28.9: Flooded Area in a Small Town and a Bamboo Path: An Economic Solution.

28.9 Towards Better Ecological services

The changes made to ecosystems improved the human well-being and the economy but have harmed the planet and the society. Biodiversity loss has occurred. Poverty is still affecting tribal communities. The degradation of ecosystem is worsening in 21st century. Forests are source for clean air and water in the area. The soil erosion is checked by mangrove plantations. Plant canopy maintains the temperature in surroundings. The changes are needed to preserve the degradation and meet the increasing demand for services. More policy interventions are required.

28.10 Practice of Work from Home

The Coronavirus crisis has drastically affected the economy of several countries. The hotel industry, aviation industry and those concerned with public transport have been facing heavy losses. People are getting unemployed and for those working, their salaries are cuts. The most affected were the migrant workers. Work from home has become the new lifestyle. People are compelled to work from home keeping social distancing in mind. It is more time efficient. While it has advantages, health and productivity wise, it has its own set of disadvantages. As people are working from home, they are finding it difficult to balance the professional and personal life. Though it is only a temporary alternative for the traditional way of working, it is practically not possible for those working in medical services, manufacturing industries *etc.*

Modern life is fast paced and competitive. The lockdown has given a break to the monotonous routine of life. It has given time to introspect and to know ourselves. People have turned to learning new skills and fruitful hobbies to beat stress like walking, cooking, reading, gardening *etc.* They are inculcating constructive hobbies which they never had time to do earlier. Some are turning to yoga and meditation to deal with anxiety and stress. Hobbies not only refresh the mind, but also provide relaxation.

For many people, the work routine gives a structure to their days. With the lockdown, the lack of routine makes life feel adrift. Sleep cycles and meal cycles have been affected. There is hardly any physical activity and people have become mere couch potatoes. This may result in obesity. Being stuck in four walls of the house may lead to stress and anxiety.

The uncertainty of life post lockdown is taking a toll on mental and emotional well being. With an overall uncertain future, job loss, pay cuts coupled with social isolation, there has been a steep rise in cases related to stress, anxiety and depression. The psychological feelings of fear, loss and helplessness may lead to increasing levels of anxiety and depression. It is getting worse for people having anxiety disorders and obsessive compulsive disorder (OCD). For people living alone, loneliness has become a major cause of concern. The most vulnerable are the senior citizens and those with pre existing medical conditions.

Since social distancing became imperative in lockdown, schools and colleges have been shut down. The students are forced to study through video conferencing and online sessions or not study at all. The pandemic has impacted the education system greatly. Online learning platforms like Google meet, Zoom have become popular. However, online classes are riddled with technical challenges. Millions of people in India do not have access to smartphone or a computer, so it is taking a toll on their studies. With 40 percent internet penetration and erratic connectivity, this way is not much feasible.

However, there are some positive consequences of the lockdown. With no domestic help available, people have managed to do the household chores themselves. We have realized that housekeeping is the most thankless, never ending job and should not be taken for granted. The malls had earlier created a problem for the small vendors and provision stores. During this pandemic, it was the local stores that have supported us. The way people are connecting has changes completely- video calls, online meetings *etc.* We have become self reliant, not depending much on other countries. We have learnt that the craze for foreign goods is not in larger interests of the country. Our own country has so much talent and potential. We have imbibed the feeling of 'Vasudhaiva Kutumbakam'- the world is a family. It has encouraged healthy eating habits and a healthy lifestyle. People are now more concious about boosting their immunity and have turned towards homemade cooking. Eatery joints and malls which were earlier crowded now bear a deserted look. The ban on spitting and littering is helping us to keep our surroundings clean. In an era where family values are dwindling, it has given time to reflect upon ourselves and to reconnect as a family.

28.11 Challenges after Covid-19 Pandemic

Nature shows her wrath in the form of such disasters. It is a constant reminder to us that it is our master. We need to realize that we are merely a toy in the hands of God and we need to learn to bow down to Him. On an ending note, let's all remember that "This too shall pass" and one day our state and our nation will be completely free from Covid-19. "India is standing strong and selflessly in these difficult times with those facing trouble in India or abroad. India's growth will always be aiding global growth."

The pandemic has given a bird eye's view of how fragile and unpredictable our lives are. We have learnt to manage with limited resources and to manage our finances. It is teaching us to be patient and to take things one at a time. A close monitoring of places like Taj mahal in India and Jema El Fina in Marrakech, Morocco (Figure 28.11), world heritage sites should be done as a large number of foreign tourists visit there daily.

Figure 28.11: UNESCO World Heritage Centre Jema el-Fina in Marrakech, Morocco with a Large Number of Foreign Tourists.

Change is something commonly feared amongst our species; a fear of the unknown future that we seem to have little to no control over. But when forced into a situation where our daily life turns on its head, we cannot help but notice how a simple difference could change our life. Just being inside our homes and not using cars to get around significantly improved the environment. Having to be more careful about health and hygiene more than ever caused a change in the sicknesses we're often ailed with. In other words, the future we thought out of our control turned out to be something that's greatly influenced by our actions. And perhaps if we were a little less afraid to do something new, we could change the world for our future generations; give them a place they deserve to grow up in.

28.12 Conclusion

The first benefit of engaging with critical thinking is developing a questioning perspective on the systems at hand. Indeed, according to Antonio Gramsci, intellectuals must question the status quo (Crehan, 2016). This challenges complacency and allows for analysis on how to improve prevailing practices. While it is projected that evolution is linear and that modern institutions surpass previous systems, many sociological thinkers have questioned this. Using such a critical approach, we can also recognize how problem-riddled current systems are. Humans had started to believe that they had become invincible with technological developments. Now, a small virus has forced about seven billion people to not only stop flying, but to also stay within the confines of their homes.

REFERENCES

Akolkar, A. B., Suresh, S., Varalaxmi, H. D., and Kumari, A. (2015). Report on Assessment of Environmental Compliance in Electroplating Units at Bengaluru. Bengaluru: Central Pollution Control Board, South Zonal Office

Anonymous 1997. White Paper on Pollution in Delhi with an Action Plan, Ministry of Environment and Forests, Government of India. 1997. [last accessed on 2011 September 20]. Available from: http: //envfor.nic.in/divisions/cpoll/delpolln.html.

Applerouth, S and L D Edles (2007): *Sociological Theory in the Contemporary Era: Text and Readings*, London: Pine Forge Press.

Balachandran S, Meena BR, Khillare PS. Particle size distribution and its elemental composition in the ambient air of Delhi. Environ Int. 2000;26: 49–54. [PubMed: 11345737]

Boni M.F. *et al.*, Preprint at bioRxiv https: //doi.org/10.1101/2020.03.30.015008; 2020; not peer reviewed before posting

Bora, M., and Joshi, N. (2014). A study on variation in biochemical aspects of different tree species with tolerance and performance index. The Bioscan, 9, 59-63.

Bradsher, Keith. "Earth-Friendly Elements, Mined Destructively." The New York Times. December 25, 2009http: //www.nytimes.com/2009/12/26/business/global/26rare.html?_r=2

Buechler, S, D Sen, N Khandekar and C Scott (2016): "Relinking Governance of Energy with Livelihoods and Irrigation in Uttarakhand, India," *Water*, Vol 8, No 10, https: //doi.org/10.3390/w8100437.

Burnham, A.; Wang, W.; Wu, Y. Energy Systems Division, Argonne National Laboratory. "Development and Application of GREET 2.7 - The Transportation Vehicle Cycle Model." November 2006.http: //www.transportation.anl.gov/pdfs/TA/378.PDF

Calhoun, C, J Gerteis, J Moody, S Pfaff and I Virk (2012): *Contemporaray Social Theory*, United Kingdon: Wiley-Blackwell.

Chambers, R (2019): "Personal Reflections on the Green Revolutions Narrative and Myths," Institute of Development Studies, https: //www.ids.ac.uk/opinions/personal-reflections-on-the-green-revolut..

Corfee-Morlot, J., Lamia K-C., Michael G. D., Ian C., Alexis R. and Teasdale, P. J. (2009), Cities, Climate Change and Multilevel Governance. OECD Environmental Working Papers N° 14, OECD publishing Dietz,

Crehan, K A F (2016): *Gramsci's Common Sense: Inequality and Its Narratives*, Durham: Duke University Press.

Cropper M. L, Simon N. B, Alberini A., and Sharma P. K (1997) The World Bank Development Research Group. 1997. Dec. The Health Effects of Air Pollution in Delhi, India, : http: //www.airimpacts.org/documents/local/WB_WPS1860_Delhi_AQ.pdf

Cushine, G. (1985). Electroplating wastewater pollution control technology. Park Ridge, N. J. Noyes Data Corporation.

Cullet, P, S Paranjape, H Thakkar, M Vani, K J Joy and M Ramesh (2012): *Water Conflicts in India: Towards a New Legal and Institutional Framework*, Pune: Forum for Policy Dialogue on Water Conflicts in India.

Freeman, H. (1988). Waste Minimization Audit Report: Case Studies of Minimization of Cyanide Waste from Electroplating Operations. EPA.

Goyal R, Khare M. Indo air quality modelling for PM 10, PM 2.5, PM 2.5, and PM 1.0 in naturally ventilated classrooms of an urban Indian school building. Environ Monit Assess. 2011;176: 501–16. [PubMed: 20635200]

Hochstetler, K. and Viola, E. (2011) Brazil and the Multiscalar Politics of Climate Change. Paper presented at the 2011 Colorado Conference on Earth Systems Governance, Colorado State University, Fort Collins, Colorado, 17-20 May 2011

Kumar A, Phadke KM, Tajne DS, Hasan MZ. Increase in inhalable particulates' concentration by commercial and industrial activities in the ambient air of a select Indian metropolis. Environ Sci Technol. 2001;35: 487–92. [PubMed: 11351718]

Kumar A, Scott Clark C. Lead loadings in household dust in Delhi, India. Indoor Air. 2009;19: 414–20. [PubMed: 19659889]

Moon, Paula. Center for Transportational Research, Argonne National Laboratory. "Vehicle-Cycle Energy and Emissions Effects of Conventional and Advanced Vehicles." 2006http: //www.transportation.anl.gov/pdfs/TA/381.pdf

Muthukaruppan, K., Sowmya, M., Prasad, M. V. R., and Satpathy. (2010). Ambient air quality at IGCAR campus, Kalpakkam. The Bioscan, 2, 383-390. www. scholink.org/ojs/index.php/ees Energy and Earth Science Vol. 3, No. 2, 2020 48 Published by SCHOLINK INC.

Ramesh, B. B., Udaya, B. S., and Seeni, M. K. (2009). Waste Minimization in Electroplating Industries: A Review. Journal of Environmental Science and Health, Part C: Environmental Carcinogenesis and Ecotoxicology Reviews, 27(3), 155-177. https: //doi.org/10.1080/10590500903124158

Ritzer, G (2010): *Sociology Theory*, Boston: McGraw-Hill.

Samaras, C; and Kyle. M,"Life Cycle Assessment of Greenhouse Gas Emissions from Plug-in Hybrid Vehicles: Implications for Policy." Environmental Science and Technology, 2008.http: //pubs.acs.org/doi/abs/10.1021/es702178s

Singh, R. K. (2014). Pollution Threat to Surface and Groundwater Quality Due to Electroplating Industries. International Journal for Research in Applied Science and Engineering Technology (IJRASET), 2(II), 50-53.

Singh, V., Ram, C., and Kumar, A. (2016). Physico-Chemical Characterization of Electroplating Industrial Effluents of Chandigrah and Haryana Region. Journal of Civil and Environmental, 6(4). https: //doi.org/10.4172/2165-784X.1000237

Strickland, Eliza. "Isn't It Ironic: Green Tech Relies on Dirty Mining in China." Discover. December 28, 2009http: //blogs.discovermagazine. com/80beats/2009/12/28/isnt-it-ironic-green-tech-relies-on-dirty-mining-in-china/

Thakar, B. K., and Mishra, P. C. (2010). Dust collection potential and air pollution tolerance index of tree vegetation around vedanta aluminium limited, Jharsuguda. The Bioscan, 3, 603612.

Turner,JH(2001): "Positivism: Sociological," *ScienceDirect*, https: //www. sciencedirect.com/topics/social-sciences/positivism.

Tiyagi S. K. Agrawal S. Kumar S. Agrawal N. Sharma S. Gupta D.K. and Tyagi S. (2020) Analysis of Ambient Air Quality Pre- and During Lockdowns due to COVID -19A Case Study of Hapur (NCR Town), Uttar Pradesh, *Indian Journal of Air Pollution Control*, 20, (1-2): 47

Umamaheshwari, R (2015): "A Visionary on Water Issues," *Hindu*, 12 September,

https: //www.thehindu.com/opinion/op-ed/obituary-on-ramaswamy-r-iyer/arti.

WHO (2005) Air quality guidelines for particulate matter, ozone, nitrogen dioxide and sulfur dioxide, Global update 2005, Summary of risk assessment.from: http: //whqlibdoc.who.int/hq/2006/WHO_SDE_PHE_OEH_06.02_eng.

Chapter 29

Emergency Preparedness Plan for the Industries to Reduce the Risk of Life and Properties

Meera Chokshi and Manthan Tailor

Department of Environmental Studies, Faculty of Science, The Maharaja Sayajirao University of Baroda, Vadodara – 390 002, Gujarat
e-mail: env.meera@gmail.com, manthantailor@gmail.com

ABSTRACT

Incident, accident, emergency as well as disaster are all undesired/ unplanned/ unexpected events with increasing risk level respectively. They have potential to cause harm to life, property and environment. Preparedness with respect to these events is the readiness at the time of occurrence of these events but involves the actions before, during and after the event. Various industries have different processes and working environments, and thus may need diverse actions which are specific to the working environment. Unlike some other industries, Chemical Industries are one of the most critical industries that are characterized by unique conditions which need critical attention. While safety procedures in some other industries are advisable, they are absolutely mandatory in the Chemical Industries. Recurrences of major process safety management (PSM) incidents like Union Carbide and Bhopal Gas leak tragedy are calamitous and must be strongly avoided. Laws are very strict for high risk industries as life and health of hundreds of citizens is directly related to them. The safety personnel take necessary actions to comply with rules with such chemical industries with high risk to ensure safety of its employees and adjacent community as well as minimization of property and environment losses.

Emergency preparedness is achieved by creating an Emergency Plan. Such plans may vary depending on the type of organization but their fundamental motive is to minimization of risk. Emergency preparedness plan involves components viz., planning and staffing, resource procurement, defining of roles and responsibilities and capacity building by regular rehearsal and frequent trainings.

The main purpose of emergency preparedness is to make the personnel ready for any type of emergency situation that may put their lives at risk and/or would cause property and environmental damage. PSM is an integral component of emergency preparedness and an essential part of process safety management (PSM) is assessing processes and making sure the people involved in those processes understand the chemicals they are working with and associated possible dangers. They need to know how the process works, what are the daily operations and maintenance protocols and how to respond when irregularities occur. Considering the legal aspects, the concerned industries must have an On-site as per Schedule 11 of MSIHCR, 1989 and for Gujarat, as per Schedule 8A of Gujarat Factories Rules, 1963. Also, details required to be furnished for Off-site Emergency Plan as per Schedule 12 of MSIHCR, 1989, for the preparation of Off-site Emergency Plan (OEP) by the district authorities. The trailing document is designed with an aim to help the manager and administrative staff for implementation of the same to their respective organizations.

Keywords: *Emergency Plan, preparedness, Industries, Chemical, Risk, Accident, Mock drill*

Abbreviations

CAEPPR [Chemical Accidents (Emergency Planning, Preparedness and Response)] DMA (Disaster Management Act), DMP (Disaster Management Plan), GFR (Gujarat Factories Rules), NDMA (National Disaster Management Authority), MSIHCR (Manufacture, Storage and Import of Hazardous Chemicals, Rules), GSDMA (Gujarat State Disaster Management Authority).

29.1 Introduction

Any workplace environment involves some or the other kind of eventualities or accident that may occur, resulting in loss which can be avoided or minimized by undertaking certain safety measures. An accident is any undesired, unexpected, unplanned unintended event and the causative agent of the same is called Hazard. Hazard is a qualitative term and it can never be zero because it exists everywhere. Existence of hazard poses risk, which in turn is governed by the probability of an event and its negative consequences. Risk can be estimated by multiplication of Probability and Severity of an accident. When the risk is zero, it is called impure risk and the resulting undesired event is termed as Incident. When risk is non-zero, it is called pure risk and the resulting event is termed as Accident. When the risk is high, the event is termed as Emergency and the event with very huge risk or catastrophic risk is known as Disaster. Thus, incident, accident, emergency and disaster are all unexpected events with different risk of increasing values.

The disasters with highest associated risk are defined as "A serious disruption of the functioning of a community or a society involving widespread human, material, economic or environmental losses and impacts, which exceeds the ability of the affected community or society to cope using its own resources." (United Nations International Strategy for Disaster Reduction, 2009)

The World Health Organization (WHO) defines a disaster as "a sudden ecological phenomenon of sufficient magnitude to require external assistance". It is also defined as any event, typically occurring suddenly, that causes damage, ecological disruption, loss of human life, deterioration of health and health services, and which exceeds the capacity of the affected community on a scale sufficient to require outside assistance (Landsman, 2001).

Emergency can be defined as any situation which pose a threat to safety of persons, property or/and environment. And preparedness means the state of readiness to deal with a threatening disaster situation or disaster and the effects thereof (NDMA, 2005).

As a whole, Emergency preparedness means the state of readiness which enables stakeholders to mobilize, organize and provide relief to deal with an impending or actual disaster or the effects of a disaster (GSDMA, 2003). In other words, Emergency Preparedness means the technical, operational and organizational measures that are planned to be implemented under the management of the emergency organization in case hazardous or accidental situations occur, in order to protect human and environmental resources and assets (NORSOK, 1998).

29.1.1 The Objectives of Emergency Preparedness

- ☆ To be in a state of readiness for any undesired/unintended event(s) and to face and minimize the adverse effects of accidents caused by hazardous substances.
- ☆ To provide following details in a document (emergency plan)
  - ❖ Measures to contain the incident,
  - ❖ Measures to minimize effects due to fire, explosion, toxic release, spillage of hazardous substances,
  - ❖ Actions to be taken before, during and after an accident.

29.1.2 The Steps for Planning of Emergencies

- ☆ Step-1: To identify and assess the hazard and its potential of an installation/department *i.e.* Risk assessment.
- ☆ Step-2: To prepare plans to prevent and control emergencies as well as mitigate consequences of an accident.
- ☆ Step-3: To ensure availability of resources on-site, mock-drill, monitoring, updating the plan and to arrange public awareness.

29.2 Classification of Emergency and Emergency Plans

There are different types of emergency situations considering an industrial pocket including the following:

1. On-site Emergency

It is caused by an accident that takes place in a chemical installation or an industry and the effects are confined to the factory premises involving the people working in the factory.

On-site Emergency Plan deals with measures to prevent and control emergencies within the factory and not affecting outside public or environment. Such eventualities are mandatorily the responsibility of the occupier. The format of On-site Emergency Plan is available as Schedule 11 of the Manufacture, Storage and Import of Hazardous Chemicals Rules, 1989 as latest amended and Sch. 8A of the Gujarat Factories Rules, 1963 as latest amended. The mock drill frequency is at least once in six months but for a chemical works it is at least once in three months.

The elements of On-Site Emergency Plan:

a) Hazardous substances and Zones (Maps/Charts)
b) Appointment of key persons and their role
c) Assigning responsibilities to the personnel
    (i) During normal working hours of the factory
    (ii) Outside the normal working hours of the factory
d) Infrastructure
    (i) Emergency Control Centre
    (ii) Safe Assembly Points
e) Operational Systems during Emergency
    (i) Communication system
    (ii) Warning system
    (iii) Emergency services
    (iv) Mutual aid
f) Declaration of off-site emergency

2. Off-site Emergency

If an accident takes place in a chemical installation/industry and its effects are felt outside its premises, the situation thus created is called an off-site emergency.

It is mandatory that the district authorities should prepare off-site emergency plan for their jurisdiction. The details required to be furnished by the Occupier for off-site emergency is available as Sch.12 of MSIHCR, 1989.

Off-site emergency plan deals with measures to prevent and control emergencies affecting public and the environment outside the premises. The manufacturer should provide the necessary information on the nature, extent and likely effects of such incidents. The mock drill frequency is at least once in a year and at any Industrial Pocket of the District. It is the responsibility of the District Collector.

The Elements of off-site emergency plan:

(i) Plan pre-requisites
    a) Quantification
    b) Hazard Analysis
    c) Scenario developments
    d) Assessment of capabilities
(ii) Plan development
    a) incidental information

b) Authority for responding
c) Basic assumption
d) Operation of the plan

(iii) Emergency Assistance
a) Name and address of key personnel
b) Communication details

(iv) Response functions

29.3 Significance of the Plan

The main purpose of Emergency Preparedness Planning in any industrial premise is to make the employees ready for any type of emergency situation like fire, explosion, toxic release, *etc.* that may arise while handling and storage of any hazardous chemical.

An Emergency Plan is majorly important as it serves as a significant document at the time of any emergency situation and would help to handle and cope the situation with minimum harm and danger. Also, regular updating of the plan and incorporating details about the steps taken and mistakes occurred for any accidents that may have happened in the past, gradually, the plan document would contain all the necessary information for all the types of emergencies.

The guidelines and legal requirements for preparation of the emergency plan differ from state to state, but the basic concept and application criteria remains the same, which is the safety of people, property and environment. In case of an On-site emergency plan, it is applicable only until the effects can be restricted within the factory's premise. Once the effects are felt beyond the factory's premise, the on-site plan no longer plays a major part, the Off-site emergency plan would have to be activated in order to stop the effects from spreading furthermore.

29.4 Recent Chemical Disaster: The Vizag Gas Leak

On 7th May, 2020, there occurred an incident of uncontrolled Styrene vapour release at LG Polymers, RR Venkatapuram, in Visakhapatnam from one of the Styrene storage-tanks (M6-Tank). The accident took lives of 12 persons in the immediate consequent period and 585 people had to be hospitalized for treatment. In addition, it also caused loss of vegetation and livestock. This Styrene vapour release, broadly referred to as "Vizag Gas Leak", is one of the major Styrene vapour release incidents from a bulk storage tank at any place in the world.

The accident is said to have occurred due to poor design of tank, insufficient refrigeration and cooling system, absence of circulation and mixing systems, insufficient measurement parameters, poor safety protocol and awareness, improper risk assessment and response, poor PSM systems, negligence of management, inadequate knowledge amongst staff, deficient knowledge of the chemical properties of Styrene, especially during storage under idle conditions. LG Polymers did not have any PSM system. On the whole, it can be said that LG Polymers did not have proper safety protocols and violated a number of safety etiquettes (Yashoda and

Teja, 2020). Similarly, many other chemical accidents have occurred worldwide in the past, took lives of many people and caused major damage to property and the environment. These accidents happened due to negligence which lead to catastrophic results. But it is a great opportunity to learn from previous incidents and develop new practices, procedures and management systems and design a safer plant. These incidents disclose many hidden facts about safety and offer efficient tools for prevention of similar incidents in the future.

Table 29.1: Industrial Hazard-wise Categorization of Districts in Gujarat

Category	*Districts*
Highly hazardous	Bharuch and Vadodara
Hazardous	Ahmedabad, Surat, Jamnagar, Kutch, Valsad, Rajkot
Less hazardous	Anand, Gandhinagar, Bhavnagar, Kheda, Panchmahal, Mehsana, Porbandar
Quite less hazardous	Amreli, Dahod, Banaskantha, Junaghadh, Dangs, Narmada, Patan, Navsari, Surendranagar, Sabarkantha

Based on GSDMA, 2013.

The above-mentioned districts of Gujarat in total have more than 400 MAH Factories that pose major threat to the state's inhabitants and property, which would also result in severe environmental harm, in case of any chemical accident.

29.5 Model for an On-site Emergency Plan of a Chemical Factory

As Per the Guidelines on Preparation of On-site and Offsite Emergency Plan by the Office of Chief Inspector of Factories, Gujarat State in April, 1989,

An on-site Emergency Plan document should include the following components.

I. Description Section

1) Introduction
2) Scope of the document
3) Objectives
4) Definitions
5) Risk and Environmental Impact Assessment
6) Emergency Organization
7) Communication System
8) Action on site
9) Off-site Emergency Plan
10) Training, rehearsal and record

II. Annexure Section

These annexure are basically designed to acquire the specific information referred from the description section. Easily readable information in table format is given which would save a considerable amount of time at the time of emergency. Also, it would be useful to surrounding factories as mutual aid and to the Government. For the purpose of having a fool-proof plan, all these annexure should be filled very precisely.

And the details for the information to be filled in the Annexure are as follows:

Identification of the Factory

It includes basic legal and contact details of the factory and its manufacturing process. It also includes details of the key personnel of the factory.

Map of the Area

Under this annexure the Key Plan of the factory premise is to be provided.

Factory Layout

The entire layout map of the factory is to be provided here.

Storage Hazard and Control

Here, all the hazardous substances with their concentration are to be listed along with their storage quantity, its location, operating conditions, associated hazards, control measures provided and details of persons in charge.

Material Safety Data Sheets

The MSDS of all the chemicals in the factory's premise is to be provided under this annexure. At the time of any chemical emergency, it will be useful for decision making.

Process and Vessel Hazards and Controls

The name of the plants or places where the hazardous processes and operations are carried out, materials used with their quantity, and all the details about operating conditions and vessels used including its maximum capacity. Also, details about persons in charge.

Other Hazards and Controls

All other hazard or directly/indirectly associated risks with its basic information like source, impacts, severity, control measures, and contact details of person in charge at source.

Trade Waste Disposal

It includes data of the trade waste like type and name of the trade waste, its physical state, generation per day, source of generation, and method of treatment and/or disposal and contact details of authorized persons.

Records of Past Incident

It includes information of occurred accidents with all minute details like date, time, place duration, time required for controlling it, type of injury, and its severity.

Gas Dispersion Concentration

Here the material dispersion concentration (in ppm) is to be given from 100m to 5km distance (downwind).

Evacuation Table

Here the radius of immediate danger, downwind and crosswind evacuation area (in km) is provided individually for each material.

Environment Impact Assessment

Here all the basic and necessary information acquired from the EIA report is to be incorporated.

Weather Condition

Here, the average weather conditions and essential meteorological data of that area is to be provided.

Incident Controllers

It includes basic details about appointed incident controllers like name, designation, qualification and contact information.

Deputy Incident Controllers

It includes basic details about appointed deputy incident controllers like name, designation, qualification and contact information.

Site Main Controller

It includes basic details about appointed site main controllers like name, designation, qualification and contact information, in order of their priorities.

Key Personnel

It includes basic details about appointed key personnel for each department like name, designation, qualification and contact information.

Essential Workers

It includes basic details about appointed essential workers, individually for every shift with their name, designation, work they are trained for, contact information along with their required equipment for first aid, firefighting, rescue, salvage, *etc.*

Safe Assembly Points

It includes Safe assembly points marked within the factory's premise along with their locations and capacity. Also, the basic contact information about the person in charge of all the assembly points at the time of emergency are to be provided.

Emergency Control Centre

Here, the location and telephone numbers of the emergency control centers is to be given along with the items that are kept in the center, their quantity, the persons who would operate/handle them and its period of operation.

Fire and Toxicity Control Arrangements

Details about all the types available extinguishers are included here. Also, the information about the PPEs, contact information about nearby fire stations are provided here.

Medical Arrangements

The basic details about first aid centers, ambulance room, hospitals, occupational health centers along with the basic and contact information of in charge person and the available facilities are to be provided here.

Transport and Evacuation Arrangements

The basic details about transport center and vehicles, if owned along with the basic and contact information of in charge person are to be provided here. Also, details of the outside shelters for evacuation along with their location and accommodation capacity are to be given.

Pollution Control Arrangements

Details about all the pollution control arrangements like effluent treatment plant, ambient air/stack monitoring, scrubbers, incinerator, land pollution control measures along with their necessary details and contact information of in-charge person are to be provided here.

Other Arrangements

Any other arrangements, if available are to be listed here along with their basic details and contact information of the key personnel and essential workers in charge of them. Also, the details about the mutual aid arrangements and contact information of the in-charge persons are to be given.

Alarms and Sirens

All the location points of the installed alarms and sirens along with their type and interpretation of each of them are to be mentioned here.

Internal Telephones

Here the telephone numbers of all the internal (within the factory) emergency services with their name and location are to be given. Also, the name, designation and the contact details of the person available on this phone are to be given.

External Telephones

The telephone numbers of all the external emergency services like police, fire brigade, hospital, ambulance, *etc.*, with their basic details and contact information of the person available on this phone and the services expected from them under on-site or offsite emergency plan are to be given.

Nominated Persons to Declare Major Emergency

Here, the location from where the major emergency would be declared and the person nominated to declare the emergency with their basic details along with their assigned duty under on-site or off-site emergency plan and contact details are to be given.

A Form to Record Emergency Telephone call

This form is used to fill the information delivered through telephonic conversation at the time of emergency. It includes essential information like, brief description, type and cause of incident and number of injured and affected people. And if the emergency is caused by vehicles then the vehicle number, name and address of the owner and if he's been informed about it or not.

Statutory Communication

The periodicity and last date for giving the statutory information to the workers, general public, neighboring firms, District emergency authority and Directorate of Industrial safety and health is given here. Also, if any suggestions are received from them, then those suggestions and the last date for implementation of them is also to be written.

Separation Distances

The safe separation distance (in meter) of the hazardous chemicals and the residing population in the area is to be provided here individually for all the hazardous chemicals in the factory with their container capacity and quantity.

Emergency Instruction Booklets

Here, the short and clear information about the duties/functions for all the roles (incident controller and his deputy, site main controller, key personnel, administration personnel, first aider, fire fighter, rescuer, salvager) assigned for the time of emergency and his reporting site at that time are given.

Additionally, if any other important information, that might be useful at the time of emergency can be added as appendix. For instance, the preventive measures like the use of appropriate PPEs individually on the basis of the chemical being handled can be provided.

Note: Followings terms are as per CAEPPR Rules, 1996 and MSIHCR, 1989 and some other National and International Standards

Table 29.2: Glossary of Terms

Term	*Definition*
Hazard	Hazard is a physical situation which may cause human injury, damage to property or the environment or some combination of these criteria.
Accident	Accident is an unplanned event, which has a probability of causing personal injury or property damage or both. It may result in physical harm (injury or disease) to person, damage to property, loss to the company, a nearby area or any combination of these effects.
Industrial Pocket	Industrial Pocket means any industrial zone ear-marked by the industrial development corporation of the State Government or by the State government
Risk	Combination of the likelihood of an occurrence of a hazardous event or exposure(s) and the severity of injury or ill health that can be caused by the event or exposure(s) (Indian Standards 18001, 2007)
Risk Assessment	The overall process of estimating the magnitude of risk and deciding whether the risk is tolerable.
Major Chemical Accident	It is an incident involving loss of life inside or outside the installation or ten or more injuries inside and/or one or more injuries outside or release of toxic chemicals or explosion or fire or spillage of hazardous chemicals resulting in on site or off site emergencies or damage to equipment leading to stopping of process or adverse effect to the environment.
Major Accident Hazards (MAH)	MAH installations means isolated storage and industrial activity at a site, handling (including transport through carrier or pipeline) of hazardous chemicals equal to or, in excess of the threshold quantities specified in column 3 of Schedules 2 and 3 respectively of MSIHCR, 1989.
Safety	State in which the risk of harm to persons or damage to property is limited to a tolerable level.

REFERENCES

GSDMA. (2003). Gujarat Act No. 20 of 2003. 20, 1–40.

GSDMA. (2013). Gujarat State Chemical Disaster Management Plan. 11.

Landesman, Y.L. (2001), Public health management of disasters: The practice guide. American Public Health Association, 800 I Street, NW, Washington, DC. 2.

NDMA. (2005). DM Act, 2005 - National Disaster Management Authority. 1–29. https: //ndma.gov.in/en/disaster.html

NORSOK. (1998). Standard Z-013, Risk and Emergency Preparedness Analysis. (Issue March).

Chemical Accidents (Emergency Planning, Preparedness and Response) Rules, 1996.

Manufacture, Storage and Import of Hazardous Chemicals Rules, 1989.

Yashoda, T., and Teja, D. (2020). EPRA International Journal of Research and Development (IJRD) VIZAG Gas Leak- A Case Study on the uncontrolled Styrene Vapour Release for the first time in India. 7838: 13–24.

Chapter 30

Phenomenon of Bioluminescence in Fireflies (*Lampyridae*) and Mushrooms

Jugal Chauhan, Nirmal Chaudhary, K. Shruti, Drashti Navlani and Arun Arya

Department of Environmental Studies, The Maharaja Sayajirao University of Baroda, Vadodara – 390 002, Gujarat
e-mail: sarojarun10arya@rediffmail.com

ABSTRACT

ATP is energy giving compound. It can be calculated in presence of luciferin and Mg^{+2}. The light emitted in night is directly proportional to Adenosine Triphosphate concentration. Usually the measurement is made through a Flourimeter. Bose was a pioneer in many areas, contributing greatly to physics, biology, botany, archaeology, and perhaps surprisingly, science fiction. Because of this, he is perhaps one of only a handful of people who could be considered a polymath in the modern period–an occupation which is usually associated with great minds of earlier centuries, in the heady days from the Renaissance to the Enlightenment, before scientific fields became more specialized and scientists found it ever more difficult to apply their work across multiple fields. Despite being rewarded a Knighthood in 1917, and being the first Indian to become a fellow for Science of the Royal Society in 1920, his work was largely overlooked at the time, and its significance has only been re-evaluated in recent years, particularly in his contributions to radio technology. In 2012–over a century after his experimental work on radio–the Institute of Electrical and Electronics Engineers (IEEE) recognized these contributions as an IEEE Milestone in Electrical and Computer Engineering, and the Indian Botanical Gardens were renamed in his honor in 2009. It is interesting to note that Sir J.C. Bose for the first time saw this phenomenon of insects in his childhood and called it a bush on fire. Different genes are cloned and efforts are on to standardize a method to detect the transfer of light genes in plants. Based on recent data, bioluminescence could make crops glow when they need water or fertilizer, or *when the plants are attacked by pests. Thus farmers could use water or chemicals more efficiently treating their fields only when Huff! Puff !*

The Kaleidoscopic fungus family comprises such a wide variety of forms, sizes, odours and life histories that it is impossible to sum up all of them. Some of these occur as solitary specimen. One may keep searching literally for years without locating a second example of a first accidental discovery. Glowing wood was used by troops in the two world wars. They stuck it in the straps of their steel helmets or on their rifles to avoid collision in darkness. Unlike light from an incandescent bulb, which glows by resistance to an electric current, bioluminescence comes from a chemical reaction between two substances organisms produce naturally: luciferin and the enzyme luciferase. Because different kinds of luciferins and luciferases occur in nature, the colour of their light varies.

Keywords: *Bioluminescence, Photinus pyralis, Fireflies, Luciferins Lampyridae, Mushrooms.*

30.1 Introduction

Imagine that instead of switching on a lamp when it gets dark, you could read by the light of a glowing plant on your desk. MIT engineers have taken a critical first step toward making that vision a reality. By embedding specialized nanoparticles into the leaves of a watercress plant, they induced the plants to give off dim light for nearly four hours. They believe that, with further optimization, such plants will one day be bright enough to illuminate a workspace. "The vision is to make a plant that will function as a desk lamp – a lamp that you don't have to plug in. The light is ultimately powered by the energy metabolism of the plant itself," says Michael Strano, the Carbon P. Dubbs Professor of Chemical Engineering at MIT and the senior author of the study. This technology could also be used to provide low-intensity indoor lighting, or to transform trees into self-powered street lights, the researchers say. MIT postdoc Seon-Yeong Kwak is the lead author of the study, which appears in the journal *Nano Letters*.

Bioluminescence, the result of a process that occurs in living organisms in which an electronically excited substance produced in a chemical reaction decays to the ground level, is a widespread phenomenon in Nature. The scientific research on this subject led to the discovery of the enzyme-substrate system, the former called luciferase and the latter luciferin (from the Latin Lucifer, ''Light-bringer''). Among all the bioluminescent organisms fireflies are the most studied and well-characterized, specially the North American firefly, *Photinus pyralis* (Order Coleoptera, Family Lampyridae). This review presents the chemical mechanism, furnishing basic information about its principal components, and some recent applications of the enzyme.

30.1.1 Firefly Luciferase

Firefly luciferase is classified as Photinus-luciferin: oxygen 4-oxidoreductase (decarboxylating, ATP-hydrolysing) (EC 1.13. 12.7). Its substrate is firefly luciferin (LH2), and the reaction requires ATP, oxygen and a metallic cation (McElroy, 1947;1951). Crystallographic studies have shown that the protein is folded into two compact domains, a large N-terminal domain and a small C-terminal domain, joined by a flexible linker peptide, which creates a wide cleft between the two domains

(Conti, *et al.*, 1996). The C-terminal portion bears the tripeptide Serine-Lysine-Leucine (SLK motif) responsible for peroxisome targeting. They have a unique flash pattern. The purpose of blinking is so that male fireflies can find potential mates.

30.1.2 Bioluminescence in Microorganisms and Fungi

The structure of bacterial luciferase has also been determined to high resolution without bound ligands (Fisher *et al.*, 1995). Bacterial luciferase is a heterodimer composed of homologous subunits, α and β (Baldwin, and Ziegler 1992). Both subunits assume the well-known $(\beta/\alpha)_8$, or TIM barrel structure, and are packed together by a parallel four-helix bundle. The active center is confined to the α subunit, and it has been proposed to reside within a large internal cavity which opens through a narrow crevice (Baldwin, *et al.*1995). The opening to the cavity lies beneath a long disordered loop which appears to undergo a conformational rearrangement upon flavin binding and catalysis(Baldwin, and Ziegler 1992; Baldwin, *et al.*1995; AbouKhair, *et al.*, 1985). It has been proposed that this conformational change may be necessary to exclude water from the intermediates and the excited state of the flavin. Thus, it appears reasonable to propose that both luciferases exclude water from their respective excited emitters by a conformational rearrangement that occurs upon or following substrate binding.

30.2 Bioluminescence in Plants

Researches are going on to develop self-repair plants having their own energy. With the help of Plant nanobionics, light genes of firefly are transferred into plants and cause them to emit light. Researcher's efforts yielded plants that can glow for about 45 minutes, which has recently been improved to 3.5 hours. *Arabidopsis thaliana*, arugula, kale, spinach and watercress plants are some on which this has been demonstrated. Plants can be designed to detect explosives and communicate that information to a smart phone, as well as plants that can monitor drought conditions. Lighting, which accounts for about 20 percent of worldwide energy consumption, seemed like a logical next target. "Plants can self-repair, they have their own energy, and they are already adapted to the outdoor environment," Strano says. "We think this is an idea whose time has come. It's a perfect problem for plant nanobionics," (Giraldo *et al.*, 2014).

To create their glowing plants, the MIT team turned to luciferase, the enzyme that gives fireflies their glow. Luciferase acts on a molecule called luciferin, causing it to emit light. Another molecule called co-enzyme A helps the process along by removing a reaction byproduct that can inhibit luciferase activity. The MIT team packaged each of these three components into a different type of nanoparticle carrier. The nanoparticles, which are all made of materials that the U.S. Food and Drug Administration classifies as "generally regarded as safe," help each component get to the right part of the plant. They also prevent the components from reaching concentrations that could be toxic to the plants. The researchers used silica nanoparticles about 10 nanometers in diameter to carry luciferase, and they used slightly larger particles of the polymers PLGA and chitosan to carry luciferin and coenzyme A, respectively. To get the particles into plant leaves, the researchers first suspended the particles in a solution. Plants were immersed in the solution and then exposed to high pressure, allowing the particles to enter the leaves through tiny pores called stomata.

Particles releasing luciferin and coenzyme A were designed to accumulate in the extracellular space of the mesophyll, an inner layer of the leaf, while the smaller particles carrying luciferase enter the cells that make up the mesophyll. The PLGA particles gradually release luciferin, which then enters the plant cells, where luciferase performs the chemical reaction that makes luciferin glow. The researchers have also demonstrated that they can turn the light off by adding nanoparticles carrying a luciferase inhibitor. This could enable them to eventually create plants that shut off their light emission in response to environmental conditions such as sunlight, the researchers say.

30.2.1 Nanobionic Plants: Supercharged Photosynthesis

Plants have many valuable functions: They provide food and fuel, release the oxygen that we breathe, and add beauty to our surroundings. Now, a team of MIT researchers have worked on "plant nanobionics." "Plants are very attractive as a technology platform," says Michael Strano, the Carbon P. Dubbs Professor of Chemical Engineering and leader of the MIT research team. "They repair themselves, they're environmentally stable outside, they survive in harsh environments, and they provide their own power source and water distribution." Plants can be used into self-powered, photonic devices such as detectors for explosives or chemical weapons.

The idea for nanobionic plants grew out of a project in Strano's lab to build self-repairing solar cells modeled on plant cells. Chloroplasts host all of the machinery needed for photosynthesis, which occurs in two stages. During the first stage, pigments such as chlorophyll absorb light, which excites electrons that flow through the thylakoid membranes of the chloroplast. The plant captures this electrical energy and uses it to power the second stage of photosynthesis – building sugars. Chloroplasts can still perform these reactions when removed from plants, but after a few hours, they start to break down because light and oxygen damage the photosynthetic proteins. Usually plants can completely repair this kind of damage, but extracted chloroplasts can't do it on their own. To prolong the chloroplasts' productivity, the researchers embedded them with cerium oxide nanoparticles, also known as nanoceria. These particles are very strong antioxidants that scavenge oxygen radicals and other highly reactive molecules produced by light and oxygen, protecting the chloroplasts from damage.

The researchers delivered nanoceria into the chloroplasts using a new technique they developed called lipid exchange envelope penetration, or LEEP. Wrapping the particles in polyacrylic acid, a highly charged molecule, allows the particles to penetrate the fatty, hydrophobic membranes that surrounds chloroplasts. In these chloroplasts, levels of damaging molecules dropped dramatically. Using the same delivery technique, the researchers also embedded semiconducting carbon nanotubes, coated in negatively charged DNA, into the chloroplasts. Plants typically make use of only about 10 percent of the sunlight available to them, but carbon nanotubes could act as artificial antennae that allow chloroplasts to capture wavelengths of light not in their normal range, such as ultraviolet, green, and near-infrared. With carbon nanotubes appearing to act as a "prosthetic photoabsorber," photosynthetic activity - measured by the rate of electron flow through the thylakoid membranes - was 49 percent greater than that in isolated chloroplasts without embedded nanotubes. When nanoceria and carbon nanotubes were delivered together, the chloroplasts remained active for a few extra hours.

The researchers then turned to living plants and used a technique called vascular infusion to deliver nanoparticles into *Arabidopsis thaliana*, a small flowering plant. Using this method, the researchers applied a solution of nanoparticles to the underside of the leaf, where it penetrated tiny pores known as stomata, which normally allow carbon dioxide to flow in and oxygen to flow out. In these plants, the nanotubes moved into the chloroplast and boosted photosynthetic electron flow by about 30 percent. Yet to be discovered is how that extra electron flow influences the plants' sugar production. "This is a question that we are still trying to answer in the lab: What is the impact of nanoparticles on the production of chemical fuels like glucose?" (Giraldo *et al.*, 2014).

30.2.2. Lean Green Machines

The researchers also showed that they could turn *Arabidopsis thaliana* plants into chemical sensors by delivering carbon nanotubes that detect the gas nitric oxide, an environmental pollutant produced by combustion. Strano's lab has previously developed carbon nanotube sensors for many different chemicals, including hydrogen peroxide, the explosive TNT, and the nerve gas sarin. When the target molecule binds to a polymer wrapped around the nanotube, it alters the tube's fluorescence. "We could someday use these carbon nanotubes to make sensors that detect in real time, at the single-particle level, free radicals or signaling molecules that are at very low-concentration and difficult to detect," (Giraldo *et al.*,2014).

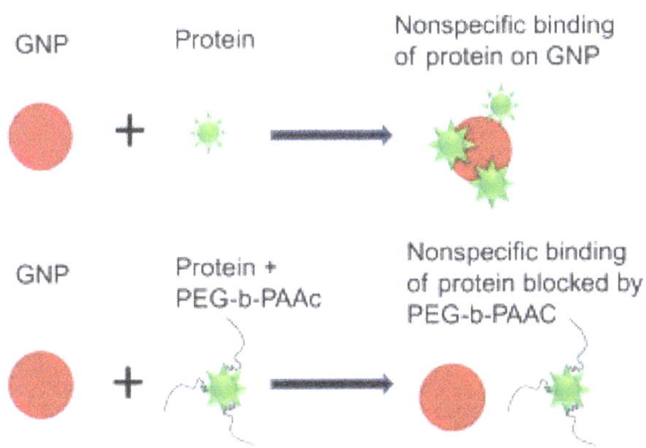

Figure 30.1: Colorimetric Detection. The strategy of protein attachement. The biofouling can be preventing using PEG-polymers.

The optical biosensor was originally developed to measure the dissolved oxygen, carbon dioxide, and pH (Galindo, 2009). Light is the output transduced signal (Monošík *et al.*, 2012). This biosensor is based on the methods, UV-vis absorption, fluorescence, chemiluminescence/bioluminescence, surface plasmon resonance, reflection spectroscopy, and laser light scattering (Korotkaya, 2014; Mello and Kubota, 2002). Colorimetric test strips are disposable cellulose pads impregnated with reagents and enzymes, a commonly availed technology for whole-blood monitoring of diabetes. Luminescence and reflection spectroscopy is very useful, especially in immunoassay (Galindo, 2009). A biosensor that involves luminescence uses firefly luciferase (Photinus-luciferin 4-monooxygenase) to identify the bacteria that presents in the samples.

"The authors nicely show that self-assembling nanoparticles can be used to enhance the photosynthetic capacity of plants, as well as serve as plant-based biosensors and stress reducers." By adapting the sensors to different targets, the researchers hope to develop plants that could be used to monitor environmental pollution, pesticides, fungal infections, or exposure to bacterial toxins. They are also working on incorporating electronic nanomaterials, such as graphene, into plants (Giraldo *et al.*, 2014).

The development of nanobionic plants is facilitated by the passive assembly of high-zeta-potential nanomaterials within the chloroplast photosynthetic machinery via kinetic trapping by lipid exchange. Semiconducting SWNTs delivered by this spontaneous mechanism have the potential for increasing chloroplast carbon capture by promoting chloroplast solar energy harnessing and electron-transport rates. Future studies will explore the effect of nanoparticle assemblies on the carbon reactions of photosynthesis and chloroplast sugar export. Nanomaterial enhancement of isolated chloroplast stability to free radicals and higher photosynthetic efficiencies opens the possibility of creating hyperstable synthetic materials that grow and repair themselves using sunlight, water and carbon dioxide. SWNT real-time sensing of NO in extracted chloroplasts and leaves could also be extended to detect a wide range of plant signalling molecules and exogenous compounds such as pesticides, herbicides and environmental pollutants. Nanomaterials offer a promising way to engineer plant function, but the absorption, transport and distribution of nanoparticles within photosynthetic organisms remain poorly understood. This nanobionics approach to engineer plant function will lead to a new research field at the interface of nanotechnology.

30.3 Why are Fireflies Disappearing?

Fireflies are fascinating creatures that light up our nights and bring a sense of magic and mystery to our environment. If they disappear, it will be a great loss to habitats and future generations of people all over the world.

30.3.1 Life Cycle of Fireflies

In mid- summer, female deposits about 100 Eggs in the soil or on vegetation. The eggs of some fireflies are bioluminescent and already glow. The eggs hatch in about three weeks. The worm-like larvae that produce light are sometimes called glowworms. The larvae hunt at night and prey on slugs, snails, worms and other insects. When it captures prey, the larva injects it's victim with enzymes to paralyze it and liquefy it's remains. Larvae live through the winter before pupating in the spring – in some species, larvae live through two winters before pupating.

During the **Pupal** stage, in a process called histolysis, the larva's body is broken down and transformative cells are activated. When the metamorphosis is complete, the adult firefly is ready to emerge, usually about three weeks after pupation.

Adult fireflies only live up to a month. Some do not eat at all during adulthood, using every precious moment to find a mate. Others are cannibalistic, but most dine on pollens and nectar. Silica nanoparticles about 10 nanoparticles in diameter to carry luciferase and slightly larger particles of the polymer PLGA and chitosan to carry luciferin and coenzyme A, respectively. Like Fourth-of-July fireworks, cool swims on hot days, and lazy vacations, fireflies are a sign of summer. Many of us have cherished memories of spotting and catching them during warm summer evenings and running through fields that sparkled as if strewn with stars.

Figure 30.2: Female Glowworm, *Lampyris noctiluca*.

Figure 30.3: Flying and Glowing Firefly, *Photinus pyralis*.

30.3.2 Changes in Habitat of Fireflies

Ecologists believe that there are signs that our kids may not grow up with the same firefly memories as we had. That's because fireflies are disappearing from marshes, fields and forests all over the country–and all over the world. And if it continues, fireflies may fade forever, leaving our summer nights a little darker and less magical. Nobody knows for sure. But most researchers blame two main factors: development and light pollution.

Most species of fireflies thrive as larvae in rotting wood and forest litter at the margins of ponds and streams. And as they grow, they more or less stay where they were born. Some species are more aquatic than others, and a few are found in more arid areas–but most are found in fields, forests and marshes. Their environment of choice is warm, humid and near standing water of some kind–ponds, streams and rivers, or even shallow depressions that retain water longer than the surrounding ground.

The problem is that in America and throughout the world, our open fields and forests are being paved over, and our waterways are seeing more development and noisy boat traffic. As their habitat disappears under housing and commercial developments, firefly numbers dwindle. Logging, pollution and increased use of pesticides may also contribute to destroying firefly habitat and natural prey. Human traffic is believed to disrupt firefly habitat as well. While scientific studies have only been done for the past few years, there's plenty of anecdotal evidence in areas that were once full of fireflies–and much of it goes back generations. Some areas once had so many fireflies that they profited from running firefly tours in marshes and forests–but since human traffic has increased, firefly populations have gone down.

30.3.3 Too Much Light at Night

Scientists don't know enough about fireflies to tell for sure. But the signs are indicating light pollution as a major factor in the disappearance of fireflies all over the world. Both male and female fireflies use their flashing lights to communicate. Some species synchronize their flashes, sometimes across large groups of thousands of insects. All species speak a language of light–scientists believe they use it to attract mates, defend their territory, and warn off predators.

Human light pollution is believed to interrupt firefly flash patterns. Scientists have observed that synchronous fireflies get out of synch for a few minutes after a car's headlights pass. Light from homes, cars, stores, and streetlights may all make it difficult for fireflies to signal each other during mating–meaning fewer firefly larvae are born next season. Where fireflies once had uninterrupted forests and fields to live and mate, homes with landscaped lawns and lots of exterior lights are taking over. The reduction of habitat and the increase in lighting at night may all be contributing to make fireflies disappear.

30.4 Developing Habitat: Suitable for Fireflies

Terrestrial animals, marine animals, freshwater animals, fungi, bacteria and many other microorganisms show bioluminescence. Fireflies were present in clean environment and are known to inhabit many bushes. Strategies for increasing fireflies, include increasing more plantation and provide suitable conditions for larvae to grow. It includes:

- ☆ **Provide habitat for females** – Mating habitat (diverse canopy heights). Places for her to lay eggs.
- ☆ **Provide habitat for larvae** – Plants that build soil/leaf litter, places to find food (snails, worms, slugs, soft-bodied insects), trap moisture.
- ☆ **Operational cover for flashing adults**

Plant List

These are plants that are described as facultative wetland plants, which occur in riparian areas but do well in wet and non-wet habitats. These are good plants to consider if you are planting for fireflies in areas that need to be restored or if you are starting from scratch to encourage fireflies to land. The plants listed below are specific to Texas but are also found in many other areas of the US. Be sure to get to know your native plants, especially wetland plants to understand what might be right for your area.

- Switch grass, Eastern gama grass, Deer grass. Lindheimer muhly, Big sacaton (Mexican firefly species like this variety), Virginia wildrye, Sideoats grama, and Vine mesquite.
- Lindheimer senna, Asters (tall aster/spiny aster), Frogfruit, Frostweed, Goldenrod, Cardinal Flower, Late boneset, River fern.
- Purple leather flower, Morning glory, Alamo vine, Passion flower.
- Sycamore. Pecan. Little Walnut. Cottonwood. American Elder. Ash (green, Mexican, velvet ash). Wafer Ash. Hackberry. Elm Trees. Mexican Buckeye. Live Oak. Chinquapin Oak. Shumard Oak. Bur Oak. Dogwood. Soapberry. Trumpet Creeper. Golden Ball Lead Tree.

30.4 Mechanism of Bioluminescence

A firefly controls the beginning and end of the chemical reaction, and thus the start and stop of its light emission, by adding oxygen to the other chemicals needed to produce light. This happens in the insect's light organ. When oxygen is available, the light organ lights up, and when it is not available, the light goes out. Insects do not have lungs, but instead transport oxygen from outside the body to the interior cells within through a complex series of successively smaller tubes known as tracheoles. For a long time it was a mystery as to how some firefly species manage such a high flash rate, considering the relatively slow speed of the muscles that control oxygen transport. Researchers fairly recently learned that nitric oxide gas plays a critical role in firefly flash control. In short, when the firefly light is off," no nitric oxide is being produced. In this situation, oxygen that enters the light organ is bound to the surface of the cell's energy-producing organelles, called the mitochondria, and is thereby not available for transport further within the light organ. The presence of nitric oxide, which binds to the mitochondria, allows oxygen to flow into the light organ where it combines with the other chemicals needed to produce the bioluminescent reaction. Because nitric oxide breaks down very quickly, as soon as the chemical is no longer being produced, the oxygen molecules are again trapped by the mitochondria and are not available for the production of light.

Fireflies appear to light up for a variety of reasons. The larvae produce short glows and are primarily active at night, even though many species are subterranean or semi-aquatic. Fireflies produce defensive steroids in their bodies that make them unpalatable to predators. Larvae use their glows as warning displays to communicate their distastefulness. As adults, many fireflies have flash patterns unique to their species and use them to identify other members of their species as well as to discriminate between members of the opposite sex. Several studies have shown that female fireflies choose mates depending upon specific male flash pattern characteristics. Higher male flash rates, as well as increased flash intensity, have been shown to be more attractive to females in two different firefly species. The adult fireflies of some species are not luminous at all, however, and instead use pheromones to locate mates. The use of pheromones as sexual signals appears to be the ancestral condition in fireflies with the use of luminous sexual signals as being a more recent development. There are species that employ both pheromonal and luminous components in their mating systems. These species appear to be evolutionarily intermediate between the pheromone-only fireflies and flash-only fireflies.

Wild-type and red mutant luciferase from *Luciola cruciata* showed that native luciferase adopts a ''closed form'' during the formation of the high-energy intermediate responsible for light emission, creating a hydrophobic pocket around the active site, and an ''open form'' in complex with the reactants and products (Nakatsu, *et al.*, 2006). The major part of the amino acid residues important to the bioluminescent activity is located in the N-terminal portion and only one in the C-terminal domain (Branchini *et al.*, 2000), and it was demonstrated that luciferase can produce light bearing only the N-domain, albeit with a luminescent output of only 0.03 per cent of the complete enzyme (Zako *et al.*, 2003). The typical emission spectrum for luciferase is in the yellow-green region (550–570 nm), with a peak at 562 nm at basic media (pH 7.5–7.8) (Seliger *et al.*, 1964a). However, luciferase is a pH-sensitive enzyme, and acid media (pH 5–6) can shift the emission to red (maximum at 620 nm), as well as higher temperatures and heavy metal cations (Seliger, *et al.*, 1964b). It is believed that conformational changes, which influence the active site microenvironment, are responsible for the different color emission.

In *Luciola cruciata* red mutant luciferase the catalytic state remains in an ''open form,'' which could allow energy loss from the excited state intermediate, leading to the emission of red light instead of the higher-energy yellow-green light (Nakatsu, *et al.*, 2006). The *in vitro* emission of light follows, under well-defined conditions, a flash pattern, with a rise in the intensity of emission that decays to low levels (about 5 per cent of the initial burst) in a few seconds, even in the presence of available substrate In Nature, however, the biochemical pathway for LH2 synthesis remains a mystery (Day *et al.*, 2004). In fact it is not even known if the fireflies contain all enzymes to synthesize LH2 or if it is obtained by ingestion of other bioluminescent organisms or by symbiosis with bacteria (Day *et al.*, 2004). Also, the evolutionary creation of LH2 is questioned. Some researchers propose that LH2 first served as an antioxidant, later being requested by the bioluminescent reaction when luciferase appeared. It has been suggested that the origin of the thiazoline ring could be a cysteine (Day *et al.*, 2004), as can be observed in the chemical synthesis in which one of the steps involves this amino acid. Cysteine has two isomers, L- and D-, and either one of them can be used to form LH2, giving rise to L-LH2 and D-LH2, respectively. Their function in the bioluminescent reaction will be highlighted in the incoming sections.

The bioluminescent reaction starts with the reaction of DLH2 with ATP. The reaction is a SN2 nucleophilic displacement, which involves the carboxylic group at the C4 carbon at the thiazoline ring of D-LH2 and the phosphate groups of ATP. After a rapid increase in the light output to a maximum an accentuated decay in the intensity of emission is verified. (RLU, relative light unit). The carboxyl group is a nucleophile, and nucleophilic attack of the electrophilic phosphorus at the a-phosphoryl group of ATP displaces inorganic pyrophosphate (PPi), an excellent leaving group, and transfers adenylate (50 -AMP) to D-LH2. The reaction is, thus, an adenylylation, and produces an enzyme-bound intermediate, luciferyl-adenylate (LH2-AMP) (Dukhovich *et al.*, 1996), a mixed anhydride. The phosphoryl group transfer is also found, for example, in the activation of fatty acids by fatty acyl-coenzyme A synthetases and in the attachment of amino acids to their correspondent tRNA in protein synthesis by aminoacyl-tRNA synthetases (Wood, 1995). Formation of ATP-Mg21 complexes partially shields the negative charges and influences the conformation of the phosphate groups, and thus explains the requirement of this divalent cation in the reaction (Nelson *et al.*, 2005). The L-LH2 isomer can also be adenylylated, giving rise to LLH2-AMP; however it is not used further in light production (Seliger, *et al.*, 1961).

Besides the light-producing pathway luciferase catalyzed bioluminescence also displays lateral reactions. In such reaction the complex luciferaseD-LH2-AMP reacts with oxygen in a dark reaction pathway leading to the oxidized product dehydroluciferyl-AMP (L-AMP) (Fontes *et al.*, 1998). The reaction involves the displacement of inorganic pyrophosphate (PPi) as a leaving group from the ATP molecule, which leads to the formation of the intermediate D-luciferyl-adenylate (D LH2 AMP). However the existence of luciferase in fireflies' peroxisomes (Gould *et al.*, 1989), an organelle rich in the oxygen detoxifying enzyme catalase, could represent a proof that this reaction is not impossible *in vivo* (Fraga *et al.*, 2006). By its turn L-AMP is capable of reacting with the PPi-Mg21 released in the light-producing pathway forming dehydroluciferin (L) and regenerating ATP. As L is an oxidative product of LH2 and ATP, through the oxidation of LH2-AMP to L-AMP and its posterior pyrophosphorolysis (Fontes *et al.*, 1998), it was originally regarded as the light emitter and equivocally called oxyluciferin. Its chemical synthesis was achieved and its role on the bioluminescent reaction, as well as of its adenylated analog L-AMP, has been clarified.

30.4.1 Color of the Bioluminescence of Fireflies

One of the most intriguing aspects of firefly bioluminescence pertains to the color of the light emitted. Recently, the luciferase from *Luciola mingrelica* fireflies, collected in the southern regions of Russia (Ugarova, 1989), as well as the luciferases of the fireflies indigenous to Japan, *Luciola cruciata* and *Luciola lateralis* (Seliger *et al.*, 1964), have been purified and characterized. The *L. mingrelica* and *L. cruciata* luciferases are similar to the *P. pyralis* luciferase, with I_{max} of 562–570 nm, whereas the reaction catalyzed by *L. lateralis* luciferase emits green light (I_{max} 552 nm). All *Luciola* luciferases inactivate rapidly at low ionic strength (Ugarova, 1989), whereas the *P. pyralis* luciferase crystallizes in active form under the same conditions (McElroy 1960).

The chemiluminescence emission spectra of a variety of luciferin analogs were studied by White *et al.* (1971). Based on the fluorescence emission spectra of these compounds in the presence and absence of proton acceptors, they suggested that the red emission arises from the keto anion form of the product molecule and the yellow-green emission from the enol anion. This hypothesis is consistent with the emission of red light in the bioluminescence reaction at lower pH, and has led to the suggestion that the various colors of light observed from different species of firefly result from a spectral mixing from these two emitters (Seliger, and McElroy 1964). However, when luciferases were isolated from different organs of the click beetle *Pyrophorus plagiophthalamus*, each emitted a different color of bioluminescence, and each emission spectrum showed a single peak, not a superimposition of two or more different spectra (Seliger *et al.*, 1961). The cDNAs of four luciferases from *P. plagiophthalamus*, chosen on the basis of differences in color of bioluminescence, have been cloned and sequenced. The amino acid sequences of these luciferases are 95–99 per cent identical, and fewer than two or three amino acid changes are needed for spectral shifts of up to 50 nm in I_{max}. The properties of these luciferases require some rethinking of the traditional explanations for spectral differences between insect luciferases, which historically had been attributed to shifts in the proposed equilibrium distribution between the keto and enol forms of the excited-state oxyluciferin product. If this were the sole explanation for the spectral variations, one would expect to see intermediate bimodal spectra, as have been reported for *P. pyralis* luciferase titrated with zinc ions (Seliger, and McElroy, 1964). However, the four recombinant click beetle luciferases emit light with sharp emission spectra with spectral maxima of 546 nm (green), 560 nm (yellow-green), 578 nm (yellow), and 593 nm (orange) (Wood, *et al.*, 1989). This finding suggests that light emission from each enzyme occurs from a single molecular species within the environment of the enzyme and that differences in color result from differences between the microenvironments of the enzyme–oxyluciferin complexes.

30.4.2 The Firefly Luciferase Superfamily of Enzymes

The number of proteins related to firefly luciferase and apparently descended from a common ancestor is growing rapidly. These enzymes are involved in the biosynthesis of iron-binding siderophores (Rusnak, *et al.*, 1989) the antibiotics gramicidin S (Kanda *et al.*, 1989), tyrocidine (Roskoski *et al.*, 1970), and penicillin(Kleinkauf, and von Döhren 1990), and pathogen defense agents in plants (Becker-Andre *et al.*, 1990), and the bio-degradation of halogenated aromatics (Scholten *et al.*, 1991). The amino acid sequence of AngR, a DNA-binding protein that modulates iron-regulated transcription, also shows homology to the firefly luciferases (Farrel *et al.*, 1990). Given the diversity and importance of the related proteins already discovered, more are likely to be found as the number of sequenced proteins increases. Each of the enzymes catalyzes the adenylation of a carboxylic acid substrate using Mg^{2+}-ATP, followed by reaction of the activated carboxylate with an acceptor and release of AMP. Several of the related enzymes use CoA or an active-site thiol as an acceptor for displacing the AMP. There does not appear to be an essential thiol in the active site of firefly luciferase(Alter, and DeLuca 1986), although there have been several reports that the addition of CoA enhances the emission of light from firefly luciferase. The role, if any, that CoA plays in the firefly luciferase reaction remains to be elucidated.

30.5 Bioluminescence in Fungi

Aristotle (384–322 BC) first described light emission from rotten wood and distinguished this living light from fire. Pliny the Elder (23–79) mentioned in his a Historia Naturalis that bioluminescent white fungi, sweet in taste and with pharmacological properties, could be found in France on decaying trees. Interestingly, G. E. Rumph (1637–1706), a Dutch physician, merchant and consul of Amboine (Moluccas, Indonesia), reported in his Herbarium Amboiense that natives were able to illuminate their path in the dark forest carrying bioluminescent fruiting bodies in their hands. Harvey pointed out uncommon uses of luminous mushrooms 200 hundred years later in Micronesia, where natives used them on their head as ornaments in ritual dances or crushed them on their face in order to scare their enemies. Curiously, these mushrooms were frequently destroyed as they were considered a bad omen. Despite Aristotle's and Pliny's writings on bioluminescent mushrooms and reports by botanists on the distribution of luminous mushrooms, early attention was focused mainly on the light emitted from rotten wood instead on the fungi. Light emission was only directly linked to the fungi in the first half of the nineteenth century. J. F. Heller (1813–1871), Professor at Vienna University, was the first to correlate cause and effect attributing to fungi and bacteria the light exhibited by decaying wood and animals, respectively (Desjardin *et al.*, 2008).

Figure 30.4: ***Panellus stipticus*, Bioluminescent Fungi.**

Figure 30.5: ***Mycena galericulata*.**

Figure 30.6: Bioluminescent Mushrooms.

Figure 30.7: Bioluminescent *Xylaria*.

A modern appraisal of this subject was provided by W. Pfeffer (1845–1920), who applied to bioluminescent fungi the terms luciferin and luciferase coined by Dubois for the thermo-stable (substrate) and labile (enzyme) factors (Dubois,1885). The number of reported bioluminescent fungus species in the world is reevaluated to include the many new species that we discovered since 2005. In addition, we discuss evolutionary aspects of bioluminescence in the fungi and raise hypotheses on the molecular mechanisms underlying the bioluminescent pathway based on the literature and our own observations. Our revision is focused on fungi bioluminescence defined as a chemical reaction that occurs in fungi leading to constant light emission with maximum intensity in the range 520–530 nm whose chemiexcitation step is catalyzed by an enzyme generally called luciferase. This phenomenon should not be confused with transient, low-level or ultraweak chemiluminescence that often increases in response to oxidative stress (*e.g.*, elevated O_2 concentrations or introduction of ROS-generating compounds) wherein light may be emitted from singlet oxygen, triplet excited states, reactions of ONOO–, lipoxygenase activity, heme protein-peroxide reactions and Fenton chemistry. The latter phenomenon, reported from some yeasts and other fungi, will not be treated herein.

Bioluminescence is everywhere in nature – species of phytoplankton, Fish, squid, bacteria, worms, fireflies and a host of other animals produce their own light, and the reasons they do it – from defense to attracting a mate – are just as varied. Mushrooms shine,the glow is the result of an interaction between a compound called luciferin and the luciferase enzyme in the presence of oxygen, they say. They examined the mushrooms with an array of techniques, including mass spectrometry and chromatography, and say the light comes when the product of this reaction, oxyluciferin, releases an oxygen molecule and returns to its ground state, thereby giving off energy in the form of light. The genes are encoding for bioluminescence in the mushrooms.

The work by Herring (1994) found that the luminescent parts of the included pileus (cap), hymenium (gills) and the mycelial threads in combination or separately) also the individual spores were also seen to be luminescent. Herring also stated that if the fruiting body (mushroom) was bioluminescent then the mycelial threads were always luminescent as well but not vice versa. Bioluminescence results because of a certain Biochemical reaction. This can be described as a chemiluminescent reaction which involves a direct conversion of chemical energy transformed to light energy. The reaction involves the following elements:

- ☆ Enzymes (Luciferase) – biological catalysts that accelerate and control the rate of chemical reactions in cells.
- ☆ Photons – packs of light energy.
- ☆ ATP – adenosine triphosphate, the energy storing molecule of all living organisms.
- ☆ Substrate (Luciferin) – a specific molecule that undergoes a chemical charge when affixed by an enzyme.
- ☆ Oxygen – as a oxidizer

A simplified formula of the bioluminescent reaction:

ATP (energy) + Luciferin (substrate)+ Luciferase (enzyme) + O_2 *(oxidizer) = Oxyluciferin + Light (photons)*

Majority of bioluminescence occurs in the Basidiomycetes and only one observation has been made involving the Ascomycetes; specifically in the Ascomycete genus *Xylaria* (Harvey 1952). At present there are 42 confirmed bioluminescent Basidiomycetes that occur worldwide and share no resemblance to each other visually, other than the ability to be bioluminescent. Of these 42 species that have been confirmed 24 of these have been identified just in the past 20 years and as such many more species may exhibit this trait but are yet to be found.

The two main genuses that display bioluminescence are the genus *Pleurotus* which have at present 12 species which occur in continental Europe and Asia. The genus Mycena have 19 species identified to date with a worldwide distribution range. In North America only 5 species of bioluminescent basiodiomycetes have been reported. These include the Honey mushroom -*Armillaria mellea*, the common *Mycena -Mycena galericulata* the Jack O'Latern – *Ophalalotus olearius, Panus styticus* and *Clitocybe illudens*.

After examining the literature and a mushroom field guide book it was evident that there was no correlation between the edibility of the mushroom and its bioluminescence. Some mushrooms such as Armillaria mellea the Honey mushroom was listed as being excellent to eat. While the Jack O'Latern – *Omphalalotus olearius* was listed as poisonous. *Mycena* were unknown and while *Panus stypticus* was listed as poisonous it was found to contain a clotting agent and useful in stopping bleeding. As it only a field guide to North American mushrooms was available, only the North American varieties were examined. If all 42 species of bioluminescent basidiomycetes were included in the search, a possible correlation may have been found.

30.5.1 Taxonomy and Evolutionary Aspects of Fungal Bioluminescence

In 1978, Wassink updated his review of luminescent fungi published in 1948 and that of Harvey, wherein he treated 42 taxa with verified or questionable luminescent properties. His review included a reevaluation of the taxonomic status, synonymy, and luminescent characteristics of all species reported before 1945 (19 species), and detailed accounts of newly described or rediscussed luminous species reported between 1946 and 1978 (23 species). In addition he provided lists of species of uncertain taxonomic position (16 epithets) and of doubtful bioluminescent capabilities (17 epithets). Many of the new accounts represented species described from Malaysia, Japan and the South Pacific, but unfortunately many of these epithets were invalidly published. In addition, the protologues provided only limited morphological data making identification of subsequent collections difficult, and consequently, vouchered reports of most of the Asian species have not been published since. Significant strides have been made in the past 30 years in augmenting our knowledge of the occurrence, distribution, ecology, taxonomy and phylogeny of luminescent fungi. For most fungi, populations are delimited into species based on a suite of shared morphological features, with distinctions between species often being subtle differences in the macro- and micromorphological characteristics of their sexual reproductive structures. Because macromorphological features are quite plastic and easily influenced by environmental factors, comprehensive descriptions that include data on all cell and tissue types that comprise the reproductive structures of individuals from numerous populations are necessary for accurate species diagnoses. These micro-data are usually absent from most descriptions published prior to the 1950s. Consequently, many synonyms have been published and many of the species listed by Wassink (1978) remained poorly known until recently. Concerted efforts by a number of fungal taxonomists to study exsiccati (type or authentic specimens) and to recollect and redescribe luminescent species has lead to the clarification of many species concepts.10–13 In addition, Desjardin and colleagues (Desjardin *et al.*, 2008) have discovered a number of new species and new luminescence accounts of species from Brazil.

A biological species concept can be applied to species that cooperate in vitro. For many saprotrophic fungal species, single spores can be isolated, mated on select media, and their ability to dikaryotize and develop into new individuals can be evaluated. Hence, interbreedability and genetic exchange can be tested. An evolutionary species concept may be applied when molecular sequences datasets are generated and clades of terminal taxa in resultant phylogenetic analyses are used to inform taxonomic decisions. The application of these new techniques of mating system studies (Petersen and Hughes1998) and phylogenetic analyses(Coetzee, *et al.*, 2001) are helpful in delimiting the taxonomic boundaries of species and have been used to clarify species concepts in a number of luminescent fungi. Phylogenetic analyses of multiple loci datasets have greatly advanced our understanding of relationships amongst the fungi and have allowed a first glimpse at the phylogenetic placement of bioluminescent fungi. After extensive searches of pertinent literature, examination of numerous exsiccati specimens, fieldwork throughout the world, and analyses of the nomenclature, taxonomy and phylogeny of reported species. If we couple these data with the seminal research published by Moncalvo *et al.* (2002) and Matheny *et al.* (2006) it is clear that all known bioluminescent fungi are Basidiomycetes and represent white-spored euagarics once placed in the polyphyletic family Trichomolataceae sensu. They are mushroom-forming, saprotrophic or rarely plant pathogenic species belonging to three distinct lineages. Twelve species belong to the Omphalotus lineage (Omphalotaceae), five species to the Armillaria lineage (Physalacriaceae), and the majority 47 species belong to the mycenoid lineages (mostly Mycenaceae). We will address each of these lineages separately and basidiomes emit light(Bermudes, *et al.*, 1991). The components of the life cycle that luminesce are variable in the mycenoid lineages and this surely influences the ecological roles and adaptive significance of bioluminescence. In many species, only the mycelium is luminescent, whereas in others both the mycelium and basidiomes emit light. Very rarely has the basidiome been reported as luminescent while the mycelium is non-luminescent (*M. irritans, M. lamprospora, M. sublucens*).

Whether the luminous propagules attract arthropod dispersal vectors thereby aiding the fungus in infecting new hosts is unknown. To avoid confusion in the use of the misapplied name *Stilbum flavida* for this stage in the life cycle, the name Decapitatus was created for the anamorphic dispersal stage of *M. citricolor*. In the most current and comprehensive phylogeny of euagaric lineages (Matheny, *et al.*, 2006) the luminescent mycenoid fungi in the Mycenaceae belong to the Tricholomatoid clade whose other members are all non-luminescent. The Tricholomatoid clade is distantly related to the Marasmioid clade in which the Omphalotaceae (Omphalotus lineage) and Physalacriaceae (*Armillaria* lineage) belong. In summary, recent multi-gene molecular sequences datasets suggest at least 3 independent origins of bioluminescence in the fungi: in the Omphalotaceae, the Physalacriaceae and the Mycenaceae. The phylogenetic placement of *G. viridilucens* and *M. lucentipes* outside of the Mycenaceae may indicate a fourth independent origin but more work needs to be done to support this contention. Within the Mycenaceae, the 45 recognized luminescent taxa belong to 18 different taxonomic groups (genera or sections within Mycena). Whether this indicates a single early origin of luminescence in the family followed by multiple losses, or multiple independent origins is currently unknown but under investigation in labs (Desjardin *et al.*, 2007).

30.5.2 Physiology of Fungal Luminescence

Wassink (1978) provided an excellent review of the literature on the physiology and biochemical aspects of luminescent fungi up to 1978. Since then, limited research has been conducted on luminescent fungi in pure culture and none in situ. A few examples are presented here. As early as 1924 it was known that there are luminescent and non-luminescent strains of *Panellus stipticus*, and interfertility studies by Macrae confirmed that non-luminescent populations from Europe were sexually compatible with luminescent populations from eastern North America. More recent research by Petersen and Bermudes (1992) indicated that populations of *P. stipticus* from eastern Russia, Japan, New Zealand, and eastern North America were all sexually compatible, and they confirmed with photometric analyses that the species is non-luminescent in Eurasia, but that both non-luminescent and luminescent strains of *P. stipticus* occur in eastern North America. The effects of various environmental and nutritional conditions on the growth and bioluminescence of mycelia of *P. stipticus* indicated that optimal conditions included: darkness; 28 æ per cent C; pH 3.8; cellobiose, glucose, maltose, pectin, or trehalose as the carbon source; and ammonia or asparagine as the nitrogen source. Studies on the localization of bioluminescent tissues in *P. stipticus* indicated a 10- to 50-fold increase in luminescence emission during basidiome development and that luminescence in the basidiomes was restricted primarily to the pileus margin and lamellar edges. In addition it was shown that in both cultures and basidiomes, the wavelength of maximum bioluminescence was at 525 nm. In *in vitro* antagonism studies, it was shown that bioluminescence emissions in P. stipticus and *O. olearius* were reduced over six orders of magnitude when grown with the mycopathogenic fungus *Trichoderma harzianum*. Cultivation of *Omphalotus japonicus* (syn. *Lampteromyces japonicus*) as a prelude to investigations on the mechanism of bioluminescence, and subsequent research resulted in the proposition of riboflavin as the putative light emitter. Other than a few papers elucidating the cultural conditions for optimal mycelial growth, and the production of degradative enzymes, toxins (illudoids), nematophagous compounds, and biomedically active (antibiotic, antitumor) compounds (not cited here), only limited research has been published on the biology of the Moon Night mushroom, *O. japonicus* (Desjardin *et al.*, 2008). In Miller's (1994) redescription on the Australian Ghost Fungus, *O. nidiformis*, he noted different color-morphs that had variable luminescent properties, with a dark form that luminesced rather weakly and a light form with very strong luminescence. Culture analyses and interfertility studies amongst the two forms indicated that they belong to a single morphologically and physiologically variable species. The dynamics of bioluminescence in three sympatric species of *Armillaria* (*A. gallica, A. mellea, A. tabescens*) was examined in response to environmental illumination and mechanical disturbance. The data revealed consistent differences in expression of bioluminescence among the three species, confirming earlier work, and among intragenet cultures. They found no evidence in support of diurnal periodicity of bioluminescence in the mycelial cultures in vitro. Similar results indicating that bioluminescence does not vary between night and day were reported recently from mycelia of an undetermined tropical fungus inhabiting natural wood substrate *in situ*.Earlier reports suggested a diurnal–nocturnal oscillation and seasonal variation of light emissions in *A. mellea* and *P. stipticus*. Weitz *et al.* (2001) studied the effects of temperature, light and pH on mycelial growth and luminescence in *A. mellea, O. olearius, M. citricolor* and *P. stipticus* and found that temperature and pH had a significant effect on both aspects but that light did not. Bermudes (1991) were some of the first to systematically use photometers to measure light emission from fungi, and as a consequence discovered that *Mycena haematopus*, previously thought to be non-luminescent, actually emitted light but at intensity not perceived by the human eye. Measurements of total luminescence of single-spore (monokaryon) or dikaryon cultures of six species of fungi revealed a significant difference in luminescence between different species and between monokaryon and dikaryon cultures of the same species (*P. stipticus, O. japonicus*). Recently, the optimal conditions for the growth of mycelia and the production of basidiomes in *Mycena chlorophos* have been published (Niitsu *et al.*, 2000). Optimal temperature for mycelial growth and primordium formation were 27°C and 21°C respectively, optimal pH of the medium was 4.0, and light was required for primordium initiation. Basidiomes of this species are typically short-lived, but luminescence at a maximum wavelength of 522 nm was emitted continuously for three days.

30.5.3 Ecological Functions of Luminescent Fungi

The potential ecological functions and adaptive significance of bioluminescence in the fungi have been reviewed recently, reiterating the ideas of numerous earlier workers. It has been suggested that bioluminescent basidiomes attract invertebrates to aid in fungal spore dispersal. In an elegant experiment, Sivinski showed that more arthropods were attracted at night to luminous forest litter containing mycelia of a *Mycena* species encapsulated in closed glass test tubes (to allow visual recognition but practically eliminate olfactory detection) than to glass test tubes containing non-luminous forest litter. In addition, subsequent research has confirmed that fungivorous insects are positively phototactic to relatively low light emissions of wavelengths in the range 300–650 nm. Hence, for those fungal species that form luminous basidiomes that sporulate at night, emitting light may have a selective advantage over non-luminous species in attracting potential spore-dispersing arthropods, especially in areas where wind dispersal is inhibited (*e.g.*, dense, closed canopy tropical forests). Given that luminous and non-luminous basidiomes can each produce millions of spores per night that are easily winddispersed, the significance of supplemental insect dispersal is unknown. Bioluminescence as an adaptation to enhance spore dispersal

30.6 Conclusion

Fascinating fireflies and bioluminescence is known to mankind for a long. Albeit much progress was done during the past decades, many mysteries remain unsolved. In fact even the mechanism of luciferase's action in all reactions is not completely clear until now. Questions as "Which amino acid residues take part in each catalytic step of the reaction?" or "What intermediate enzyme-substrate states are formed during reaction?" are examples of problems that should be elucidated in firefly bioluminescent catalysis. The phenomenon is seen in fungi and genes are manipulated in other eukaryotic organisms including some higher plants. Various novel applications based on luciferase are currently being developed and new discoveries are likely to occur in the near future. The unification of so diverse fields as molecular biology, genetics and chemistry will certainly bring to light a new era of bioluminescence research.

REFERENCES

AbouKhair, N.K., Ziegler, M.M. and Baldwin, T.O. (1985). Bacterial luciferase: demonstration of a catalytically competent altered conformational state following a single turnover. Biochemistry 24, 3942–3947.

Alter, S.C. and DeLuca, M. (1986). The sulfhydryls of firefly luciferase are not essential for activity. Biochemistry 25, 1599–1605.

Baldwin, T.O. and Ziegler, M.M. (1992). The biochemistry and molecular biology of bacterial bioluminescence. In Chemistry and Biochemistry of Flavoenzymes. (Müller, F, ed), vol. III, pp. 467–530, CRC Press, Boca Raton, FL.

Baldwin, T.O., *et al.*, and Rayment, I. (1995). Structure of bacterial luciferase. Curr. Opin. Struct. Biol. 5, 798–809.

Baldwin, T.O. (1996)Firefly luciferase: the structure is known, but the mystery remains. Structure 15 4: 223–228 © Current Biology Ltd ISSN 0969-2126.

Becker-Andre, M., Schulze-Lefert, P. and Hahlbrock, K. (1991). Structural comparison, modes of expression, and putative cis-acting elements of the two 4-coumarate: CoA ligase genes in potato. J. Biol. Chem. 266, 8551–8559.

Bermudes, D., Boraas, M. E. and Nealson, K. H.)1991). In vitro antagonism of bioluminescent fungi by Trichoderma harzianum, Mycopath, 115, 19

Branchini, B. R., Murtiashaw, M. H., Magyar, R. A., and Anderson, S. M. (2000) The role of lysine 529, a conserved residue of the acyl-adenylate-forming enzyme superfamily, in firefly luciferase. Biochemistry 39, 5433–5440.

Conti, E., Franks, N. P., and Brick, P. (1996) Crystal structure of firefly luciferase throws light on a superfamily of adenylate-forming enzymes. Structure 4, 287–298.

Coetzee, P.A. Wingfield, B. D. Bloomer, P Ridley, GS Kile G. A. and M. J. Wingfield, M.J. (2001) Phylogenetic relationships of Australian and New Zealand Armillaria species, Mycologia, 2001, 93, 887.

Desjardin D.E. Oliveira A.G. and StevaniC.V. (2008) Fungi bioluminescence revisited. Photochem. Photobiol. Sci., 2008, 7, 170–182. DOI: 10.1039/b713328f

Dubois, R. (1887) Note sur la fonction photogenique chez le Pholas dactylus, C. R. Seances Soc. Biol. Paris, 1887, 39, 564.

Dubois, R. (1885) Note sur physiologie des Pyrophores, C. R. Seances Soc. Biol. Paris, 1885, 2, 559. 4.

Dukhovich, A., Sillero, A., and Sillero, M. A. G. (1996) Time course of luciferyl adenylate synthesis in the firefly luciferase reaction. FEBS Lett. 395, 188–190.

Farrel, D.H., Mikesell, P., Actis, L.A. and Crosa, J.H. (1990). A regulatory gene, angR, of the iron uptake system of Vibrio anguillarium: similarity with phage P22 cro and regulation by iron. Gene 86, 45–51.

Fisher, A.J., Raushel, F.M., Baldwin, T.O. and Rayment, I. (1995). The three-dimensional structure of bacterial luciferase from Vibrio harveyi at 2.4 Å resolution. Biochemistry 34, 6581–6586.

Fontes, R., Ortiz, B., de Diego, A., Sillero, A., and Sillero, M. A. G. (1998) Dehydroluciferyl-AMP is the main intermediate in the luciferin dependent synthesis of Ap4A catalyzed by firefly luciferase. FEBS Lett. 438, 190–194.

Fraga, H., Fernandes, D., Novotny, J., Fontes, R., and Esteves da Silva, J. C. G. (2006) Firefly luciferase produces hydrogen peroxide as a coproduct in dehydroluciferyl adenylate formation. ChemBioChem. 7, 929–935.

Garcý´a-Campan˜a, A. M. and Baeyens, W. R. G., eds. (2001) Chemiluminescence in Analytical Chemistry. Marcel Dekker, New York.

Giraldo JP, Landry M.P., Faltermeier, SM, McNicholas.TP, Iverson NM, Boghossian,AA, Reuel,NF, Hilmer,AJ, Sen, F., Brew JA and Strano,MS (2014) Plant nanobionics approach to augment photosynthesis and biochemical sensing. *Nature Materials*, 13: 400-408.Pub online: 16 March 2014 | DOI: 10.1038/NMAT3890.

Gould, S. J., Keller, G.-A., Hosken, N., Wilkinson, J., and Subramani, S. (1989) A conserved tripeptide sorts proteins to peroxisomes, J. Cell. Biol. 108, 1657–1664.

Herring, P.J. (1994) Luminous fungi, Mycologist, 1994, 8, 181.

Kanda, M., Hori, K., Kurotsu, T., Miura, S. and Saito, Y. (1989). Reaction mechanism of gramicidin synthetase 1, phenylalanine racemase of Bacillus brevis. J. Biochem. (Tokyo) 105, 653–659.

Kleinkauf, H. and von Döhren, H. (1990) Nonribosomal biosynthesis of peptide antibiotics. Eur. J. Biochem. 192, 1–15.

Kirchmair, M. Poder, R. Huber CG. and O. K. Miller Jr., O.K. (2002) Chemotax- ¨ onomical and morphological observations in the genus Omphalotus Fayod (Omphalotaceae), Persoonia, 17: 583.

Lapota D. Galt C. LoseeJR Huddell H.D. Orzech JK. Nealson KH (1988) Observations and measurements of planktonic bioluminescence in and around a milky sea, 119, (1): 55-81

Matheny, P.B. Curtis, J.M. V. Hofstetter,M. C. Aime, J.-M.Moncalvo, Z.-W. Ge, Z.-L. Yang, J. C. Slot, J. F. Ammirati, T. J. Baroni, N. L. Bougher, K.W. Hughes, D. J. Lodge, R.W. Kerrigan,M. T. Seidl, D. K. Aanen,M. DeNitis, G.M. Daniele, D. E. Desjardin, B. R. Kropp, L. L. Norvell, A. Parker, E. C. Vellinga, R. Vilgalys and D. S. Hibbett,(2006) Major clades of Agaricales: a multilocus phylogenetic overview, Mycologia, 98: 982.

McElroy, W. D. (1947) The energy source for bioluminescence in an isolated system. Proc. Natl. Acad. Sci. USA 33, 342–345.

McElroy, W. D. (1951) Properties of the reaction utilizing adenosinetriphosphate for bioluminescence. J. Biol. Chem. 191, 547–557.

McElroy, W.D. (1960). Crystalline firefly luciferase. Methods Enzymol. 5, 445–448.

Moncalvo, J.M. Vilgalys, R. S. A. Redhead,SA. ? J. E. Johnson, T. Y. James, M. C. Aime, V. Hofstetter, S. J. W. Verduin, E. Larsson, T. J. Baroni, R. G. Thorn, S. Jacobsson, H. Clemenc ´ ¸on and O. K. Miller Jr., One hundred and seventeen clades of euagarics, Mol. Phylogenet. Evol., 2002, 23, 357.

Money N.P. (2017) Mushrooms: A natural and cultural history. Reaktion Books ltd. London 197pp.

Nakatsu, T., Ichiyama, S., Hiratake, J., Saldanha, A., Kobashi, N., Sakata, K., and Kato, H. (2006) Structural basis for the spectral difference in luciferase bioluminescence. Nature 440, 372–376.

Nelson, D. L., and Cox, M. M. (2005) Lehninger Principles of Biochemistry, 4th edn. W. H. Freeman, New York.

Niitsu, H. Hanyuda N. and Sugiyama, Y. (2000)Cultural properties of a luminous mushroom, Mycena chlorophos, Mycoscience, 2000, 41, 551.

Oliveira, A. G., Desjardin, D.E., Perryc, B.A., and Stevani C.V. (2012) Evidence that a single bioluminescent system is shared by all known bioluminescent fungal lineages. Photochemical and Photobiological Sciences. 11, 848-852.

Petersen R.H. and Bermudes, D. (1992) *Panellus stypticus*: geographically separated interbreeding populations, Mycologia, 1992, 84, 209.

Petersen R.H. and Hughes, K.W. (1998) Mating systems in Omphalotus (Paxillaceae, Agaricales), Plant Syst. Evol., 211, 217.

Rusnak, F., Faraci, W.S. and Walsh, C.T. (1989). Subcloning, expression, and purification of the enterobactin biosynthetic enzyme 2,3- dihydroxybenzoate-AMP ligase: demonstration of enzyme-bound (2,3- dihydroxybenzoyl) adenylate product. Biochemistry 28, 6827–6835.

Roskoski, R., Jr., Gevers, W., Kleinkauf, H. and Lipmann, F. (1970). Tyrocidine biosynthesis by three complementary fractions from Bacillus brevis (ATCC 8185). Biochemistry 9, 4839–4845.

Seliger, H. H., Buck, J. B., Fastie, W. G., and McElroy, W. D. (1964a) The spectral distribution of firefly light. J. Gen. Physiol. 48, 95–104.

Seliger, H. H., McElroy, W. D., White, E. H., and Field, G. F. (1961) Stereospecificity and firefly bioluminescence, a comparison of natural and synthetic luciferins. Proc. Natl. Acad. Sci. USA 47, 1129–1134.

Seliger, H.H. and McElroy, W.D. (1964). The colors of firefly bioluminescence: enzyme configuration and species specificity. Proc. Natl. Acad. Sci. USA 52, 75–81.

Scholten, J.D., Chang, K.H., Babbitt, P.C., Charest, H., Sylvestre, M. and Dunaway-Mariano, D. (1991). Novel enzymic hydrolytic dehalogenation of a chlorinated aromatic. Science 253, 182–185.

Wood, K. V. (1995) The chemical mechanism and evolutionary development of beetle bioluminescence. Photochem. Photobiol. 62, 662–673.

Wassink, W.C. (1978) Luminescence in fungi, in Bioluminescence in Action, ed. P. J. Herring, Academic Press, New York, 1978, pp. 171.

Wassink, E.C. (1948) Observations on the luminescence in fungi, I, including a critical review of the species mentioned as luminescent in literature, Recl. Trav. Bot. Neerl. ´, 1948, 41, 150

Ugarova, N.N. (1989). Luciferase of Luciola mingrelica fireflies. kinetics and regulation mechanism. J. Biolumin. Chemilumin. 4, 406–418.

Weitz, H.J. (2004) Naturally bioluminescent fungi, Mycologist, 2004, 18,

White, E.H., Rapaport, E., Seliger, H.H. and Hopkins, T.A. (1971). The chemi- and bioluminescence of firefly luciferin: an efficient chemical production of electronically excited states. Bioorg. Chem. 1, 92–122.

Wood, K.V., Lam, Y.A., Seliger, H.H. and McElroy, W.D. (1989). Complementary DNA coding click beetle luciferases can elicit bioluminescence of different colors. Science 244, 700–702.

Wood, K.V. (1992). Improved kinetics of light production by beetle luciferase using thiol reagents. Chem. Abstr. 116, 230750n.

Wood, K.V. (1991). Recent advances and prospects for use of beetle luciferases as genetic reporters. In Bioluminescence and Chemiluminescence: Current Status. (Stanley, P.E. and Kricka, L.J., eds), pp. 543–546, John Wiley and Sons, Chichester.

Zako, T., Ayabe, K., Aburatani, T., Kamiya, N., Kitayama, A., Ueda, H., and Nagamune, T. (2003) Luminescent and substrate binding activities of firefly luciferase N-terminal domain. Biochim. Biophys. Acta. Proteins Proteom. 1649, 183–189.

Chapter 31

Geospatial Applications for Macro Level Air Pollution Transport and Monitoring: A Review

*Shashikant Patel[1], Janak P. Joshi[2] and Bindu Bhatt[3]**

[1]Scientist SG, Punjab Remote Sensing Centre
[2]GIS Expert, Smart City Project, Vadodara, Gujarat
[3]Department of Geography, The Maharaja Sayajirao University of Baroda, Vadodara – 390 002, Gujarat
**e-mail: bindoobhatt@gmail.com*

ABSTRACT

Air pollution is one of the major environmental concerns globally and in developing countries. It is one of the challenge to attain development and balancing the resource quality. India is also one of the countries in the pursuit to attain the developmental need, judicial usage of the resource, and maintaining the quality. Biomass burning, industrial growth, transportation, and power production in country and global scale have been increasing in recent times, resulting in air pollution and air quality degradation issues throughout India. Emissions from agricultural rice residue burning, forest biomass burning as well as industrial sources have all been linked to long and medium range transport of air pollution in different regions of the world. As severe air pollution in North India has become an annual affair, especially in metro cities and surroundings, this issue rises every year and the warms political debates for health issues. The emissions are also at the higher rate in comparison to the nature's balancing act. Thus, a requirement of the monitoring capability is required to quantify the issues pertaining to the contaminants that are contributed at any area of origin but has the wide spread effect due to the global dispersion.

The present practice of the using data from the local sensors may help to quantify and act on the impact locally, although it fails to give the regional and global impacts understanding.

Over the time, the development of network of earth observation satellites (EOS) has given us the capabilities of continuous synoptic understanding regional scale transportation of air pollution, Improve the data coverage, and air pollution trend awareness for forecasting. Green house and other Gases Emissions Inventories and regulatory effectiveness can also be monitored, in the present context of attaining sustainable development goals.

The use of remotely sensed (RS) data as source of input data in air pollution and for environmental processes modelling has taken its due popularity in recent years. With the availability of remotely-sensed data from different sensors of various platforms with a wide range of spatiotemporal, radiometric and spectral resolutions has made remote sensing as the best source of data for geographically large scale environmental analysis and study. In this review, we summarize some of the most commonly used applications of the geospatial technologies in pollution mapping and analysis. The review also summarizes the capability of remote sensing to monitor air quality with available remote sensing dataset at various scales.

This review paper also aims to discuss some of the developed geospatial enabled air pollution monitoring systems.

Keywords: *Air pollution, Geospatial, Remote sensing, Public health, Monitoring, PM2.5.*

31.1 Introduction

"Air quality" refers to the condition of the air within environment or surroundings. Cleanliness of air is measured in the form of Air Quality Index (AQI) that pertains the Good air quality, which the air is clean, clear and free from pollutants such as smoke, dust and smog among other gaseous impurities in the air (Michaelides, *et al.*, 2017). Air quality has attained considerable scientific interest during the last decades because it directly or indirectly impacts a large number of social and economic activities but above all it adversely affects human health and well-being.

The AQI is an index for reporting daily air quality. It tells us how clean or polluted air is, and what associated health effects might be a concern for living lives. The AQI focuses on health effects people or living lives may experience (https: // www.airnow.gov/).

The air in our atmosphere is mostly made up of two gases that are essential for life on Earth: nitrogen and oxygen. However, the air also contains smaller amounts of many other gases and particles. AQI tracks five major air pollutants:

1. Ground level ozone
2. Carbon monoxide
3. Sulphur dioxide
4. Nitrogen dioxide
5. Airborne particles, or aerosols

Ground level ozone and airborne particles are the two air pollutants that pose the greatest risk to human health. They are also the two of the main ingredients in smog, a type of air pollution that reduces visibility.

31.1.1 Understanding of the AQI

Table 31.1: Air Quality Index Values and their Health Concern

Sl.No.	Air Quality Index (AQI) Values	Levels of Health Concern
1	0-50	Good
2	51-100	Moderate
3	101-150	Unhealthy for Sensitive Groups
4	151-200	Unhealthy
5	201-300	Very Unhealthy
6	301-500	Hazardous

31.1.2 Trace Gases

Remote Sensing of trace gases uses the signature of gas absorption. The research in atmospheric deposition has increased a great deal over the past years, because of its increasing significant contribution to the explanation of pollution phenomena in many environmental compartments along with the possibility to evaluate the impacts of pollution sources at long and short distance (as in fugitive emissions) and the possibility to carry out long-term studies aimed at performing health impact assessment on exposed population. The atmosphere is the carrier on which some natural and anthropogenic organic and inorganic chemicals are transported, and deposition events are the most important processes that remove those chemicals, depositing it on soil and water surfaces.

31.2 Air Pollution Impacts on Public Health

Transportation related emissions influence changes in the natural balance of the atmospheric air and cause adversely high level of air pollutants. Several megacities of the world are currently affected by major air pollutants, such as ground-level O_3, NO_2, particulate matter (PM), and carbon monoxide (CO). Particulate matter 10 microns in aerodynamic diameter (PM_{10}) or smaller are considered inhalable and have the greatest impact on public health. Particles 2.5 microns or smaller ($PM_{2.5}$) are thought to be the most damaging (www.epa.gov/oar/aqtrnd97).

The Copernicus Sentinel-5P (S5P) data is available for download since July 2018 to monitor air quality and changes in ozone over Antarctica. The TROPO spheric Monitoring Instrument (TROPOMI) is the single sensor on board of the S5P satellite.

The S5P is the first of the atmospheric composition Sentinels (operational satellite missions supporting the Copernicus programme), launched in 2017, for a nominal lifetime of 7 years. S5P, is a gap-filler and a preparatory programme covering products and applications for Sentinel-5. The S5P mission will fill the gap between the end of the Ozone Monitoring Instrument (OMI) and SCIAMACHY exploitation and the Sentinel-5 mission (credit: ESA).

This high spatial resolution data is useful for air pollution to locate origin of key pollutants (trace gases such as sulphur dioxide in the atmosphere) and finding pollution hotspots. Measurements of atmospheric ozone from the Copernicus S5P satellite are now being used in daily forecasts of air quality. scihub https: //scihub.copernicus.eu/. I downloaded a level 2 NO_2 file in netCDF format (.nc files).

31.3 Sources of Available Remote Sensing Data Set

Table 31.2: Different Sources of Remote Sensing Data

Satellite	*Altitude*	*Coverage Cycle*	*Spatial Resolution*
Landsat-7 Multispectral	705 km	16 days	30 m
SPOT-5 Multispectral	822 km	26 days	10 m
IRS 1C/1D Multispectral		24 days	23m
IKONOS Mulitspectral	681km	2.9 days	4m
QuickBird Multispectral	450 km	6 days	825m
GOES-8 and -10	35,790 km	Geostationary	1,4,8km
MODIS (Terra and Aqua)		Daily, 8 Day, 30 Day	250 m – 1km
MISR (Terra)		Monthly, 3-Month, Annual	275 m – 1.1km
OMI (Aura)		Daily, Monthly	13x24 km
VIIRS (NPP)		Daily, Monthly	750 m
INSAT-3D	~35,786 km		1,4, 8 km
Sentinel 5P	-	Daily	-

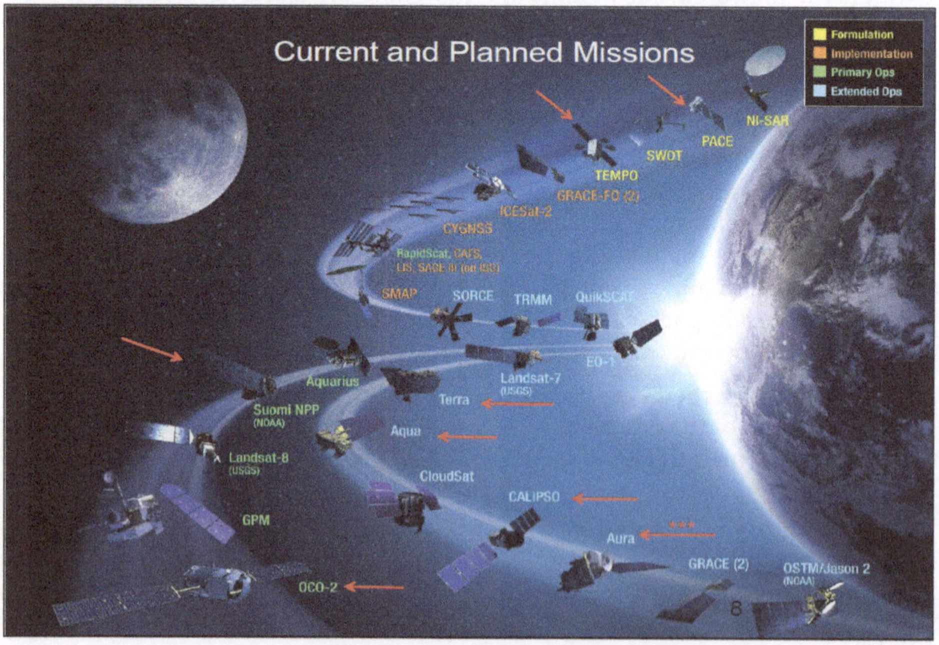

Figure 31.1: Picture Depicting Ongoing and Future Planned Missions (*Source*: ARSET: https: //arset.gsfc.nasa.gov/).

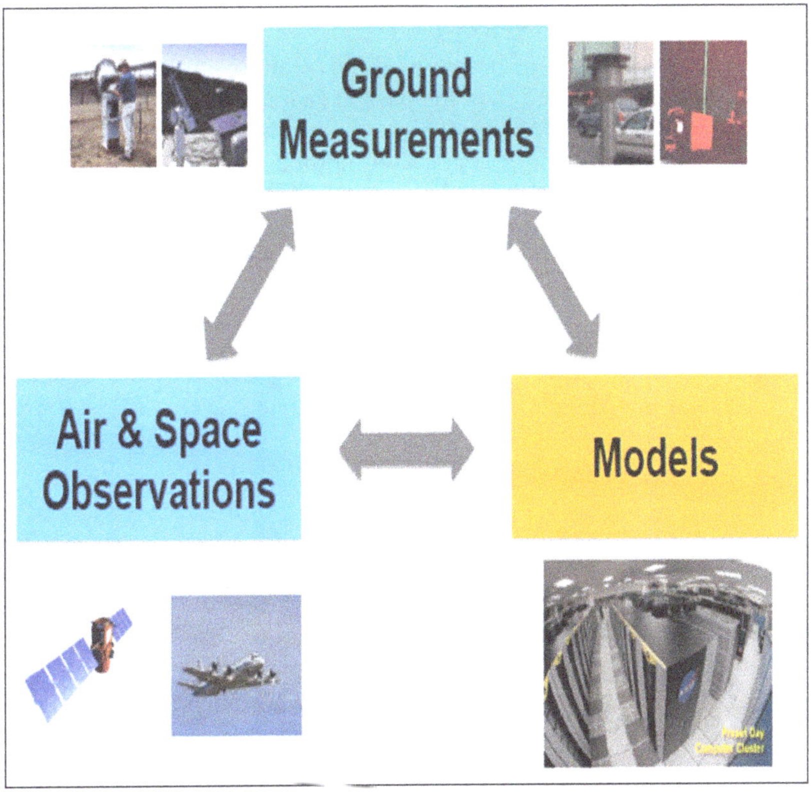

Figure 31.2: Conceptual Block Diagram of Data Processing.

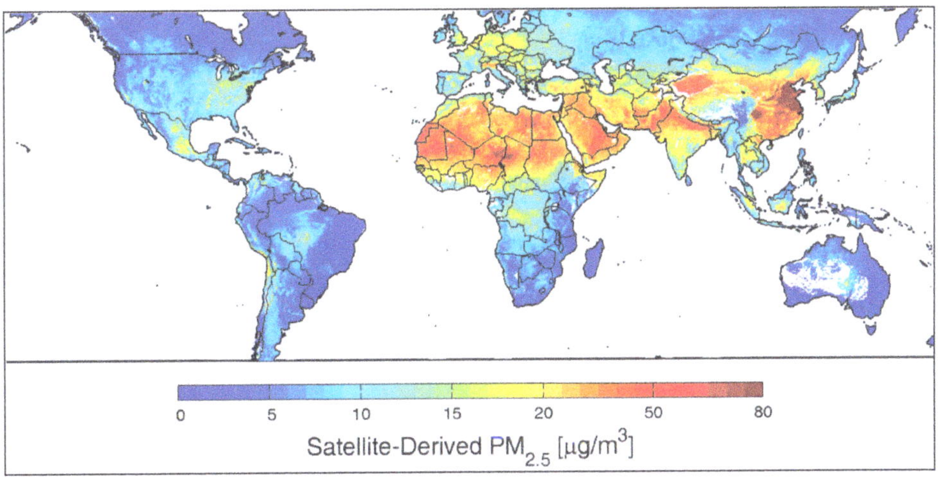

Figure 31.3: World Map Showing $PM_{2.5}$ Pollution Measured through Remote Sensing Data (*Source*: van Donkelaar *et al.*, 2006, 2009).

31.4 Methods and Products

Aerosol thickness is an indicator of the overall pollution of an area. There have been many methods used over the past 30 years to monitor aerosol thickness, the major four of which are (1) ocean method, (2) brightness method, (3) contrast-reduction method, and (4) dark vegetation method (Sifakis *et al.*, 1998). Aerosol Optical Depth (AOD) indicates the level at which particles in the air (aerosols) prevent light from travelling through the atmosphere. Aerosols scatter and absorb incoming sunlight, which reduces visibility.

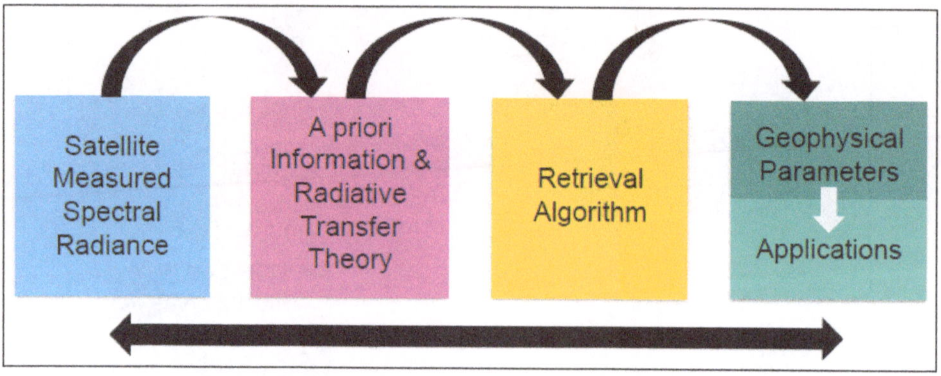

Process of Remote Sensing

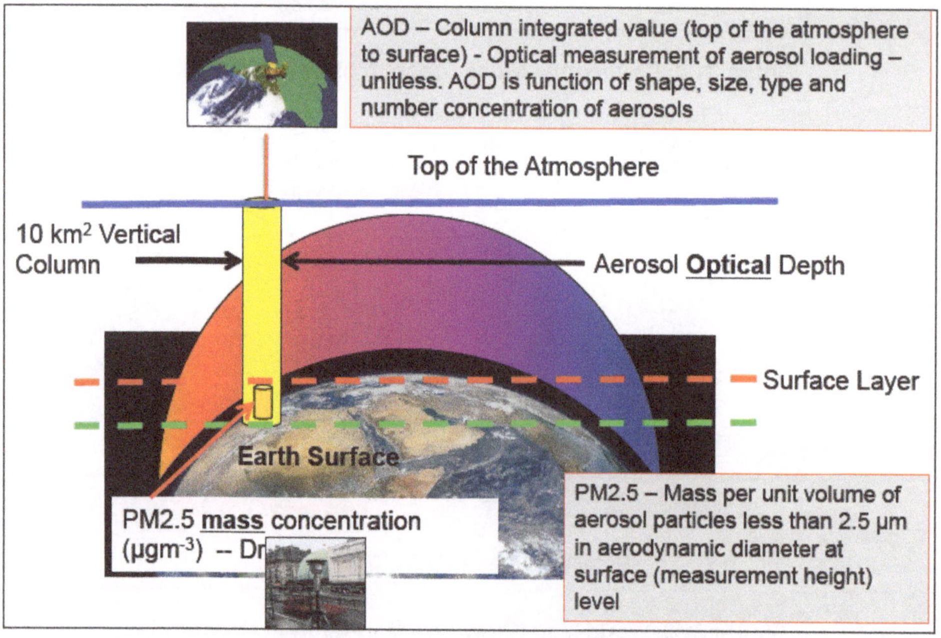

Figure 31.3: Diagram of Remote Sensing Data Processing for Aerosols (*Source*: van Donkelaar *et al.*, 2006, 2009).

Aerosol optical depth (AOD) retrieval algorithms use cloud screening techniques to disregard AOD in cloudy regions due to difficultly in separating visible reflectance of aerosols and clouds.

31.5 Conclusion

Data products from both Low orbits (Sun synchronous) and high orbit (Geostationary platform) will add valuation to data with planned Mission for the betterment of regional and global monitoring of the transport of air pollution. This paper presents the potential of satellite dataset for continuous mapping and monitoring of air pollution. Satellite dataset shows its capability towards the spatial coverage, multiple spatial resolution and Macro level transport monitoring.

Major limitations of Remote Sensing air pollution analysis is the cloud interaction, spectral limits of interferences and adequate working skills along with the validation facilities.

This paper also shows the potential capabilities of geospatial technologies to analyse air quality using GIS and remote sensing. The expansion of the ground sample station, training and the collaboration from local to national authority can aid in understanding the disruption cause of the natures harmony with the development.

REFERENCES

ARSET Applied Remote Sensing Training, Accessed via https: //arset.gsfc.nasa. gov/

Michaelides, S., Paronis,D., Retalis, A., and Filippos Tymvios, F (2017). Monitoring and Forecasting Air Pollution Levels by Exploiting Satellite, Ground-Based, and Synoptic Data, Elaborated with Regression Models, *Advances in meteorology*, Vol.2017, | https: //doi.org/10.1155/2017/2954010

Sifakis, N. I., N. A. Soulkellis, and D. K. Paronis, (1998). Quantitative mapping of air pollution density using earths observations: a new process method and application to an urban area, International Journal of Remote Sensing, 19(17): 3289-3300.

U.S. Environmental Protection Agency (1997): National air quality and emissions trends. Report, EPA-454/R-98- 016, 1997. www.epa.gov/oar/aqtrnd97.

van Donkelaar A, Martin RV, and Park RJ. (2006) Estimating ground-level $PM_{2.5}$ using aerosol optical depth determined from satellite remote sensing. *J Geophys Res*. 111: 21201. doi: 10.1029/2005JD006996.

van Donkelaar A, Martin RV, and Brauer M, (2009) Global estimates of ambient fine particulate matter concentrations from satellite-based aerosol optical depth: development and application. *Environ Health Perspect*. 118(6): 847–855. doi: 10.1289/ehp.0901623

Chapter 32

Environmental Pollution: Conferences and Treaties

Priyanka Kanungo

Department of Environmental Studies, Faculty of Science, The Maharaja Sayajirao University of Baroda, Vadodara – 390 002, Gujarat
e mail: priyanka.ujjain@gmail.com

ABSTRACT

The industrialization in 1980s in Europe changed the phase of planet earth and pollution was result of this massive activity. Anthropogenic causes like deforestation, modern agricultural practices, fertilizer, cement industries, coal run power plants, biomass burning, the land use changes like urbanization, industrialization, creation of special economic zones, and increase in number of vehicles on road are major causes of pollution. Global warming is caused due to increased green house gases. To reduce the environmental pollution, effect of climate change a large number of conventions and treaties are held in past. First major event was the United Nations Conference on Environment, at Stockholm in 1972. We celebrate the world Environment day every year on 5th June to mark the opening of this event. Smt. Indira Gandhi then Prime Minister of India participated in the event. Another major event was Earth summit in 1992. The paper describes few important of these events and their objectives in brief.

Keywords: *Environmental pollution, Conferences, Treaties, Convention, IPCC, Climate change.*

32.1 Introduction: Green Economy

The United Nations is now promoting "green economy in the context of sustainable development and poverty eradication" aspires to catalyse the changes needed in the region. Under the principle of shared but differentiated responsibilities, the green economy is understood in opposition to a brown economy, which compartmentalizes, pollutes, excludes and destroys. A green economy is one which augments and affords priority to human well-being and social equity, while significantly reducing environmental risks and ecological scarcities. This proposal advocates the redesign of specific public policies that promote a low-carbon development pattern resistant to disasters and climate change, create green jobs and factor into decision-making the economic costs and benefits associated with the use of ecosystem services and materials. An economy for sustainable development reduces negative environmental impacts, such as carbon emissions and pollution, promotes efficient use of energy and resources and avoids the loss of biological diversity and ecosystem services, thus improving well-being now and in the future.

A biogeochemical model of vegetation using observed climate data predicts the high northern latitude greening trend over the past two decades observed by satellites and a marked setback in this trend after the Mount Pinatubo volcano eruption in 1991. The observed trend toward earlier spring budburst and increased maximum leaf area is produced by the model as a consequence of biogeochemical vegetation responses mainly to changes in temperature. High-latitude CO_2 uptake during these years is predicted as a consequence of the differential response of heterotrophic respiration and net primary production (Lucht *et al.*, 2002, Beer *et al.*, 2010). In Palestine Rabi *et al.* (2004) had found that within a period of 41 years the mean annual average temperature has risen by 2.3 °C. They were cultivating earlier wheat and barley which is changed now. Increasing the tree plantation and green cover helps to improve the environmental conditions (Rathore and Jasrai, 2015). Agricultural activities like growing more paddy and raising cattle has had a large influence in the 151 per cent rise of atmospheric methane, and it increases the atmospheric temperature (Bhavnani and Makwana, 2015).

International agreements, at both the global and regional levels, can and must contribute to these objectives, by fostering an environment and incentives to guarantee greater social inclusion, access to fair and sustainable exchanges of ecosystem goods and services and sound stewardship of global environmental public goods. United Nations system, in particular the United Nations Environment Programme (UNEP), United Nations Development Programme (UNDP), United Nations Population Fund (UNFPA), Food and Agriculture Organization of the United Nations (FAO), United Nations Human Settlements Programme (UN-Habitat), United Nations Educational, Scientific and Cultural Organization (UNESCO), United Nations Entity for Gender Equality and the Empowerment of Women(UN-Women), United Nations Children's Fund (UNICEF), are helping for development of states without pollution and destruction. League of Nations was formed after World War I and after World War II the United Nations came into existence. With the promotion of peace among nations a large number of conventions and treaties are held time to time to promote pollution free environment and development of society in a sustainable way. The difference between Treaty, Convention and protocol is:

- ☆ **Treaties** - Treaty is an agreement in written form between nation- states (or international agencies, that have been given treaty- making capacity by the states that created them) that is intended to establish a relationship governed by International law.

- **Convention –** It is a meeting or formal assembly, as of representatives or delegates, for discussion of and action on particular matters of common concern.
- **Protocols** - It is an official procedure or system of rules governing affairs of state or diplomatic occasions.

32.2 The UN Conference on the Human Environment, Stockholm 1972

The United Nations Conference on the Human Environment, having met at Stockholm from 5 to 16 June 1972, having considered the need for a common outlook and for common principles to inspire and guide the peoples of the world in the preservation and enhancement of the human environment.

1. Man is both creator and moulder of his environment, which gives him physical sustenance and affords him the opportunity for intellectual, moral, social and spiritual growth. In the long and tortuous evolution of the human race on this planet a stage has been reached by man and the power to transform his environment in countless ways.
2. The protection and improvement of the human environment is a major issue which affects the well-being of peoples and economic development throughout the world; it is the urgent desire of the peoples of the whole world and the duty of all Governments.
3. Man has constantly to sum up experience and go on discovering, inventing, creating and advancing. In our time, man's capability to transform his surroundings, if used wisely, can bring to all peoples the benefits of development and the opportunity to enhance the quality of life.
4. In the developing countries most of the environmental problems are caused by under-development. Millions continue to live far below the minimum levels required for a decent human existence, deprived of adequate food and clothing, shelter and education, health and sanitation. Therefore, the developing countries must direct their efforts to development, with their priorities to safeguard and improve the environment.
5. The natural growth of population continuously presents problems for the preservation of the environment, and adequate policies and measures should be adopted, as appropriate, to face these problems. Of all things in the world, people are the most precious. It is the people that propel social progress, create social wealth, develop science and technology and, through their hard work, continuously transform the human environment.
6. A point has been reached in history when we must shape our actions throughout the world with a more prudent care for their environmental consequences. Through ignorance or indifference we can do massive and irreversible harm to the earthly environment on which our life and well-being depend. Conversely, through fuller knowledge and wiser action, we can achieve for ourselves and our posterity a better life in an environment more in keeping with human needs and hopes.

7. To achieve this environmental goal will demand the acceptance of responsibility by citizens and communities and by enterprises and institutions at every level.

Principle 1 Man has the fundamental right to freedom, equality and adequate conditions of life, in an environment of a quality that permits a life of dignity and well-being, and he bears a solemn responsibility to protect and improve the environment for present and future generations. In this respect, policies promoting or perpetuating apartheid, racial segregation, discrimination, colonial and other forms of oppression and foreign domination stand condemned and must be eliminated.

Principle 2 The natural resources of the earth, including the air, water, land, flora and fauna and especially representative samples of natural ecosystems, must be safeguarded for the benefit of present and future generations through careful planning or management, as appropriate.

Principle 3 The capacity of the earth to produce vital renewable resources must be maintained and, wherever practicable, restored or improved.

Principle 4 Man has a special responsibility to safeguard and wisely manage the heritage of wildlife and its habitat, which are now gravely imperilled by a combination of adverse factors. Nature conservation, including wildlife, must therefore receive importance in planning for economic development.

Principle 5 The non-renewable resources of the earth must be employed in such a way as to guard against the danger of their future exhaustion and to ensure that benefits from such employment are shared by all mankind.

Principle 6 The discharge of toxic substances or of other substances and the release of heat, in such quantities or concentrations as to exceed the capacity of the environment to render them harmless, must be halted in order to ensure that serious or irreversible damage is not inflicted upon ecosystems. The just struggle of the peoples of all countries against pollution should be supported.

Principle 7 States shall take all possible steps to prevent pollution of the seas by substances that are liable to create hazards to human health, to harm living resources and marine life, to damage amenities or to interfere with other legitimate uses of the sea.

Principle 8 Economic and social development is essential for ensuring a favourable living and working environment for man and for creating conditions on earth that are necessary for the improvement of the quality of life.

Principle 9 Environmental deficiencies generated by the conditions of under-development and natural disasters pose grave problems and can best be remedied by accelerated development through the transfer of substantial quantities of financial and technological assistance as a supplement to the domestic effort of the developing countries and such timely assistance as may be required.

Principle 10 For the developing countries, stability of prices and adequate earnings for primary commodities and raw materials are essential to environmental management since economic factors as well as ecological processes must be,taken into account.

Principle 11 The environmental policies of all States should enhance and not adversely affect the present or future development potential of developing countries.

Principle 12 Resources should be made available to preserve and improve the environment, taking into account the circumstances and particular requirements of developing countries and any costs which may emanate from their incorporating environmental safeguards into their development planning and the need for making available to them, upon their request, additional international technical and financial assistance for this purpose.

Principle 13 In order to achieve a more rational management of resources and thus to improve the environment, States should adopt an integrated and co-ordinated approach to their development planning so as to ensure that development is compatible with the need to protect and improve environment for the benefit of their population.

Principle 14 Rational planning constitutes an essential tool for reconciling any conflict between the needs of development and the need to protect and improve the environment.

Principle 15 Planning must be applied to human settlements and urbanization with a view to avoiding adverse effects on the environment and obtaining maximum social, economic and environmental benefits for all. In this respect, projects which are designed for colonialist and racist domination must be abandoned.

Principle 16 Demographic policies which are without prejudice to basic human rights and which are deemed appropriate by Governments concerned should be applied in those regions where the rate of population growth or excessive population concentrations are likely to have adverse effects on the environment of the human environment and impede development.

Principle 17 Appropriate national institutions must be entrusted with the task of planning, managing or controlling the environmental resources of States with a view to enhancing environmental quality.

Principle 18 Science and technology, as part of their contribution to economic and social development, must be applied to the identification, avoidance and control of environmental risks and the solution of environmental problems and for the common good of mankind.

Principle 19 Education in environmental matters, for the younger generation as well as adults, giving due consideration to the underprivileged, is essential in order to broaden the basis for an enlightened opinion and responsible conduct by individuals, enterprises and communities in protecting and improving the environment in its full human dimension.

Principle 20 Scientific research and development in the context of environmental problems, both national and multinational, must be promoted in all countries, especially the developing countries.

Principle 21 States have, in accordance with the Charter of the United Nations and the principles of international law, the sovereign right to exploit their own resources pursuant to their own environmental policies, and the responsibility to ensure that activities within their jurisdiction or control do not cause damage to the environment of other States or of areas beyond the limits of national jurisdiction.

Principle 22 States shall co-operate to develop further the international law regarding liability and compensation for the victims of pollution and other environmental damage caused by activities within the jurisdiction or control of such States to areas beyond their jurisdiction.

Principle 23 Without prejudice to such criteria as may be agreed upon by the international community, or to standards which will have to be determined nationally, it will be essential in all cases to consider the systems of values prevailing in each country, and the extent of the applicability of standards which are valid for the most advanced countries but which may be inappropriate and of unwarranted social cost for the developing countries.

Principle 24 International matters concerning the protection and improvement of the environment should be handled in a co-operative spirit by all countries, big and small, on an equal footing. Co-operation through multilateral or bilateral arrangements or other appropriate means is essential to effectively control, prevent, reduce and eliminate adverse environmental effects resulting from activities conducted in all spheres, in such a way that due account is taken of the sovereignty and interests of all States.

Principle 25 States shall ensure that international organizations play a co-ordinated, efficient and dynamic role for the protection and improvement of the environment.

Principle 26 Man and his environment must be spared the effects of nuclear weapons and all other means of mass destruction. States must strive to reach prompt agreement, in the relevant international organs, on the elimination and complete destruction of such weapons.

32.3 Rio de Janeiro Earth Summit 1992

The United Nations Conference on Environment and Development (UNCED), also known as the Rio de Janeiro Earth Summit, the Rio Summit, the Rio Conference, and the Earth Summit (Portuguese: ECO92), was a major United Nations conference held in Rio de Janeiro from 3 to 14 June, in 1992. Earth Summit was held as a platform for other Member States to collaborate. Since the creation, many others in the field of sustainability show a similar development to the issues discussed in these conferences, including non-governmental organizations (NGOs). The issues addressed included:

- ☆ Systematic scrutiny of patterns of production – particularly the production of toxic components, such as lead in gasoline, or poisonous waste including radioactive chemicals
- ☆ Alternative sources of energy to replace the use of fossil fuels which delegates linked to global climate change

- New reliance on public transportation systems in order to reduce vehicle emissions, congestion in cities and the health problems caused by polluted air and smoke
- The growing usage and limited supply of water

An important achievement of the summit was an agreement on the Climate Change Convention which in turn led to the Kyoto Protocol and the Paris Agreement. Another agreement was to "not to carry out any activities on the lands of indigenous peoples that would cause environmental degradation or that would be culturally inappropriate". Forests contribute to the welfare of the entire planet, they have a variety of trees and medicinal herbs (Sabnis 2020). The Convention on Biological Diversity was opened for signature at the Earth Summit, and made a start towards redefinition of measures that did not inherently encourage destruction of natural ecoregions and so-called uneconomic growth. Although President George H.W. Bush signed the Earth Summit's Convention on Climate, his EPA Administrator William K. Reilly acknowledges that U.S. goals at the conference were difficult to negotiate and the agency's international results were mixed, including the U.S. failure to sign the proposed Convention on Biological Diversity.

The Earth Summit resulted in the following documents:

- Rio Declaration on Environment and Development
- Agenda 21
- Forest Principles

Moreover, important legally binding agreements (Rio Convention) were opened for signature:

- Convention on Biological Diversity
- Framework Convention on Climate Change (UNFCCC)
- United Nations Convention to Combat Desertification

In order to ensure compliance to the agreements at Rio (particularly the *Rio Declaration on Environment and Development* and *Agenda 21*), delegates to the Earth Summit established the *Commission on Sustainable Development* (CSD). In 2013, the CSD was replaced by the *High-level Political Forum on Sustainable Development* that meets every year as part of the ECOSOC meetings, and every fourth year as part of the General Assembly meetings.

32.4 World Summit on Sustainable Development, Johannesburg, 2002

Resolutions adopted by the Summit Resolution 1* Political Declaration The World Summit on Sustainable Development, Having met in Johannesburg, South Africa, from 26 August to 4 September 2002,

1. Adopts the Johannesburg Declaration on Sustainable Development, which is annexed to the present resolution;

2. Recommends to the General Assembly that it endorse the Johannesburg Declaration on Sustainable Development as adopted by the Summit. Annex Johannesburg Declaration on Sustainable Development From our origins to the future
    1. We, the representatives of the peoples of the world, assembled at the World Summit on Sustainable Development in Johannesburg, South Africa, from 2 to 4 September 2002, reaffirm our commitment to sustainable development.
    2. We commit ourselves to building a humane, equitable and caring global society, cognizant of the need for human dignity for all.
    3. At the beginning of this Summit, the children of the world spoke to us in a simple yet clear voice that the future belongs to them, and accordingly challenged all of us to ensure that through our actions they will inherit a world free of the indignity and indecency occasioned by poverty, environmental degradation and patterns of unsustainable development.
    4. As part of our response to these children, who represent our collective future, all of us, coming from every corner of the world, informed by different life experiences, are united and moved by a deeply felt sense that we urgently need to create a new and brighter world of hope.
    5. Accordingly, we assume a collective responsibility to advance and strengthen the interdependent and mutually reinforcing pillars of sustainable development ó economic development, social development and environmental protection ó at the local, national, regional and global levels.
    6. From this continent, the cradle of humanity, we declare, through the Plan of Implementation of the World Summit on Sustainable Development and the present Declaration, our responsibility to one another, to the greater community of life and to our children.

Adopted at the 17th plenary meeting, on 4 September 2002; for the discussion, see chap. VIII. 2 A/CONF.199/20 7.

Recognizing that humankind is at a crossroads, we have united in a common resolve to make a determined effort to respond positively to the need to produce a practical and visible plan to bring about poverty eradication and human development. From Stockholm to Rio de Janeiro to Johannesburg thirty years ago, in Stockholm, it was agreed on the urgent need to respond to the problem of environmental deterioration. Ten years ago, at the United Nations Conference on Environment and Development, held in Rio de Janeiro, we agreed that the protection of the environment and social and economic development are fundamental to sustainable development, based on the Rio Principles. To achieve such development, we adopted the global programme entitled Agenda 213 and the Rio Declaration on Environment and Development, to which we reaffirm our commitment. The Rio Conference was a significant milestone that set a new agenda for sustainable development. Between Rio and Johannesburg, the world nations have met in several major conferences under the auspices of the United Nations, including the International Conference on Financing for Development, as well as the Doha Ministerial Conference. These conferences defined for the world a comprehensive vision for the future of humanity.

At the Johannesburg Summit, we have achieved much in bringing together a rich tapestry of peoples and views in a constructive search for a common path towards a world that respects and implements the vision of sustainable development. The Johannesburg Summit has also confirmed that significant progress has been made towards achieving a global consensus and partnership among all the people of our planet.

32.5 The Kyoto Protocol 1997

The Kyoto Protocol was adopted on 11 December 1997. Owing to a complex ratification process, it entered into force on 16 February 2005. Currently, there are 192 Parties to the Kyoto Protocol. The Kyoto Protocol operationalizes the United Nations Framework Convention on Climate Change by committing industrialized countries and economies in transition to limit and reduce greenhouse gases (GHG) emissions in accordance with agreed individual targets. The Convention itself only asks those countries to adopt policies and measures on mitigation and to report periodically. It only binds developed countries, and places a heavier burden on them under the principle of "common but differentiated responsibility and respective capabilities", because it recognizes that they are largely responsible for the current high levels of GHG emissions in the atmosphere. In its Annex B, the Kyoto Protocol sets binding emission reduction targets for 37 industrialized countries and economies in transition and the European Union. Overall, these targets add up to an average 5 per cent emission reduction compared to 1990 levels over the five year period 2008–2012 (the first commitment period).

The Kyoto Mechanisms

One important element of the Kyoto Protocol was the establishment of flexible market mechanisms, which are based on the trade of emissions permits. Under the Protocol, countries must meet their targets primarily through national measures. However, the Protocol also offers them an additional means to meet their targets by way of three market-based mechanisms:

- International Emissions Trading
- Clean Development Mechanism (CDM)
- Joint implementation (JI)

These mechanisms ideally encourage GHG abatement to start where it is most cost-effective, for example, in the developing world. It does not matter where emissions are reduced, as long as they are removed from the atmosphere. This has the parallel benefits of stimulating green investment in developing countries and including the private sector in this endeavor to cut and hold steady GHG emissions at a safe level. It also makes leap-frogging–that is, the possibility of skipping the use of older, dirtier technology for newer, cleaner infrastructure and systems, with obvious longer-term benefits–more economical.

Monitoring Emission Targets

The Kyoto Protocol also established a rigorous monitoring, review and verification system, as well as a compliance system to ensure transparency and hold Parties to account. Under the Protocol, countries' actual emissions have to be monitored and precise records have to be kept of the trades carried out. Registry systems track and record transactions made by Parties under the mechanisms. The UN Climate Change Secretariat, based in Bonn, Germany, keeps an international transaction log to verify that transactions are consistent with the rules of the Protocol.

32.6 The World Climate Conferences

A series of international meetings were organized by the World Meteorological Organization (WMO), about global climate issues principally global warming in addition to climate research and forecasting. Created in 1988 the objective of the IPCC is to provide governments at all levels with scientific information that they can use to develop climate policies. IPCC reports are also a key input into international climate change negotiations. The IPCC currently has 195 members. Thousands of people from all over the world contribute to the work of the IPCC. For the assessment reports, IPCC scientists volunteer their time to assess the thousands of scientific papers published each year to provide a comprehensive summary of what is known about the drivers of climate change, its impacts and future risks, and how adaptation and mitigation can reduce those risks. An open and transparent review by experts and governments around the world is an essential part of the IPCC process, to ensure an objective and complete assessment and to reflect a diverse range of views and expertise. Through its assessments, the IPCC identifies the strength of scientific agreement in different areas and indicates where further research is needed. The IPCC does not conduct its own research.

The **First World Climate Conference** was held on 12–23 February 1979 in Geneva and sponsored by the WMO. It was one of the first major international meetings on climate change. Essentially a scientific conference, it was attended by scientists from a wide range of disciplines. In addition to the main plenary sessions, the conference organized four working groups to look into climate data, the identification of climate topics, integrated impact studies, and research on climate variability and change. The Conference led to the establishment of the World Climate Programme and the World Climate Research Programme. It also led to the creation of the Intergovernmental Panel on Climate Change (IPCC) by WMO and UNEP in 1988.

The **Second Climate Conference** was held on 29 October to 7 November 1990, again in Geneva. It was an important step towards a global climate treaty and somewhat more political than the first conference. The main task of the conference was to review the WCP set up by the first conference. The IPCC first assessment report had been completed in time for this conference. The scientists and technology experts at the conference issued a strong statement highlighting the risk of climate change. The conference issued a Ministerial Declaration only after hard bargaining over a number of difficult issues; the declaration disappointed many of the participating scientists as well as some observers because it did not offer a high level of commitment. Eventually, however, developments at the conference led to the establishment of the United Nations Framework Convention on Climate Change (UNFCCC), of which the Kyoto Protocol is a part, and to the establishment of the Global Climate Observing System (GCOS), a global observing system of systems for climate and climate-related observations.

World Climate Conference-3 (WCC-3) was held in Geneva, Switzerland, 31 August - 4 September 2009. Its focus was on climate predictions and information for decision-making at the seasonal to multi-decadal timescales. The goal was to create a global framework that will link scientific advances in these climate predictions and the needs of their users for decision-making to better cope with changing conditions. Key users of climate predictions include food producers, water managers, energy developers and managers, public health workers, national planners, tourism managers and others, as well as society at large. Participants in WCC-3 included these users, as well as climate service providers and high-level policy-makers. The Conference also aimed to increase commitment to, and advancements in, climate observations and monitoring to better provide climate information and services worldwide that will improve public safety and well-being.

WCC-3 outcomes also intended to contribute to the achievement of the United Nations Millennium Development Goals and broader UN climate goals, including the Hyogo Framework for Action on Disaster Risk Reduction. The Conference theme complemented global work under way to help societies adapt to climate change in line with Bali Action Plan, especially the Nairobi Work Programme. The outcomes formed part of WMO input to the 2009 UNFCC COP-15 meeting for climate mitigation in Copenhagen in the December following WCC-3.

32.7 The Major Environmental Events

To protect nature and make a balance between development and degradation of environment following major conventions, treaties and actions are taken by United Nations and member countries.

Table 32.1: Major Environment Related Conventions/Treaties and their Purpose

Event Name	*Year*	*Purpose*	*Organization*
Agenda 21	1992	Action agenda for the UN that can be executed at local, national and global levelAlso termed as Rio Earth Summit	Action Plan of UN
Barcelona convention	1976	Protection of Mediterranean Sea from pollution	UNEP
Basel convention	1989	To reduce movements of hazardous waste	
Benzene convention	1971	Prevention from toxic substances arising from Benzene	

Event Name	*Year*	*Purpose*	*Organization*
Bonn convention	1983	Conservation of migratory species of wild animals	United Nations Env. Program
Brundtland report	1987	Gave concept of Sustainable Development	
Cartagena Protocol	2000	Protect Biological Diversity from the risks posted by Genetically modified organisms	First protocol on CBD
Convention on Desertification	1994	To combat desertification and mitigate the effects of drought through national Action Programs	United Nations
Convention on Biological Diversity	1993	Convention on Biological Diversity and sustainable use of components of Biological diversity	
Convention on International Trades in Endangered Biodiversity	1963	Protection of wild animals and plants from illegal trade to save their life	International union of conservation of Nature
Geneva Protocol	1925	Prohibition of the use in war of Asphyxiating, poisonous or other gases, and of Bacteriological methods of Warfare.	
Helsinki Declaration	1964	Set of ethics for Human Experimentation developed for medical community	
Kyoto Protocol	1997	Set emission Target for Development countries	UNFCC
Minamata Convention	2013	Protection of Human health and Environment from the harmful effect of mercury	UN Treaty
Montreal Protocol	1987	Agreement to protect the ozone layer by phasing out production and consumption of ozone depleting substances.	
Nagoya Protocol	2010	Access to genetic resources and fair and equitable sharing of benefits arising from their utilization to the convention on biological diversity.	
Nairobi Declaration	1985	Agreement between the parties to work for the prosperity of Western Indian Ocean Region	UN Environment's Regional seas programme
Paris Agreement	2015	Aim is to strengthen the global response to the threat of climate change by keeping a global temperature rise this century well below 2 degree Celsius above pre-industrial levels	Action Plan of the UNFCC
Ramsar Convention on wetlands	1971	Framework for conservation of wetlands and their Resources	
Rotterdam Convention	1998	The Prior informed consent procedure for certain hazardous chemicals and pesticides in International Trade	UN Treaty

Event Name	Year	Purpose	Organization
Stock holm Convention on Persistent organic	2001	To eliminate or Restrict the use and production of persistent organic pollutants	Stockholm, Sweden
UNConvention on the Law of the Sea	1982	For Rights and responsibilities of Nations for the Use of Ocean water and established guidelines for Marine Natural Resources	
UNFramework convention on Climate Change	1992	To stabilize greenhouse gas concentration in Atmosphere	
Vienna Convention	1988	Framework on Protection of Ozone Layer	
World Heritage Convention	1988	Defines the kind of natural and cultural sites which can be considered as World Heritage site under UNESCO	Multilateral Environment agreement.

32.8 Conclusion

Anthropogenic pollution like industrialization, change in agricultural practices, biomass burning, electric power generation with coal and increase in number of vehicles in roads has resulted into problems of pollution, ozone hole and climate change. Various action plans are suggested to reduce the carbon emission and plans like tree plantation, reduction in CFCs and GHGs, carbon sequestration are suggested. This will result in stabilizing the increasing pollution on earth and will lead a path of sustainable developmet. The paper highlights some action plans initiated by United Nations and member nations.

REFERENCES

Beer C, Reichstein M, Tomelleri E, Ciais P, Jung M, Carvalhais N, Rödenbeck C, Arain MA, Baldocchi D, Bonan GB, Bondeau A, Cescatti A, Lasslop G, Lindroth A, Lomas M, Luyssaert S, Margolis H, Oleson KW, Roupsard O, Veenendaal E, Viovy N, Williams C, Woodward FI, and Papale D. (2010) Terrestrial gross carbon dioxide uptake: global distribution and covariation with climate. Science. 13;329 (5993): 834-8. doi: 10.1126/science.1184984.

Bhavnani H.V. and Makwana A.R. (2015) Energy Conservation and Global warming, In; Impact of global warming an climate change on human and plant health (eds. Arya A. and Patel V.S.) Daya Pub. House New Delhi. 81-100pp.

IPCC (2007) Intergovernmental Panel on Climatic Change working group I. Climate change 2007: The Physical Science Basis IPPC working group I

Lucht, W. Prentice,I.C. Myneni, R.B. Sitch, S. Friedlingstein, P. Cramer, P. Philippe P. Bousquet,P. Buermann,W. SmithB. (2002) Climatic control of the high-latitude vegetation greening trend and Pinatubo effect. Science, 31;296(5573): 1687-9. doi: 10.1126/science.1071828.

Rabi A. Carmi N. Abu Jamus, NO and Abu Saadah M. (2004) expected impact of climate change of population and livelihood in Arid and semi-arid areas: Case studies from Palestine.

Rathore A. and Jasrai Y.T. (2015) Air pollution, climate change and mitigation strategies. In; Impact of global warming an climate change on human and plant health (eds. Arya A. and Patel V.S.) Daya Pub. House New Delhi. 81-100pp.

Sabnis S. (2020) Forests and biodiversity conservation: a religious perspective, In: Environment at Crossroads, challenges and green solutions. Arya A. (ed.) Scientific Publishers (India)233-235.

www.glf.co.in

www.ipcc.ch/data assessed on 30/10/2020.

Index

C

D

K

L

M

N

O

P

R

S

T

U

V

COLOR PLATES

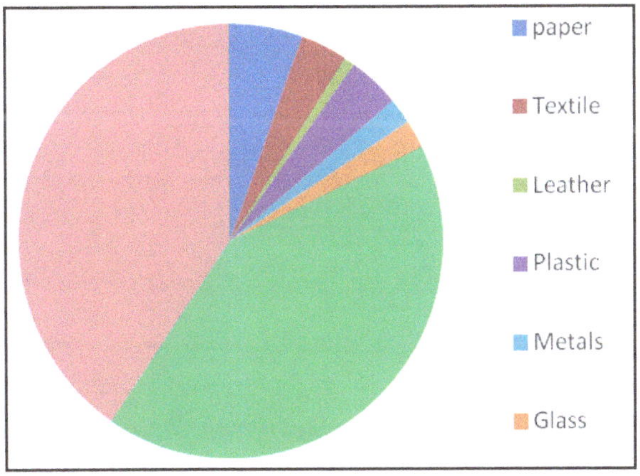

Page: 39

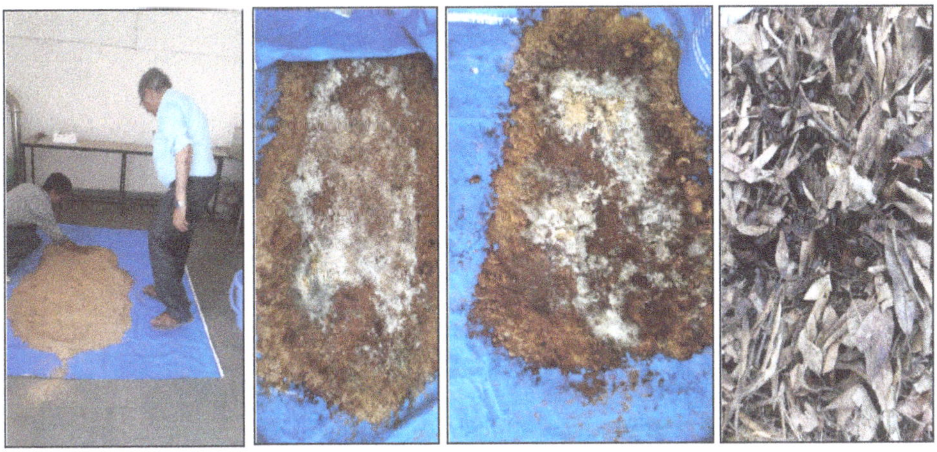

Page: 52

Page: 53

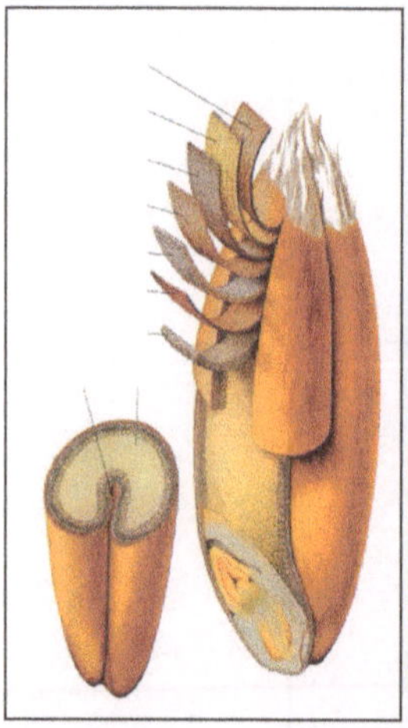

Figure 4.2: Effect of SO_2 Pollution on Wheat Leaves.

Figure 4.3: Reduced Fruit Setting in Drumstick due to SO_2 Pollution.

Figure 4.4: Black Tip Occurs in Orchards Located Close to Brick-kilns.

Page: 70

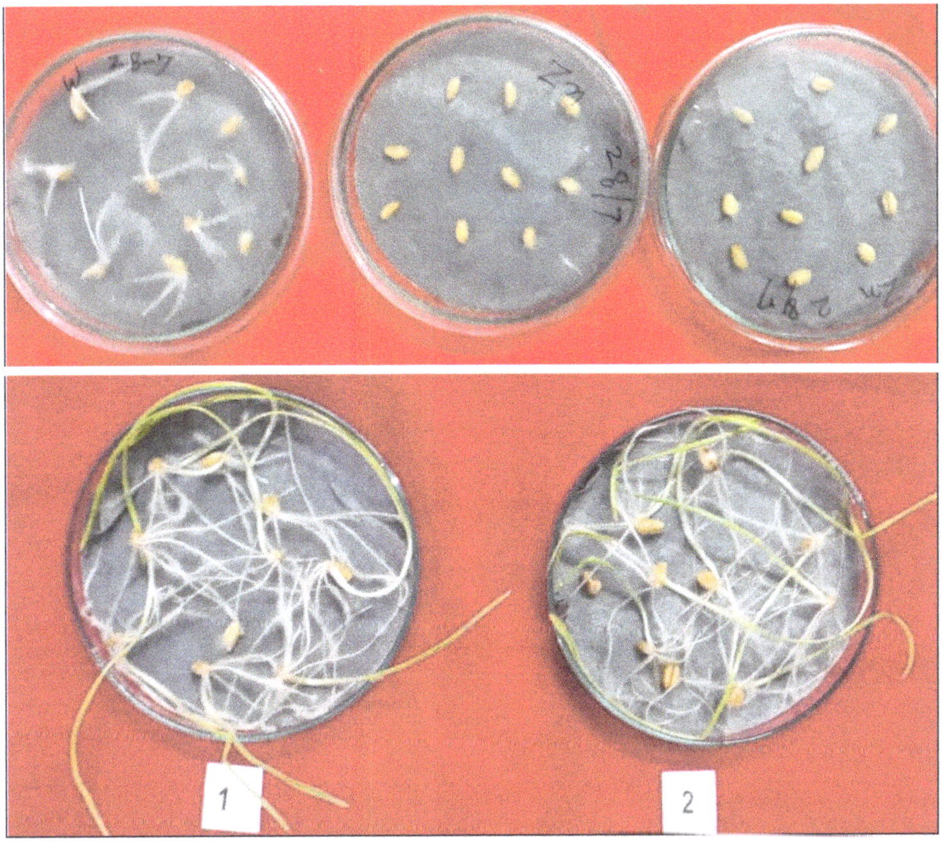

Page: 85

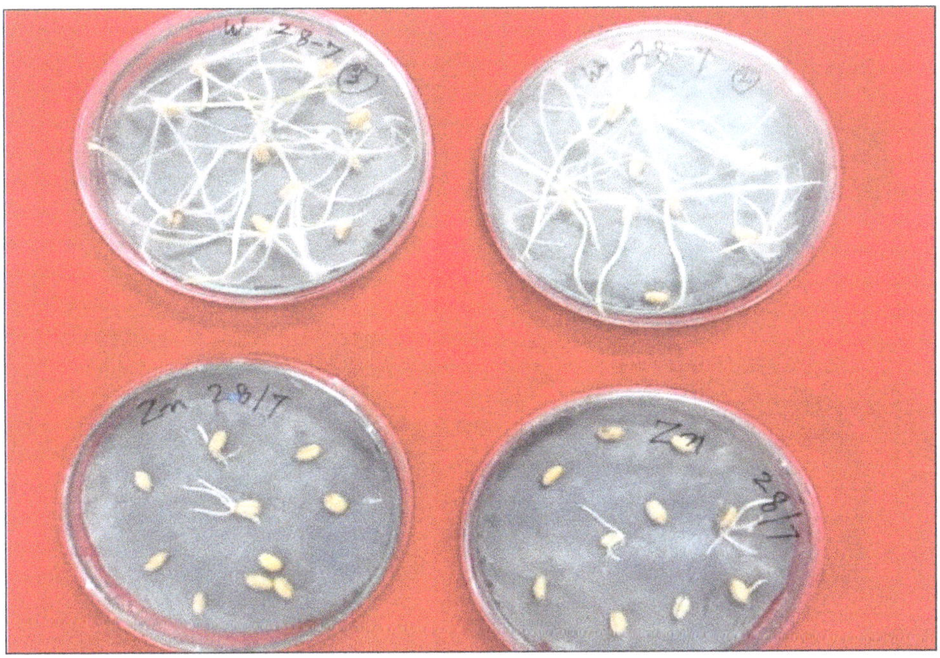

Page: 86

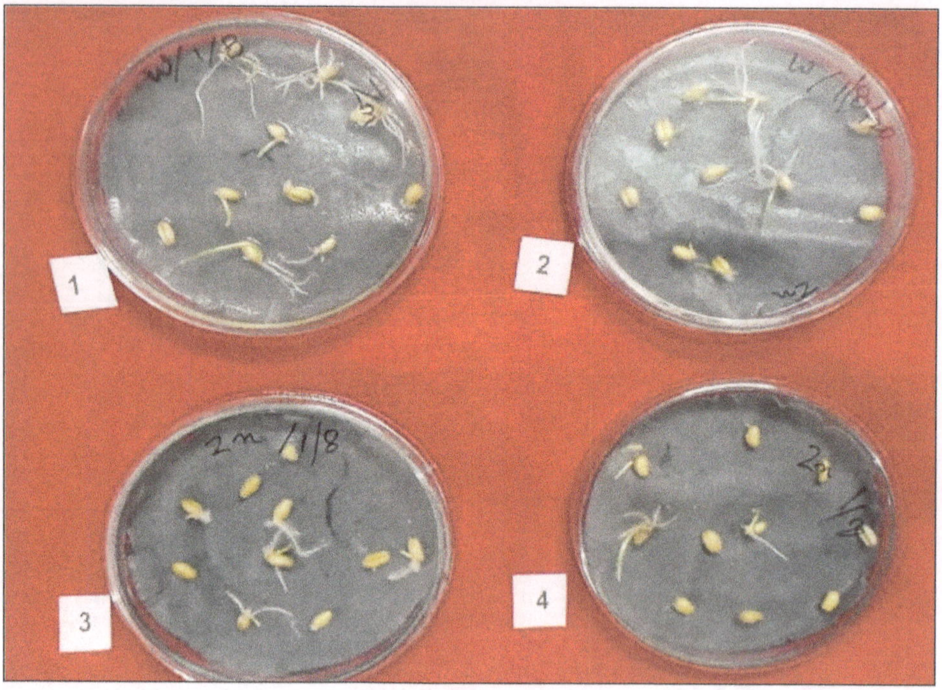

Page: 86

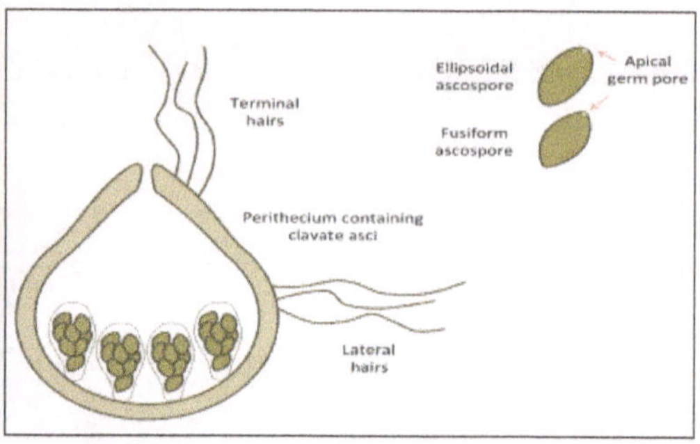

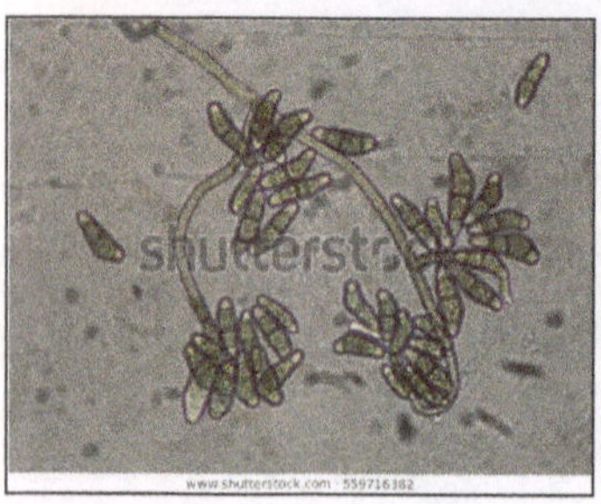

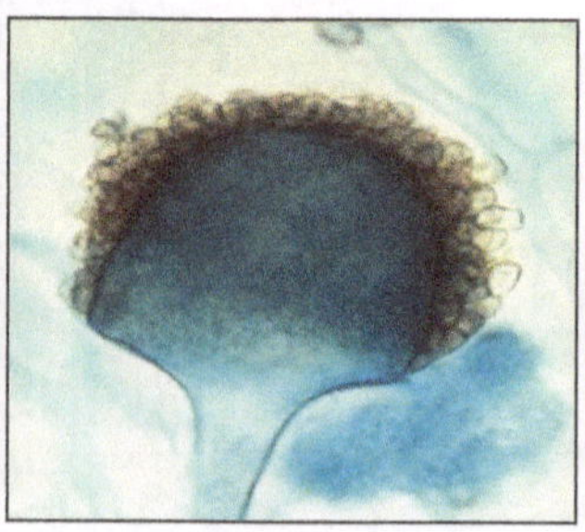

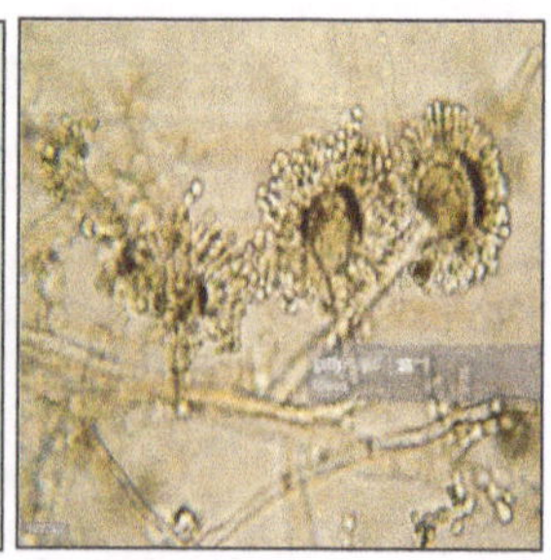

Page: 102

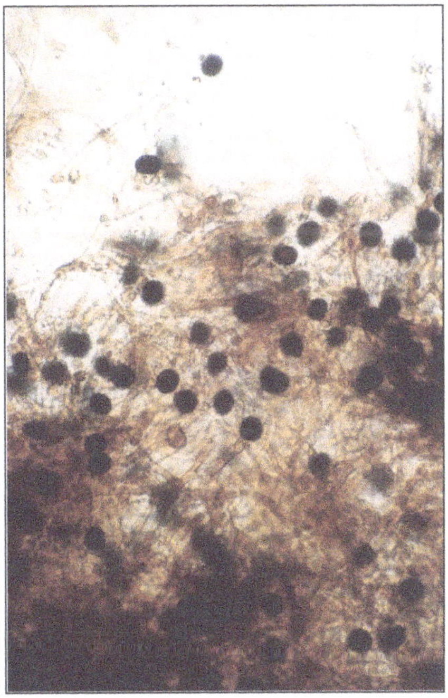

Page: 106

Page: 106

(a) Dust deposition on leaves of ***Ficus hispida***

(b) Leaves of ***Ficus hispida*** from controlled sites

(c) Dust deposition on leaves of ***Calotropis procera***

(d) Leaves of ***Calotropis procera*** from controlled sites

(e) Dust deposition on leaves of ***Datura metel***

(f) Leaves of ***Datura metel*** from controlled sites

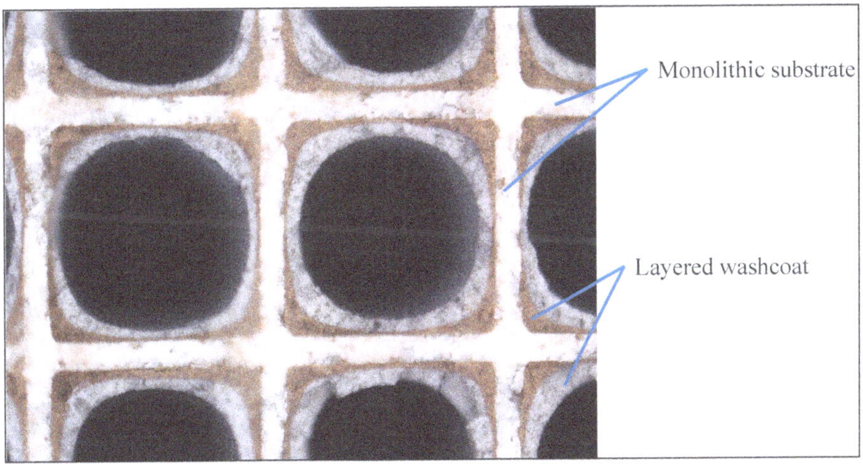

Page: 154

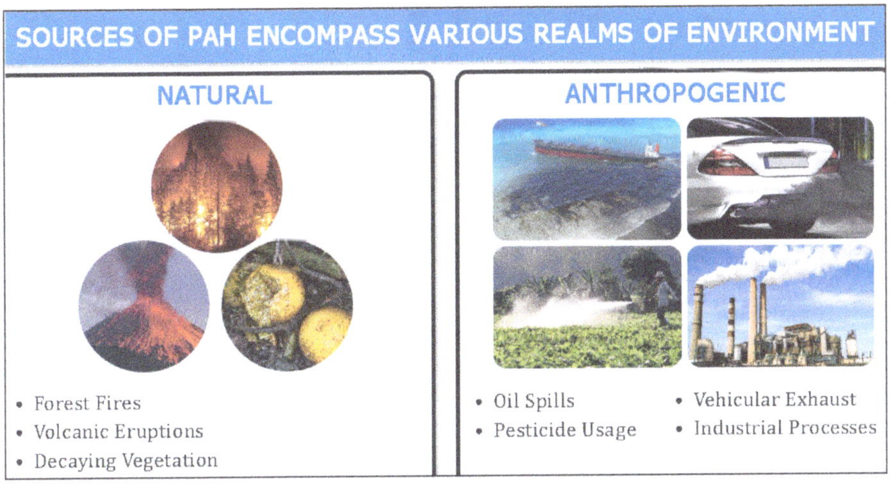

Page: 177

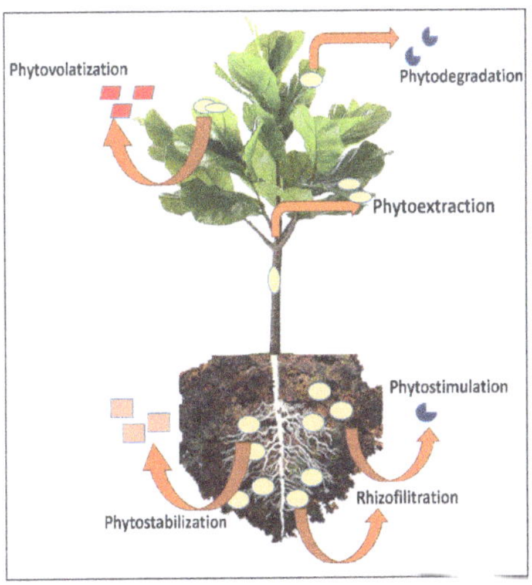

Page: 201

Page: 412

Page: 222

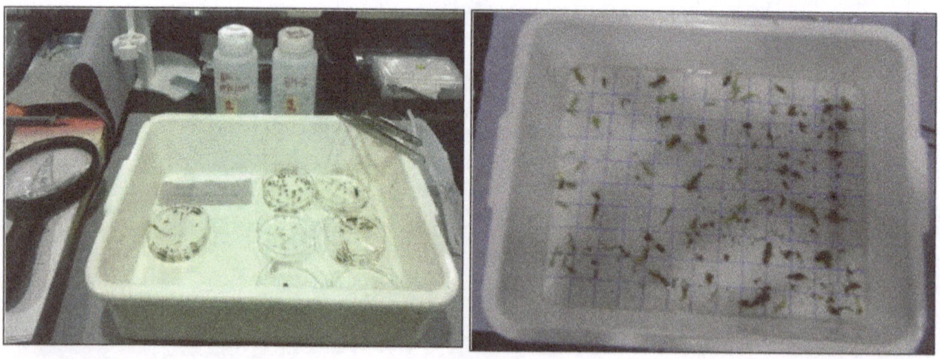

Page: 223

Plate 20.1: Birds Occurring in Pratapgunj

(a) Black-rumped Flameback (***Dinopium benghalense***); (b) Grey Hornbill (***Ocyceros birostris***); (c) White-throated Kingfisher (***Halcyon smyrnensis***); (d) Green Bee-eater (***Merops orientalis***); (e) Shikra (***Accipiter badius***); (f) Rufous Treepie (***Dendrocitta vagabunda***).

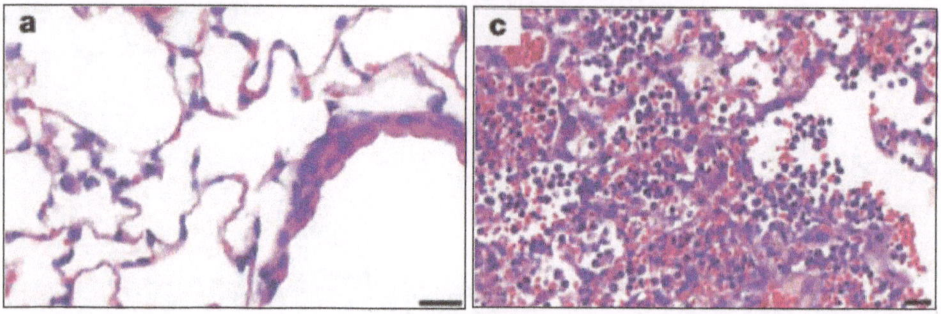

Figure 22.3: The 1918 Virus was Extremely Virulent. Pg. 362

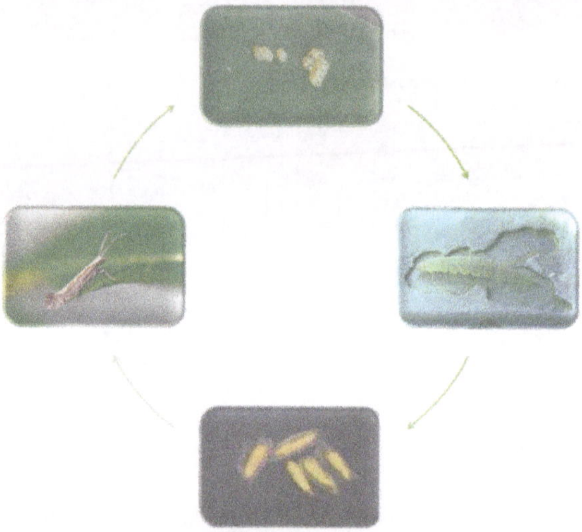

Page: 413

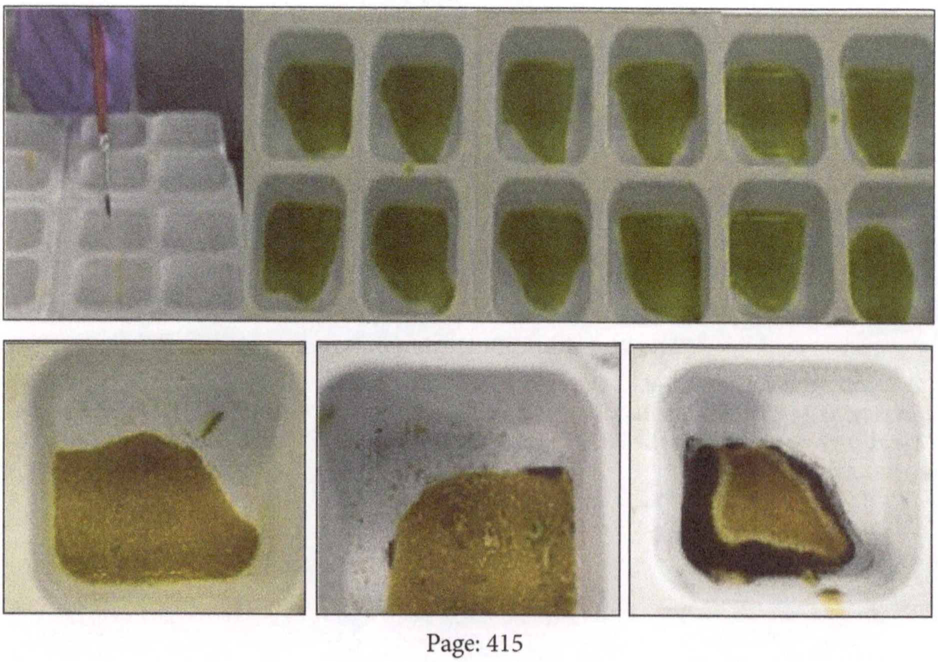

Page: 415

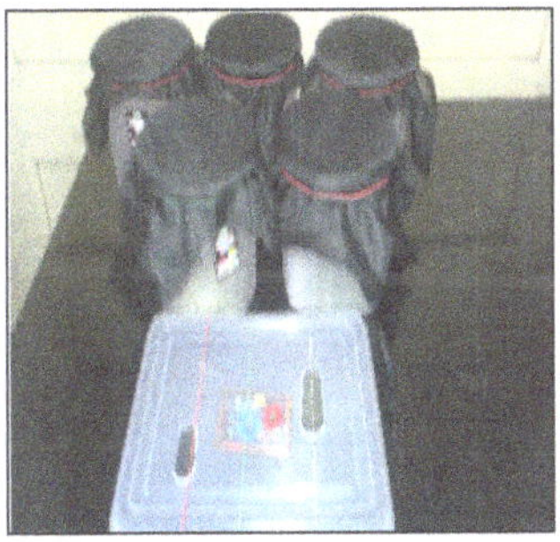

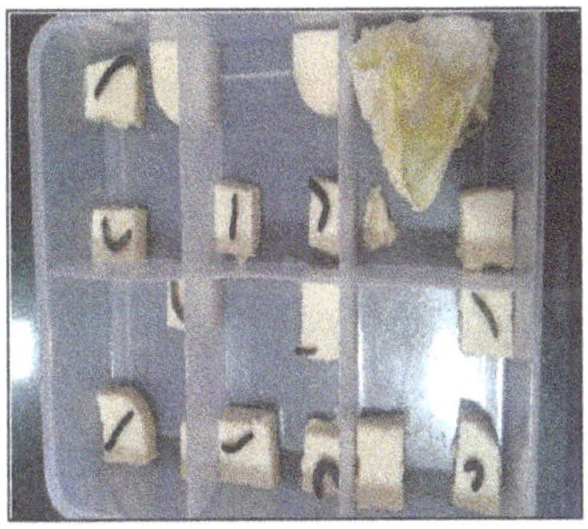

Page: 423

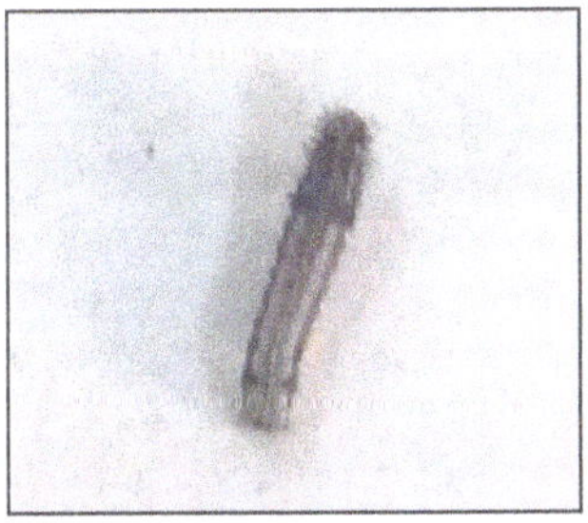

Live Larvae

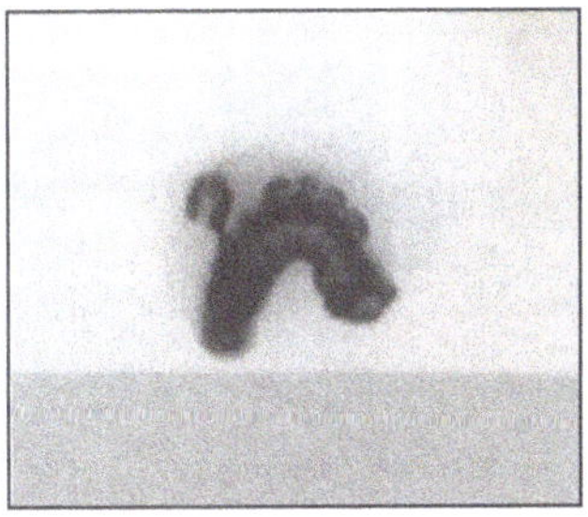

Moribund Larvae

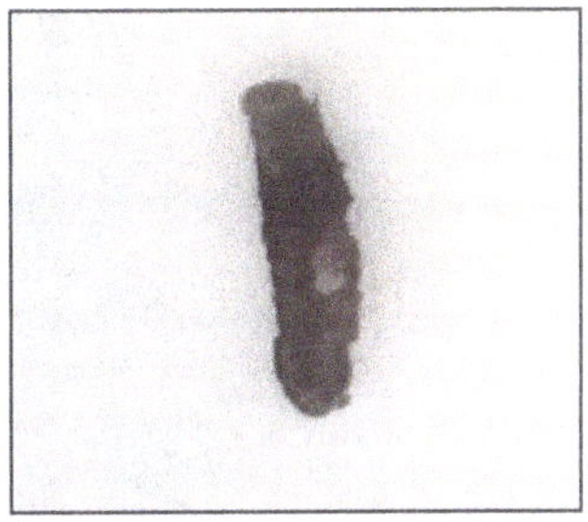

Dead Larvae

Page: 426

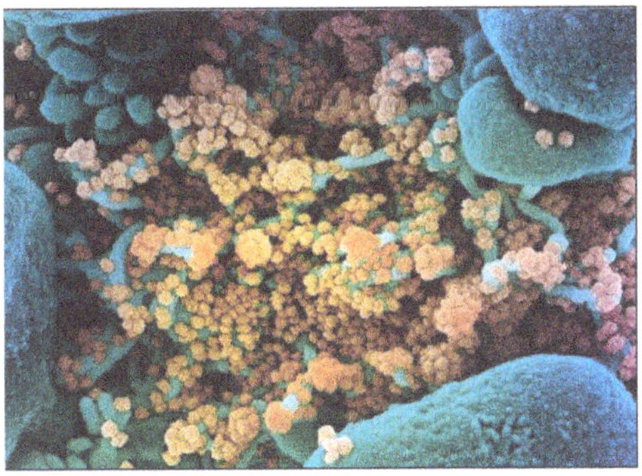

Figure 28.1: Corona Virus Particles in Lung Tissues. Pg. 458

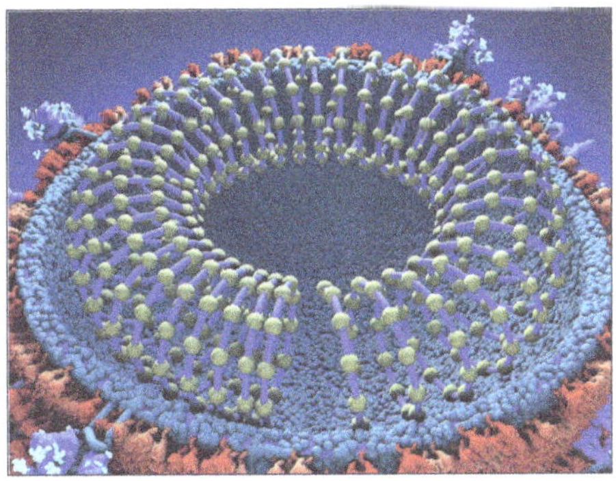

Figure 28.2: Architecture of CORONA Virus Particle Showing Spikes.

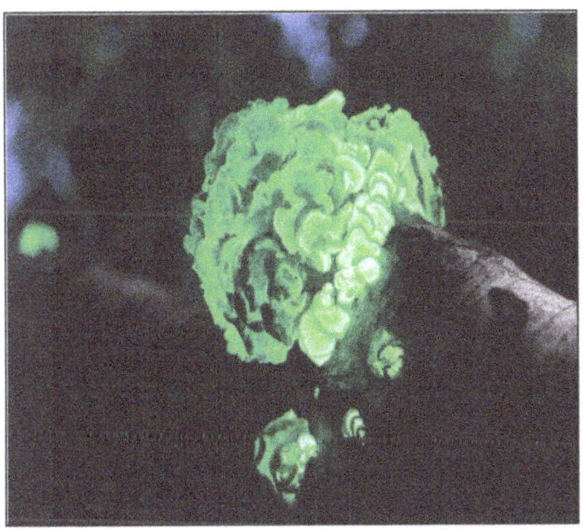

Page: 505

Page: 505

Figure 30.2: Female Glowworm, *Lampyris noctiluca*.

Figure 30.3: Flying and Glowing Firefly, *Photinus pyralis*.

Page: 499

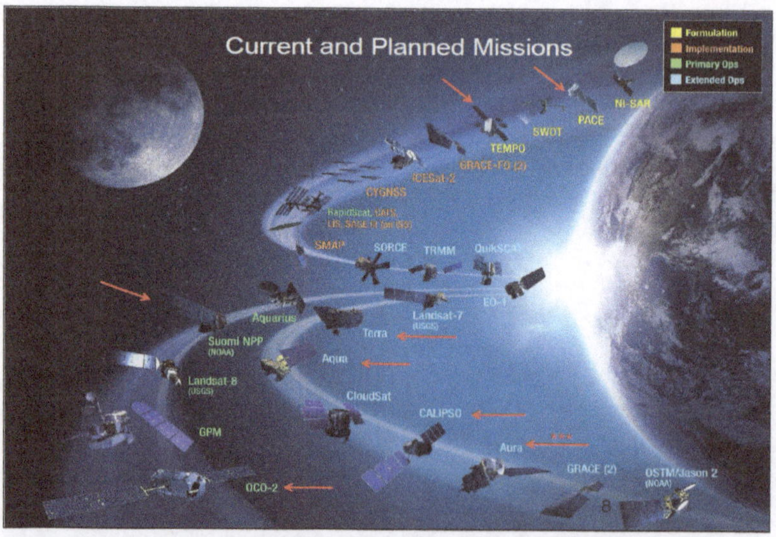

Figure 31.1: Picture Depicting Ongoing and Future Planned Missions (*Source*: ARSET: https: //arset.gsfc.nasa.gov/). Pg. 518

Printed at Repro India Ltd.